AF847120

Studies in Systems, Decision and Control

Volume 30

Series editor

Janusz Kacprzyk, Polish Academy of Sciences, Warsaw, Poland
e-mail: kacprzyk@ibspan.waw.pl

About this Series

The series "Studies in Systems, Decision and Control" (SSDC) covers both new developments and advances, as well as the state of the art, in the various areas of broadly perceived systems, decision making and control- quickly, up to date and with a high quality. The intent is to cover the theory, applications, and perspectives on the state of the art and future developments relevant to systems, decision making, control, complex processes and related areas, as embedded in the fields of engineering, computer science, physics, economics, social and life sciences, as well as the paradigms and methodologies behind them. The series contains monographs, textbooks, lecture notes and edited volumes in systems, decision making and control spanning the areas of Cyber-Physical Systems, Autonomous Systems, Sensor Networks, Control Systems, Energy Systems, Automotive Systems, Biological Systems, Vehicular Networking and Connected Vehicles, Aerospace Systems, Automation, Manufacturing, Smart Grids, Nonlinear Systems, Power Systems, Robotics, Social Systems, Economic Systems and other. Of particular value to both the contributors and the readership are the short publication timeframe and the world-wide distribution and exposure which enable both a wide and rapid dissemination of research output.

More information about this series at http://www.springer.com/series/13304

Viktor A. Sadovnichiy · Mikhail Z. Zgurovsky
Editors

Continuous and Distributed Systems II

Theory and Applications

Springer

Editors
Viktor A. Sadovnichiy
Lomonosov Moscow State University
Moscow
Russia

Mikhail Z. Zgurovsky
Kiev Polytechnic Institute
National Technical University of Ukraine
Kiev
Ukraine

ISSN 2198-4182 ISSN 2198-4190 (electronic)
Studies in Systems, Decision and Control
ISBN 978-3-319-19074-7 ISBN 978-3-319-19075-4 (eBook)
DOI 10.1007/978-3-319-19075-4

Library of Congress Control Number: 2013953260

Springer Cham Heidelberg New York Dordrecht London
© Springer International Publishing Switzerland 2015
This work is subject to copyright. All rights are reserved by the Publisher, whether the whole or part of the material is concerned, specifically the rights of translation, reprinting, reuse of illustrations, recitation, broadcasting, reproduction on microfilms or in any other physical way, and transmission or information storage and retrieval, electronic adaptation, computer software, or by similar or dissimilar methodology now known or hereafter developed.
The use of general descriptive names, registered names, trademarks, service marks, etc. in this publication does not imply, even in the absence of a specific statement, that such names are exempt from the relevant protective laws and regulations and therefore free for general use.
The publisher, the authors and the editors are safe to assume that the advice and information in this book are believed to be true and accurate at the date of publication. Neither the publisher nor the authors or the editors give a warranty, express or implied, with respect to the material contained herein or for any errors or omissions that may have been made.

Printed on acid-free paper

Springer International Publishing AG Switzerland is part of Springer Science+Business Media (www.springer.com)

This volume is dedicated to the distinguished mathematician Prof. Viktor A. Sadovnichiy on the occasion of his 75th birthday.

This volume is dedicated to the distinguished mathematician Prof. Mikhail Z. Zgurovsky on the occasion of his 65th birthday.

Preface

The given collected articles have been organized as a result of joint academic panels of research workers from the Faculty of Mechanics and Mathematics of Lomonosov Moscow State University and Institute for Applied System Analysis of the National Technical University of Ukraine "Kyiv Polytechnic Institute," devoted to applied problems of mathematics, mechanics, and engineering, which attracted the attention of researchers from leading scientific schools of Brazil, France, Germany, Poland, Russian Federation, Spain, Mexico, Ukraine, USA, and other countries. Modern technological applications require development and synthesis of fundamental and applied scientific areas, with a view to reducing the gap that may still exist between theoretical basis used for solving complicated technical problems and implementation of obtained innovations. To solve these problems mathematicians, mechanics, and engineers from wide research and scientific centers have been working together. Results of their joint efforts, including applied methods of modern algebra and analysis, fundamental and computational mechanics, nonautonomous and stochastic dynamical systems, and optimization, control and decision sciences for continuum mechanics problems, are partially presented here. In fact, serial publication of such collected papers to similar seminars is planned.

This is the sequel of an earlier volume.

The book is addressed to a wide circle of mathematical, mechanical, and engineering readers.

Moscow
Kiev
November 2014

Viktor A. Sadovnichiy
Mikhail Z. Zgurovsky

International Editorial Board of this Volume

Editors-in-Chief

V.A. Sadovnichiy, Lomonosov Moscow State University, Russian Federation
M.Z. Zgurovsky, National Technical University of Ukraine "Kyiv Polytechnic Institute," Ukraine

Associate Editors

V.N. Chubarikov, Lomonosov Moscow State University, Russian Federation
D.V. Georgievskii, Lomonosov Moscow State University, Russian Federation
P.O. Kasyanov, Institute for Applied System Analysis, National Technical University of Ukraine "Kyiv Polytechnic Institute" and World Data Center for Geoinformatics and Sustainable Development, Ukraine
J. Valero, Universidad Miguel Hernandez de Elche, Spain

Editors

Tomás Caraballo, Universidad de Sevilla, Spain
N.M. Dobrovol'skii, Tula State Lev Tolstoy Pedagogical University, Russian Federation
E.A. Feinberg, State University of New York at Stony Brook, USA
D. Gao, Virginia Tech, Australia
María José Garrido-Atienza, Universidad de Sevilla, Spain
O.V. Kapustyan, Institute for Applied System Analysis, National Technical University of Ukraine "Kyiv Polytechnic Institute," Ukraine
D. Korkin, University of Missouri, Columbia, USA

Acknowledgments

We express our gratitude to editors of the Springer Publishing House who worked with collection, and to everybody who took part in the preparation of the manuscript.

We want to express our special gratitude to Olena L. Poptsova for technical support of our collection.

Contents

Contributors

Vladimir V. Alexandrov Mechanics and Mathematics Faculty, Department of Applied Mechanics and Control, Lomonosov Moscow State University, Moscow, Russian Federation

Raul Temoltzi Avila Mathematics Faculty, Universidad Autonoma Del Estado de Hidalgo, Pachuca, Mexico

Dmitry I. Bugrov Mechanics and Mathematics Faculty, Lomonosov Moscow State University, Moscow, Russian Federation

Érika Capelato Departamento de Economia, Faculdade de Ciências e Letras, Universidade Estadual Paulista, Araraquara, SP, Brazil

Tomás Caraballo Dpto. Ecuaciones Diferenciales y Análisis Numérico, Universidad de Sevilla, Sevilla, Spain

Vladimir Nikolaevich Chubarikov Faculty of Mechanics and Mathematics, Lomonosov Moscow State University, Moscow, Russian Federation

Gleb Vladimirovich Fedorov Mechanics and Mathematics, Faculty of Lomonosov Moscow State University, Research Associate of Scientific Research Institute of System Development, Moscow, Russia

A.T. Fomenko Lomonosov Moscow State University, Moscow, Russia

V.V. Galatenko Moscow State University, Moscow, Russia

María J. Garrido-Atienza Dpto. Ecuaciones Diferenciales Y Análisis Numérico, Universidad de Sevilla, Sevilla, Spain

D.V. Georgievskii Moscow State University, Moscow, Russia

Mark O. Gluzman Institute for Applied System Analysis, National Technical University of Ukraine "Kyiv Polytechnic Institute", Kiev, Ukraine

I.M. Gorban Institute of Hydromechanics, National Academy of Sciences of Ukraine, Kyiv, Ukraine

Nataliia V. Gorban Institute for Applied System Analysis, National Technical University of Ukraine "Kyiv Polytechnic Institute", Kiev, Ukraine

Xiaoying Han Department of Mathematics and Statistics, Auburn University, Auburn, AL, USA

Alexander O. Ivanov Chair of Differential Geometry and Applications, Faculty of Mechanics and Mathematics, Lomonosov Moscow State University, Moscow, Russia; Chair of Mathematical Modelling, Bauman Moscow State Technical University, Moscow, Russia

Piotr Kalita Faculty of Mathematics and Computer Science, Jagiellonian University, Kraków, Poland

E.O. Kantonistova Lomonosov Moscow State University, Moscow, Russia

O.A. Kapustian Taras Shevchenko National University of Kyiv, Kiev, Ukraine

O.V. Kapustyan Taras Shevchenko National University of Kyiv, National Technical University of Ukraine "KPI", Kyiv, Ukraine

V.O. Kapustyan NTUU "KPI", Kiev, Ukraine

Pavlo O. Kasyanov Institute for Applied System Analysis, National Technical University of Ukraine "Kyiv Politechnic Institute", Kiev, Ukraine

O.V. Khomenko Institute for Applied System Analysis, National Technical University of Ukraine "Kyiv Polytechnic Institute", Kyiv, Ukraine

Peter E. Kloeden School of Mathematics and Statistics, Huazhong University of Science and Technology, Wuhan, China; Felix-Klein-Zentrum Für Mathematik, TU Kaiserslautern, Kaiserslautern, Germany

Olha P. Kupenko Dnipropetrovsk Mining University, Dnipropetrovsk, Ukraine; Institute for Applied System Analysis, National Technical University of Ukraine "Kiev Polytechnic Institute", Kiev, Ukraine

Stephan S. Lemak Mechanics and Mathematics Faculty, Lomonosov Moscow State University, Moscow, Russian Federation

Günter Leugering Department Mathematik, Friedrich-Alexander-Universität Erlangen-Nürnberg Lehrstuhl AMII, Erlangen, Germany

Kening Lu 346 TMCB Brigham Young University, Provo, UT, USA

T.P. Lukashenko Moscow State University, Moscow, Russia

Grzegorz Łukaszewicz Faculty of Mathematics, Informatics and Mechanics, University of Warsaw, Warsaw, Poland

O.K. Mazur National University of Food Technologies, Kiev, Ukraine

Alain Miranville Laboratoire de Mathématiques et Applications, UMR CNRS 7348, SP2MI, Université de Poitiers, Chasseneuil Futuroscope Cedex, France

Alexander S. Mishchenko Chair of High Geometry and Topology, Faculty of Mechanics and Mathematics, Lomonosov Moscow State University, Moscow, Russia

Liliia S. Paliichuk Institute for Applied System Analysis, National Technical University "Kyiv Polytechnic Institute", Kiev, Ukraine

V.V. Palin Lomonosov Moscow State University, Moscow, Russian Federation

B.E. Pobedria Moscow State University, Moscow, Russia

I.O. Pyshnograiev NTUU "KPI", Kiev, Ukraine

E.V. Radkevich Lomonosov Moscow State University, Moscow, Russian Federation

Alain Rapaport UMR INRA/SupAgro MISTEA, Montpellier, France; MODEMIC Team, INRIA, Sophia-Antipolis, France

Victor A. Sadovnichiy Lomonosov Moscow State University, Moscow, Russian Federation

Björn Schmalfuß Institut Für Stochastik, Friedrich Schiller Universität Jena, Jena, Germany

Jacson Simsen Instituto de Matemática e Computacção, Universidade Federal de Itajubá, Itajubá, MG, Brazil

Katerina V. Tikhonova Mechanics and Mathematics Faculty, Lomonosov Moscow State University, Moscow, Russian Federation

Alla M. Tkachuk National University of Food Technologies, Kiev, Ukraine

Alexey A. Tuzhilin Chair of Differential Geometry and Applications, Faculty of Mechanics and Mathematics, Lomonosov Moscow State University, Moscow, Russia

B. Wang Department of Mathematics, New Mexico Institute of Mining and Technology, Socorro, NM, USA

Mikhail Z. Zgurovsky National Technical University of Ukraine "Kyiv Politechnic Institute", Kiev, Ukraine

Part I
Applied Methods of Modern Algebra and Analysis

Chapter 1
The Absolute Stability of Orthorecursive Expansions in Redundant Systems of Subspaces

V.V. Galatenko, T.P. Lukashenko and V.A. Sadovnichiy

Abstract Orthorecursive expansions in redundant systems of subspaces are considered and the results on the absolute stability of these expansions are presented. Problems of stability with respect to projection errors and stability with respect to system perturbations are studied. The presented results are the generalization of theorems on absolute stability of orthorecursive expansions in redundant systems of Hilbert space elements.

1.1 Introduction

Orthorecursive expansion in a system of subspaces [1] is a natural generalization of orthorecursive expansion in a system of elements of Hilbert space [2], which is in turn a generalization of classical orthogonal expansions.

Let us recall the definition of orthorecursive expansions in a system of subspaces. Let H be a Hilbert space (here we consider spaces over $\mathbb{R}$, however, the case of spaces over $\mathbb{C}$ is similar). Let $\{H_n\}_{n=1}^{\infty}$ be a system of closed subspaces of H and let $f \in H$ be an approximated element. We inductively define the elements of the expansion $\{\widetilde{f}_n\}_{n=1}^{\infty}$ and the sequence of remainders $\{r_n(f)\}_{n=0}^{\infty}$:

$$r_0(f) = f;$$

$$\widetilde{f}_{n+1} = P_{n+1}(r_n(f)), \quad r_{n+1}(f) = r_n(f) - \widetilde{f}_{n+1} = P_{n+1}^{\perp}(r_n(f))$$

V.V. Galatenko (✉) · T.P. Lukashenko · V.A. Sadovnichiy
Moscow State University, Moscow 119991, Russia
e-mail: vgalat@imscs.msu.ru

T.P. Lukashenko
e-mail: lukashenko@mail.ru

V.A. Sadovnichiy
e-mail: info@rector.msu.ru

© Springer International Publishing Switzerland 2015

V.A. Sadovnichiy and M.Z. Zgurovsky (eds.), *Continuous and Distributed Systems II*, Studies in Systems, Decision and Control 30,
DOI 10.1007/978-3-319-19075-4_1

(here P_{n+1} is the orthogonal projector onto H_{n+1}, $P_{n+1}^{\perp}$ is the orthogonal projector onto the orthogonal complement of H_{n+1}).

Definition 1.1 The series $\sum_{n=1}^{\infty} \widetilde{f}_n$ is called *an orthorecursive expansion* of f in a system of subspaces $\{H_n\}_{n=1}^{\infty}$.

If all H_n are one-dimensional and equal to $< e_n >$, this definition is equivalent to the definition of orthorecursive expansion in a system of elements $\{e_n\}_{n=1}^{\infty}$.

As one can easily see, the remainder of the orthorecursive expansion $r_N(f)$ equals the difference $\left(f - \sum_{n=1}^{N} \widetilde{f}_n\right)$, and hence the convergence of $\sum_{n=1}^{\infty} \widetilde{f}_n$ to f is equivalent to the convergence of the remainders $r_n(f)$ to zero. In [1] it is noted that due to Bessel's identity

$$\left\| f - \sum_{n=1}^{N} \widetilde{f}_n \right\|^2 = \|r_N(f)\|^2 = \|f\|^2 - \sum_{n=1}^{N} \|\widetilde{f}_n\|^2$$

this convergence is also equivalent to Parseval's identity

$$\|f\|^2 = \sum_{n=1}^{\infty} \|\widetilde{f}_n\|^2 .$$

In the current research we consider the case of redundant systems of subspaces and address the problem of absolute stability of orthorecursive expansions in such systems with respect to projection errors and to perturbations of the system. In case of orthorecursive expansions in a system of elements this problem was studied in [3]. For pure greedy algorithms [4] that utilize orthorecursive scheme for coefficients computation similar problems were considered in [5, 6].

In order to study the problem of stability of orthorecursive expansions with respect to projection errors, we first introduce the model of the expansion process that contains these errors. Let $\{H_n\}_{n=1}^{\infty}$ be a system of closed subspaces of H and let $f \in H$ be an approximated element. Let us additionally consider the sequence of elements $\{\xi_n\}_{n=1}^{\infty} \subset H$. We inductively define the elements of the expansion $\{\widetilde{f}_n^{err}\}_{n=1}^{\infty}$ and the sequence of remainders $\{r_n^{err}(f)\}_{n=0}^{\infty}$:

$$r_0^{err}(f) = r_0(f) = f;$$

$$\widetilde{f}_{n+1}^{err} = P_{n+1}(r_n^{err}(f)) + \xi_{n+1}, \quad r_{n+1}^{err}(f) = r_n^{err}(f) - \widetilde{f}_{n+1}^{err}.$$

Definition 1.2 The series $\sum_{n=1}^{\infty} \widetilde{f}_n^{err}$ is called *an orthorecursive expansion* of f in a system of subspaces $\{H_n\}_{n=1}^{\infty}$ *with projection errors* $\{\xi_n\}_{n=1}^{\infty}$.

In practical implementations of orthorecursive expansions the exact values of projection errors $\{\xi_n\}_{n=1}^{\infty}$ remain unknown, however, their norms can be controlled (e.g., by taking an appropriate number of nodes in numerical integration methods in case of functional Hilbert spaces). In some applications where the projection is evaluated as a linear combination of certain elements that belong to the subspace, projection errors also automatically satisfy the condition $\xi_n \in H_n$ $(n = 1, 2, 3, \ldots)$.

Similar to [3], we also introduce the notion of strongly redundant systems of closed subspaces with respect to orthorecursive expansions.

Definition 1.3 A system of closed subspaces $\{H_n\}_{n=1}^{\infty}$ is called *strongly redundant with respect to orthorecursive expansions* if for every $N \in \mathbb{N}$ and every $f \in H$ an orthorecursive expansion of f in the system $\{H_{N+n}\}_{n=1}^{\infty}$ converges to f.

A natural and simple example of a system of closed subspaces that is strongly redundant with respect to orthorecursive expansions is an expanding system [7]: $H_n \subset H_{n+1}$ $(n = 1, 2, 3, \ldots)$, $\overline{\bigcup_n H_n} = H$.

In order to state the result on the absolute stability of orthorecursive expansions with respect to perturbations of a system of subspaces, we also generalize the notion of quadratically close systems [3, 8] to the case of systems of closed subspaces.

Let $\{H_n\}_{n=1}^{\infty}$ and $\{L_n\}_{n=1}^{\infty}$ be systems of closed subspaces of H, and let P_n and Q_n be orthogonal projectors onto H_n and L_n respectively $(n = 1, 2, 3, \ldots)$.

Definition 1.4 Systems of subspaces $\{H_n\}_{n=1}^{\infty}$ and $\{L_n\}_{n=1}^{\infty}$ are called *quadratically close* if

$$\sum_{n=1}^{\infty} \|P_n - Q_n\|^2 < \infty.$$

The norm here is the standard (strong) operator norm.

1.2 Main Results

The main results on the absolute stability of orthorecursive expansions in a system of subspaces with respect to projection errors can be stated as follows.

Theorem 1.1 *Let H be a Hilbert space over $\mathbb{R}$ and let $\{H_n\}_{n=1}^{\infty}$ be an arbitrary system of closed subspaces of H that is strongly redundant with respect to orthorecursive expansions. Then for every element $f \in H$ and every sequence $\{\xi_n\}_{n=1}^{\infty} \subset H$ for which $\sum_{n=1}^{\infty} \|\xi_n\| < \infty$ the orthorecursive expansion of f in the system $\{H_n\}_{n=1}^{\infty}$ with projection errors $\{\xi_n\}_{n=1}^{\infty}$ converges to f.*

Theorem 1.2 *Let H be a Hilbert space over $\mathbb{R}$ and let $\{H_n\}_{n=1}^{\infty}$ be an arbitrary system of closed subspaces of H that is strongly redundant with respect to orthorecursive* *expansions. Then for every element* $f \in H$ *and every sequence* $\{\xi_n\}_{n=1}^{\infty} \subset H$ *for which* $\xi_n \in H_n$ $(n = 1, 2, 3, \ldots)$ *and* $\sum_{n=1}^{\infty} \|\xi_n\|^2 < \infty$ *the orthorecursive expansion of* f *in the system* $\{H_n\}_{n=1}^{\infty}$ *with projection errors* $\{\xi_n\}_{n=1}^{\infty}$ *converges to* f.

The construction from [3, S 3] shows that the condition $\sum_{n=1}^{\infty} \|\xi_n\|^2 < \infty$ in Theorem 1.2 cannot be weakened even in case of one-dimensional subspaces generated by the repeats of elements of an orthogonal basis. A similar construction shows that the condition $\sum_{n=1}^{\infty} \|\xi_n\| < \infty$ in Theorem 1.2 is also sharp.

The main result on the absolute stability of orthorecursive expansions in a system of subspaces with respect to system perturbations can be stated as follows.

Theorem 1.3 *Let* H *be a Hilbert space over* $\mathbb{R}$, $\{H_n\}_{n=1}^{\infty}$ *be an arbitrary system of closed subspaces of* H *that is strongly redundant with respect to orthorecursive expansions, and let* $\{L_n\}_{n=1}^{\infty}$ *be a system of closed subspaces of* H *that is quadratically close to* $\{H_n\}_{n=1}^{\infty}$. *Then the system* $\{L_n\}_{n=1}^{\infty}$ *is also strongly redundant with respect to orthorecursive expansions.*

The construction from [3, S 4] shows that the condition of quadratic closeness in Theorem 1.3 cannot be weakened even in case of $H = \mathbb{R}^2$.

Theorems 1.1 and 1.3 can be combined in the following way.

Theorem 1.4 *Let* H *be a Hilbert space over* $\mathbb{R}$, $\{H_n\}_{n=1}^{\infty}$ *be an arbitrary system of closed subspaces of* H *that is strongly redundant with respect to orthorecursive expansions, and let* $\{L_n\}_{n=1}^{\infty}$ *be a system of closed subspaces of* H *that is quadratically close to* $\{H_n\}_{n=1}^{\infty}$. *Then for every element* $f \in H$ *and every sequence* $\{\xi_n\}_{n=1}^{\infty} \subset H$ *for which* $\sum_{n=1}^{\infty} \|\xi_n\| < \infty$ *the orthorecursive expansion of* f *in the system* $\{L_n\}_{n=1}^{\infty}$ *with projection errors* $\{\xi_n\}_{n=1}^{\infty}$ *converges to* f.

Theorems 1.2 and 1.3 can be combined in a similar way.

The results above are a generalization of the results on absolute stability of orthorecursive expansions in a system of elements of a Hilbert space [3].

1.3 Proofs

The proofs of the theorems are based on the following lemmas.

Lemma 1.1 *Let* H *be a Hilbert space over* $\mathbb{R}$ *and let* $\{H_n\}_{n=1}^{\infty}$ *be an arbitrary system of closed subspaces of* H. *Then for every element* $f \in H$ *and every sequence*

$\{\xi_n\}_{n=1}^{\infty} \subset H$ *the estimate*

$$\left\| r_N^{err}(f) - r_N(f) \right\| \le \sum_{n=1}^{N} \|\xi_n\|$$

(where $r_N(f)$ *is the remainder of the orthorecursive expansion of* f *in* $\{H_n\}_{n=1}^{\infty}$, *and* $r_N^{err}(f)$ *is the remainder of the orthorecursive expansion of* f *in* $\{H_n\}_{n=1}^{\infty}$ *with projection errors* $\{\xi_n\}_{n=1}^{\infty}$*) holds for all* $N \in \mathbb{N}$.

Lemma 1.2 *Let* H *be a Hilbert space over* $\mathbb{R}$ *and let* $\{H_n\}_{n=1}^{\infty}$ *be an arbitrary system of closed subspaces of* H. *Then for every element* $f \in H$ *and every sequence* $\{\xi_n\}_{n=1}^{\infty} \subset H$ *with* $\xi_n \in H_n$ $(n = 1, 2, 3, \ldots)$ *the estimate*

$$\left\| r_N^{err}(f) - r_N(f) \right\|^2 \le \sum_{n=1}^{N} \|\xi_n\|^2$$

(where $r_N(f)$ *is the remainder of the orthorecursive expansion of* f *in* $\{H_n\}_{n=1}^{\infty}$, *and* $r_N^{err}(f)$ *is the remainder of the orthorecursive expansion of* f *in* $\{H_n\}_{n=1}^{\infty}$ *with projection errors* $\{\xi_n\}_{n=1}^{\infty}$*) holds for all* $N \in \mathbb{N}$.

Lemma 1.3 *Let* H *be a Hilbert space over* $\mathbb{R}$, *and let* $\{H_n\}_{n=1}^{\infty}$ *and* $\{L_n\}_{n=1}^{\infty}$ *be arbitrary systems of closed subspaces of* H. *Then for every element* $f \in H$ *the estimate*

$$\|\rho_N(f) - r_N(f)\|^2 \le \|f\|^2 \left(\sum_{n=1}^{N} \|P_n - Q_n\|^2 + 4 \left(\sum_{n=1}^{N} \|P_n - Q_n\|^2 \right)^{\frac{1}{2}} \right)$$

(where $r_N(f)$ *is the remainder of the orthorecursive expansion of* f *in* $\{H_n\}_{n=1}^{\infty}$, $\rho_N(f)$ *is the remainder of the orthorecursive expansion of* f *in* $\{L_n\}_{n=1}^{\infty}$, *and* P_n, Q_n *are orthogonal projectors onto* H_n, L_n *respectively) holds for all* $N \in \mathbb{N}$.

The theorems can be proved using these lemmas in a standard way [3]. For e.g., let us show how Theorem 1.1 can be proved using Lemma 1.1.

Let ε be an arbitrary positive number. There exists such a number $M \in \mathbb{N}$ that $\sum_{n=M+1}^{\infty} \|\xi_n\| < \frac{\varepsilon}{2}$. Let ρ_n denote the remainders of the orthorecursive expansion in the truncated system of subspaces $\{H_{M+n}\}_{n=1}^{\infty}$, and let ρ_n^{err} denote the remainders of the orthorecursive expansions in the truncated system of subspaces $\{H_{M+n}\}_{n=1}^{\infty}$ with projection errors $\{\xi_{M+n}\}_{n=1}^{\infty}$. Due to the strong redundancy of $\{H_n\}_{n=1}^{\infty}$ with respect to orthorecursive expansions, $\rho_n(g) \to 0$ $(n \to \infty)$ for every $g \in H$, hence there exists such a number $K \in \mathbb{N}$ that $\|\rho_k(r_M^{err}(f))\| < \frac{\varepsilon}{2}$ for all $k > K$. Then for all $n > M + K$ we have

$$\|r_n^{err}(f)\| = \|\rho_{n-M}^{err}(r_M^{err}(f))\| \leq \|\rho_{n-M}(r_M^{err}(f))\| + \sum_{m=1}^{n-M} \|\xi_{M+m}\| < \frac{\varepsilon}{2} + \frac{\varepsilon}{2} = \varepsilon.$$

Thus $r_n^{err}(f) \to 0$ $(n \to \infty)$ or, equivalently, the orthorecursive expansion $\sum_{n=1}^{\infty} \widetilde{f}_n^{err}$ converges to f.

Therefore we need only to prove the above lemmas. We start with Lemmas 1.1 and 1.2 and apply the induction. Let $\theta_N(f)$ denote the difference $\left(r_N^{err}(f) - r_N(f)\right)$ $(N = 0, 1, 2, \ldots)$. For $N = 0$ both remainders $r_N(f)$ and $r_N^{err}(f)$ equal f, so estimates in Lemmas 1.1 and 1.2 hold. Let us assume that the estimates hold for $N = M$ and prove them for $N = M + 1$. The definition of orthorecursive expansions implies the equalities

$$\begin{aligned} r_{M+1}^{err}(f) &= r_M^{err}(f) - \widetilde{f}_{M+1}^{err} \\ &= r_M(f) + \theta_M(f) - (P_{M+1}(r_M(f) + \theta_M(f)) + \xi_{M+1}) \\ &= P_{M+1}^{\perp}(r_M(f)) + P_{M+1}^{\perp}(\theta_M(f)) - \xi_{M+1} \\ &= r_{M+1}(f) + P_{M+1}^{\perp}(\theta_M(f)) - \xi_{M+1}. \end{aligned}$$

Hence,

$$r_{M+1}^{err}(f) - r_{M+1}(f) = P_{M+1}^{\perp}(\theta_M(f)) - \xi_{M+1},$$

so

$$\left\|r_{M+1}^{err}(f) - r_{M+1}(f)\right\| \leq \|P_{M+1}^{\perp}(\theta_M(f))\| + \|\xi_{M+1}\| \leq \|\theta_M(f)\| + \|\xi_{M+1}\|,$$

which completes the proof of Lemma 1.1.

In case of Lemma 1.2 the projection error ξ_{M+1} belongs to H_{M+1}, so it is orthogonal to $P_{M+1}^{\perp}(\theta_M(f))$, and hence

$$\left\|r_{M+1}^{err}(f) - r_{M+1}(f)\right\|^2 = \|P_{M+1}^{\perp}(\theta_M(f))\|^2 + \|\xi_{M+1}\|^2 \leq \|\theta_M(f)\|^2 + \|\xi_{M+1}\|^2.$$

This inequality completes the proof of Lemma 1.2.

The proof of Lemma 1.3 is less straightforward. We start with the introduction of a number of notations. For $n = 1, 2, 3, \ldots$ let $D_n = P_n - Q_n$ (hence $Q_n = P_n - D_n$ and $Q_n^{\perp} = Id - Q_n = P_n^{\perp} + D_n$). For $n = 0, 1, 2, \ldots$ let $\theta_n(f) = \rho_n(f) - r_n(f)$. Note that for all n the trivial estimate $\|\theta_n(f)\| \leq \|\rho_n(f)\| + \|r_n(f)\| \leq 2\|f\|$ holds.

Let $\sum_{n=1}^{\infty} \widetilde{f}_n^L$ be the orthorecursive expansion of f in the system $\{L_n\}_{n=1}^{\infty}$.

Inductively, we show that the inequality

$$\|\theta_N(f)\|^2 \le \|f\|^2 \cdot \sum_{n=1}^{N} \|D_n\|^2 + 4\|f\| \cdot \sum_{n=1}^{N} \left\| \widetilde{f_n^L} \right\| \|D_n\|$$

holds for every $N = 0, 1, 2, \ldots$. The estimate of Lemma 1.3 directly follows from this inequality, Cauchy-Schwarz inequality and Bessel's inequality $\sum_{n=1}^{\infty} \| \widetilde{f_n^L} \|^2 \le \|f\|^2$, which is an immediate corollary of the Bessel's identity [1].

For $N = 0$ the inequality trivially holds as $\theta_0(f) = 0$. Let us assume that the inequality holds for $N = M$ and prove it for $N = M+1$. According to the definition of orthorecursive expansions

$$\rho_{M+1}(f) = Q_{M+1}^{\perp}(\rho_M(f)) = \left(P_{M+1}^{\perp} + D_{M+1}\right)(r_M(f) + \theta_M(f)) =$$

$$= r_{M+1}(f) + P_{M+1}^{\perp}(\theta_M(f)) + D_{M+1}(\rho_M(f)),$$

so

$$\theta_{M+1}(f) = P_{M+1}^{\perp}(\theta_M(f)) + D_{M+1}(\rho_M(f)).$$

Thus,

$$\|\theta_{M+1}(f)\|^2 = \|P_{M+1}^{\perp}(\theta_M(f))\|^2 + \|D_{M+1}(\rho_M(f))\|^2 +$$

$$+ 2\left(P_{M+1}^{\perp}(\theta_M(f)), D_{M+1}(\rho_M(f))\right) \le$$

$$\le \|\theta_M\|^2 + \|f\|^2\|D_{M+1}\|^2 + 2\left|\left(P_{M+1}^{\perp}(\theta_M(f)), D_{M+1}(\rho_M(f))\right)\right|.$$

The inner product can be estimated as follows:

$$\left|\left(P_{M+1}^{\perp}(\theta_M(f)), D_{M+1}(\rho_M(f))\right)\right| = \left|\left(P_{M+1}^{\perp}(\theta_M(f)), (P_{M+1} - Q_{M+1})(\rho_M(f))\right)\right| =$$

$$= \left|\left(P_{M+1}^{\perp}(\theta_M(f)), Q_{M+1}(\rho_M(f))\right)\right| = \left|\left(\left(P_{M+1}^{\perp} - Q_{M+1}^{\perp}\right)(\theta_M(f)), Q_{M+1}(\rho_M(f))\right)\right| =$$

$$= \left|\left(D_{M+1}(\theta_M(f)), \widetilde{f_{M+1}^L}\right)\right| \le \|\theta_M(f)\| \cdot \|D_{M+1}\| \cdot \left\|\widetilde{f_{M+1}^L}\right\| \le$$

$$\le 2\|f\| \cdot \|D_{M+1}\| \cdot \left\|\widetilde{f_{M+1}^L}\right\|.$$

This estimate completes the proof of Lemma 1.3.

Acknowledgments This research was supported by the Ministry of Education and Science of the Russian Federation (agreement 02.G25.31.0030), the Russian Foundation for Basic Research (project 14–01–00417), and the President grants for the support of the leading scientific schools of the Russian Federation (grants NSh–1096.2014.1, NSh–3682.2014.1).

References

1. Lukashenko, T.P., Sadivnichii, V.A.: Orthorecursive expansions with respect to subspaces. Dokl. Math. **445**(2), 135–138 (2012)
2. Lukashenko, T.P.: Properties of orthorecursive expansions in nonorthogonal systems. Moscow Univ. Math. Bull. **56**(1), 5–9 (2001)
3. Galatenko, V.V.: On orthorecursive expansions with errors in the calculation of coefficients. Izv. Math. **69**(1), 1–14 (2005)
4. Temlyakov, V.N.: Weak greedy algorithms. Adv. Comput. Math. **12**(2–3), 213–227 (2000)
5. Gribonval, R., Nielsen, M.: Approximate weak greedy algorithms. Adv. Comput. Math. **14**(4), 361–378 (2001)
6. Galatenko, V.V., Livshits, E.D.: Generalized approximate weak Greedy algorithms. Math. Notes **78**(1–2), 170–184 (2005)
7. Lukashenko, T.P., Sadivnichii, V.A.: On orthorecursive expansions in terms of a chain of systems. Dokl. Math. **425**(6), 741–746 (2009)
8. Bari, N.K.: On the stability of completeness properties of systems of functions. Dokl. Akad. Nauk SSSR **37**, 99–103 (1942). (in Russian)

Chapter 2
Topological Classification of Geodesic Flows on Revolution 2-Surfaces with Potential

A.T. Fomenko and E.O. Kantonistova

Abstract The paper is devoted to a short explanation of the topological classification (up to Liouville equivalence) of the integrable geodesic flows of two-dimensional surfaces of revolution with potential. The classification is given in the terms of so-called "marked molecules," i.e., Fomenko–Zieschang invariants for integrable systems with two degrees of freedom on three-dimensional isoenergy sufraces.

2.1 "Atoms" and Morse Functions

The critical point is called *nondegenerate* if the second differential

$$d^2f = \Sigma \frac{\partial^2 f}{\partial x_i \partial x_j} dx_i dx_j$$

is nondegenerate at this point. There are three possible types of nondegenerate critical points for functions on two-dimensional surfaces: maximum, minimum, and saddle (see Fig. 2.1).

A smooth function is called a *Morse function* if all its critical points are nondegenerate. The Morse functions are everywhere dense in the space of all smooth functions on a smooth manifold.

By c we shall denote critical values of f, i.e., those in whose preimage there is at least one critical point. By arbitrary small perturbation, one can do so that, on every critical level c there is exactly one critical point. In other words, the critical points which occur in the same level can be moved close but in different levels. If each critical level $f^{-1}(c)$ contains exactly one critical point, then f is called *simple Morse function*.

A.T. Fomenko (✉) · E.O. Kantonistova
Lomonosov Moscow State University, Leninskie Gory, 1, Moscow, Russia
e-mail: atfomenko@mail.ru

E.O. Kantonistova
e-mail: kysin@rambler.ru

© Springer International Publishing Switzerland 2015

V.A. Sadovnichiy and M.Z. Zgurovsky (eds.), *Continuous and Distributed Systems II*, Studies in Systems, Decision and Control 30,
DOI 10.1007/978-3-319-19075-4_2

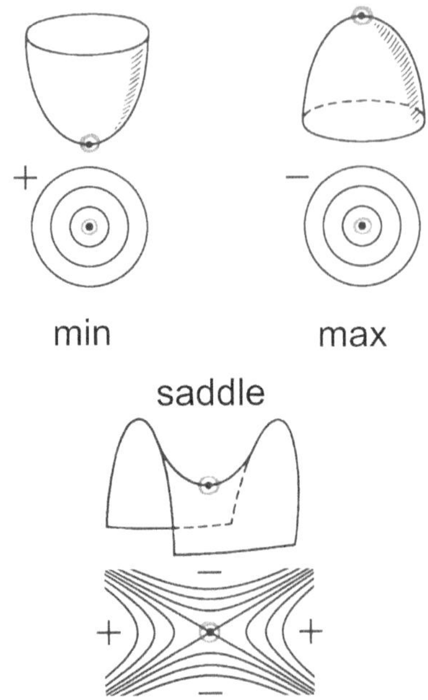

Fig. 2.1 Types of critical points

Let f be a Morse function on a compact smooth manifold X^n. For any $a \in R$, consider the level surface $f^{-1}(a)$ and its connected components which can be as called *fibers*. As a result, on the manifold there appears the structure of a foliation with singularities. By declaring each fiber to be a point and introducing the natural quotient topology in the space Γ of fibers, we obtain some quotient space. It can be considered as the base of the foliation. For Morse function, the space Γ is a finite graph. The graph is called the *Reeb graph* of the Morse function f on manifold X^n. Consider, for instance, the two-dimensional torus in R^3 embedded as shown in Fig. 2.2, and take the natural height function to be a Morse function on the torus.

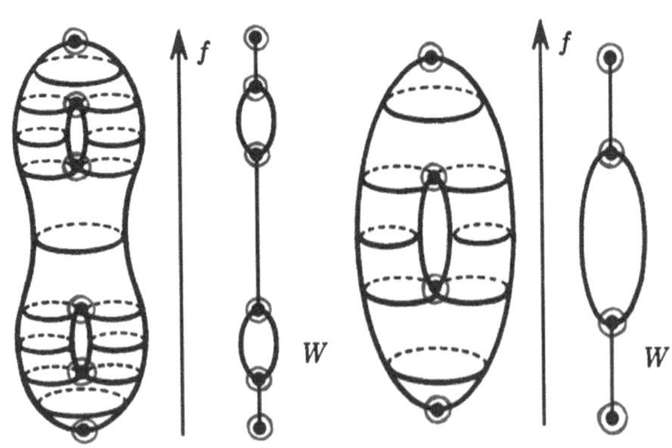

Fig. 2.2 Reeb graphs for height functions

Then its Reeb graph has the form as shown in the same figure.

It is a natural problem to give the classification of Morse functions on two-dimensional surfaces up to the fiber equivalence. To solve it, first we need to study the local question, namely, to describe the local topological structure of singular fibers.

We begin with the informal definition. An *atom* is defined to be the topological type of a two-dimensional Morse singularity. In other words, this is the topological type of singular fiber of the foliation defined on a two-dimensional surface by a Morse function. More precisely, we can reformulate this as follows.

Definition 2.1 An *atom* is a neighborhood P^2 of a critical fiber (which is defined by inequality $c - \varepsilon \leq f(x) \leq c + \varepsilon$ for sufficiently small ε), foliated into level lines of f and considered up to the fiber equivalence. In other words, an atom is the germ of the foliation on a singular fiber.

The atom P^2 is called *simple*, if the Morse function f in the pair (P^2, f) is simple. The other atoms are called *complicated*. The *complexity* of an atom is a number of critical points on its critical level $f(x) = c$. The atom is called *orientable (oriented)* or *nonorientable* depending on whether the surface P^2 is orientable (oriented) or nonorientable.

First, consider a nonsingular level line which is close to a local maximum point. This line is a circle. As the regular value tends to the local maximum, the circle shrinks into a point (Fig. 2.3). Let us represent this evolution and the bifurcation in the following conventional, but in quite visual manner. Every regular level line (a circle) we represent as one point is located on the level a (Fig. 2.3). As a changes, this point moves running through a segment. At the moment, when the value of the function becomes critical (equal to c), a circle has shrunk into a point. Denote this event by the letter A with a segment going out of it. This segment is directed downwards. In the case of minimum, we proceed the similar way (Fig. 2.3).

If c is a critical saddle value, then the singular level line looks like a figure eight curve. As a tends to c, two circles are getting closer and, finally, touch at a point. After this, the level line bifurcation happens and, instead of two, we obtain just one circle, see Fig. 2.4.

Let f be a simple Morse function on a compact closed surface X^2 (orientable or nonorientable). Consider Reeb graph Γ. The vertices of Γ correspond to critical fibers of f. Let us replace these vertices by corresponding atoms (either A, or B, or $\tilde{B}$, which is nonorientable version of B). The graph obtained is called a *simple molecule* W. In fact, the notion of the simple molecule does not differ yet from that of the Reeb graph. However, for complicated Morse functions the molecule W will carry more information than the Reeb graph Γ (Fig. 2.5).

A minimal simple Morse function on the pretzel, i.e., on the sphere with two handles, is realized as the height function on the embedding of the pretzel, presented in Fig. 2.8. The corresponding simple molecule is also shown here.

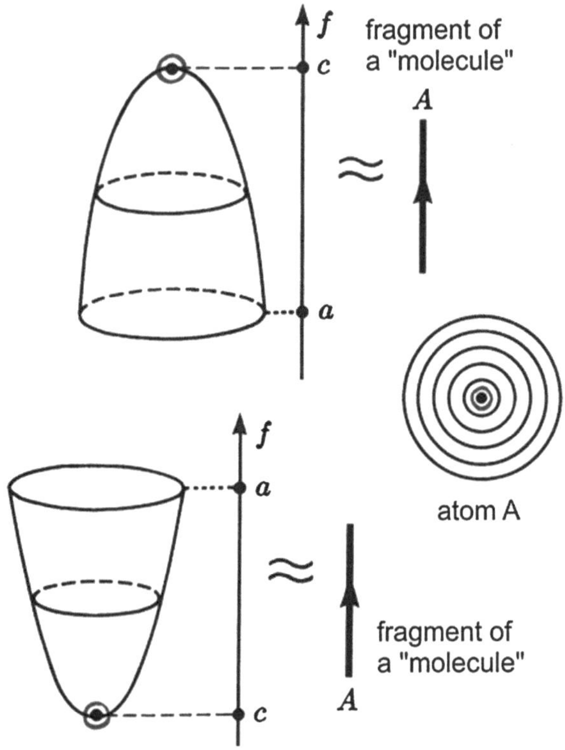

Fig. 2.3 Atom *A*

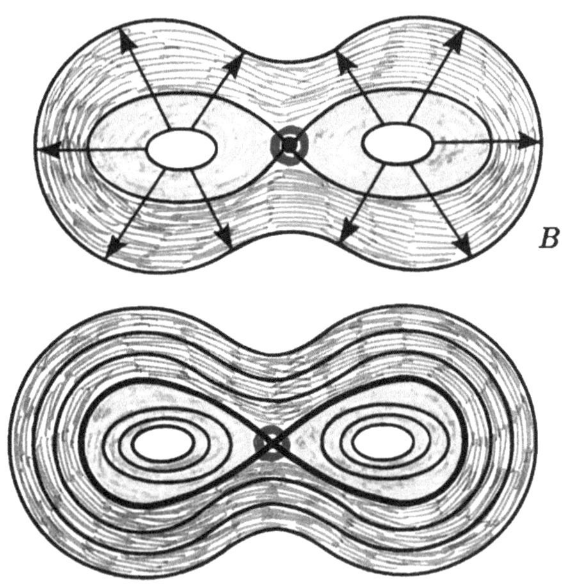

Fig. 2.4 Atom *B*. Level *curves* of a simple function *f* are shown

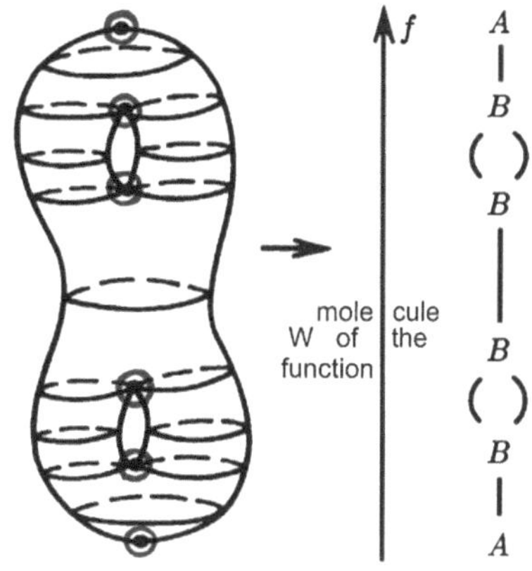

Fig. 2.5 Minimal simple Morse function on the pretzel and its simple molecule

2.2 Complicated Atoms and Molecules

Recall that an atom is complicated if critical connected level surface of function f contains several critical points. Such objects naturally arise in many problems in geometry and physics (see Fig. 2.6).

We now give a simple example. Suppose that a finite group G acts smoothly on a surface X^2, and let f be a G-invariant Morse function; then, as a rule, such function will be complicated. Indeed, if, for instance, the orbit of a critical point x entirely belongs to a connected component of the level line $\{f(x) = \text{const}\}$, then this level contains several critical points.

Of course, a small perturbation can make the function simple by moving critical points into different levels. However, this destroys the symmetry. Thus, in the problems that require studying symmetries of different kinds, one has to investigate complicated Morse functions as an independent object. The molecules W classify Morse functions on M^2 up to fiber equivalence, see [1].

It is convenient to denote every atom (P^2, K) by some letter with number of incoming and outgoing edges. The end of each edge corresponds to a certain boundary

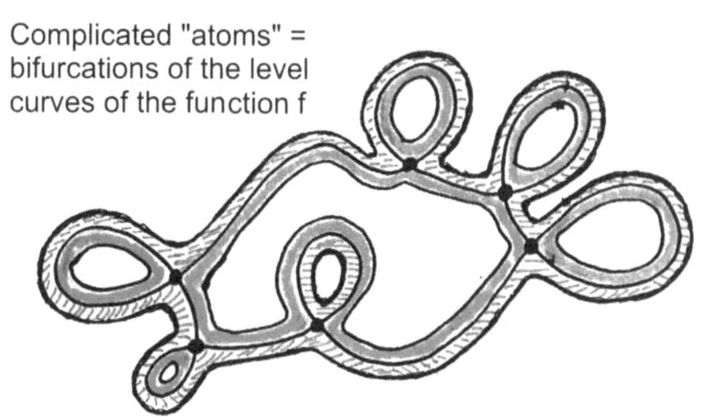

Fig. 2.6 Complicated Morse functions

circle of the surface P. It is important to emphasize that, generally speaking, the ends of an atom (P, K) are not equivalent; because the boundary circles of the surface P are not equivalent in the sense that not every two of them can be matched by a homeomorphism of the pair (P, K) onto itself. Some atoms of low complexity (both orientable and nonorientable) are listed in Fig. 2.7. In the same table, one can see the corresponding pairs of f-graphs, as well as the surface $\tilde{P}$ obtained from P by gluing disks to all of its boundary circles. See details in [1].

2.3 Topology of Integrable Hamiltonian Systems with Two Degrees of Freedom

Consider symplectic manifold M^{2n} with closed nondegenerate skew-symmetric two-form ω. It defines on the space of smooth functions on M^{2n} Poisson bracket

$$\{f, g\} = \omega^{-1}(df, dg).$$

The system of ordinary differential equations defined by vector field v is called *Hamiltonian system* if there exists such function H (called *Hamiltonian*) that for any function g on M^{2n} the equality holds $v(g) = \{g, H\}$. We denote $v = \mathrm{sgrad} H$.

The system is integrable in Liouville sense, if there exists exactly n functionally independent commuting integrals $f_1, \ldots, f_n$, and all vector fields $\mathrm{sgrad} f_i$ are complete. They define the so-called Liouville foliation (see Fig. 2.8). The Liouville theorem states, that if the regular (i.e., f_i are functionally independent on it) common level surface of these functions is compact and connected, then it is a torus T^n. The solution (integral trajectory) in general case determines almost periodic motion on this torus.

This class of Hamiltonian systems contains many important examples from physics an classical mechanics: the different cases of motion of rigid body (Euler case, Lagrange top), geodesic flow on ellipsoid, interaction of the material points, located on the line or on the circle S^1. The classification of such systems is a very difficult task. It turns out, however, that in case of $n = 2$ the full topological classification exists.

Consider a symplectic manifold M^4 with an integrable Hamiltonian system $v = \mathrm{sgrad} H$; let Q_h^3 be a nonsingular compact connected isoenergy 3-surface in M^4. Let f be an additional integral of the system v that is independent of H. We denote its restriction to Q_h^3 by the same letter f. It is assumed to be a Bott function (nondegenerate, see [1]) on Q_h^3. Our aim is to investigate the topology of the Liouville foliation on Q_h^3 defined by the given integrable system. Its nonsingular leaves are Liouville tori, and the singular ones correspond to critical levels of the integral f on Q_h^3.

Now consider a topologically stable (see [1]) integrable system with Bott integral f on an isoenergy 3-surface Q_h^3 and take some singular leaf L of the corresponding Liou-

No	ATOM	f-GRAPHS	CODE	GENUS
	Complexity 1, orientable			
1	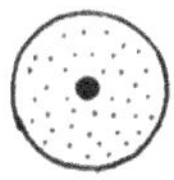		$-A$	S^2
2	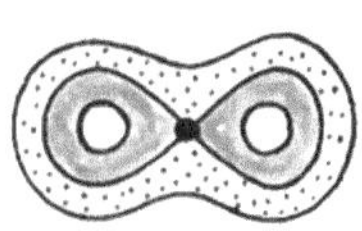	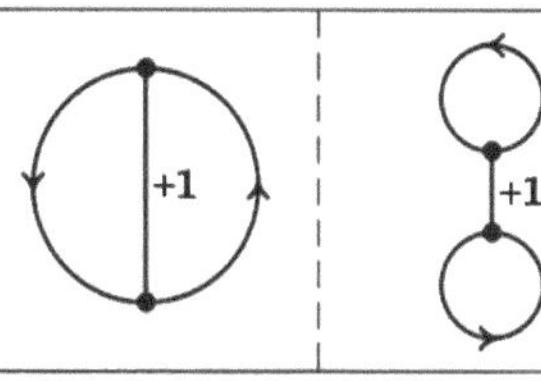	$-B<$	S^2
	Complexity 1, non-orientable			
$\widetilde{1}$	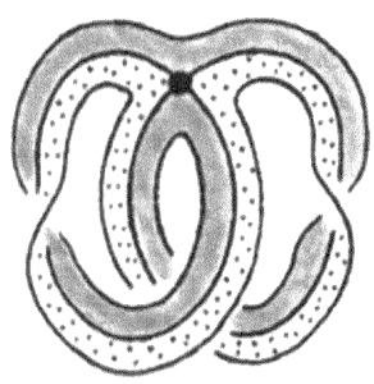	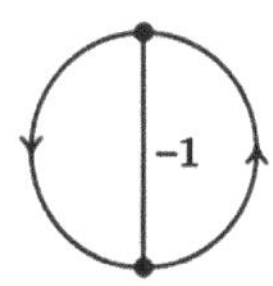	$-\widetilde{B}-$	$\mathbb{R}P^2$
	Complexity 2, orientable			
1		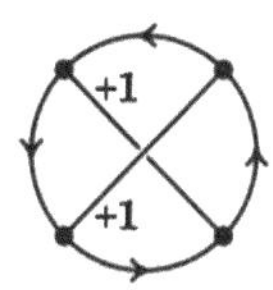	$-C_1-$	T^2
2	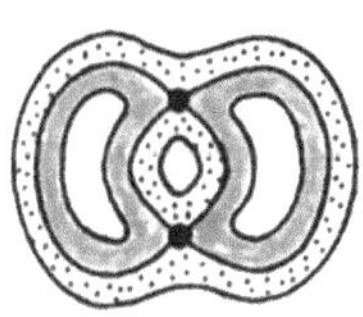	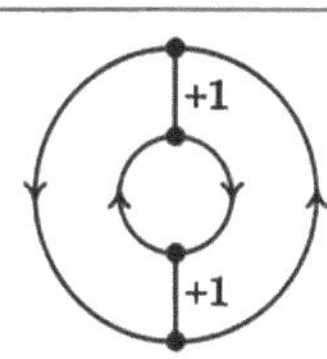	$>C_2<$	S^2
3	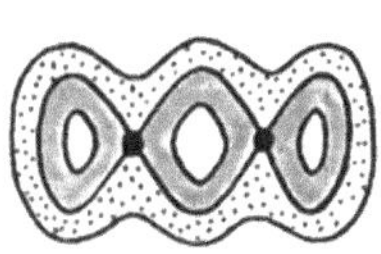	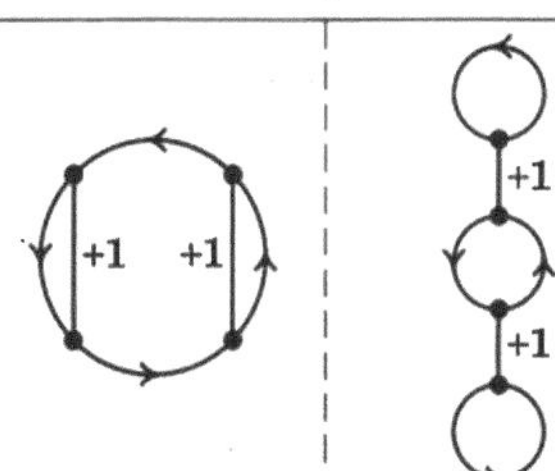	$-D_1\equiv$	S^2

Fig. 2.7 Several atoms of low complexity

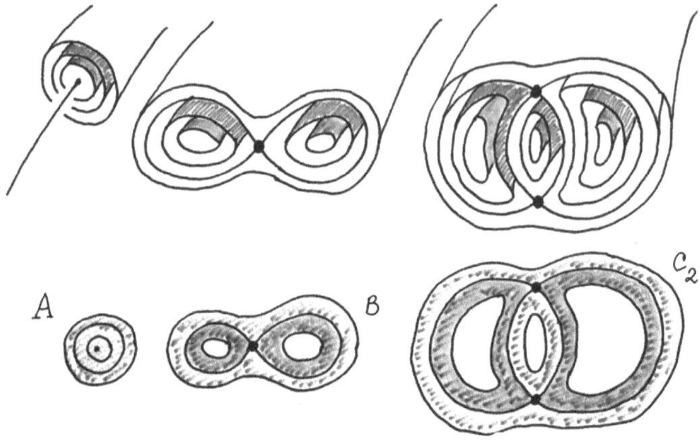

Fig. 2.8 The two-dimensional "cross-section" of Liouville foliation in three-dimensional invariant submanifold around singular leaf

ville foliation on Q_h^3. Consider a neighborhood of this leaf, i.e., a three-dimensional manifold $U(L)$ with the Liouville foliation structure and fixed orientation. By analogy with the two-dimensional case, as neighborhood $U(L)$, we take the connected component of the set $c - \varepsilon \leq f(x) \leq c + \varepsilon$ that contains the singular leaf L (same as in previous section $f(L) = c$ is a critical value of f). Such an object is naturally called a 3-*atom*. However, from the formal viewpoint, we have to be more careful. We shall assume two such 3-manifolds $U(L)$ and $U'(L)$ with the structure of the Liouville foliation to be fiberwise equivalent if
(1) there exists a diffeomorphism between them that maps the leaves of the first Liouville foliation into those of the second one,
(2) this diffeomorphism preserves both the orientation on 3-manifolds and the orientation on the critical circles defined by the Hamiltonian flows.

Definition 2.2 The equivalence class of the three-dimensional manifold $U(L)$ is called a 3-*atom*. The number of critical circles in the 3-atom is called its *atomic weight* or *complexity* (for details, see [10, 11]).

Consider 3-atom $U(L)$ with the structure of a Seifert fibration on it. Let

$$\pi : U(L) \to P^2$$

denote its projection onto a two-dimensional bas P^2 with the embedded graph $K = \pi(L)$. Let us mark those points on the base P^2, into which the singular fibers of the Seifert fibration (i.e., the fibers of type (2, 1)) are projected. Recall that the base P^2 has a canonical orientation. The point is that an orientation is already fixed on $U(L)$, as well as on the fibers of the Seifert fibration. It is clear that, as a result, we obtain some oriented 2-atom (P^2, K).

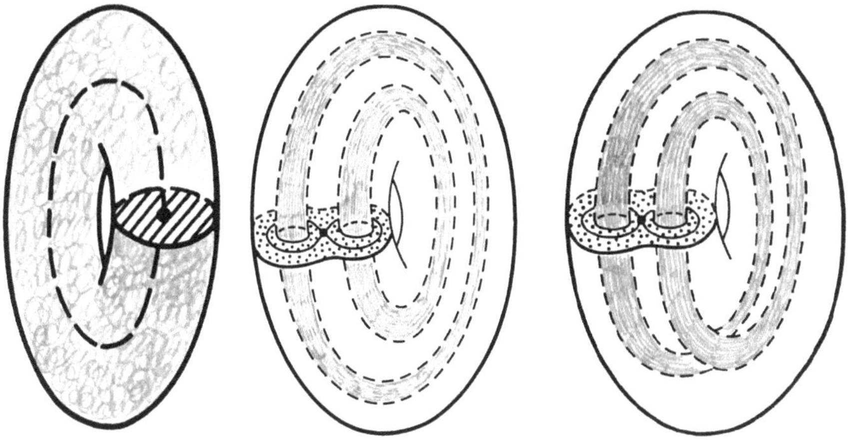

Fig. 2.9 3-atoms A, B and B^*

Theorem 2.1 (Fomenko). *Under the projection $\pi : U(L) \to P^2$, the 3-atom $U(L)$ turns into the 2-atom (P^2, K); and moreover, the singular fibers of the Seifert fibration on the 3-atom are in one-to-one correspondence with the star-vertices of the 2-atom. This correspondence between 2-atoms and 3-atoms is bijection.*

The example of 3-atom is shown in the Fig. 2.9.

Let us describe 3-atom A (see Fig. 2.9). This 3-atom is presented as a solid torus foliated into concentric tori, shrinking into the axis of the solid torus. In other words, the 3-atom A is the direct product of a circle and a disk foliated into concentric circles. From the viewpoint of the corresponding dynamical system, A is a neighborhood of a stable periodic orbit. The examples of saddle 3-atoms are presented in the Fig. 2.9.

Now we make more precise the definition of the Liouville equivalence for integrable Hamiltonian systems. From now on, we shall assume that two Liouville foliations are Liouville equivalent if and only if there exists a diffeomorphism that sends the leaves of the first foliation to those of the second one and satisfies two conditions related to the orientation. Namely, it preserves the orientation of 3-manifolds Q_h^3 and Q'^3_h; and moreover, it also preserves the orientation on the critical circles given by the Hamiltonian flows. The molecule W contains a lot of essential information on the structure of the Liouville foliation on Q_h^3. However, this information is not quite complete. Indeed, the molecule of the form $A{-}A$, for example, informs us that the manifold Q_h^3 is glued from two solid tori foliated into concentric tori in a natural way. However, it does not tell us how this gluing is made, and what three-dimensional manifold is obtained as a result. Therefore, we have to add some additional information to the molecule W, namely, the rules that clarify how to glue the isoenergy surface Q_h^3 from individual 3-atoms. As it was discovered by A.T. Fomenko and H. Zieschang, the molecule W, which corresponds to integrable Hamiltonian system, can be endowed by some numerical marks in such a way, that these *marked molecules* W^* will classify such a systems up to Liouville equivalence. In short, the *marked molecule* W^* is molecule W, equipped with three sets of numbers r_i, ε_i and n_k called *marks*, see [1].

Theorem 2.2 (Fomenko, Zieschang). *Two integrable systems* (v, Q_h^3) *and* (v', Q'^3_h) *are Liouville equivalent if and only if their marked molecule* W^* *and* $W^{*\prime}$ *coincide.*

The marks cannot be chosen arbitrary, as there are several important conditions on them. Any marked molecule W^* with marks, which satisfy these conditions, is called *abstract marked molecule*.

Theorem 2.3 *Any abstract marked molecule* W^* *is realized as a marked molecule of some integrable Hamiltonian system.*

Corollary 2.1 *(1) There exist a one-to-one correspondence between the Liouville equivalence classes of integrable systems and marked molecules. In particular, the set of Liouville equivalence classes of integrable systems is discrete (countable) and has no continuous parameters.*
(2) There exists an enumeration algorithm for marked molecules (i.e., classes of integrable systems).
(3) There exists an algorithm for comparison of marked molecules, i.e., the algorithm that gives answer to the question whether two integrable systems corresponding to given molecule are Liouville equivalent or not.

2.4 Geodesic Flows with Potential on the Surfaces of Revolution

Consider a manifold M, which is diffeomorphic to a cross product of finite interval (a, b) and circle S^1. Let us define a metrics $ds^2 = dr^2 + f^2(r)d\varphi^2$ on M, where r is a natural parameter on $f(r)$. $f(r)$ is a smooth positive function on (a, b), and $f(a) = f(b) = 0$; and, moreover, we assume that a surface of revolution obtained by rotation of this function around the axis Oz is a smooth manifold. φ is a conventional angle coordinate on a circle S^1. Let $V(r)$ be a smooth function on the surface of revolution, obtained by the rotation of $f(r)$. It plays a role of a *potential* of the system.

Definition 2.3 We call a system, defined by pair $(f(r), V(r))$, a *system on the surface of revolution*.

The phase space of this system has a dimension of four: it has coordinates $(r, \varphi, p_r, p_\varphi)$.

Lemma 2.1 *The system given by a pair* $(f(r), V(r))$ *is integrable Hamiltonian system on* T^*M *(on the cotangent bundle of* M*) for any smooth potential function* $V(r)$.

This statement is clear since the investigated system has two integrals of motion: H — energy, and p_φ — linear first integral, the mechanical sense of the later being the projection of kinetic moment on axis of rotation. Hence, according to the definition of Liouville integrability (see [1–3, 12]), this system is integrable.

The Hamilton function of this system has a form

$$H = \frac{p_r^2}{2} + \frac{p_\varphi^2}{2f^2(r)} + V(r). \tag{2.1}$$

We can rewrite the function H in the form

$$H = g^{ij}(r)p_r p_\varphi + V(r). \tag{2.2}$$

According to a famous Maupertuis principle, there exists a value of energy $H = h_0 > \max(V(r))$, such that for any $h > h_0$ the integral trajectories of the vector field $v = \text{sgrad}\, H$ on the isoenergetic level $Q^3 = \{H(r, \varphi, p_r, p_\varphi) = h\}$ coincide with the integral trajectories of the vector field $\widetilde{v} = \text{sgrad}\, \widetilde{H}$, where the Hamitonian $\widetilde{H}$ has a form

$$\widetilde{H} = \frac{g^{ij}(r)}{h - V(r)} p_r p_\varphi. \tag{2.3}$$

It is clear that the vector field $\widetilde{v}$ defines a *geodesic flow* of a Riemannian metrics $\widetilde{g}^{ij}$ on the manifold T^*M, where $\widetilde{g}^{ij}(r) = (h - V(r))g^{ij}(r)$.

So, we can consider the system on the surface of revolution as a geodesic flow with the potential on the surface of revolution.

2.5 The Case of Gravitational Potential: Topological Classification

Now let us examine the systems on the surfaces of revolution, which are defined by pairs $(f(r), V(r))$, where the function $f(r)$ gives us a smooth surface of revolution on $(0, \pi)$, and the potential function $V(r) = \cos r$, which means that we have the action of the gravitational field in such systems.

In this chapter, we want to give the topological classification of such systems. For this purpose, it is necessary to define some useful notions.

Definition 2.4 We will call the map

$$\Phi : M \to R^2 : (r, \varphi, p_r, p_\varphi) \mapsto (H(r, \varphi, p_r, p_\varphi), p_\varphi(r, \varphi, p_r, p_\varphi)) \tag{2.4}$$

the *momentum map*.

Definition 2.5 If the rank of $d\Phi(x)$ is less than 2, then x is a *critical point* of the momentum map, and $\Phi(x)$ is its *critical value*. A set of critical values $\Sigma = \{\xi = \Phi(x),\ x \text{ is a critical point}\}$ is called a *bifurcation diagram*.

For more details, see [7–9].

Let us define an *effective potential* function:

$$U_{eff}(p_\varphi, r) = \frac{p_\varphi^2}{2f^2(r)} + V(r). \tag{2.5}$$

Assume that the following conditions are satisfied:

1. In all points r_i, which are the solutions of the system

$$\begin{cases} \frac{\partial U_{eff}(r,p_\varphi)}{\partial r} = 0 \\ \frac{\partial^2 U_{eff}(r,p_\varphi)}{\partial r^2} = 0 \end{cases} \tag{2.6}$$

we have

$$\frac{\partial^3 U_{eff}(r_i, p_\varphi)}{\partial r^3} \neq 0 \tag{2.7}$$

(it is a condition of the existence of semicubical point of return on the bifurcation diagram);

2. In all points r_i where the function $U_{eff}(r, p_\varphi)$ has a local minima, all the values of $U_{eff}(r, p_\varphi)$ are different for the fixed p_φ. (this condition provides the existence of atoms of only two types A and B in the system).

Theorem 2.4 *If the system satisfies the conditions above, then the bifurcation diagram of this system is constructed of the curves of three types:*

(i) The curve of a "parabole" type (see Fig. 2.10)

(ii) Two points of the rank 0 with coordinates $(H, p_\varphi) = (\pm 1, 0)$. The point $(-1, 0)$ has a center–center type, the point $(1, 0)$ has a focus–focus type (see Fig. 2.11)

(iii) The curve of a "beak" type (see Fig. 2.11: a, b, c particular cases, d the general case)

And besides, the bifurcation diagram may consist of only one curve of type (i) and of any number of the curves of type (iii).

Example 2.1 Let us consider a function $f(r) = \sin r$ (this system called a "spherical pendulum system" (this system was studied in [13])). In this case, the bifurcation diagram has a very simple form: it has no curves of type (iii). The function $f(r)$ and the bifurcation diagram are shown in the Fig. 2.12.

Example 2.2 Consider a function $f(r)$, which gives us a surface with two local maxima at the points a and b (see Fig. 2.13a)). In this case, the bifurcation diagram also has the curves of type (iii) (see Fig. 2.13b)).

With the help of the bifurcation diagrams, we can construct the molecules. All the molecules consist of atoms of only two types A and B.

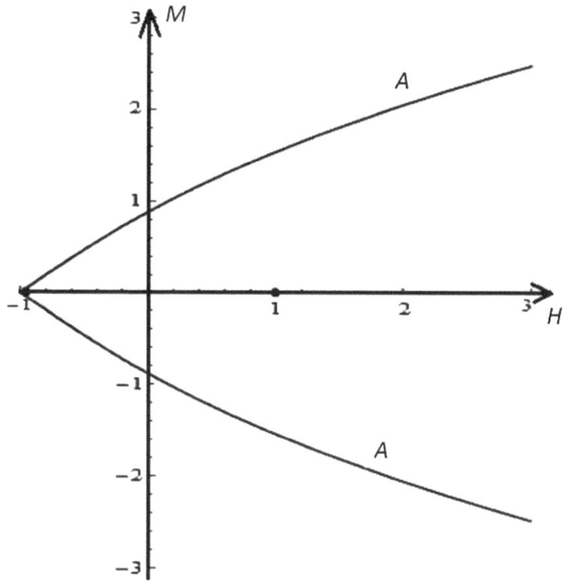

Fig. 2.10 The curve of a "parabole" type and critical points of rank 0

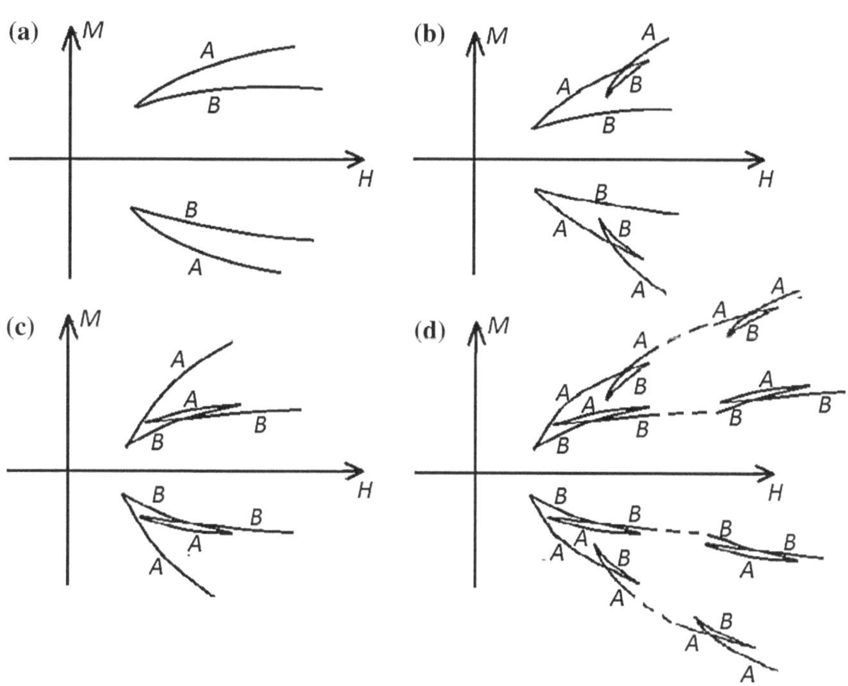

Fig. 2.11 Curves of a "beak" type

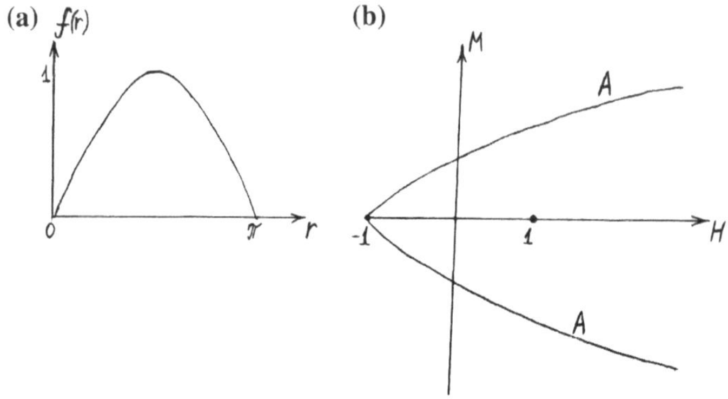

Fig. 2.12 The bifurcation diagram for "spherical pendulum" system

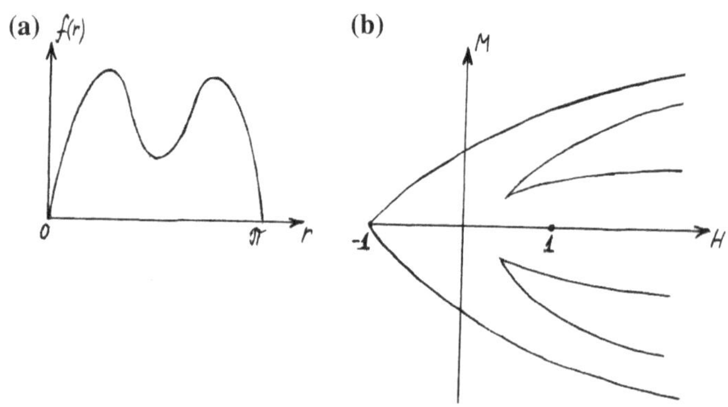

Fig. 2.13 The bifurcation diagram for the system with $f(r) = \sin r - a \cdot \sin^3 r$ and $V(r) = \cos r$

2.6 Topological Equivalence Between Different Integrable Systems

Now we continue to study the equivalence in the sense of Liouville. As soon as we have constructed the molecules, we have some information about the Liouville fibration of the system. But this information is not full (see Chap. 3). Let us consider an arbitrary rib of the molecule. In the preimage of any regular point from this rib, we have one regular torus in T^*M. So we can choose two regular tori, which correspond to the regular points of the rib near both edges of this rib.

Then, we choose a basis on each of these two tori, and we deform these bases toward each other. As a result, we obtain two bases on the torus, which correspond to the central (regular) point of the rib. We can write the transition matrix from one coordinate system to another, and this matrix gives us information about the way of gluing two tori in the ends of the rib (see Fig. 2.14).

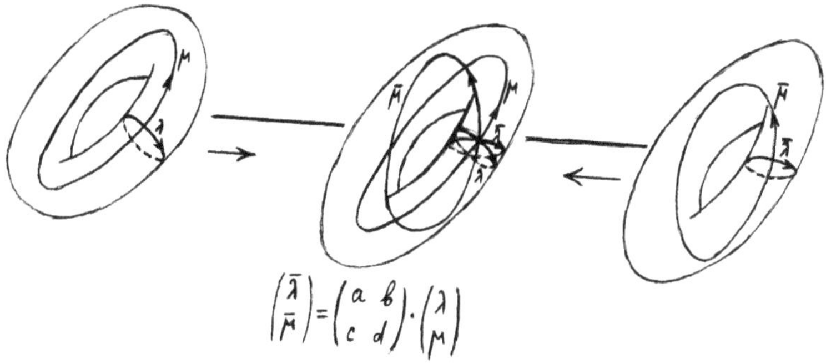

Fig. 2.14 The way of gluing two tori defined by matrix

We can choose the basis on a torus in different ways, therefore gluing matrix depends on the coordinate systems. But we can calculate some numbers given by this matrix, which do not depend on the coordinate system. These invariants are called the *marks* on the ribs of the molecule. The molecule with the marks is called a *Fomenko–Zieschang invariant* (or simply a *marked molecule*).

Theorem 2 says that the Fomenko–Zieschang invariant is the invariant of topological equivalence of the system. Namely, if two integrable Hamiltonian systems, defined by their Hamiltonian vector fields $v = \text{sgrad}H$ and $\widetilde{v} = \text{sgrad}\widetilde{H}$, have the same Fomenko–Zieschang invariants on the isoenergetic surfaces Q^3 and $\widetilde{Q}^3$, these systems are equivalent in the Liouville sense.

Hence, it is reasonable to calculate the marked molecules of integrable systems.

There exist three types of marks: r, ε and n. The mark r is defined by the rule

$$r = \begin{cases} \frac{a}{b} \bmod 1 \in Q/Z, & b \neq 0 \\ \infty, & b = 0 \end{cases} \tag{2.8}$$

The mark ε can be calculated in the following way:

$$\varepsilon_i = \begin{cases} \operatorname{sign} b, & b \neq 0 \\ \operatorname{sign} b, & b = 0 \end{cases} \tag{2.9}$$

The mark n has a more complicated definition (see [4–6, 14]).

Theorem 2.5 *If all the conditions of the Theorem 4 are satisfied, then the marked molecules of the systems with gravitational potential consist of the ribs of the following types: (a) the rib A–A with the mark $r = 0$, if $Q^3 = \{H = h < 1\}$; and with the mark $r = 1/2$, if $Q^3 = \{H = h > 1\}$. The mark $\varepsilon = +1$ in both cases;*
(b) the rib A–B with the mark $r = 0$ and the mark $\varepsilon = +1$;
(c) the rib B–B with the mark $r = \infty$ and the mark $\varepsilon = -1$, if the rib is symmetric relative to the axis OH; and the mark $\varepsilon = +1$ in other case;
(d) if the system admits the atoms of type B, then there exist marks of type n. If

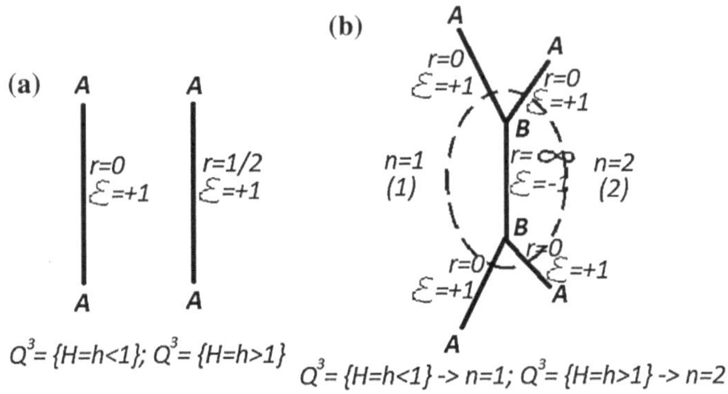

Fig. 2.15 The marked molecules for the systems from the examples 1 and 2

$Q^3 = \{H = h < 1\}$, *then the mark* $n = 1$, *and if* $Q^3 = \{H = h > 1\}$, *the mark* $n = 2$.

The marked molecules for the examples 1 and 2 are shown in Fig. 2.15.

Corollary 2.2 *The molecule of the example* 2 *(when the energy* $H > 1$*) coincides with a molecule in the classical integrable Hamiltonian system called Zhukovskii system. Almost all marks on the ribs of these molecules coincide except for one mark* ε *on the rib* $B{-}B$*. So, the systems are different in the Liouville sense, but their invariants almost coincide.*

References

1. Fomenko, A.T.: Symplectic Geometry, Second revised edn. Gordon and Breach, New York (1995)
2. Bolsinov, A.V., Fomenko, A.T.: Integrable Geodesic Flows on Two-Dimensional Surfaces. Consultants Bureau, New York (2000). (Kluwer Academic/Plenum Publishers, New York)
3. Fomenko, A.T., Konyaev, A.Y.: Algebra and geometry through Hamiltonian systems. In: Zgurovsky, M.Z., Sadovnichiy, V.A. (eds.) Continuous and Distributed Systems. Theory and Applications. Solid Mechanics and Its Applications, pp. 3–21. Springer, Berlin (2014)
4. Bolsinov, A.V., Fomenko, A.T.: Integrable Hamiltonian Systems: Geometry, Topology, Classification. Chapman and Hall/CRC, (A CRC Press Company) Boca Raton (2004)
5. Bolsinov, A.V., Fomenko, A.T.: Integrable Hamiltonian Systems: Geometry, Topology, Classification, 2. Chapman and Hall/CRC, (A CRC Press Company) Boca Raton
6. Fomenko., A.T.: A bordism theory for integrable nondegenerate Hamiltonian systems with two degrees of freedom. A new topological invariant of higher-dimensional integrable systems. Math. USSR Izvestiya **39**(1), 731–759 (1992)
7. Bolsinov, A.V., Fomenko, A.T.: Integrable geodesic flows on the sphere, generated by Goryachev-Chaplygin and Kowalewskaya systems in the dynamics of a rigid body. Math. Notes **56**(1–2), 859–861 (1994)
8. Kudryavtseva, E.A., Nikonov, I.M., Fomenko, A.T.: Maximally symmetric cell decompositions of surfaces and their coverings. Sb. Math. **199**(9), 3–96 (2008)

9. Fomenko, A.T., Konyaev, AYu.: New approach to symmetries and singularities in integrable Hamiltonian systems. Topol. Appl. **159**, 1964–1975 (2012)
10. Fomenko, A.T.: Hidden symmetries in integrable Hamiltonian systems. In: Progress in Analysis. Proceedings of the 8th Congress of the International Society for Analysis, its Applications, and Computation. pp. 22–27 August 2011. Moscow, Peoples' Friendship University of Russia, 2012. Vol. 1, 26–41
11. Kudryavtseva, E.A., Fomenko, A.T.: Every finite group is a symmetry group of a map (of atom-bifurcation), Vestnik MGU, 1. Matem. Mech. **3**, 21–29 (2013)
12. Fomenko, A.T., Kudryavtseva, E.A.: Each finite group is a symmetry group of some map (an atom-bifurcation). Mosc. Univ. Math. Bull. **68**(3), 148–155 (2013)
13. Kantonistova, E.O.: Integer lattices of action variables for generalized Lagrange case. Mosc. Univ. Math. Bull. **1**, 54–58 (2012). ISSN 0201-7385
14. Kantonistova, E.O.: Integer lattices of action variables for spherical pendulum system. Mosc. Univ. Math. Bull. **69**(4), 135–147 (2014)

Chapter 3
Multiplicative and Additive Problems of Partitions of Natural Numbers

Vladimir Nikolaevich Chubarikov and Gleb Vladimirovich Fedorov

Abstract This article deals with the problem of multiplicative factorization of the natural number n on k factors and the additive partition of a natural number n on k terms provided that parameter k is a function of n and $k \to \infty$ with $n \to \infty$. We obtained asymptotic formulas for limiting cases of the order of growth of parameter k, which are characterized by the fact that the form of the asymptotic formula changes when k passes corresponding critical values $k = k_{cr}(n)$. This feature occurs in additive and multiplicative problems of the partition (factorization) of natural numbers. As an application, it is noted the point of maximum of the function of additive partition into unordered terms of interest to the critical state in Maslov's model of Bose-condensate, which built a new distribution corresponding to the real noble gas and the equation of state for him. Another application is the new fast algorithms for computing multiplicative and additive functions of partitions with different conditions on parameters n and k.

3.1 Additive Problems

Euler laid the fundamentals of the theory of partitions of numbers; already the "Introduction to the analysis of the infinite" (1748) [1] is presented partitio numerorum (partition numbers into terms) in a separate chapter. The problem of the partition of natural numbers to the natural term has many interpretations, in particular, the arithmetic basic of statistics Bose–Maslov.

V.N. Chubarikov
Faculty of Mechanics and Mathematics, Lomonosov Moscow State University, GSP-1, Leninskie Gory, Moscow 119991, Russian Federation
e-mail: chubarik2014@live.ru

G.V. Fedorov (✉)
Mechanics and Mathematics, Faculty of Lomonosov Moscow State University, Research Associate of Scientific Research Institute of System Development, Moscow, Russia
e-mail: glebonyat@mail.ru

© Springer International Publishing Switzerland 2015

V.A. Sadovnichiy and M.Z. Zgurovsky (eds.), *Continuous and Distributed Systems II*, Studies in Systems, Decision and Control 30,
DOI 10.1007/978-3-319-19075-4_3

We call *ordered partition* of a natural number n for natural terms its representation of a sum of natural numbers $n = x_1+x_2+\cdots+x_k$, where k is the number of parts of the partition, and solution $(x_1, x_2, \ldots, x_k)$ is a set of natural numbers. The number of solutions of the Diophantine equation is the number of ways in which you can place $k-1$ vertical lines in $n-1$ gaps between n points, that is, there is $\binom{n-1}{k-1}$.

Let $p_k(n)$ denote the number of *disordered partitions* of a natural number n on k of positive terms. Usually partition of n on k parts is presented as a record of these parts in nonincreasing order. Consequently, the number of $p_k(n)$ is equal to the number of solutions in positive integers of the equation

$$x_1 + x_2 + \cdots + x_k = n, \qquad x_1 \geq x_2 \geq \cdots \geq x_k \geq 1.$$

We have the following recursive Euler's formula (Sect. 318, c. 242) [1]:

$$p_k(n) = p_k(n-k) + p_{k-1}(n-1).$$

If k parts $x_1, \ldots, x_k$ are different in the disordered partition, it gives $k!$ ordered partitions. Then for $n \to \infty$ and for fixed k it is to be expected that

$$p_k(n) \sim \frac{\binom{n-1}{k-1}}{k!}. \tag{3.1}$$

In 1941 P. Erdös and J. Lehner [2] proved a more general result.

Theorem 3.1 (Erdös–Lehner) *Let $n \to \infty$ and $k = o(n^{1/3})$. Then the asymptotic relation (3.1) holds.*

In 1942, F.C. Auluck, S. Chowla, and H. Gupta gave a simple proof of this theorem [3].

G.I. Arkhipov and V.N. Chubarikov [4] proved in Theorem 3.1 that $k = O(n^{1/3})$.

In 1941, P. Erdös and J. Lehner [2] proved that the "normal" number of terms k of the partitions of n with $n \to \infty$ is asymptotically equal to $k_{cr} \sim c^{-1}n^{1/2}\log n$, where $c = \pi\sqrt{2/3}$. More precisely, they proved the following assertion.

Theorem 3.2 (Erdös–Lehner) *Let $p(n)$ denotes the number of partitions of n on an arbitrary number of terms and $k = c^{-1}n^{1/2}\log n + xn^{1/2}$. Then the asymptotic relation*

$$\lim_{n\to\infty} \frac{p_k(n)}{p(n)} = \exp\left(-\frac{2}{c}e^{-\frac{cx}{2}}\right)$$

holds and the right-hand side of this equation is a function of the probability distribution.

Proof of theorem based on the application of the principle of inclusion–exclusion, which results in this case the identity

$$p_k(n) = p(n) - \sum_{1\le r\le n-k} p(n-(k+r))+$$

$$+ \sum_{\substack{0<r_1<r_2\\1<r_1+r_2\le n-2k}} p(n-(k+r_1)-(k+r_2))-$$

$$- \sum_{\substack{0<r_1<r_2<r_3\\1<r_1+r_2+r_3\le n-3k}} p(n-(k+r_1)-(k+r_2)-(k+r_3))+\cdots=$$

$$= p(n)(1-S_1+S_2-S_3+\ldots).$$

Next, we use the inequalities Bruno sieve. For any natural number ν we get

$$1-S_1+S_2-S_3+\cdots-S_{2\nu-1}\le\frac{p_k(n)}{p(n)}\le 1-S_1+S_2-S_3+\cdots-S_{2\nu}.$$

From the asymptotic formula of Hardy–Ramanujan for $p(n)$ we deduce

$$S_1\sim\frac{2}{c}\exp\left(-\frac{cx}{2}\right).$$

Moreover, for any ν we have

$$S_\nu\sim\frac{1}{\nu!}\left(\frac{2}{c}\exp\left(-\frac{cx}{2}\right)\right)^\nu.$$

Hence,

$$\frac{p_k(n)}{p(n)}\sim 1+\sum_{\nu=1}^{\infty}(-1)^\nu S_\nu\sim\exp\left(-\frac{2}{c}\mathrm{e}^{-\frac{cx}{2}}\right).$$

Lastly, the asymptotic expression completes the proof of Theorem 3.2.

F.C. Auluck et al. [3] formulated the following hypotheses.

Conjecture 1 *For any fixed n let $p_{k_0}(n)$ denote the largest value of $p_k(n)$, i.e., for every k we have $p_k(n)\le p_{k_0}(n)$. Then, for $n\to\infty$ asymptotic relation $k_0\sim c^{-1}n^{1/2}\log n$ holds.*

Conjecture 2 *For any fixed n* let *$p_{k_0}(n)$ denotes the largest value of $p_k(n)$, i.e. for every k we obtain $p_k(n)\le p_{k_0}(n)$. Then for $k_1<k_2\le k_0$ we have $p_{k_1}(n)\le p_{k_2}(n)$ and for $k_0<k_1<k_2$ we have $p_{k_2}(n)<p_{k_1}(n)$.*

They proved that for any $\delta>0$ and for sufficiently large n, we have $n^{1/2}<k_0\le(1+\delta)c^{-1}n^{1/2}\log n$. In 1945, P. Erdös [5] found that $k_0=c^{-1}n^{1/2}\log n+an^{1/2}+o(n^{1/2})$, where the constant a is determined from the equation $c/2=\exp(-ca/2)$. Thus, Conjecture 1 was solved.

Currently, Conjecture 2 of the intervals of monotonicity of $p_k(n)$ is not solved. The point of maximum of $p_k(n)$ is interesting in critical states in the Maslov model of Bose-condensate. Maslov marks [6] that we obtained from this data "identical with the data for noble gases as far as the noble gases according to the law of corresponding states identical with each other."

We present A.G. Postnikov theorem [7], derived from Ingham Tauberian theorem.

Let there be a given sequence $0 = \lambda_0 < \lambda_1 < \lambda_2 < \lambda_2 < \dots$ of real numbers, and let $\rho(\lambda)$—the number of integers λ_ν, not exceeding λ, i.e.,

$$\rho(\lambda) = \#\{\nu : \lambda_\nu \le \lambda\}$$

and let there exist $B > 0$ and $\beta > 0$ such that

$$\rho(\lambda) = B\lambda^\beta + R(\lambda),$$

and for some constant b, c with $\lambda \to \infty$ asymptotic relation

$$\int_0^\lambda \frac{R(v)}{v}\,dv = b \ln \lambda + c + o(1)$$

holds. Suppose, further, $P_k(\lambda)$—the number of solutions in nonnegative integers $r_1, r_2, \dots,$ of inequality

$$r_1\lambda_1 + r_2\lambda_2 + \dots \le \lambda$$

provided $r_1 + r_2 + \dots = k$. The number of decisions of the previous inequality on an unlimited number of terms k is denoted by $P(\lambda)$.

Theorem 3.3 (Postnikov) *For $\lambda \to \infty$ the following asymptotic relation*

$$P(\lambda) \sim \left(\frac{1-\alpha}{2\pi}\right)^{1/2} e^c M^{-(b+\frac{1}{2})\alpha} \lambda^{(b+\frac{1}{2})(1-\alpha)-\frac{1}{2}} e^{\frac{1}{\alpha}(M\lambda)^\alpha}$$

holds and constant α, M are defined as

$$\alpha = \frac{\beta}{\beta+1}, \qquad M = (B\beta\Gamma(\beta+1)\zeta(\beta+1))^{1/\beta}.$$

In particular, it follows from a theorem of Hardy–Ramanujan on the asymptotics of the number of partitions of a natural number for an unlimited number of positive terms. We set $\lambda_\nu = \nu (\nu = 1, 2, 3, \dots)$. Then

$$\rho(\lambda) = [\lambda], \quad R(\lambda) = [\lambda] - \lambda, \quad B = \beta = 1,$$

$$\alpha = 1/2, \qquad M = \zeta(2) = \pi^2/6.$$

Furthermore, according to Stirling's formula, we have

$$\int_0^\lambda \frac{R(v)}{v}\,dv = \int_0^\lambda \frac{[v]-v}{v}\,dv = -\frac{1}{2}\ln\lambda - \frac{1}{2}\ln 2\pi + o(1).$$

Hence, $b = -1/2$, $c = -\ln(2\pi)/2$. From here we find

$$P(\lambda) \sim \left(\frac{1}{4\pi}\right)^{1/2} \lambda^{-1/2} e^{2\sqrt{\zeta(2)\lambda}} = \frac{1}{2\sqrt{\pi\lambda}} e^{\pi\sqrt{\frac{2\lambda}{3}}}.$$

In 2012, V.E.Nazaykinsky [8] found an elementary method to derive the asymptotic behavior of the entropy of the gas Bose–Maslov $\ln P(\lambda)$. If $\lambda \to \infty$, it has the form

$$\ln P(\lambda) \sim (\beta+1)\left(\frac{B\Gamma(\beta+1)\zeta(\beta+1)}{\beta^\beta}\right)^{\frac{1}{\beta+1}} \lambda^{\frac{\beta}{\beta+1}},$$

which is consistent with the asymptotic formula of Postnikov for $P(\lambda)$.

3.2 Multiplicative Problems

We define the *multidimensional divisor function* $\tau_k(n)$ in a standard way as the number of representations of a positive integer n in the form of a product $x_1 \cdot x_2 \cdot \ldots \cdot x_k = n$, where $x_1, x_2, \ldots, x_k$ are positive integers; we assume that $\tau_k(0) = 0$, $\tau_k(1) = 1$, and $\tau_1(n) = 1$. In the case $k = 2$, the value $\tau_2(n) = \tau(n)$ equals the number of different divisors of the positive integer n. In the general case, given a multidimensional divisor function $\tau_k(n)$, we refer to the number of factors k in the factorization of n as the dimension of this function and the number n itself as the argument of the divisor function.

In 1907, S. Wigert showed in [9] that

$$\max_{n\le x} \tau(n) = \exp\left(\ln 2 \frac{\ln x}{\ln\ln x} + O\left(\frac{\ln x \cdot \ln\ln\ln x}{(\ln\ln x)^2}\right)\right).$$

In 1915, S. Ramanujan [10] gave a rather interesting simpler proof of this relation. Sharper upper bounds for this function were obtained in [11, 12]. A generalization for any fixed dimension $k \ge 2$ is the upper limit

$$\limsup_{n\to\infty} \frac{\ln \tau_k(n) \ln\ln n}{\ln n} = \ln k. \tag{3.2}$$

In [13] G.V. Fedorov obtained the upper order of growth for the ratio of the number of divisors of neighboring binomial coefficients. For integers $k \geq 1$ and $n > \prod_{p \leq k+1} p^{[\log_p k]} + k$, the exact inequality

$$T_k(n) = \frac{\tau\left(C_n^k\right)}{\tau\left(C_n^{k+1}\right)} \leq \frac{\tau(k+1)}{2}, \tag{3.3}$$

was proved; moreover, for each integer $k \geq 1$, there exist infinitely many positive integers n for which relation (3.3) is an equality.

It turns out that if $k = k(n) \to \infty$ as $n \to \infty$, then the maximum value (in the sense of the upper limit) of the divisor function $\tau_k(n)$ differs from the classical upper limit (3.2), provided that the dimension k increases rapidly enough; moreover, the upper limit is attained for a power sequence rather than for the sequence of products of consecutive primes.

In [14] we proved the following theorem.

Theorem 3.4 (Fedorov) *Suppose $k = k(n) \to \infty$ as $n \to \infty$. If $\frac{k}{\log_2 n} \to 0$, then the upper limit*

$$\limsup_{n\to\infty} \frac{\log_2 \tau_k(n) \cdot \log_2 \log_2 n}{\log_2 n \cdot \log_2 k} = 1$$

is attained. If $\frac{\log_2 n}{k} \to 0$, then the upper limit

$$\limsup_{n\to\infty} \frac{\log_2 \tau_k(n)}{\log_2 n \cdot \log_2 \frac{k}{\log_2 n}} = 1$$

is attained.

Note that, for fixed values k of the dimension, the upper limit is the same as in the first case of this theorem.

The critical case, when $k \sim \ln n$, was analyzed in article [15].

Theorem 3.5 (Fedorov) *If $\frac{\ln n}{k} = A + o(1)$, then*

$$\limsup_{n\to\infty} \frac{\ln \tau_k(n)}{\ln n} = \log_2 \left(\frac{\ln 2}{A} + 1\right) + \frac{1}{A} \cdot \ln \left(\frac{A}{\ln 2} + 1\right),$$

that is,

$$\limsup_{n\to\infty} \frac{\log_2 \tau_k(n)}{\log_2 n \cdot \log_2 \left(\frac{k}{\log_2 n} + 1\right) + k \cdot \log_2 \left(\frac{\log_2 n}{k} + 1\right)} = 1.$$

In 2001, A.I. Pavlov [16] discovered a similar effect for the mean value of the multidimensional divisor function with growing dimension. For a fixed value of the parameter k, the following relation holds (see, e.g., [17]):

$$D_k(x) = \sum_{n \le x} \tau_k(n) = x P_k(\ln x) + O\left(x^{\alpha_k + \epsilon}\right), \tag{3.4}$$

where P_k is a $(k-1)$th degree polynomial, $\epsilon > 0$ is an arbitrary small fixed real number, and $\alpha_k < 1$. If the parameter k increases as $x \to \infty$ so that $k \succ \sqrt{\ln x}$, then the principal term of the asymptotic expression for $D_k(x)$ differs from the expected principal term, which can be extracted from (3.4):

$$x \frac{(\ln x)^{k-1}}{(k-1)!}.$$

In [18], G.V. Fedorov continued Pavlov's study.

Theorem 3.6 (Fedorov) *Suppose k is an integer parameter, $k = k(x) \to \infty$ as $x \to \infty$, and $k \le (\ln x)^{\frac{4}{5+\rho}}$ for some fixed $0 < \rho < \frac{1}{3}$. Then the asymptotic formula*

$$D_k(x) = x \frac{(\ln x)^{k-1}}{(k-1)!} \exp\{q_k(x)\} L_k(x) \left(1 + O\left(\frac{k^{5+\rho}}{\ln^4 x}\right) + O\left(k^{3\rho - 1}\right)\right),$$

takes place, where the functions $q_k(x)$ and $L_k(x)$ are defined as

$$q_k(x) = \gamma_0 \frac{k^2}{\ln x} - \left(\gamma_0^2 + \gamma_1\right) \frac{k^3}{\ln^2 x} + \left(\frac{5}{3}\gamma_0^3 + 3\gamma_0\gamma_1 + \frac{\gamma_2}{2}\right) \frac{k^4}{\ln^3 x},$$

$$L_k(x) = 1 - \frac{k}{\ln x}\left(\gamma_0 + \frac{3}{2}\right) + \frac{k^2}{\ln^2 x}\left(\frac{3\gamma_0^2}{2} + \gamma_1 + 3\gamma_0 + \frac{7}{4}\right) -$$

$$- \frac{k^3}{\ln^3 x}\left(\frac{21}{4}\gamma_0^2 + \frac{5}{2}\gamma_0^3 + 3\gamma_0\gamma_1 + \frac{3}{2}\gamma_1 + \frac{21}{4}\gamma_0 + \frac{15}{8}\right),$$

and the Stieltjes constants are defined as

$$\gamma_n = \lim_{m \to \infty} \left(\sum_{k=1}^{m} \frac{(\ln k)^n}{k} - \frac{(\ln m)^{n+1}}{n+1}\right),$$

in particular, $\gamma_0 = \gamma$ is the Euler constant.

This theorem implies that the principal terms in the asymptotic expressions for the mean value of the multidimensional divisor function differ for k in each of the ranges $(\ln x)^{\frac{m}{m+1}} \le k \le (\ln x)^{\frac{m+1}{m+2}}$, where m takes integer values such that $0 \le m \le 3$.

Acknowledgments The second author was supported in part by RFFI grant 13-01-12402.

References

1. Euler, L.: Introductio in Analysin Infinitorum T.I., Chapter 16.—Losannae, Marcum—Michaelem Bouscuet 1748
2. Erdös, P., Lehner, J.: The distribution of the number of summands in the partitions of a positive integer. Duke Math. J. **8**, 335–345 (1941)
3. Auluck, F.C., Chowla, S., Gupta, H.: On the maximum value of the number of partitions of n into k parts. J. Indian Math. Soc. (N.S.) **6**, 105–112 (1942)
4. Archipov, G.I., Chubarikov, V.N.: Arithmetic questions of the Bose-Maslov statistics. Uchen. Zap. OGU. **6**, 1–10 (2012) (in Russian)
5. Erdös, P.: On some asymptotic formulas in the theory of partitions. Bull. Am. Math. Soc. **46**, 185–188 (1946)
6. Maslov, V.P.: Mathematical solution of the Gibbs paradox. Math. Notes **89**(2), 266–276 (2011)
7. Postnikov, A.G.:Vvedenie v analiticheskuu teoriu chisel M.: Naika, 1971, 416 p. (in Russian)
8. Nazaikinskii, V.E.: On the Entropy of a Bose-Maslov Gas. Dokl. Math. **87**(1), 50–52 (2013)
9. Wigert, S.: Sur l'ordre de grandeur du nombre des diviseurs d'un entier. Arkiv för Matematik, Astronomi och Fysik **3**, 1–9 (1907)
10. Ramanuman, S.: Highly composite numbers. Proc. Lond. Math. Soc. **14**(2), 347–409 (1915)
11. Nicolas, J.L., Robin, G.: Majorations explicites pour le nombre de diviseurs de n. Can. Math. Bull. **26**, 485–492 (1983)
12. Robin, G.:Thése d'état Université de Limoges, France (1983)
13. Fedorov, G.V.: On the number of divisors of binomial coefficients. Math. Notes **93**(2), 308–316 (2013)
14. Fedorov, G.V.: The upper limit value of the divisor function with growing dimension. Dokl. Math. **88**(2), 529–531 (2013)
15. Fedorov, G.V.: The greatest order of the divisor function with increasing dimension. Mathematica Montisnigri **XXVIII**, 17–25 (2013)
16. Pavlov, A.I.: Asymptotics of an Arithmetic Sum. Dokl. Math. **63**, 48–51 (2001)
17. Karacuba, A.A.: Uniform approximation of the remainder term in the Dirichlet divisor problem. Math. USSR-Izvestiya **6**(3), 467–475 (1972)
18. Fedorov, G.V.: On a theorem of A.I. Pavlov. Dokl. Math. **86**(2), 648–649 (2012)

Chapter 4
Critical Analysis of Amino Acids and Polypeptides Geometry

Alexander O. Ivanov, Alexander S. Mishchenko and Alexey A. Tuzhilin

At ubi materia, ibi Geometria.
Johannes Kepler

Abstract We present several geometrical and statistical methods analyzing information contained in Protein Data Bank for consistency and general theory fitting. In addition, we include a list of examples demonstrating the necessity of careful analysis and sampling of the PDB items. In particular, we show some examples visualizing the distribution of disturbance level of the famous Pauling plane low. Besides, we present a method of automatical helices finding based on discrete analogues of curvature and torsion of smooth curve.

A.O. Ivanov (✉) · A.A.Tuzhilin
Chair of Differential Geometry and Applications, Faculty of Mechanics and Mathematics, Lomonosov Moscow State University, Moscow 119991, Russia
e-mail: aoiva@mech.math.msu.su

A.O. Ivanov
Chair of Mathematical Modelling, Bauman Moscow State Technical University, Moscow 119991, Russia

A.A.Tuzhilin
e-mail: tuz@mech.math.msu.su

A.S. Mishchenko
Chair of High Geometry and Topology, Faculty of Mechanics and Mathematics, Lomonosov Moscow State University, Moscow 119991, Russia
e-mail: asmish.prof@gmail.com

© Springer International Publishing Switzerland 2015
V.A. Sadovnichiy and M.Z. Zgurovsky (eds.), *Continuous and Distributed Systems II*, Studies in Systems, Decision and Control 30,
DOI 10.1007/978-3-319-19075-4_4

4.1 Introduction

The investigation of polypeptides confirmation is a well-known and very important problem. In particular, its great importance is related to the fact that conformation changes of proteins in a living cell motivated by some mutation could lead to serious diseases. Notice that experimental determination of such mutations consequences is time-consuming and expensive process. Therefore, it is naturally and important to perform those experiments by means of computer. But the modeling problem itself turns out to be very hard and in spite of many efforts of the worldwide community the problem is very far from its solution.

Notice that this problem stands at the interfaces of many sciences such as biology, chemistry, physics, geometry, variational calculus, probability, computational mathematics. Therefore, it is necessary to attract attention of specialists from very different branches of science and to create a possibility for the interested novices to get abreast of and to start their own experiments with the data collected.

Central place in collecting the information concerning the structure of polypeptides is taken by the Protein Data Bank (PDB). The opening section of the present paper is devoted to PDB. We describe how to extract necessary geometrical information on proteins from PDB in a reasonable time. Then we tell about several disappointing obstacles we faced with during our work with PDB.

The initial aim of the paper was to demonstrate an efficiency of some geometrical analogies for conformation investigations. But at some instant we have decided to verify the PDB data for "typicalness" to imagine more exactly the subject we deal with. The results of this research are presented in the first part of the paper. After several "clean outs," we have extracted a part of the base which probably contains "typical" polypeptides. We hope that our methods of geometrical data analysis could be useful for creating more reliable databases.

In final sections we show how to visualize "mobility" of amino acids in polypeptides, and describe how to investigate the space form of polypeptides by means of two functions which are discrete analogues of curvature and torsion of a space curve in the sense of classical differential geometry. Notice that the space structure of polypeptides is investigated in a huge number of papers. Here we list some of them most related to our research: [1–5, 9–13].

The present paper is based on the materials published in [6, 7]. New results are contained in Sect. 4.5 mostly.

The work was supported by Russian Scientific Foundation (RNF), Project 14-50-00029.

4.2 Protein Data Bank

In the present section, we would like to share our impression of using Protein Data Bank (http://www.rcsb.org/pdb/home/home.do) for geometry investigation of amino acids in polypeptides. As a matter of fact, we wanted to extract from PDB an information on the three-dimensional structure of polypeptides. We decided to use the files with `pdb`-extension: those files contain lines of the following form:

```
ATOM 1 N ASN A 32     65.950  -13.181  61.696     1.00  0.00 N
```

where 7–9 positions contain the space coordinates of the atom (in the example, they are `65.950 -13.181 61.696`) whose name is located on the third position (in the example it is `N`). The fourth position presents a reference to the amino acid containing this atom, the second position is the number of the amino acid in the polypeptide under consideration. All details concerning the structure of `pdb`-format can be found in more than two hundred pages file

ftp://ftp.wwpdb.org/pub/pdb/doc/format_descriptions/Format_v33_Letter.pdf

To work with the database, we have written our own programs in *Mathematica* (Wolfram Research Group). Users having fast Internet need not pre-download `pdb`-files. For example, to use the file describing the polypeptide with the short name `2LTR` it suffices to run the following command:

pdbFile=Import["http://www.rcsb.org/pdb/files/2LTR.pdb", "List"]

and after that the variable `pdbFile` will be the list of strings forming the file `2LTR.pdb`. Those strings are ready for further processing by means of *Mathematica* commands to extract necessary information. To work with another polypeptide, one just needs to change `2LTR` in the above command by the short name of the new polypeptide. It remains to find out how to get a list of the polypeptides short names.

4.2.1 Extracting Information from PDB

Here we list our algorithm:

(1) open the cite `http://www.rcsb.org/pdb/home/home.do`;
(2) at the left upper corner press the button `Advanced` located under the button `Search`;
(3) at the appeared tab `Advanced Search Interface` do:
   (a) at the tab `Choose a Query Type` select `All/Experimental Type/Molecule Type`; then the following two fields appear: `Experimental Method` and `Molecule Type`;
   (b) at the tab `Molecule Type` choose `Protein`;
   (c) press the right button `Result Count`; the number of chosen objects appears under the button: in our case it was written `89790 PDB Entries (Structures)`;
   (d) click this inscription, and below we see the list of all polypeptides selected together with their short names and many other information; on default all the polypeptides are chosen (all checkmarks are marked).

Of course, it is convenient to save the list of all short names somewhere. To do that click the field `Filter` (just over the list of polypeptides in `Menu` tab) and choose the tab `Download Checked`. This opens the window `Structure Downloaded` which contains the field `Enter PDB IDs:`, where all the short names of polypeptides are listed (they coincide with the corresponding names of `pdb`-files). Click this field, select all elements from the list (press `Ctrl-A` under Windows), copy it to clipboard (press `Ctrl-C` under Windows), create or open some text file, and copy from the clipboard into it (press `Ctrl-V` under Windows). Then save the file. Now it contains the short names of all polypeptides currently present in PDB.

Remark 4.1 Verify if there is a "space" after the last name in the last line of the resulting file, and delete it (the presence of the "space" led to errors while running our programs in *Mathematica*).

By the way, similar algorithm provides an opportunity to download all the `pdb`-files with the descriptions of polypeptides. To do that, after the appearance of the tab `Structure Downloaded`, instead of the list saving, do the following:

(1) in the column `Download Type` remove the checkmark in the field `mmCIF Format` and mark the field `PDB Format`;
(2) in the column `Compression Type` mark the field `uncompressed`;
(3) press `Next`;
(4) at this point you need JAVA: it calls the button `Browse` which gives an opportunity to choose a path for files saving; after choosing the path, press `Start Download`. After some waiting you get the set of the `pdb`-files describing the polypeptides chosen.

We start with visualization of the polypeptides and some of their fragments.

4.2.2 The First Steps in Polypeptides Visualization

It is not difficult to choose the strings of the form `"ATOM ..."` (see above); group the strings in accordance with consecutive amino acids residues (below we call them *amino acids* for shortness) using the sixth position of the string containing the number of the amino acid in the polypeptide chain under consideration; and extract from each such string the name of the atom, its space coordinates, and the name of the corresponding amino acid. To complete our preparation, we need to know the bonds structure. Unfortunately, the `pdb`-file does not contain it. Of course, the structure of amino acids bonds is well known, but what is the correspondence between the consecutive strings `"ATOM ..."` describing an amino acid in `pdb`-file and atoms from known structural formula of this amino acid? (Some atoms of each amino acid belong to the same type, that is an obstacle to construct the correspondence directly.) We did not find this information in the file `Format_v33_Letter.pdf`, but we found an old version, namely, http://www.wwpdb.org/documentation/PDB_format_1992.pdf

involving the corresponding pictures, see p. 27. However, the latter information does not cover the hydrogens. Nevertheless, the bonds with hydrogens can be restored from their notations. For example, if a hydrogen is written as `HD`, then we need to find in the current amino acid an atom of either carbon, or nitrogen, or oxygen, or sulfur which is denoted by `CD`, or `ND`, or `OD`, or `SD`, respectively, and to bond it with this HD. As a result we constructed the following bonds table:

```
N→CA, CA→C, C→O, CA→CB, C→OXT,
CB→CG, CB→CG1, CB→CG2, CB→OG, CB→OG1, CB→SG,
CG→CD, CG→CD1, CG→CD2, CG→ND1, CG→ND2, CG→OD1, CG→OD2,
CG→SD,
CG1→CD1,
CD→CE, CD→NE, CD→OE1, CD→OE2,
CD1→CE1, CD1→NE1,
CD2→CE2, CD2→CE3, CD2→NE2,
CE→NZ, CE→SD,
CE1→CZ, CE1→ND1, CE1→NE2,
CE2→CZ, CE2→CZ2, CE2→NE1,
CE3→CZ3,
CH2→CZ2, CH2→CZ3,
CZ→NE, CZ→NH1, CZ→NH2, CZ→OH,
CA→HA, CA→HA2, CA→HA3,
CB→HB, CB→HB1, CB→HB2, CB→HB3,
CG→HG, CG→HG2, CG→HG3,
CG1→HG11, CG1→HG12, CG1→HG13,
CG2→HG21, CG2→HG22, CG2→HG23,
CD→HD2, CD→HD3, CD→N, CD→NE2,
CD1→HD1, CD1→HD11, CD1→HD12, CD1→HD13,
CD2→HD2, CD2→HD21, CD2→HD22, CD2 →HD23,
CE→HE1, CE→HE2, CE→HE3,
CE1→HE1,
CE2→HE2,
CE3→HE3,
CH2→HH2,
CZ→HZ,
CZ2→HZ2,
CZ3→HZ3,
N→H, N→H1, N→H2, N→H3,
ND1→HD1,
ND2→HD21, ND2→HD22,
NE→HE,
NE1→HE1,
NE2→HE21, NE2→HE22,
NH1→HH11, NH1→HH12,
NH2→HH21, NH2→HH22,
NZ→HZ1, NZ→HZ2, NZ→HZ3,
OG→HG,
OG1→HG1,
OH→HH,
SG→HG
```

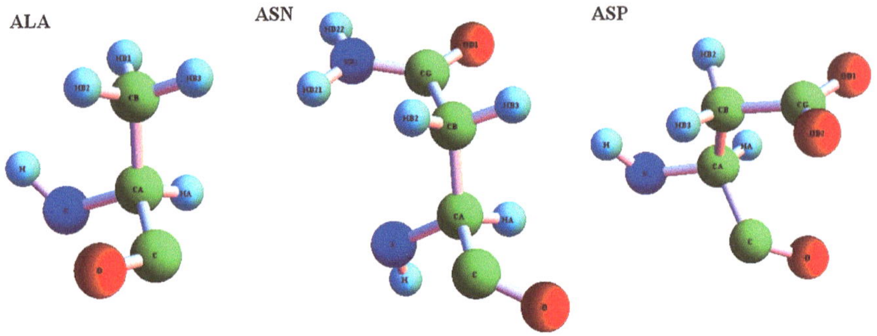

Fig. 4.1 Amino acids visualization, examples

Using *Mathematica* capabilities related to graphs visualization we created our own functions to visualize various parts of polypeptides, in particular, their amino acids. Several examples of amino acids visualization are shown in Fig. 4.1.

However, it turns out that some amino acids images contain evident errors, see Fig. 4.2.

What is the reason? Notice the bond `CD`→`N` empathized in the above Table. It turns out that this bond spoils everything. We found out that if we omit this bond in all amino acids except proline, then we get no errors. But to visualize proline, see Fig. 4.3, we need to add the bond `CD`→`N`.

Is it possible to solve this problem by changing the notations of the atoms in amino acids? It suffices to re-denote the `CD`-atoms in proline to `CF` or, say, `CW` (i.e., to find the notation that does not appear among other atoms names of other amino acids). Apparently, PDB designers do not face this problem.

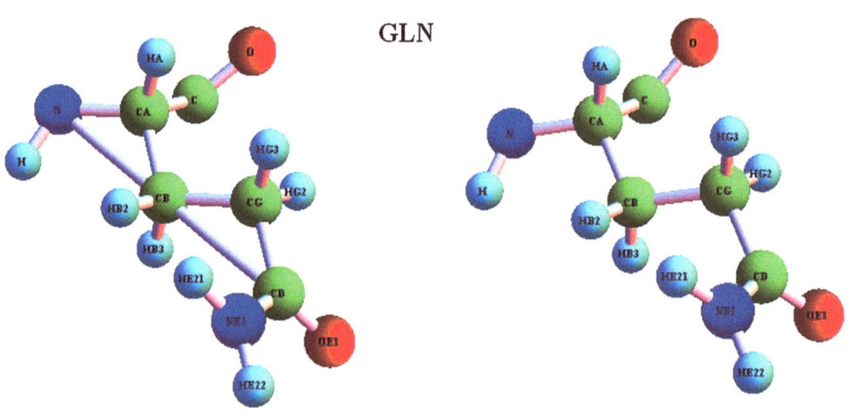

Fig. 4.2 Incorrect structural formula of glutamine (*left*) and the correct one (*right*)

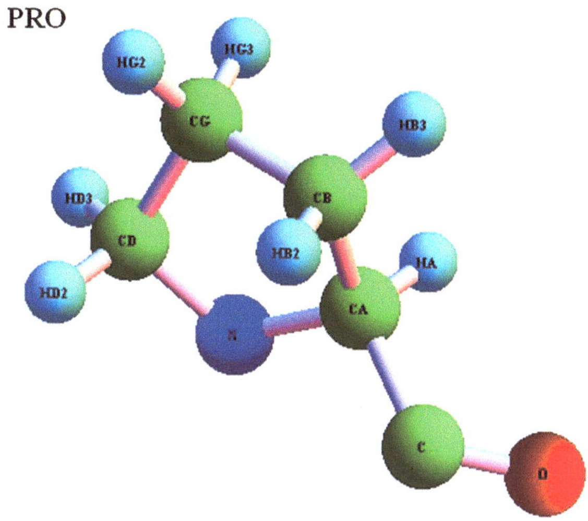

Fig. 4.3 Proline

After unraveling the bonds, we decide to compare the geometry of the same type amino acids placed in the same polypeptide or in different ones. Again, we faced several problems.

4.2.3 Some Difficulties in PDB-Files Treatment

(1) Not all `pdb`-files contain information on hydrogens. It turns out, that some methods of polypeptides' space structure investigation, for example, X-ray analysis, give no information about hydrogens. We find out that for our purposes the best method is the nuclear magnetic resonance (NMR). To choose the files obtained by this method, we go back to PDB and in the above algorithm creating the list of short names we mark `NMR` in the field `Experimental Method` (before that we did not modify this field). Now we get ten times less files in the `Result Count` (about 9000).

(2) The next files do not contain strings `"ATOM ..."`:

`1R9V.pdb, 1S1O.pdb, 1S4A.pdb, 2KVJ.pdb`

(they contain only strings of the form `"HETATM ..."`). We deleted those files from our list.

(3) Usually a `pdb`-file contains several variants of the same polypeptide, which are referred as *models*. Each model starts with string `MODEL N`, where `N` is the number of the model, and ends by string `ENDMODEL ....` We decide to choose the first model in each file. For some details on the other models see Sect. 4.5.

Notice that each consecutive `MODEL N` and `ENDMODEL ...` restrict a block of strings `"ATOM ..."` completed by terminal string `"TER ..."` (the `"TER ..."` does not contain coordinates). Hence, proceeding a pdb-file, we delete all the beginning up to the first string "ATOM ...", then we form the list of consecutive strings until "TER ...".

(4) Creating the algorithm described above, we supposed that between MODEL 1 and the first "TER ..." all strings look like "ATOM ...". However, it turned out that it was incorrect, because in some files such blocks may contain strings of the form "HETATM ..." (those strings correspond to so-called *heterogeneous groups*, the residues distinct from the ones of the standard 20 amino acids). Here we write down an example from the file 1A13.pdb:

```
.........................
ATOM     235 HD22 LEU A  14       -7.905  -4.241   0.457  1.00  0.00           H
ATOM     236 HD23 LEU A  14       -9.150  -4.650   1.637  1.00  0.00           H
HETATM   237  N   NH2 A  15       -4.520  -3.275   5.538  1.00  0.00           N
HETATM   238  HN1 NH2 A  15       -4.461  -2.351   5.201  1.00  0.00           H
HETATM   239  HN2 NH2 A  15       -3.975  -3.556   6.306  1.00  0.00           H
TER      240      NH2 A  15
.........................
```

We found all such files (604 ones) and excluded them from our list.

(5) Sometimes the description of an amino acid contains several copies of the same atom (so-called *alternative location*). The information about the location type is contained in the fifth position of the corresponding string "ATOM ...". We decided to take only the strings containing the symbol "A" at the fifth position. By the way, if such *alternative fragments* are too long, then they are gathered in a separate list after the terminal string. So, we improved our algorithm to take into account the fifth position.

(6) In several files (e.g., 1PNJ.pdb, 2PNI.pdb) some consecutive amino acids are marked by the same number but distinguished by the next position which is referred as iCode (code for insertion of residues). Here we give an example from the file 1PNJ.pdb:

```
.........................
ATOM       1  N   GLY A   1A      18.846  -2.578 -21.181  1.00  0.00           N
ATOM       2  CA  GLY A   1A      19.477  -3.527 -22.142  1.00  0.00           C
ATOM       3  C   GLY A   1A      18.401  -4.379 -22.820  1.00  0.00           C
ATOM       4  O   GLY A   1A      18.321  -4.433 -24.032  1.00  0.00           O
ATOM       5  H1  GLY A   1A      17.815  -2.713 -21.185  1.00  0.00           H
ATOM       6  H2  GLY A   1A      19.216  -2.760 -20.227  1.00  0.00           H
ATOM       7  H3  GLY A   1A      19.069  -1.602 -21.461  1.00  0.00           H
ATOM       8  HA2 GLY A   1A      20.022  -2.968 -22.888  1.00  0.00           H
ATOM       9  HA3 GLY A   1A      20.157  -4.175 -21.609  1.00  0.00           H
ATOM      10  N   SER A   1B      17.604  -5.017 -22.000  1.00  0.00           N
ATOM      11  CA  SER A   1B      16.455  -5.834 -22.505  1.00  0.00           C
ATOM      12  C   SER A   1B      15.466  -5.012 -23.344  1.00  0.00           C
ATOM      13  O   SER A   1B      14.698  -5.571 -24.101  1.00  0.00           O
ATOM      14  CB  SER A   1B      15.737  -6.447 -21.291  1.00  0.00           C
ATOM      15  OG  SER A   1B      16.667  -7.386 -20.765  1.00  0.00           O
ATOM      16  H   SER A   1B      17.763  -4.960 -21.035  1.00  0.00           H
ATOM      17  HA  SER A   1B      16.848  -6.626 -23.126  1.00  0.00           H
ATOM      18  HB2 SER A   1B      15.523  -5.698 -20.541  1.00  0.00           H
ATOM      19  HB3 SER A   1B      14.832  -6.958 -21.575  1.00  0.00           H
```

```
ATOM     20  HG  SER A   1B      17.396  -7.490 -21.384  1.00  0.00           H
ATOM     21  N   MET A   1C      15.515  -3.708 -23.185  1.00  0.00           N
ATOM     22  CA  MET A   1C      14.651  -2.769 -23.978  1.00  0.00           C
ATOM     23  C   MET A   1C      13.159  -2.991 -23.648  1.00  0.00           C
ATOM     24  O   MET A   1C      12.312  -3.029 -24.519  1.00  0.00           O
ATOM     25  CB  MET A   1C      14.942  -3.005 -25.496  1.00  0.00           C
ATOM     26  CG  MET A   1C      14.656  -1.721 -26.304  1.00  0.00           C
ATOM     27  SD  MET A   1C      15.658  -0.251 -25.959  1.00  0.00           S
ATOM     28  CE  MET A   1C      17.288  -0.860 -26.468  1.00  0.00           C
ATOM     29  H   MET A   1C      16.138  -3.336 -22.527  1.00  0.00           H
ATOM     30  HA  MET A   1C      14.912  -1.756 -23.706  1.00  0.00           H
ATOM     31  HB2 MET A   1C      15.976  -3.283 -25.625  1.00  0.00           H
ATOM     32  HB3 MET A   1C      14.325  -3.806 -25.874  1.00  0.00           H
ATOM     33  HG2 MET A   1C      14.777  -1.962 -27.348  1.00  0.00           H
ATOM     34  HG3 MET A   1C      13.620  -1.444 -26.162  1.00  0.00           H
ATOM     35  HE1 MET A   1C      17.204  -1.841 -26.912  1.00  0.00           H
ATOM     36  HE2 MET A   1C      17.720  -0.181 -27.189  1.00  0.00           H
ATOM     37  HE3 MET A   1C      17.934  -0.915 -25.603  1.00  0.00           H
........................
```

If we ignore this fact, then we get an "amino acid" of the following structure:

N, CA, C, O, H1, H2, H3, HA2, HA3, N, CA, C, O, CB, OG, H, HA, HB2, HB3, HG, N, CA, C, O, CB, CG, SD, CE, H, HA, HB2, HB3, HG2, HG3, HE1, HE2, HE3.

After taking into account the `iCode` field, this "amino acid" falls apart to three ones: `GLY`, `SER`, and `MET` (glycine, serene, and methionine), see Fig. 4.4.

(7) Amino acids having the same name could have different atomic structure. Of course, the ending amino acids of a polypeptide must be different from the nonending ones, namely, the N-end contains a pair of "extra" hydrogens, and the C-end contains an "extra" oxygen. Thus, we decided to restrict ourselves by the *interior* (nonending) amino acids.

However, it turns out that interior amino acids with the same name also can have different structures! For example, the sulfur atom in cysteine sometimes bonds with hydrogen and sometimes does not, see Fig. 4.5.

We take all the files from out list, and for each amino acid find all possible *atomic structures* in all polypeptides. Also, for each structure found, we calculated the number of its appearances. For example, for glycine `GLY` we obtained the following result:

```
55705: N,CA,C,O,H,HA2,HA3,1576,N,CA,C,O,H1,HA2,HA3;
  870: N,CA,C,O,H1,H2,H3,HA2,HA3,399,N,CA,C,O,OXT,H,HA2,HA3;
  390: N,CA,C,O,384,N,CA,C,O,H,47,N,CA,C,O,H,HA2;
   37: CA,37,N,CA,C,O,HA2,HA3,26,N,CA,C,O,H2,HA2,HA3;
   15: N,CA,C,O,H1,H2,HA2,HA3,15,N,CA,C,O,HA2,HA3,H1,H2,H3;
   13: N,CA,C,O,H,HA,13,N,CA,C,O,HA2, HN,HA1,9,N,CA,C;
    8: N,CA,C,O,H1,H2,H3,7,C,CA,H,HA2,HA3,N,O;
    5: N,CA,C,O,OXT,H,5,N,CA,C,H,HA2,HA3;
    4: N,CA,C,O,H,HA3,HA2,2,N,CA,C,O,H1,H2,H3,HA3;
    2: N,CA,C,O,HA2,HA3,H1,H2,2,N,CA,C,O,OXT,2,CA,C;
    1: N,CA,C,O,OXT,H,HA2,HA3,HXT,1,N,CA,C,O,H,HA2,HA3,H2,H3;
    1: N,CA,C,O,H2,H,H3,HA2,HA3,1,N,CA,C,O,H2,H,HA2,HA3,1,N;
    1: N,CA,C,OXT,H,HA2,HA3,1,N,CA,C,O,HA2,1,N,CA,C,O,H,HA3.
```

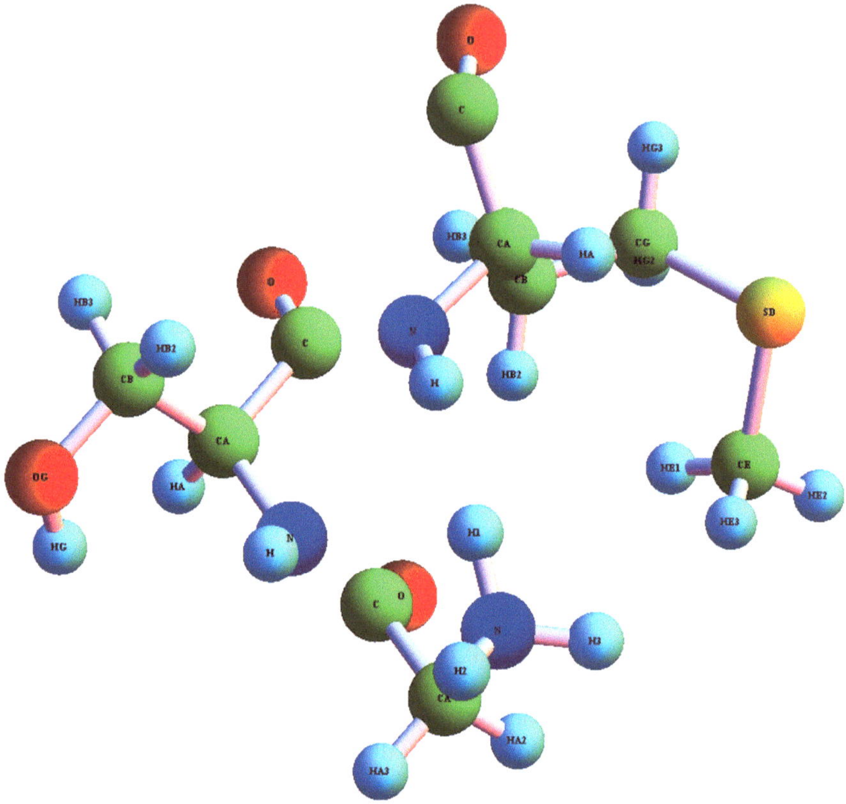

Fig. 4.4 Three different amino acids recognized as a single one, due to error

CYS

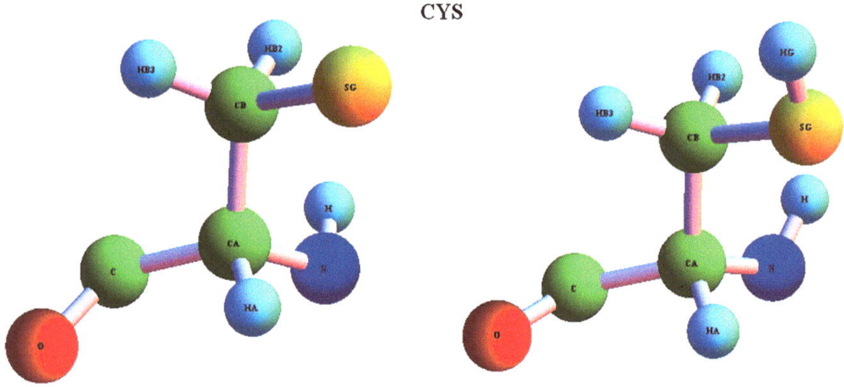

Fig. 4.5 Different atomic structures of interior cysteine

Then we choose the structures that appear most often and call them *standard*. We get the following list of the standard atomic structures of all amino acids (together with the number of their appearances):

```
(GLY:55705) N,CA,C,O,H,HA2,HA3;
(ALA:50673) N,CA,C,O,CB,H,HA,HB1,HB2,HB3;
(SER:52599) N,CA,C,O,CB,OG,H,HA,HB2,HB3,HG;
(CYS:11784) N,CA,C,O,CB,SG,H,HA,HB2,HB3;
(PRO:32529) N,CA,C,O,CB,CG,CD,HA,HB2,HB3,HG2,HG3,HD2,HD3;
(VAL:48533) N,CA,C,O,CB,CG1,CG2,H,HA,HB,HG11,HG12,HG13,
                                             HG21,HG22,HG23;
(THR:39499) N,CA,C,O,CB,OG1,CG2,H,HA,HB,HG1,HG21,HG22,HG23;
(ILE:37003) N,CA,C,O,CB,CG1,CG2,CD1,
              H,HA,HB,HG12,HG13,HG21,HG22,HG23,HD11,HD12,HD13;
(LEU:60276) N,CA,C,O,CB,CG,CD1,CD2,
                H,HA,HB2,HB3,HG,HD11,HD12,HD13,HD21,HD22,HD23;
(ASP:40010) N,CA,C,O,CB,CG,OD1,OD2,H,HA,HB2,HB3;
(ASN:29276) N,CA,C,O,CB,CG,OD1,ND2,H,HA,HB2,HB3,HD21,HD22;
(GLU:53111) N,CA,C,O,CB,CG,CD,OE1,OE2,H,HA,HB2,HB3,HG2,HG3;
(GLN:29973) N,CA,C,O,CB,CG,CD,OE1,NE2,H,HA,HB2,HB3,HG2,
                                              HG3,HE21,HE22;
(MET:14570) N,CA,C,O,CB,CG,SD,CE,H,HA,HB2,HB3,
                                        HG2,HG3,HE1,HE2,HE3;
(LYS:50343) N,CA,C,O,CB,CG,CD,CE,NZ,
             H,HA,HB2,HB3,HG2,HG3,HD2,HD3,HE2,HE3,HZ1,HZ2,HZ3;
(ARG:35509) N,CA,C,O,CB,CG,CD,NE,CZ,NH1,NH2,
          H,HA,HB2,HB3,HG2,HG3,HD2,HD3,HE,HH11,HH12,HH21,HH22;
(HIS: 8040) N,CA,C,O,CB,CG,ND1,CD2,CE1,NE2,H,HA,HB2,HB3,
                                                HD1,HD2,HE1;
(PHE:26312) N,CA,C,O,CB,CG,CD1,CD2,CE1,CE2,CZ,H,HA,HB2,HB3,
                                         HD1,HD2,HE1,HE2,HZ;
(TYR:22457) N,CA,C,O,CB,CG,CD1,CD2,CE1,CE2,CZ,
                             OH,H,HA,HB2,HB3,HD1,HD2,HE1,HE2,HH;
(TRP: 8998) N,CA,C,O,CB,CG,CD1,CD2,NE1,CE2,CE3,CZ2,CZ3,CH2,
                            H,HA,HB2,HB3,HD1,HE1,HE3,HZ2,HZ3,HH2.
```

Remark 4.2 Notice that all the structures chosen correspond to interior amino acids. This is natural since the ending ones appear much more rarely.

Paying attention to the number of appearances of different structures of some amino acids could be quite comparable.

For example, the type of interior cysteine shown in the left part of Fig. 4.5 appear two times often than the one depicted there in the right. Here we present the structures and the appearances numbers of these amino acids:

```
11784: N,CA,C,O,CB,SG,H,HA,HB2,HB3;
 6254: N,CA,C,O,CB,SG,H,HA,HB2,HB3,HG.
```

Remark 4.3 It is easy to see that the atom H (which bonds with nitrogen from the polypeptide backbone) appears at all amino acids except the proline PRO, see Fig. 4.3. And it is natural, due to the presence of the bond CD→N that we discussed above.

We decided to look more carefully at the discarded structures of proline and found out several ones containing hydrogen:

```
10: N,CA,C,O,CB,CG,CD,H,H3,HA,HB2,HB3,HG2,HG3,HD2,HD3;
```

```
1: N,CA,C,O,H,HA,HB2;
1: N,CA,C,O,CB,CG,CD,H3,H,HA,HB2,HB3,HG2,HG3,HD2,HD3;
1: N,CA,C,O,CB,CG,CD,H,HA,HB2,HB3,HG2,HG3,HD2,HD3;
1: N,CA,C,O,CB,CG,CD,HA,HB2,HB3,HG2,HG3,HD2,H.
```

Here the first number in each string stands for the appearances number of the amino acids having the corresponding atomic structure. Notice that such structure appears very rarely, namely, 14 times, versus 32,529 appearances of the standard one.

In the final stage, we tested all `pdb`-files and extracted all such ones for which all their interior amino acids (in the first model) are standard. Our list was reduced by about three times (now it contains 2892 files). After that we transformed each file in a "short format" taking only the necessary information. Now our files consist of the lists of the following entries:

```
{
{Name-of-the-first-Interior-amino-acid,
    {Coordinates-of-the-first-atom,
                    Coordinates-of-the-second-atom, ...}},
...
{Name-of-the-last-Interior-amino-acid,
    {Coordinates-of-the-first-atom,
                    Coordinates-of-the-second-atom, ...}},
}
```

Thus, if the initial base of polypeptides obtained by NMR was more than 18 Gb, then the resulting base was reduced down to 300 Mb. As a consequence, the new base may be treated much faster.

4.3 Metric Analysis of PDB

So, we have obtained the list of polypeptides for which all their interior amino acids with the same name have the same atomic structures. Then we go back to the problem of comparing geometry of the amino acids with the same name. We started with analysis of possible variation of covalent bonds lengths, and of the angles between the bonds.

4.3.1 Estimation of Spread in Lengths of Covalent Bonds in Amino Acids

To estimate the spread of covalent bonds in the set of polypeptides chosen, we proceed as follows:

(1) choose in all polypeptides the interior amino acids of fixed type, for example, all glycines;
(2) calculate the mean value of the length of each covalent bond over all chosen amino acids;
(3) for each type of amino acids (for example, for glycines) and each covalent bond in amino acids of this type (for example, for the bond `N-H`) we calculate maximal relative deviation from the mean value over all the amino acids of this type (for convenience, we measured the deviation in percents).

For polypeptides chosen randomly from our base, those deviations turn out to be sufficiently small (few percents). But after processing the whole base, we found out enormous deviations. The largest one appears in alanine (705.819 % for the bond `N-H`):

$$\frac{454.736}{\mathrm{N-CA}}, \frac{362.668}{\mathrm{CA-C}}, \frac{77.8317}{\mathrm{C-O}}, \frac{244.119}{\mathrm{CA-CB}}, \frac{328.258}{\mathrm{CA-HA}}, \frac{105.323}{\mathrm{CB-HB1}}, \frac{185.751}{\mathrm{CB-HB2}}, \frac{152.231}{\mathrm{CB-HB3}}, \frac{705.819}{\mathrm{N-H}},$$

here the numerator is the maximal deviation from the mean value (in percents) of the covalent bond standing in the denominator.

We "caught" the file leading to this deviation. It turns out that it was the file `2PDE.pdb`. We visualized the amino acids from this file, and the results exceeded all our expectations, see Fig. 4.6.

Notice that the last eight amino acids from this file look standard.

Next "record-holder in alanine deviations" is the file `2HQ3.pdb`, see Fig. 4.7.

After excluding the file `2PDE.pdb`, the deviations in glycine became more than in alanine. The "offender" is the file `2I2J.pdb`, see Fig. 4.8.

We found all the files containing deviations (for some amino acids, not necessary for glycine) more than 15 %. It turns out that there are only few such files, namely 30. Here we give the complete list of them:

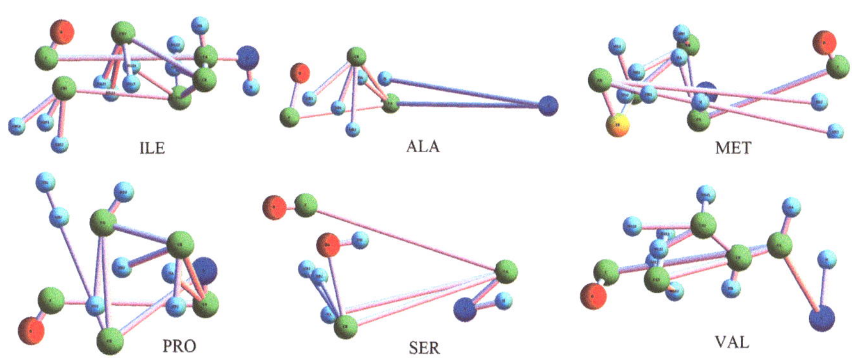

Fig. 4.6 Some "amino acids" from the file `2PDE.pdb`

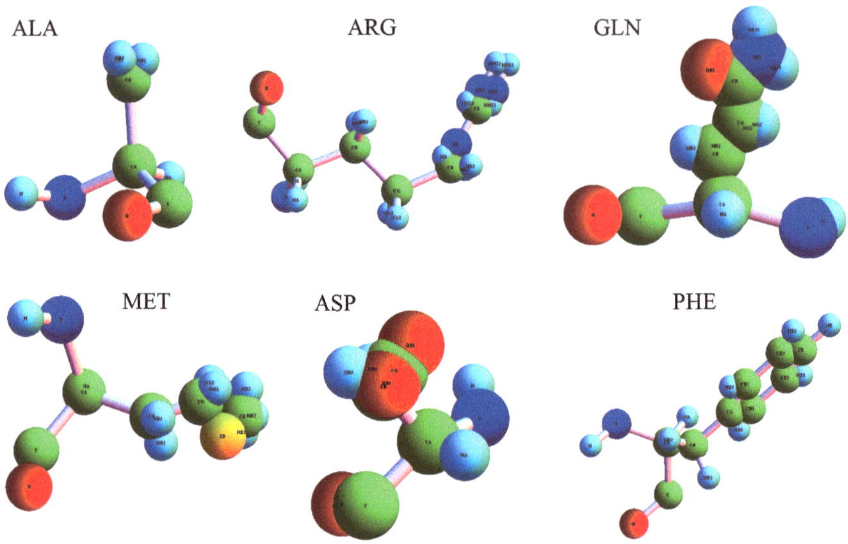

Fig. 4.7 Some "amino acids" from the file `2HQ3.pdb`

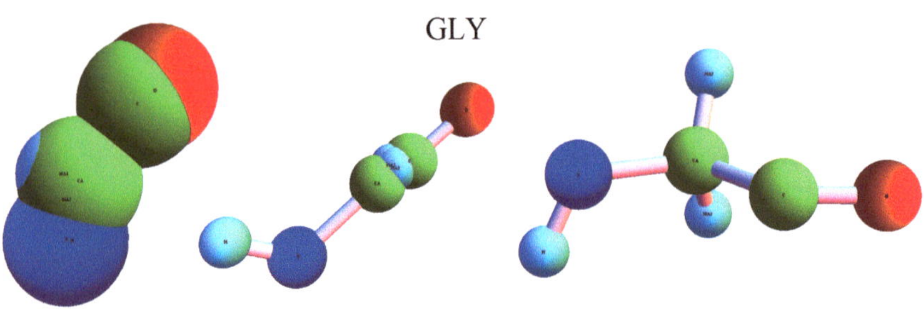

Fig. 4.8 Two "glycines" from the file `2I2J.pdb` (*left*), and the "standard" glycine (*right*)

`1AC0, 1BBA, 1COD, 1DOQ, 1GXX, 1J4M, 1K76, 1KWD, 1KWE, 1L0M, 1L8Z, 1MZI, 1OT4, 1PGY, 1XEE, 1Y7J, 2CYK, 2EVZ, 2HQ3, 2I2J, 2JMM, 2K1W, 2KZG, 2L9V, 2NR1, 2PDE, 3ZGK, 3ZOB, 3ZPK, 4BXU.`

We excluded these files and obtained maximal deviation of 14.6 % (for valine):

$$\frac{8.89612}{\mathtt{N{\to}CA}}, \frac{6.43247}{\mathtt{CA{\to}C}}, \frac{4.27047}{\mathtt{C{\to}O}}, \frac{8.92551}{\mathtt{CA{\to}CB}}, \frac{5.34567}{\mathtt{CB{\to}CG1}}, \frac{7.08376}{\mathtt{CB{\to}CG2}}, \frac{14.6298}{\mathtt{CA{\to}HA}},$$
$$\frac{9.57366}{\mathtt{CB{\to}HB}}, \frac{7.72194}{\mathtt{CG1{\to}HG11}}, \frac{7.7138}{\mathtt{CG1{\to}HG12}}, \frac{7.71296}{\mathtt{CG1{\to}HG13}}, \frac{7.70222}{\mathtt{CG2{\to}HG21}},$$
$$\frac{7.69787}{\mathtt{CG2{\to}HG22}}, \frac{7.7378}{\mathtt{CG2{\to}HG23}}, \frac{10.1827}{\mathtt{N{\to}H}}.$$

It is interesting to see the distribution of the amount of files with various deviations. To do that we partitioned the segment $[-15, 15]$ of the real axis into small segments

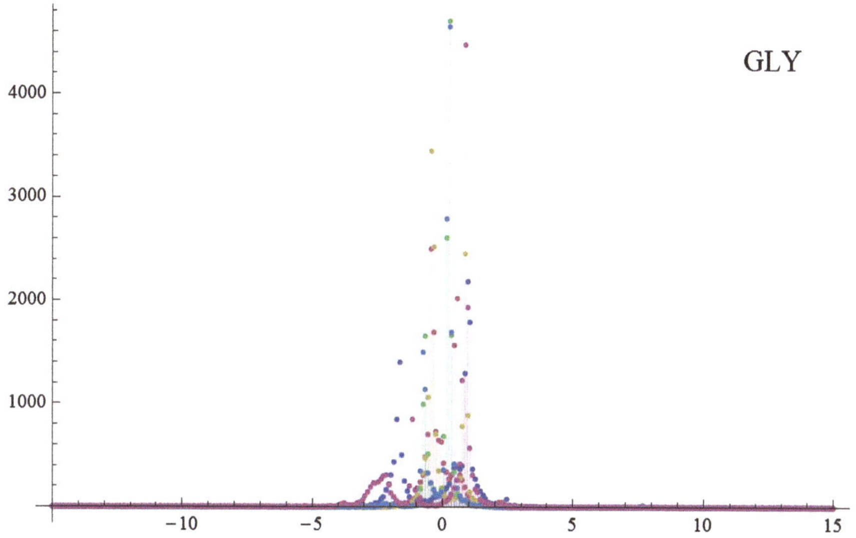

Fig. 4.9 The distribution of numbers of files with respect to the maximal deviations of the lengths of covalent bonds in glycine

of the length 0.1 and calculated the amount of files with maximal deviation in these small segments (in *Mathematica* it is convenient to use the function `BinCounts`). We visualized the results on the following graph (on abscissa axis we mark the numbers of the small segments, and on ordinate axis we mark the amount of files whose maximal deviations lie in the corresponding small segment). All these graphics turn out to be very similar, for example, see the graphic for glycine in Fig. 4.9.

The direct analysis of all 20 graphs shows that the deviations in the most files do not exceed 5 %. For each amino acid we calculated the number of files with deviations more than 5 %. Here is the answer:

```
GLY: 28; ALA: 26; SER: 47; CYS: 10; PRO: 157; VAL: 37;
THR: 43; ILE: 31; LEU: 38; ASP: 29; ASN: 33; GLU: 41;
GLN: 30; MET: 15; LYS: 53; ARG: 48; HIS: 122; PHE: 26;
TYR: 42; TRP: 183.
```

It is not difficult to see that all amino acids in this table can be divided into two groups: 17 of them have maximal deviations in 10–53 polypeptides, and 3 of them with deviations in 122–183 polypeptides (namely, `HIS, PRO, TRP`). In Fig. 4.10 we show the graph for tryptophan.

Notice that the total number of files with "large deviations" in the above table (1039) is more than the true number of such files (some files contain large deviations in several amino acids). The true number of these files equals 442. Since the number of files in our current list is 2862, the number of files with large deviations is quite large. In the following table, we present the amounts of files with maximal deviations

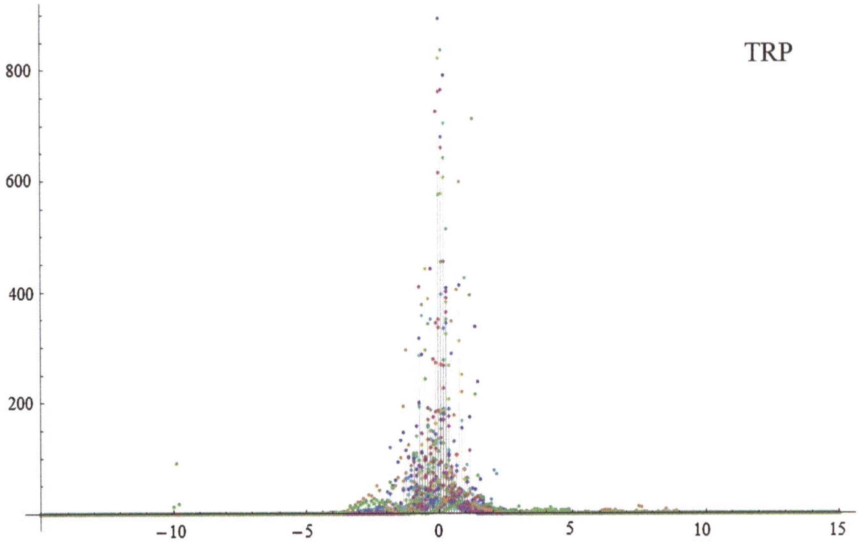

Fig. 4.10 The distribution of amounts of files with various maximal deviations of the lengths of covalent bonds in tryptophan

more than $n\%$, for $n = 6, \ldots, 12$:

$$6\,\% : 371;\ 7\,\% : 279;\ 8\,\% : 162;\ 9\,\% : 128;\ 10\,\% : 20;\ 11\,\% : 8;\ 12\,\% : 6.$$

Since we are initially interested in typical metric properties of polypeptides, we decided to exclude 20 files corresponding to 10 % deviations, so, we removed the following files

```
1ALE, 1ALF, 1C7V, 1E0Q, 1E2B, 1IFY, 1LVQ, 1MEQ, 1ODP,
1ODR, 1OEF, 1OEG, 1OPP, 1PFT, 1SOL, 1SVR, 1TXA, 1ZRP,
2BAU, 2KOX
```

and obtained the database of 2842 files such that the covalent bonds' lengths deviation from the corresponding mean values do not exceed 10 %.

The next step was to investigate the deviations from the mean value of the distances between the alpha carbons in consecutive interior amino acids over all polypeptides chosen above. *In what follows we denote the alpha carbons in two consecutive amino acids by* `CA-CA`.

4.3.2 Spread Estimation of Distances Between Consecutive Alpha Carbons (Beginning)

For each polypeptide we calculate the maximal deviation from the mean value of the distances CA-CA and plot these deviations, see Fig. 4.11. As above, the abscissa axis corresponds to the polypeptide number (in the current base), and the ordinate axis to the maximal deviation.

Notice that there are just a few "spikes." We found the corresponding polypeptides: 1BH7, 1JJS, 2KSE, 2L6F, 2L6G, 2L6H. The reason turned out to be very simple: some amino acids have not been recognized in the experiment. Notice that these files contain correspondent indications. First, these files contain comments of the following type (this example was taken from the file 1BH7.pdb):

```
REMARK 465 MISSING RESIDUES
REMARK 465 THE FOLLOWING RESIDUES WERE NOT LOCATED IN THE
REMARK 465 EXPERIMENT. (RES=RESIDUE NAME; C=CHAIN IDENTIFIER;
REMARK 465 SSSEQ=SEQUENCE NUMBER; I=INSERTION CODE.)
REMARK 465     RES C SSSEQI
REMARK 465     LEU A    17
REMARK 465     VAL A    28
```

Second, the sixth position in strings "ATOM ..." may be also used to catch the omitted amino acids. For example, in the same file 1BH7.pdb the value 18 of the sixth position follows immediately after the value 16.

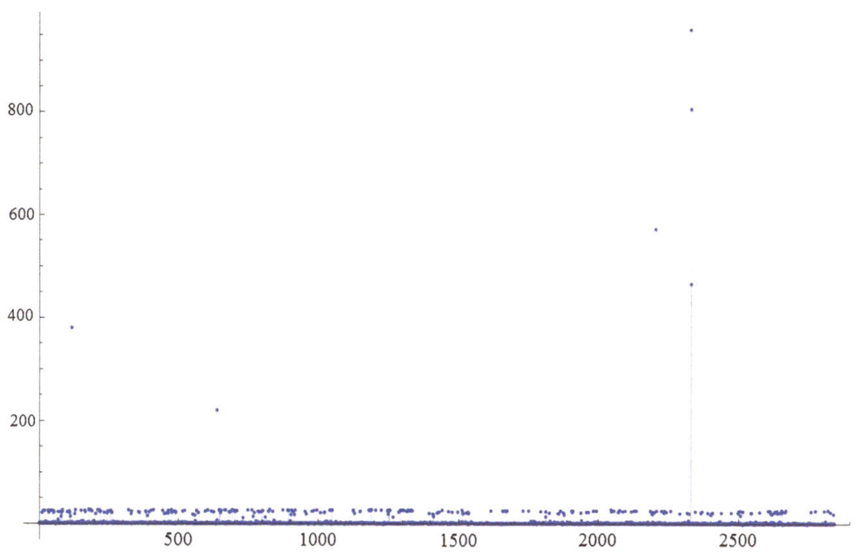

Fig. 4.11 Maximal relative deviation from the mean value between consecutive α-carbons in the polypeptides chosen above

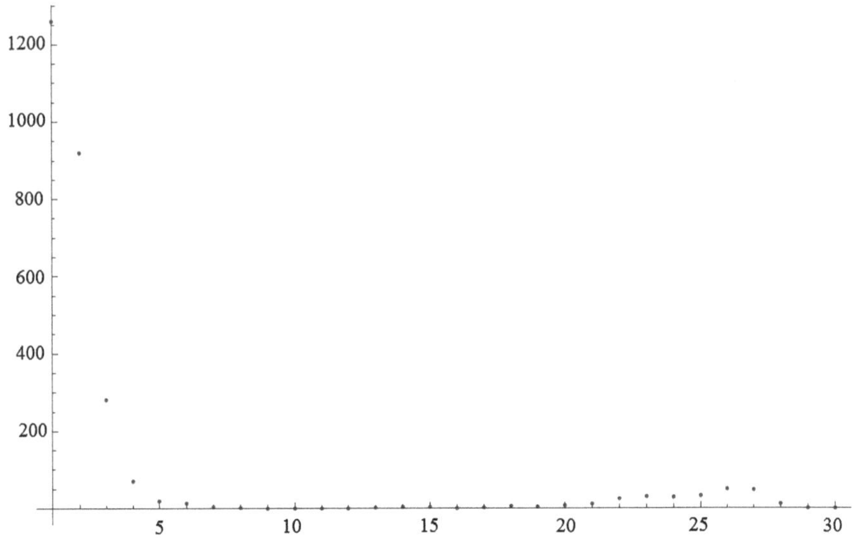

Fig. 4.12 The numbers of files (ordinate axis) with various maximal deviations (abscissa axis) in CA-CA bonds

We excluded these six files from our list, and obtained the maximal deviation of at most 30 %. Figure 4.12 demonstrates the distribution of the amount of files (ordinate axis) with various maximal deviations (abscissa axis): 1 tick mark for abscissa axis corresponds to maximal deviation of 1 %.

Here we list the corresponding values:

1260, 920, 282, 70, 20, 13, 4, 2, 1, 0, 0, 0, 2, 3, 3, 1, 2, 5, 3, 8, 12, 24, 30, 28, 33, 49, 48, 12, 1, 0

Remark 4.4 It is interesting that the most of these files have the deviation less than 5–6 %, but there is another accumulation value, namely, 25 %. In what follows, we shall explain the reason.

Of course, one of the evident reasons for such "accumulations" could be large deviation in the peptide bonds C-N.

4.3.3 Spread Estimation of the Lengths of Peptide Bonds

We investigated the deviations in the peptide bonds C-N (between consecutive amino acids). It turns out that these deviations are less than 7.8 %. Figure 4.13 shows the distribution of the numbers of polypeptides with various maximal deviations; the abscissa axis step equals 0.2 %.

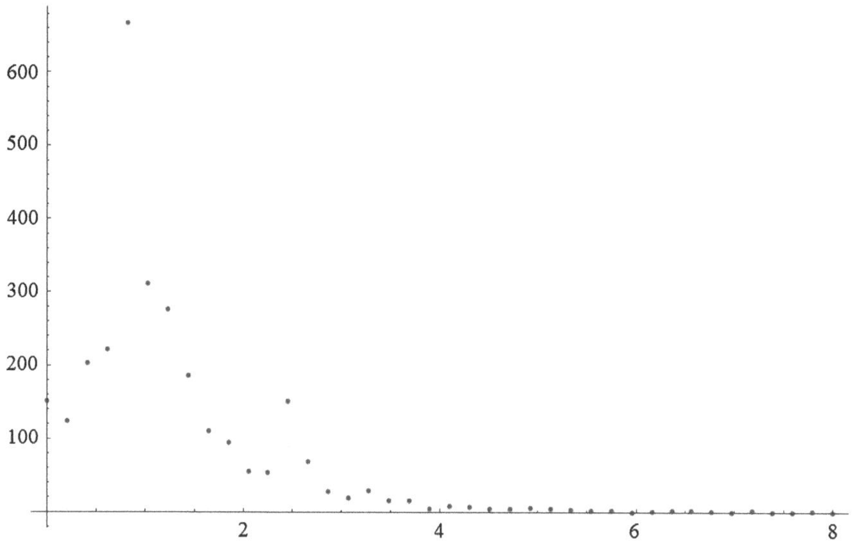

Fig. 4.13 The numbers of files (ordinate axis) with various maximal deviations (abscissa axis) in C–N bonds

Thus, the accumulations in the spread of CA–CA distances are not caused by the peptide bonds.

4.3.4 Pauling Plane Law

Usually in biochemistry, the following six atoms in the peptide bond are marked out: the atoms CA, C, and O from $(i-1)$th amino acid, and the atoms CA, N, and H from ith amino acid. Those six atoms are referred as the *peptide groups*. In 1960s, see [8], Pauling formulated the rule that each peptide group is located in a single two-dimensional plane. We will refer to this statement as to *plane law*. Violations of this law also influence to the spread of the CA–CA distances.

Let us verify the plane law taking a random polypeptide, say 1CKR, and choosing a random pair of consecutive amino acids in it, say the 91st (VAL) and the 92nd (THR), see Fig. 4.14.

Figure 4.14 clearly shows that this peptide group forms almost planar configuration, hence, the plane law holds in this case.

If the second amino acid is proline, we do not have the peptide group described above, because proline has no the hydrogen H. Figure 4.15 illustrates the situation with proline as the second amino acid.

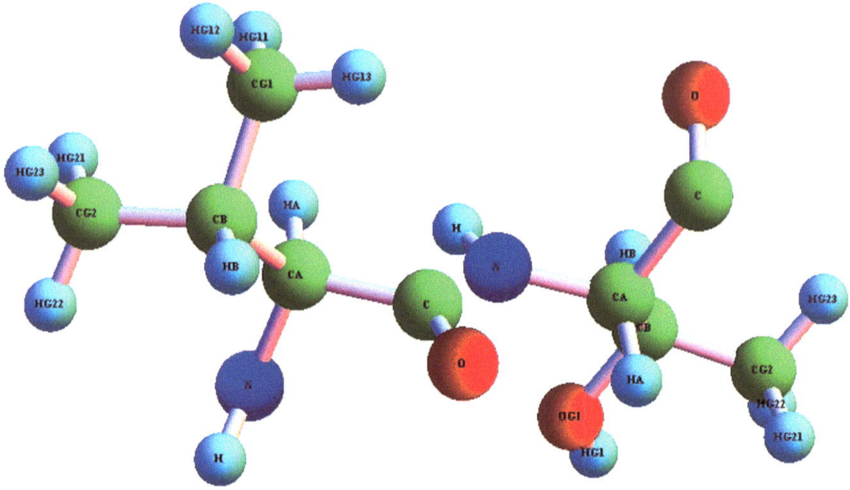

Fig. 4.14 The atoms CA, C, and O of the $(i - 1)$th amino acid and the atoms CA, N, and H of the ith amino acid lie in a single plane

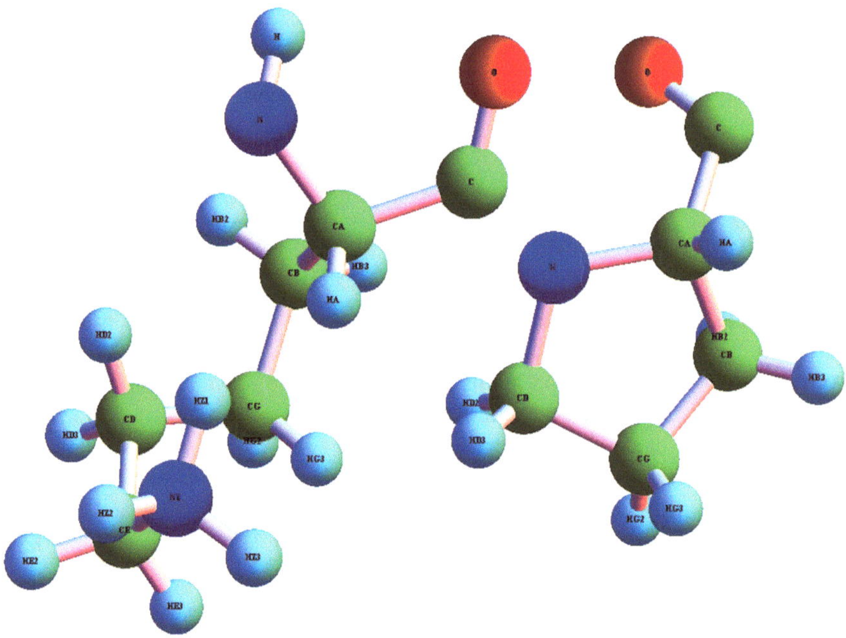

Fig. 4.15 Consecutive amino acids, the second one is proline

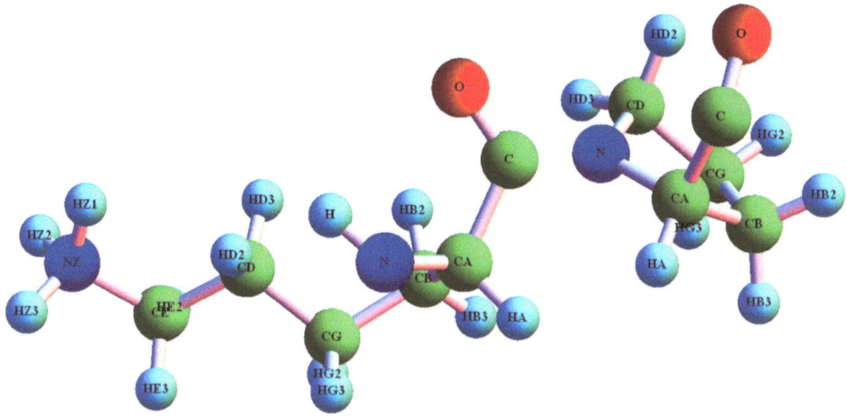

Fig. 4.16 "Counter Example" to the plane law ("singular" configuration)

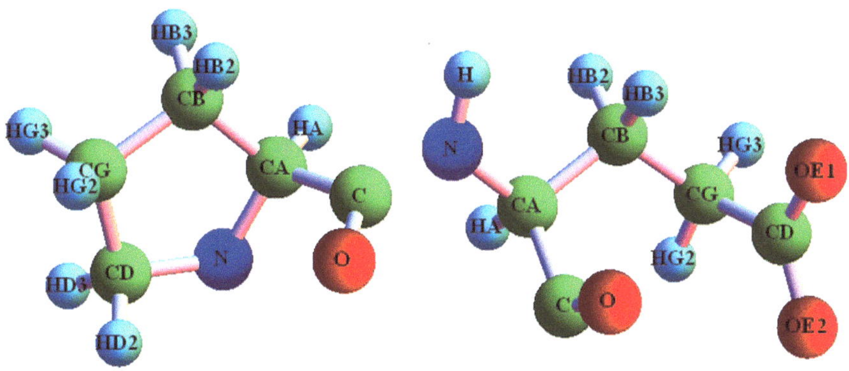

Fig. 4.17 "Counter Example" to the plane law ("nonsingular" configuration)

It is evident from this figure how to modify the plane law in this case: it suffice to take the atom `CD` instead of the atom `H`. *In what follows, by the plane law we shall always mean this generalization.*

To verify the plane law in all the remaining cases, we decided to use the volume of convex hull of peptide group atoms as the measure of planarity. For several random polypeptides the maximal volume turned out to be less than 0.5 (such deviation from the plane is almost invisible). But the maximal value for the entire database is equal to 4.74761. We found the corresponding polypeptide and the pair of its consecutive amino acids: `1PBA`, (38th `LYS`, 39th `PRO`). They are depicted in Fig. 4.16.

The example depicted in Fig. 4.16 is *singular* in the sense that the second amino acid is proline (recall that we generalized the plane law just for such configurations). Generally speaking, the values of volumes could measure nonplanarity of the peptide group for such singular configurations not in the same way as for nonsingular ones. We found out that the volumes of nonsingular configurations with similar deviations

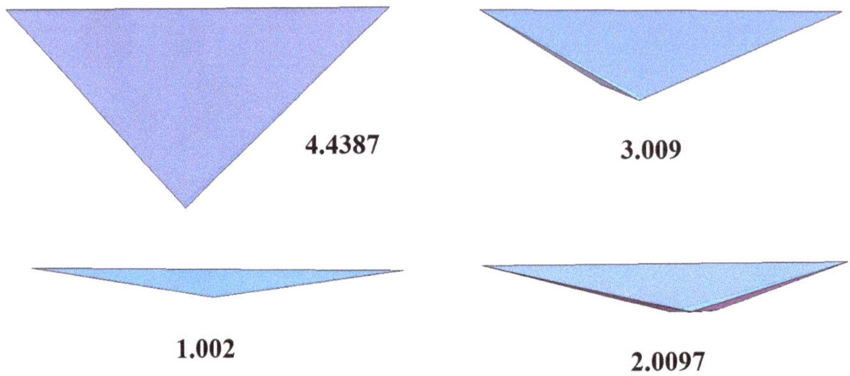

Fig. 4.18 Frontal projections of the convex hulls having the least heights

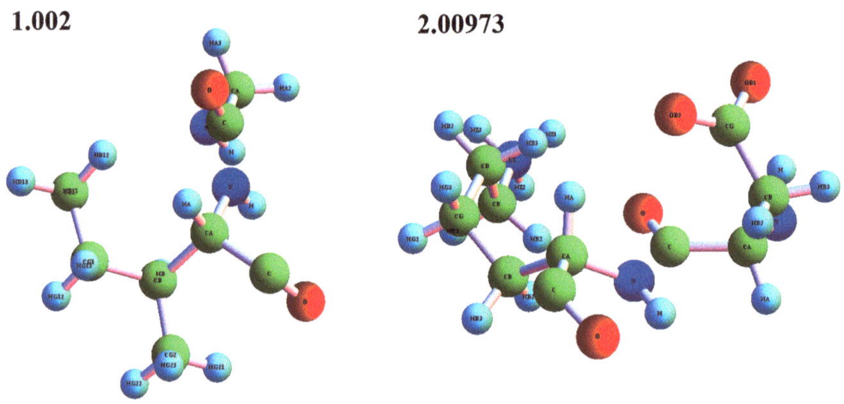

Fig. 4.19 Pairs of amino acids corresponding to the volumes ∼1 and ∼2, see Fig. 4.18

are less. We demonstrate that in Fig. 4.17: here we see an example where the maximal volume (4.4387) among nonsingular configurations is attained.

To estimate the relations between the volumes and the planarity of peptide groups, we draw the convex hulls of some polypeptide groups with different volumes and try to orient them in such a way that the height of their frontal projection becomes minimal (this height estimates the planarity of the corresponding configuration as follows: less height corresponds to more planar configuration). Some nonsingular examples, together with the corresponding volumes, are depicted in Fig. 4.18. One can see that even for volume 1 the configuration does not look quite planar.

Figure 4.19 shows the pairs of amino acids corresponding to the volumes ∼1 and ∼2, see Fig. 4.18.

We calculated the distribution of the numbers of polypeptides with respect to the maximal values of the volumes. To do that we partitioned the segment [0, 4, 8] into small segments of length 0.05, and for each resulting small segment calculated the

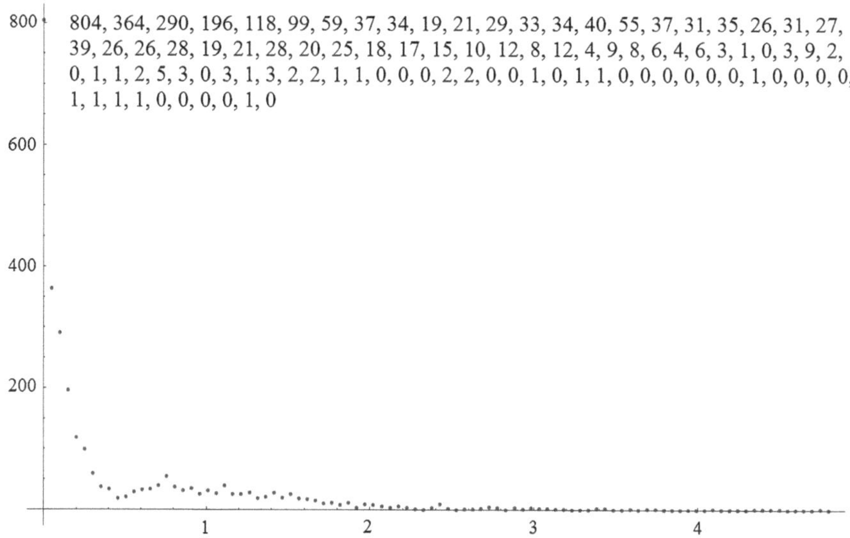

Fig. 4.20 Distribution of the numbers of polypeptides (ordinate axis) with respect to the maximal volumes of the convex hulls of the peptide groups (abscissa axis)

number of polypeptides having the maximal volumes of the convex hulls of their polypeptide groups in this interval. The corresponding graph is shown in Fig. 4.20.

Remark 4.5 It is not difficult to calculate that the maximal volume is more than 1 for 475 polypeptides (of 2836), i.e., about 16.7 %. As we have already mentioned, for such volumes the nonplanarity of configuration can be easily seen from the visualization. Thus, each polypeptide of those 16.7 % contains a pair of amino acids with visual violation of the plane law.

On the other hand, we considered all the pairs of consecutive amino acids in all (2836) polypeptides. There are 190,185 of such pairs. Usually a law is considered as a reliable one, if it is valid for more than 95 % cases. We get the following result: for 95 % of pairs of consecutive amino acids, the maximal volume is less than 0.575. An example with the volume 0.575008 is shown in Fig. 4.21.

Notice that here the plane law is valid with a fine precision. Thus, we can conclude that the plane law passes our "validity test." But to get the 95.1 % reliability we need to exclude the files with maximal volume more than 1.67. There are 138 of such files.

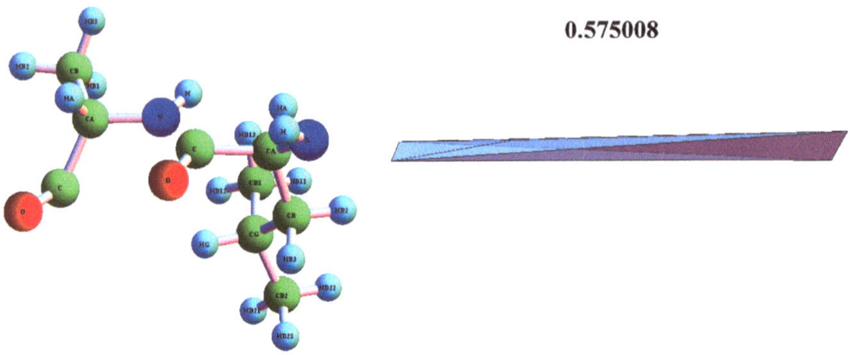

Fig. 4.21 Pair of amino acids with volume 0.575008

4.3.5 Spread Estimation of Distances Between Consecutive Alpha Carbons (End)

We decided to make the following "radical" experiment: we exclude all the files with maximal volume more than 0.5 (in the remaining files the plane law holds with very good precision) and then verify again the deviations of the `CA-CA` distances. We excluded 815 files (it remains 2021). The deviation of the `CA-CA` distances in these 2021 files turns out to be still large, about 28 %, i.e., the violation of the plane law is not the reason of these deviations.

The real reason, as the readers have already guessed, is the existence of so-called *cis* and *trans* configurations of the peptide groups, see Fig. 4.22.

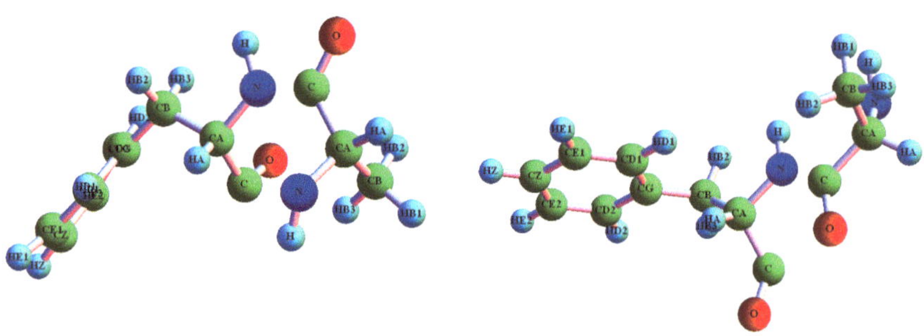

Fig. 4.22 Cis configuration (*left*) and trans configuration (*right*) of polypeptide groups. In cis configuration the atoms `CA` are located "at one side", and in trans configuration they are "from different sides" (of the line NC, if the plane law holds; see below for more details)

4.3.5.1 Cis and Trans Configurations of Peptide Groups

To define these two configurations formally, we consider two vectors: the vector products `[N-CA, N-H]` and `[C-CA, C-O]` for a peptide group (in the case of proline, we take `CD` instead of `H`). The peptide group is said to be in *trans configuration*, if the angle between these two vectors is acute (and their scalar product is positive). Otherwise, we say that the polypeptide group is in *cis configuration*.

We calculated the number of polypeptides containing at least one pair of consecutive amino acids in cis configuration. It turns out that the number of such polypeptides is relatively small, namely, 263. At first we tried just to exclude these 263 files from our list and recalculate the spread of the `CA-CA` distances. It turns out that the spread becomes better, but not too much, namely, it turns out to be equal to 17.67 %. Then we also decided to remove the files with the most significant deviations in the plane law (with the volume more than 3). Recall that for the latter polypeptides the plane law is hardly broken. It turns out that we need to exclude the following 19 polypeptides:

```
1AW3, 1BLV, 1EGO, 1ESK, 1FDF, 1I1S, 1I6C, 1L6H, 1MEA, 1MED,
1MNY, 1OZO, 1PBA, 1W9R, 1YUS, 2ADX, 2H3J, 2ITH, 7HSC.
```

The resulting list became really nice: now the spread of `CA-CA` is equal to 7.4 %. The resulting list consists of 2563 files.

Now, we pass to verification of the spread of the angles between covalent bonds.

4.3.6 Spread Estimation of Angles Between Covalent Bonds in Polypeptides

We really hoped that the covalent bonds angles analysis would be "simply formal," because we have chosen the polypeptides with small spread of covalent bonds. Formally speaking, this can guarantee nothing, but we believed that our strong "clear out" has already excluded all large deviations. Unfortunately, we were disappointed: the maximal deviation of the angles from their mean values turn out to be 71.12 %. In Fig. 4.23 there are three examples corresponding to three largest deviations of the angles (we also give the names of the amino acids and the ambient polypeptides).

Notice that the deviations in the remaining amino acids with less spread appear due to essentially nonuniform distribution of the covalent bonds for some atoms of valency 4: the neighbors of such atoms could be located in a half-space with respect to some planes passing through these atoms, see Fig. 4.24.

We found out that such strong deviations start from the spread about 26 %. Our current list contains 59 polypeptides with such large deviations. We excluded all the corresponding files from our database, and it remains 2532 files. In the resulting database the angles deviations in amino acids do not exceed 26 %, the metric deviations do not exceed 10 %, and the deviations in `CA-CA` distances do not exceed 7.5 %. Moreover, all the polypeptide groups form trans configuration, and almost all the polypeptide groups sufficiently satisfy the plane law. The further investigations were proceeded just for this database (we call the polypeptides from this base the "typical" ones).

4.4 Amino Acids' Mobility

Looking at different figures showing amino acids with the same name located in different places (in the same polypeptide or in different ones) we have found out that the space structures of these amino acids can be essentially dissimilar. To compare the geometry of amino acids with the same name we decided to draw them in the same figure. However, we had to understand how to align different amino acids with the same names. Taking in mind the above metric analysis of the geometry of amino acids we introduced the concept of the *amino acid frame*. Namely, we put the *origin* at the alpha carbon `CA`, and chose the nitrogen `N`, carbon `C`, and beta carbon `CB` as the ends of the three basis vectors (the `CB` does not appear in glycine, so in this case we took the hydrogen `HA3` instead of the `CB`).

Remark 4.6 Notice that in the order of atoms of amino acids used in the `PDB`-files the atoms `N`, `CA`, `C`, and `CB` have the numbers 1, 2, 3, and 5, respectively; in the case of glycine the 5th position contains an "other" hydrogen, but the hydrogen we need stands at the 7th position.

We used the frames to construct the alignment transformations: (1) the affine transformation taking one frame onto the other, or (2) an orthogonal transformation minimizing the total distances between the corresponding atoms, see example in Fig. 4.25 (in our illustrations we show only the first-type transformation).

Now we can easily see that the hydrogen adjacent to the nitrogen `N` and the oxygen adjacent to the carbon `C` can rotate around the corresponding vector of the frame.

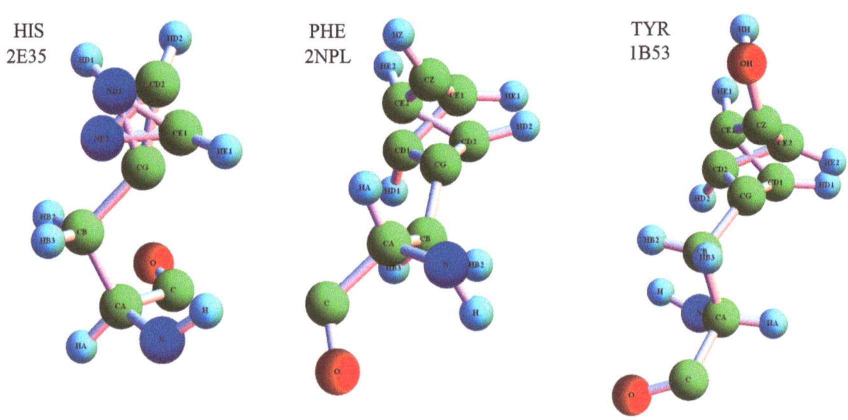

Fig. 4.23 Amino acids with maximal deviations of the angles

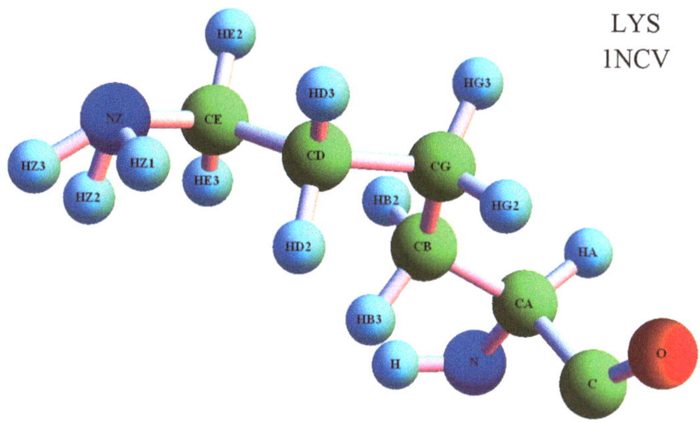

Fig. 4.24 The atoms neighboring to NZ lie in a single half-space with respect to some plane passing through NZ

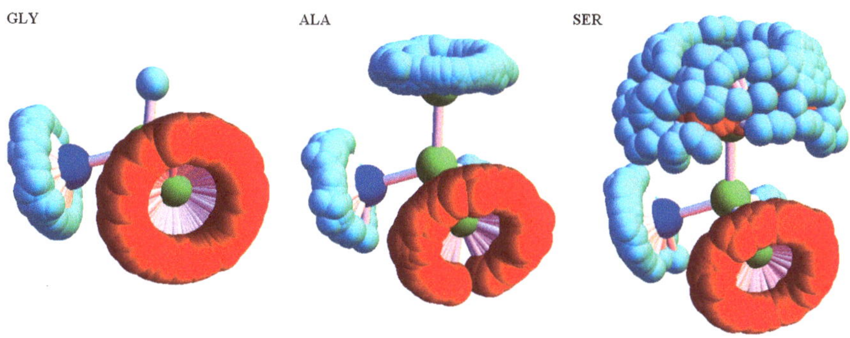

Fig. 4.25 Glycine, alanine, and serine: aligning the amino acids' frames

Similarly, in the case of alanine the hydrogen adjacent to the carbon CB also can rotate. In the case of serine the situation is more complicated: the carbon CB branches into two hydrogens and the oxygen OG also adjacent with hydrogen. The fork triplet adjacent to CB can rotate around the CA-CB axis, and besides the hydrogen adjacent to the oxygen also can rotate around the CB-OG axis.

More complicated situations are shown in Figs. 4.26 and 4.27.

The above figures lead to an idea that it may be more appropriate to represent amino acids in polypeptides in the form of "clouds." The hydrogen bonds, however, must decrease the configuration space of each amino acid. It would be very interesting to understand the stricture of these configuration spaces.

Remark 4.7 During the computer experiment with the aligning we initially used transformations preserving the orientations of the amino acids' frames. As a result, we found out that not all the frames have the same orientation. We shall discuss this problem in the next section.

4.4.1 Orientation of Amino Acids

We call an amino acid *positively oriented*, if the orientation of its frame coincides with the standard one, i.e., if the determinant of the matrix composed of the vector columns of the frame is positive. Otherwise we call the amino acid *negatively oriented*.

In our database of 2563 files we found all the polypeptides that contain at least one negatively oriented amino acid. It turns out to be 140 such polypeptides. One of such negatively oriented amino acids, namely glycine, is shown in Fig. 4.28, left.

However, if we swap the hydrogens HA2 and HA3, then the negative (left) glycine becomes positively oriented, see Fig. 4.29.

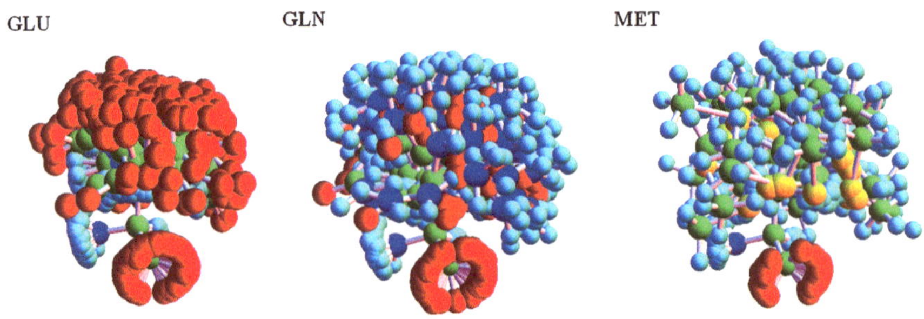

Fig. 4.26 Glutamic acid, glutamine, and methionine: aligning the amino acids' frames

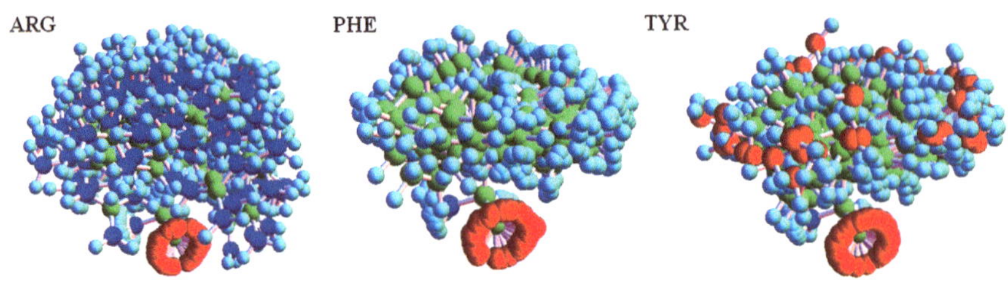

Fig. 4.27 Arginine and tyrosine: aligning the amino acids' frames

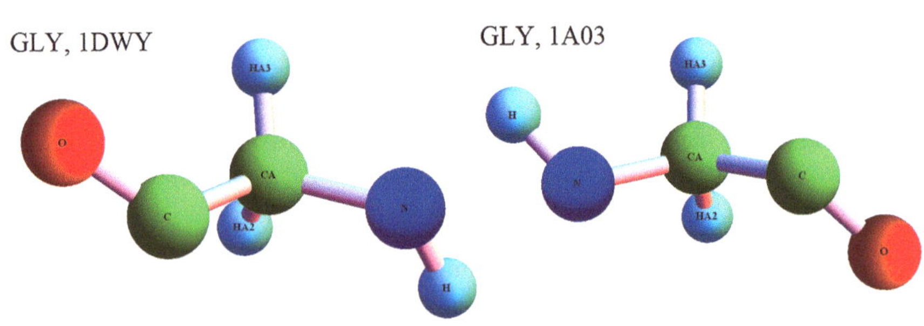

Fig. 4.28 Negatively (*left*) and positively (*right*) oriented glycine

Notice that in the whole database the frame of glycine with the hydrogen HA3 is positively oriented much more often than the frame of glycine with the hydrogen HA2.

Thus, it is not difficult to fix the problem with two orientations in glycine. Is glycine the unique amino acid with orientation problem? The answer is negative. Fortunately, the number of polypeptides containing negatively oriented amino acids which are not glycine is very small, namely, there are four such polypeptides. Here we list them:

```
1NMJ, 1ZDB, 2B8F, 2NS4
```

Notice that each of this polypeptide contains exactly one negatively oriented amino acid. In three of them it stands at the last position, and in the rest one it is located at the second place. We include here the pictures of these amino acids together with their positively oriented analogues taken from the (Figs. 4.30, 4.31, 4.32 and 4.33).

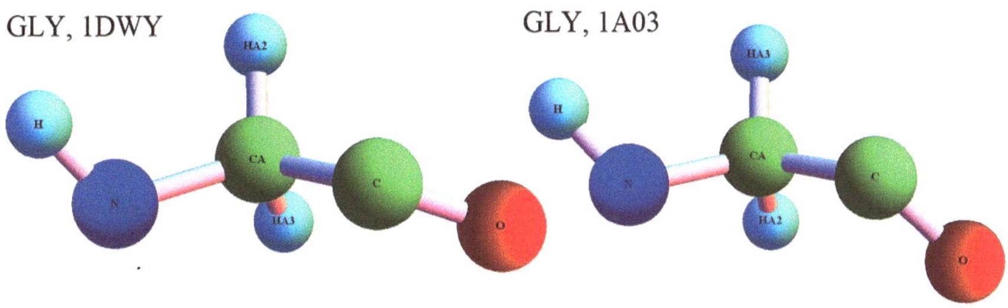

Fig. 4.29 The same amino acids after swapping

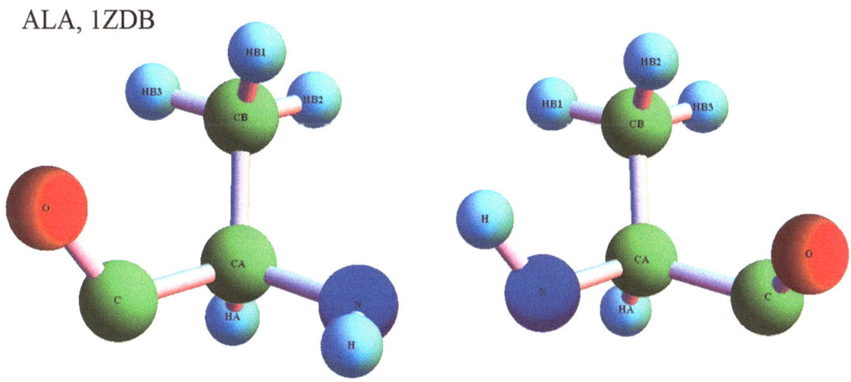

Fig. 4.30 Negatively (*left*) and positively (*right*) oriented ALA

4.5 Addendum (in collaboration with E. A. Vilkul)

We discussed the above ideas with specialists who raised a series of questions. In this Addendum we shall present some of our considerations. Recall that previously for our calculations we always chose the first model; also, we selected those pdb-files which have small enough deviations of the metric parameters reputed to be more or less stable. The specialists noticed that the first model could not be representative which could be one of the reasons of the "pathologies" discussed above. We decided to look through the other models and try to find the more representative one. Here we list just several examples, postponing the detailed report on the experiments with different models to our next publication.

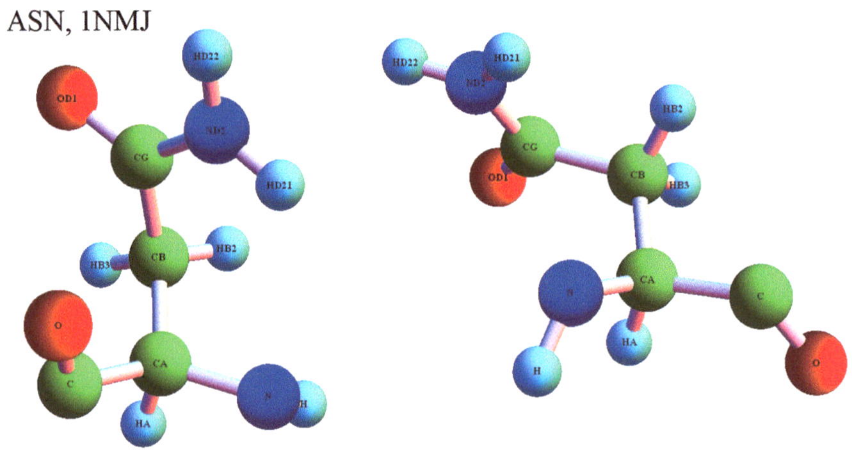

Fig. 4.31 Negatively (*left*) and positively (*right*) oriented ASN

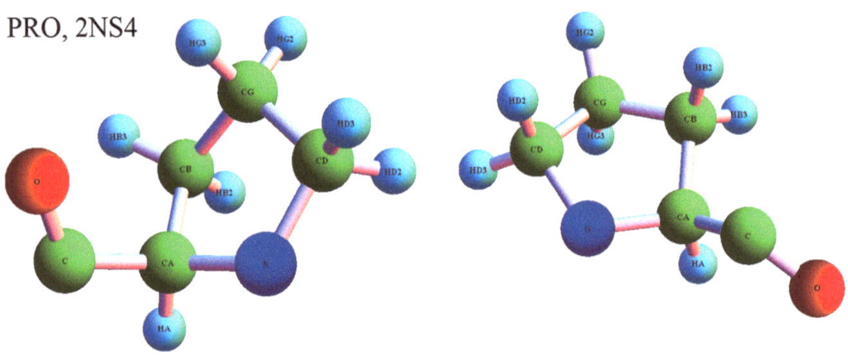

Fig. 4.32 Negatively (*left*) and positively (*right*) oriented PRO

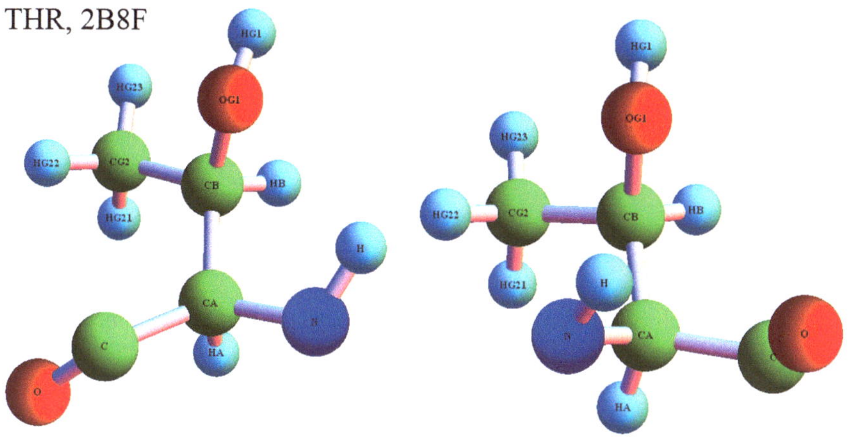

Fig. 4.33 Negatively (*left*) and positively (*right*) oriented THR

4.5.1 The Number of Models' Distribution

To start with we found out how many models appear in pdb-files, analyzed the number of the files containing *k* models, and discussed the relation of those numbers with the year of the corresponding research. "People believe" that as a rule the pdb-files describing the results of NMR investigations of the polypeptides usually contain quite enough models, about 15–20. Besides, the files with just a few models are usually dated by the previous century.

We verified those statements. To do that we take the 2892 files selected in Sect. 4.3 and calculated the number of models in each of them and also found out the year of the corresponding research.

Here we list all the numbers of models appearing in these files:

```
1, 2, 3, 4, 5, 6, 7, 8, 9, 10, 11, 12, 13, 14, 15, 16, 17, 18,
19, 20, 21, 22, 23, 24, 25, 26, 27, 28, 29, 30, 31, 32, 33, 34,
35, 36, 37, 38, 39, 40, 41, 42, 43, 44, 45, 46, 47, 48, 49, 50,
52, 54, 55, 56, 58, 59, 60, 63, 64, 68, 70, 72, 76, 80, 81, 84,
90, 92, 96, 100, 108, 116, 120, 128, 140, 144, 160, 240, 290,
304, 308, 320, 640
```

Notice that the maximal number is 640, and it appears in only one file, namely in 2KOX.pdb. This file was created in 2009. The next "champions" contain 320 models. There are two such files, 2K0F.pdb and 2M5N.pdb, the first one was created in 2008, and the second one appeared in 2013.

Then we calculated the distribution of the number of models appearance (in this list the number of models is located at the first place of each pair, and the second place shows the number of files containing such number of models):

```
{1, 315}, {2, 50}, {3, 7}, {4, 9}, {5, 22}, {6, 5}, {7, 3},
{8, 3}, {9, 12}, {10, 240}, {11, 20}, {12, 24}, {13, 9},
```

```
{14, 21}, {15, 82}, {16, 22}, {17, 9}, {18, 24}, {19, 25},
{20, 1291}, {21, 40}, {22, 17}, {23, 14}, {24, 23}, {25, 58},
{26, 10}, {27, 5}, {28, 7}, {29, 11}, {30, 118}, {31, 13},
{32, 22}, {33, 8}, {34,  10}, {35, 16}, {36, 13}, {37, 1},
{38, 8}, {39, 4}, {40, 164}, {41, 4}, {42, 11}, {43, 2},
{44, 7}, {45, 2}, {46, 5}, {47, 3}, {48, 3}, {49, 2}, {50, 18},
{52, 2}, {54, 1}, {55, 2}, {56, 3}, {58, 2}, {59, 1}, {60, 20},
{63, 1}, {64, 1}, {68, 2}, {70, 3}, {72, 1}, {76, 1}, {80, 15},
{81, 1}, {84, 1}, {90, 1}, {92, 1}, {96, 1}, {100, 3}, {108, 1},
{116, 1}, {120, 4}, {128, 1}, {140, 1}, {144, 1}, {160, 1},
{240, 1}, {290, 1}, {304, 1}, {308, 1}, {320, 2}, {640, 1}.
```

Thus we found that for each $n > 80$ the number of files containing exactly n models does not exceed 4 and it is equal to 1 almost always. We excluded all these files, and for the remaining ones we drew the following graph, see Fig. 4.34, where at the abscissa axis we represent the number of models (from 1 to 80), and ordinate axis corresponds to the number of files.

As we can see from the graph in Fig. 4.34, the maximal number of files falls to the 20 models. There are 1291 such files. Besides, the number of files containing only one model is also sufficiently large, namely, there are 315 such files (it is the next maximum).

We also constructed the distribution of the files with 20 models with respect to the year of research, see Fig. 4.35. We can see that the maximal value corresponds to 2007 (201 file), but in 1993 such files (4 ones) were also produced.

At last, we also constructed the distribution of the number of files containing only one model with respect to the year, see Fig. 4.36 (recall that there are 315 such files).

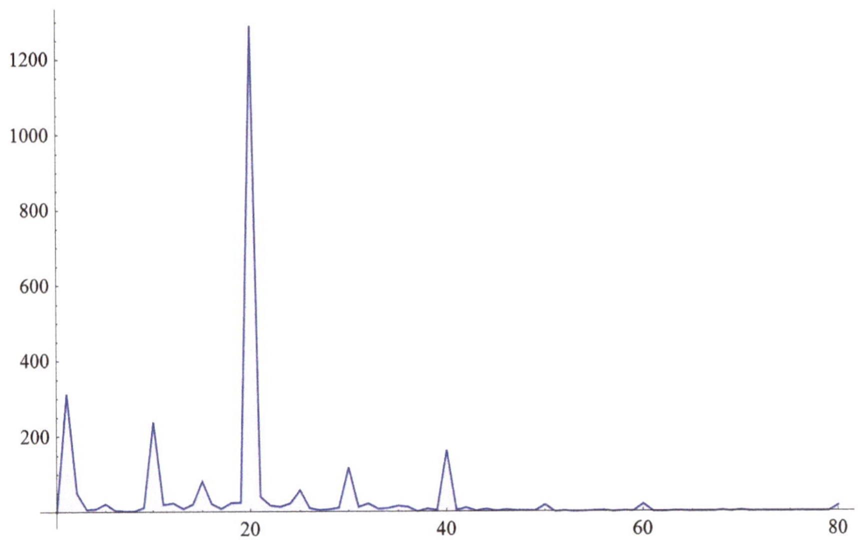

Fig. 4.34 The number of files' distribution with respect to the number of models (in the range from 1 to 80)

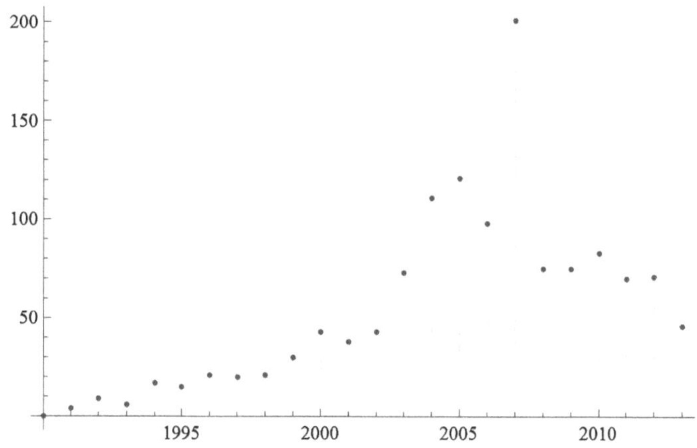

Fig. 4.35 The distribution of the files' number containing 20 models with respect to the year of research

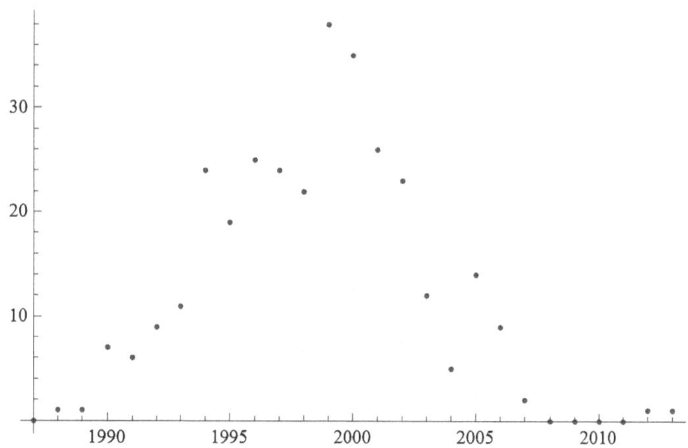

Fig. 4.36 The distribution of the files' number containing only 1 model with respect to the year of research

We can see that the maximal number of 1-model files appeared in 1999–2000 (73 of them), and from 2005 only 27 such files were created.

4.5.2 "Representativity" of the First Model from the Pathologies Point of View

In the previous sections we found out several "pathological" `pdb`-files, where the coordinates of some atoms are certainly wrong. Besides, we found some other devi-

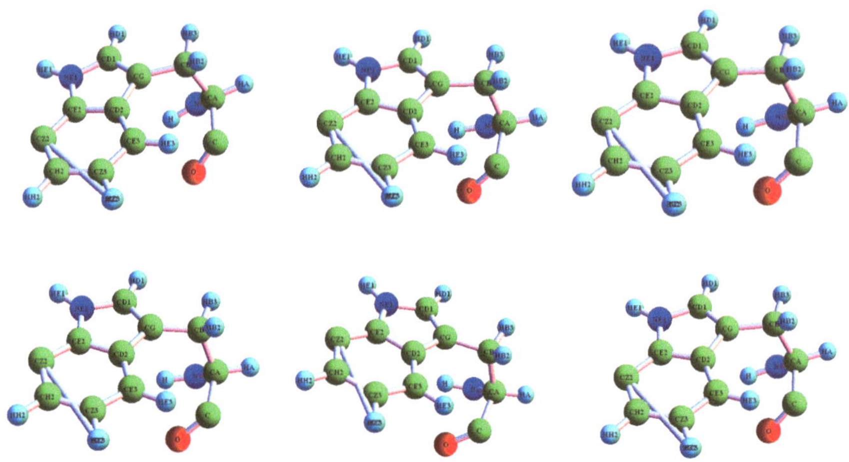

Fig. 4.37 The "pathological" tryptophan in the first six models (it is the same in the other 29)

ations, for example, in the plane law. As we already mentioned, these problems may be inspired by the fact that the first model could be nonrepresentative. We decided to verify this statement.

Recall that in Sect. 4.3.1 we investigated the relative deviations of the covalent bonds lengths from their mean values. We decided to investigate all the 30 files for which the deviations are more than 15 % (the list of these files is given in Sect. 4.3.1). It turns out that twelve of them contain only one model, so we cannot improve the data using another model in these 12 cases. For example, the "record" file `2PDE.pdb` with deviation more than 700 % belongs to this list.

The numbers of models in the remaining 18 files are distributed as follows (the first position of each pair means the number of models, the second one shows the amount of files with such number of models):

```
{9,1},{10,3},{11,1},{16,2},{20,7},{21,1},{34,1},{35,1},{81,1}
```

Let us start "from the end," i.e., consider the files containing maximal number of models, and choose among them the ones containing evident pathologies. The first such file is `1MZI` containing 81 models, but the deviations of its first model do not lead to visible deviations in the form of its amino acids.

The second file is `1OT4`. It has 35 models, and its tryptophan (`TRP`) on the 83rd position (the 82nd interior amino acid) contains an evident error. We looked through all the other models in `1OT4.pdb` and found out that this error appears in all the 35 models, see Fig. 4.37.

Notice that the tryptophan looks very similar in all pictures. May be the first model was copied? We compared the coordinates of the corresponding atoms in different models, and it turns out that they are different.

The third example is the file `2I2J` containing 34 models. All the amino acids in its first model look wrong, see Fig. 4.8. We looked through all the other models, and

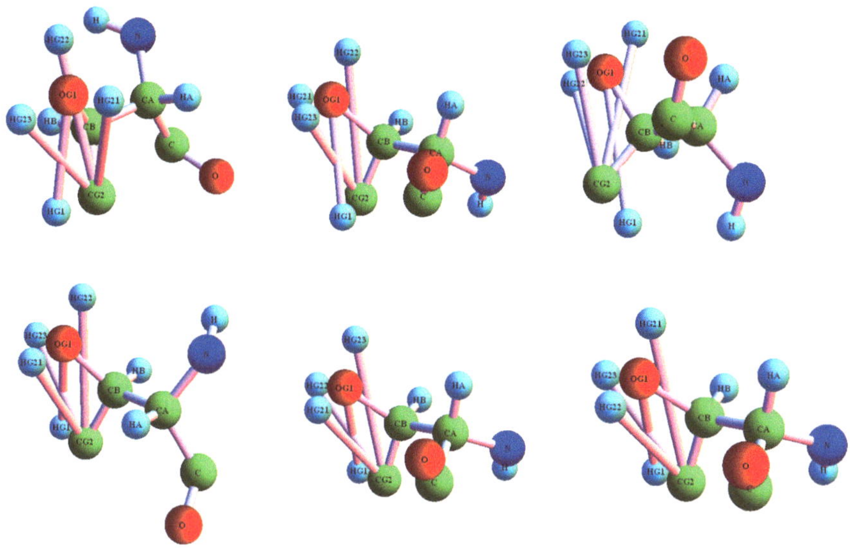

Fig. 4.38 The "pathological" alanine in the first six models

found out that they are correct (all the amino acids look standard), so this example confirms the above statement.

The fourth example turns out to be similar to the second one. The file `1KWD` has 16 models, and the 9th amino acid is wrong in all these models, see Fig. 4.38.

In the next of our papers we suppose to make a detailed analysis of all "stable" pathologies (i.e., the pathologies which appear in all models).

4.5.3 "Representativity" of the First Model from the Plane Law Point of View

In Sect. 4.3.4 we found out that in the first model the plane law is often violated. Let us investigate what happens in other models.

Recall that to measure the deviations from the plane law we calculated the volume of the convex hull of the atoms forming the peptide group. The volumes less than 0.5 indicate good correspondence to the law; if the values are more than 1, then the deviations become quite visible; the maximal values are about 4.5, see above.

We decided to analyze the "champions" in the plane law breaking. Recall that the first place here belongs to the file `1PBA`. It contains 20 models. For each model, we plot the convex hull volumes for consequent pairs of amino acids. The graph for the first model is shown in Fig. 4.39.

We can evidently see that the first model of the polypeptide contains exactly one serious deviation from the plane law. We call it *big peak*. It appears at the pair of amino acids with the numbers (38, 39).

Here we also present the graph for the second model, see Fig. 4.40.

Now we can clearly see two peaks in the graph: besides the big peak, there appears another one, we call it the *small peak*. The respective value is more than 2, hence the deviation is essential. The small peak corresponds to the pair (46, 47).

Let us draw some other graphs, see Fig. 4.41.

Notice that the deviations in these two graphs are less than 1, so the plane law holds more or less. Notice that the left graph contains both the peaks, and the right one contains only the small one.

Also notice that in this example 14 models have the deviations more than 2, and 5 models have the deviations less than 1. Besides, the big peak always appears either at the pair (38, 39) or at the pair (46, 47).

Let us consider another example. It is interesting because it contains essentially more deviations with the volume more than 4. We construct the corresponding graphs for the file `1I1S` containing 20 models, see Fig. 4.42.

Let us try to find some regularity in appearances of the plane law breaks. It is well known that there are two typical elements of the secondary structure of polypeptides, namely, so-called helices and sheets. The information concerning their location in a polypeptide can be found in the corresponding `pdb`-file, namely in the strings starting with the words `HELIX` and `SHEET`, respectively.

For convenience, we mark in red and blue the segments of the abscissa axis corresponding to the helices and sheets, respectively. Notice that the unique peak in the first model of `1PBA` corresponding to the (38, 39) pair of amino acids is located outside the marked domains, see Fig. 4.43.

The situation for the second model looks similar: the small peak appearing at the pair (46, 47) is located at the boundary of such a segment. Also, one can see some "attenuation" going from the big peak to the middle of the second red segment, see Fig. 4.44.

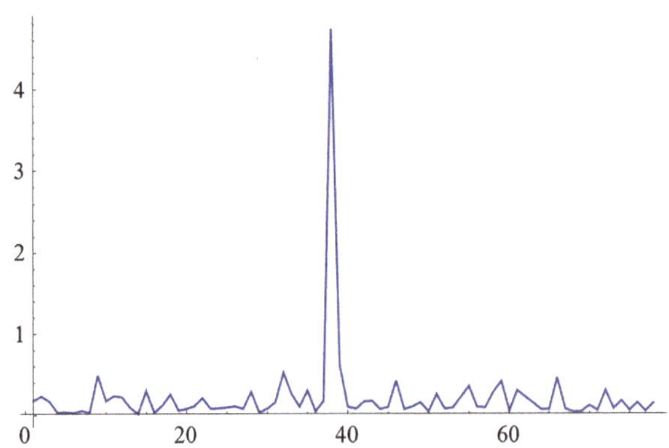

Fig. 4.39 The volumes measuring the deviations from the plane law for consequent pairs of the amino acids in the first model of the polypeptide `1PBA`

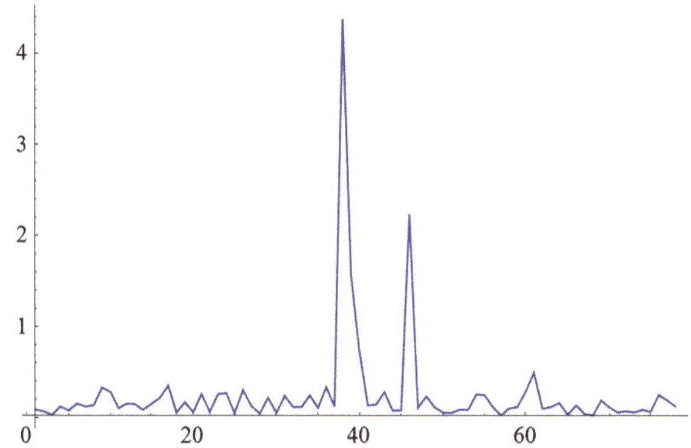

Fig. 4.40 The volumes measuring the deviations from the plane law for consequent pairs of the amino acids in the second model of the polypeptide `1PBA`

We also present the graphs constructed for the polypeptide `1I1S` (they contain much more peaks). Under careful analysis one can see the same effects: the largest deviations from the plane law appear either at the boundaries of the segments corresponding to sheets and helices, or outside of those segments. Also one can see the "attenuation" from the end of a segment to its middle point, see Fig. 4.45.

This phenomena can be explained by the fact that the helices and sheets are more or less stable configurations, as distinct from the parts between them that are movable, so the deviations appearing at movable segments attenuate along the stable elements of the secondary structure.

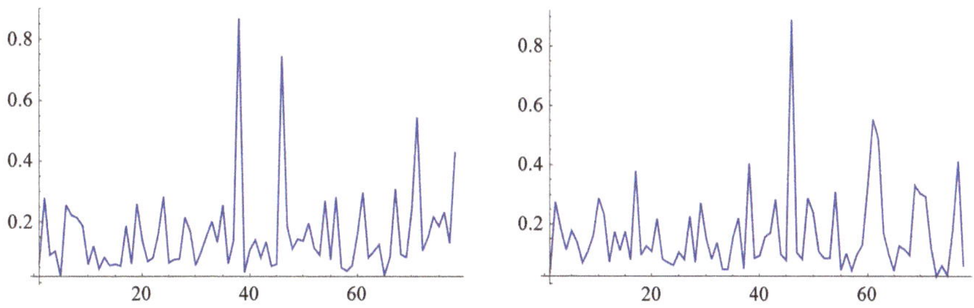

Fig. 4.41 The volumes measuring the deviations from the plane law for consequent pairs of the amino acids in the 6th and 12th models of the polypeptide `1PBA`

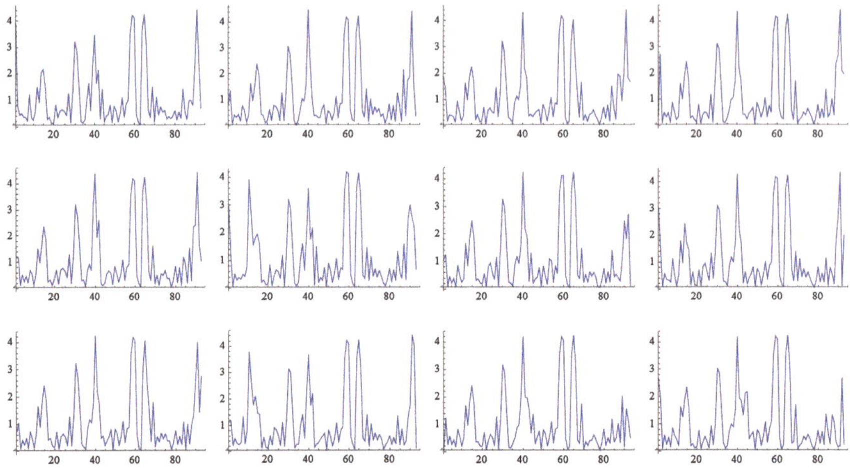

Fig. 4.42 The volumes measuring the deviations from the plane law for consequent pairs of the amino acids in the first twelve models of the polypeptide `1I1S`

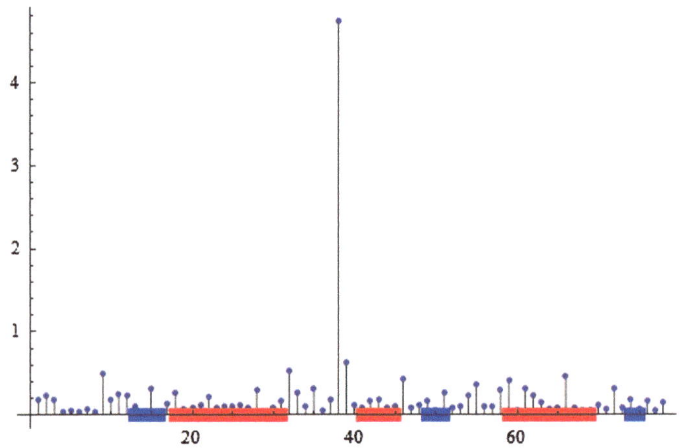

Fig. 4.43 The graph of volumes for consequent pairs of amino acids in the first model of the polypeptide `1PBA` with marked segments of helices and sheets

4.6 Geometry of Planar and Space Polygonal Lines: Spirals Detecting

In this section we discuss some other geometrical ideas that could be useful in investigation of polypeptides conformations. Consider an arbitrary *geometrical realization* of polypeptide, i.e., its special location in the three-space. We start with investigation of the alpha carbons. If we place them in accordance to the order of the corresponding amino acids in the polypeptide, we represent them as consecutive vertices of a

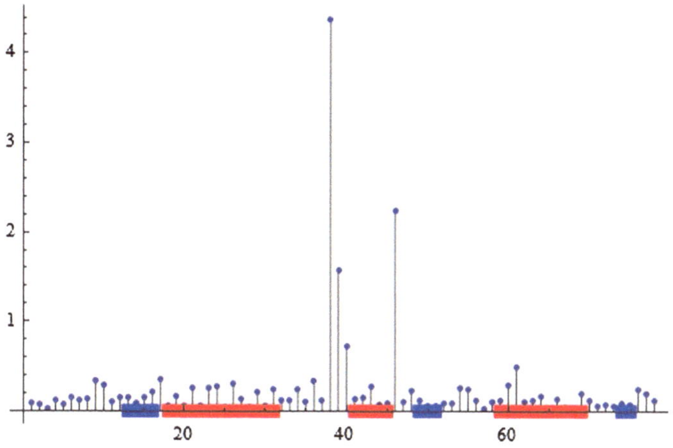

Fig. 4.44 The graph for the second model of 1PBA

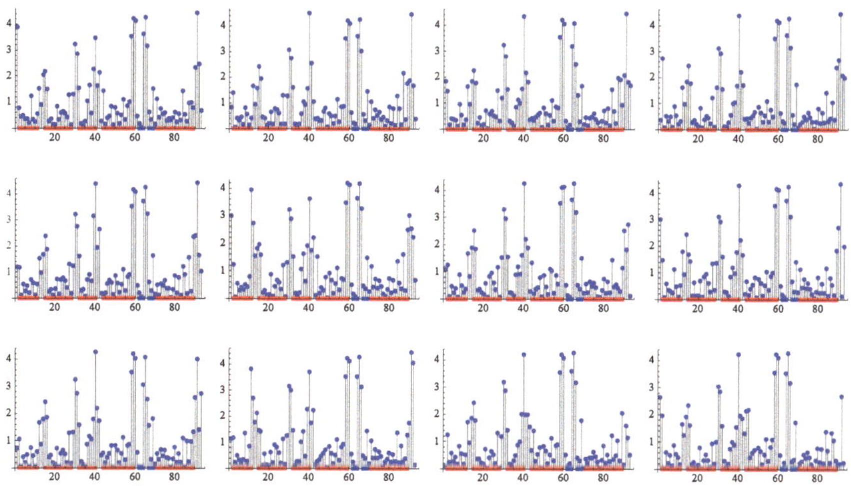

Fig. 4.45 The volumes measuring the deviations from the plane law for consequent pairs of the amino acids in the first twelve models of the polypeptide 1I1S with marked segments of helices and sheets

polygonal line L. What do we need to know about a polygonal line to reconstruct it from these data uniquely up to an orientation preserving motion of the ambient three-space? The answer is evident: we need to know the edges' lengths, the rotation angles between the consecutive edges, and the rotation angles between each pair of consecutive planes, where the first plane is spanned on the $(i-1)$th and ith edges and the second plane on the ith and $(i+1)$th edges.

Define the above angles formally. Let $L = A_0 A_1 \cdots A_n$, $A_i \in \mathbb{R}^3$, be the *vertices* of the polygonal line L. By the *edge* e_i *of the polygonal line* L we call the pair $A_{i-1}A_i$.

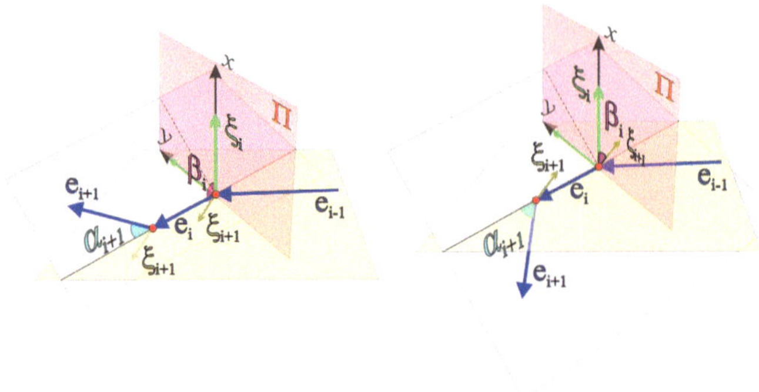

Fig. 4.46 Angular characteristics of polygonal lines

It is covenant to consider the edges e_i as the corresponding vectors $A_i - A_{i-1}$. We always assume that consecutive vectors $e_i,\ e_{i+1}$ are noncollinear. Let α_i be the value of the angle between the vectors $e_i,\ e_{i+1}$, i.e.,

$$\alpha_i = \arccos \frac{\langle e_i, e_{i+1}\rangle}{\|e_i\| \cdot \|e_{i+1}\|},$$

where $\langle v, w\rangle$ stands for the standard scalar product of the vectors v and w, and $\|v\|$ is the corresponding norm, i.e., $\|v\| = \sqrt{\langle v, v\rangle}$. The above assumptions imply that $0 < \alpha_i < \pi$.

Now consider an arbitrary *interior* edge (i.e., nonending one) e_i of the polygonal line L, then there are two neighboring edges e_{i-1} and e_{i+1}. We want to define the value β_i of the angle between two planes, the first one is spanned on the edges e_{i-1}, and e_i, and the second plane is generated by the edges e_i, and e_{i+1}. More exactly, we orient the planes in a natural way and define β_i taking into account these orientations.

Namely, we consider the vector products $\xi_i = [e_{i-1}, e_i]$ and $\xi_{i+1} = [e_i, e_{i+1}]$. The first vector is orthogonal to the first plane, and the second one is orthogonal to the second plane. Consider a plane Π perpendicular to e_i. The both vectors ξ_i and ξ_{i+1} are parallel to it. Define Cartesian coordinates in Π as follows, see Fig. 4.46. Put the origin at the intersection point of the plane Π and the straight line passing through the edge e_i; emit the axis Ox in the direction e_x of the vector ξ_i; emit the axis Oy in the direction e_y such that the triplet ξ_i, e_y, e_i is positively oriented. In the other words, $e_y = \big[e_i/\|e_i\|, \xi_i/\|\xi_i\|\big]$. Consider the direction $e = \xi_{i+1}/\|\xi_{i+1}\|$ of the vector ξ_{i+1}. Then the coordinates of the vector e in the basis $(e_x,\ e_y)$ have the form $(\cos\beta_i, \sin\beta_i)$ for some uniquely defined angle $\beta_i \in [-\pi, \pi)$. Just this β_i is taken for the value of the angle between the planes under consideration.

Let us calculate the angle β_i explicitly. Notice that $\cos\beta_i = \langle e, e_x\rangle$, $\sin\beta_i = \langle e, e_y\rangle$, therefore

$$\beta_i = \operatorname{sign}(\sin\beta_i)\arccos(\cos\beta_i) = \operatorname{sign}\big(\langle e, e_y\rangle\big)\arccos\big(\langle e, e_x\rangle\big) =$$
$$= -\operatorname{sign}\big[\det(\xi_i, e_i, \xi_{i+1})\big]\arccos\frac{\langle \xi_i, \xi_{i+1}\rangle}{\|\xi_i\|\,\|\xi_{i+1}\|}.$$

Show that the lengths s_i of the consecutive edges of the polygonal line L together with its angles α_i, β_j define the polygonal line L uniquely up to a motion.

(1) choose an arbitrary location of the initial vertex A_0 of the polygonal line L.
(2) Put the second vertex A_1 of L on the distance s_1 from A_0 in arbitrary way.
(3) From the vertex A_1 emit a ray r_1 forming the angle α_1 with the ray $r_0 = A_0A_1$ (it can be done in infinitely many ways) and put the point A_2 at the ray r_1 forming the segment A_1A_2 of the length s_2.

Notice that the steps (1)–(3) contain some "freedom of choice," but all subsequent steps are uniquely defined. Namely, assume that the vertices $A_0, \dots, A_i$, $i \geq 2$, are already constructed. In particular, the edges e_{i-1} and e_i are already constructed.

$(i+2)$ Calculate the vector $\xi_i = [e_{i-1}, e_i]$ and put $e_x = \xi_i/\|\xi_i\|$.
$(i+3)$ Calculate the vector $e_y = [e_i, e_x]/\|e_i\|$.
$(i+4)$ Then, by definition, we have $\xi_{i+1}/\|\xi_{i+1}\| = \cos\beta_i\, e_x + \sin\beta_i\, e_y$.
$(i+5)$ To construct the point A_{i+1}, we need to consider the plane Π passing through the edge e_i and orthogonal to ξ_{i+1}, and in this plane construct the vector e_{i+1} beginning at A_i having the length s_{i+1} and forming the angle α_{i+1} with the vector e_i. There are two such vectors, but, by definition, the vector ξ_{i+1} is equal to $[e_i, e_{i+1}]$, so there remains only one opportunity, namely, the following one:

$$e_{i+1} = s_{i+1}\left(\cos\alpha_{i+1}\frac{e_i}{\|e_i\|} + \sin\alpha_{i+1}\left[\cos\beta_i\, e_x + \sin\beta_i\, e_y,\ \frac{e_i}{\|e_i\|}\right]\right).$$

Now assume that all the edges of the polygonal line L have the same length. Then the angles α_i and β_i are constants (distinct from 0 and π), if and only if the vertices lie in some spiral curve. This idea gives an opportunity to find out spirals in polypeptides. Recall that in our resulting database the distances between consecutive alpha carbons deviate from the mean value at most by 7.5 %, therefore we can consider them as almost constant. Thus, if for some part of the polygonal line L constructed on the consecutive alpha carbons the angles α_i and β_i turn out to be almost constant (and distinct from 0 and π), then this fragment has spiral form. We demonstrate this in the following example.

Consider the polypeptide `1A03`. For visuality we draw a curve through its consecutive alpha carbons. Figure 4.47 (left) demonstrates that this polypeptide contains four spirals.

Also in Fig. 4.47 (right) we visualized the sequences of α- and β-angles of the polygonal line with vertices which are consecutive alpha carbons of the polypeptide `1A03`. It is easy to see the locally constant fragment on the graphs which were obtained. There are four such fragments and they correspond to the spirals of the polypeptide. It is not difficult to write a procedure that automatically finds fragments where the angles are almost constant and, hence, finding the spirals without visualization.

To make the picture more evident we drew both the graphs in one figure (we stretched one of them to obtain appropriate scales), see Fig. 4.48.

Remark 4.8 It would be interesting to find a criterion for an arbitrary polygonal line (not necessary with equal edges) describing the polygonal lines inscribed into spirals. Such condition would be useful for similar algorithms of automatic spirals finding in the polypeptides having large deviations in the distances between consecutive alpha carbons.

What is an advantage of the approach described above? For short polypeptides, of course, it is not difficult to find "by sight" the structural elements, such as spiral fragments. But for sufficiently large polypeptides it is a hard job to investigate their form directly. Instead we suggest to control two or three functions (the third one is the consecutive distances). Besides, standard properties of the functions such as monotony or convexity could lead to consideration of new types of fragments of polypeptides and, hence, to deeper understanding of their geometrical properties.

Also, let us notice that these ideas concerning description of space curves in terms of two functions have deep analogues in classical differential geometry (space curves theory). In the next section, we suggest some other topics to stimulate the reader to search analogues.

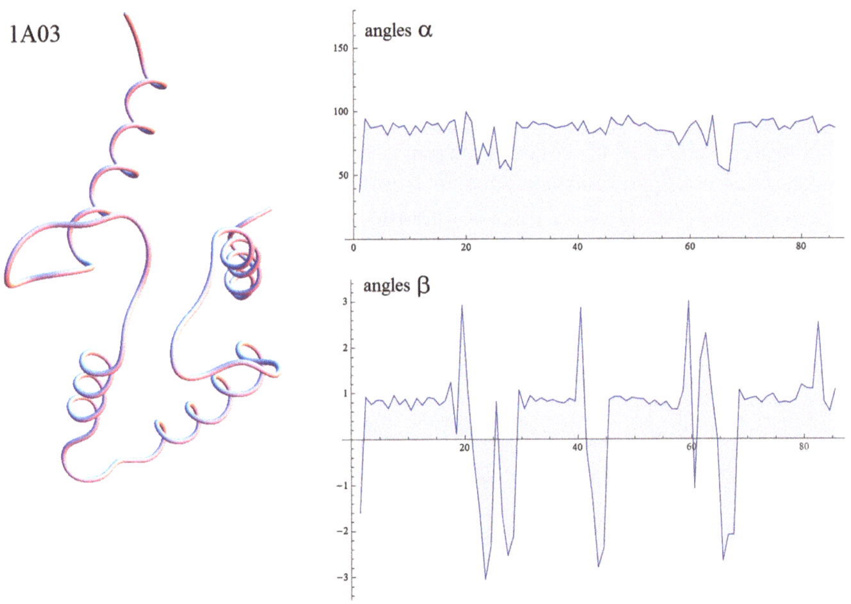

Fig. 4.47 Approximation of the polypeptide bond of 1A03 by a curve (*left*) and graphs of the consecutive angles α_i and β_i (*right*)

4.7 Torsion and Curvature of Space Curves

Consider a smooth curve $\gamma(t) = \big(x_1(t),\ x_2(t),\ x_3(t)\big)$ in the three-space $\mathbb{R}^3$. Here the smoothness means that the coordinate functions $x_i(t)$ are continuously differentiable as much as we need to. The *velocity of the curve* γ *at the point* $\gamma(t)$ is the vector $\dot{\gamma}(t) = \big(\dot{x}_1(t),\ \dot{x}_2(t),\ \dot{x}_3(t)\big)$ consisting of the derivatives of the coordinate functions. The vector $\ddot{\gamma}(t) = \big(\ddot{x}_1(t),\ \ddot{x}_2(t),\ \ddot{x}_3(t)\big)$ consisting of the second derivatives is referred as the *acceleration vector*. A curve with everywhere nonzero velocity vector is called *regular*, and if the velocity and the acceleration vectors are linear independent, then the curve is called *biregular*.

If the length of the velocity vector of the curve $\gamma(t)$ is constant and equals to 1, then the parameter t is called *natural*. If the parameter of a curve is natural, then its velocity and acceleration are perpendicular to each other at each point of the curve. For such curve the absolute value of the acceleration is used as a characteristic of the curvature of the curve, large turning of the curve correspond to large acceleration. The length of the acceleration vector of a naturally parameterized curve is called the *curvature* and is denoted by k. If the curvature is not zero, i.e., the curve is biregular, then the direction $\ddot{\gamma}/\|\ddot{\gamma}\|$ of the acceleration vector $\ddot{\gamma}$ is called the *principal normal of the curve* γ and is usually denoted by ν.

So, let $\gamma(t)$ be a biregular curve with natural parameter. Put $\tau = \dot{\gamma}$, then τ and ν are unit pairwise orthogonal vectors. Their vector product $\beta = [\tau, \nu]$ is called the *binormal*. The triplet (τ, ν, β) of pairwise orthogonal vectors is called the *Frenet frame*.

Frenet formuals shows how the Frenet frame changes along the curve. It is not difficult to verify, that the velocity $\dot{\beta}$ of the binormal β changing is collinear to the vector ν. The *torsion* $\varkappa$ of the curve is defined as the following proportionality factor: $\dot{\beta} = -\varkappa\nu$.

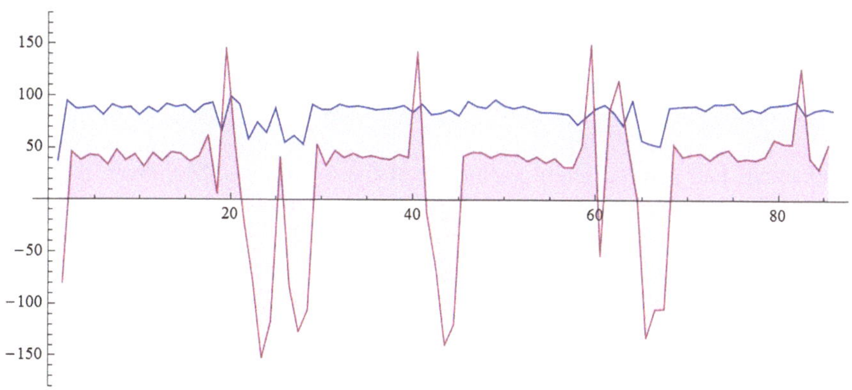

Fig. 4.48 The both graphs of α- and β-angles

We give several examples demonstrating the relations between the properties of the curvature and the torsion of a curve (considered as smooth functions) and the geometry of the curve.

Example 4.1 The straight lines are exactly the curves of zero curvature (notice that in the case of zero curvature the torsion is not defined).

Example 4.2 The curvature of the circle of radius r is equal to $1/r$. A planar curve is a circle, if and only if its curvature is a nonzero constant.

Example 4.3 A *helix* or *spiral* is a curve having in some Cartesian coordinates the representation $(a\cos t,\ a\sin t,\ bt)$, where a and b are some constants. Its curvature and torsion are nonzero constants. Conversely, each biregular curve with nonzero constant curvature and torsion is a helix.

Example 4.4 A biregular curve lies in some plane, if and only if its torsion is equal to zero.

It is well known that the functions $k(t)$ and $\varkappa(t)$ define completely the geometrical form of naturally parameterized biregular curve $\gamma(t)$: if two such curves have the same curvature and torsion, then these curves coincide up to a motion. Moreover, each two smooth functions $f(t)$ and $g(t)$, where $f(t)$ is positive, are the curvature and the torsion of some naturally parameterized biregular space curve. The procedure of the corresponding curve construction can be reduced to solution of a system of ODE.

The characterization of polygonal lines in terms of angles described above is a "discrete analogue" of the characterization of smooth curves by their curvatures and torsions. And the condition on the edges of the equal length is an analogue of the natural parametrization.

If a parameter t of a curve $\gamma(t)$ is an arbitrary one (not necessary natural), then the curvature and the torsion of the curve $\gamma(t)$ can be calculated by the following formulas:

$$k(t) = \frac{\left\|[\dot\gamma, \ddot\gamma]\right\|}{\|\dot\gamma\|^3}, \quad \varkappa(t) = \frac{\det(\dot\gamma, \ddot\gamma, \dddot\gamma)}{\left\|[\dot\gamma, \ddot\gamma]\right\|^2}.$$

One can find some discrete analogues of these formulas. Moreover, there are many different formulas for "curvature" and "torsion" of polygonal lines that sustain the passage to the limit in the following sense: if we inscribe polygonal lines into a biregular curve reducing the lengths of the edges to zero, then the curvature torsions of the polygonal lines tend to the curvature and torsion of the initial curve, respectively. In due time, the authors used this approach and test a series of functions, but the graphs of these functions do not have qualitative differences from the above graphs of the α- and β-angles (we considered the polygonal lines with almost constant length of the edges).

Working with Frenet formulas in conformational investigations of polypeptides is quite popular, see for example [3, 4], or [5].

In our opinion, serious using of mathematical methods to extract real mechanisms regulating conformations of polypeptides is possible only after a clarification of reliability of the databases such as PDB. To disjoint the real phenomena from phantoms in such databases becomes a long-felt need.

In our opinion, serious using of mathematical methods to extract real mechanisms regulating conformations of polypeptides is possible only after radical cleaning of PDB base from phantoms that appeared as a result of various bugs.

References

1. Balabaev, N.K., Shaitan, K.V.: Computer modelling of molecular dynamics. Computer-aided modelling of molecular dynamics. In: Ivanov, V.A., Rabinovich, A.L., Khokhlov, A.R. (eds.) Computer Modelling of Polymer and Biopolymer Systems, pp. 35–62. Librokom, Moscow (2009)
2. Bojenko, V.K., Ivanov, A.O., Mishchenko, A.S., Tuzhilin, A.A., Schischkin, A.M.: Geometrical models in biology. What and how can we simulate. Sci. Vis. **1**(1), 66–99. http://sv-journal.com/2009-1/ (2009) (in Russian)
3. Glasel, J.A., Deutscher, M.P. (eds.): Introduction to Biophysical Methods for Protein and Nucleic Acid Research. Academic Press, London (1995)
4. Goriely, A., Hausrath, A., Neukirch, S.: The differential geometry of proteins and its applications to structure determination. Biophys. Rev. Lett. **3**(1, 2), 77–101 (2008)
5. Hu, S., Lundgren, M., Niemi, A.J.: The Discrete Frenet frame, infection point solitons and curve visualization with applications to folded proteins. arXiv:1102.5658v1 (2011)
6. Ivanov, A.O., Mishchenko, A.S., Tuzhilin, A.A.: Geometry of polygonal lines and polypeptides. Nanostructures. Math. Phys. Model. **10**(1), 39–47. http://nano-journal.ru (2014) (in Russian)
7. Ivanov, A.O., Mishchenko, A.S., Tuzhilin, A.A.: Geometry of amino acids and polypeptides. Nanostructures. Math. Phys. Model. **10**(1), 49–76. http://nano-journal.ru (2014) (in Russian)
8. Pauling, L.: The Nature of the Chemical Bond, 3rd edn. Cornell University Press, Ithaca (1960)
9. Pauling, L., Corey, R.B., Branson, H.R.: The structure of proteins; two hydrogen-bonded helical configurations of the polypeptide chain. Proc. Natl. Acad. Sci. **37**, 205–211 (1951)
10. Shaitan, K.V.: Energy surface and conformational dynamics of molecules. Electrochemical **39**(2), 212–219 (2003) (in Russian)
11. Shaitan, K.V., Belyakov, A.A., Leontiev, K.M., Saraikin, S.S., Mikhailyuk, M.G., Egorova, K.B., Orlov, M.V.: Geometry of energy surface and conformational dynamics: from carbons to proteins and peptides. Chem. Phys. **22**(2), 57–68 (2003)
12. Shaitan, K.V., Tourleigh, Y.V., Golik, D.N., Tereshkina, K.B., Levtsova, O.V., Fedik, I.V., Shaitan, A.K., Li, A., Kirpichnikov, M.P.: Dynamic molecular design of bio- and nanostructures. J. Rus. Chem. Soc. **50**(2), 53–65 (2006) (in Russian)
13. Shaitan, K.V., Fedic, I.V.: Dynamics simulation of self-organization of the structure of model biomimetic polymers. Biophysics. **60**(3), 421 (2015)

Chapter 5
On the Riemann's Problem for One Nonstrictly Hyperbolic System

V.V. Palin and E.V. Radkevich

Abstract We consider the Riemann's problem for the multicomponent Euler system. In this paper, we describe possible types of shockwave and rarefaction wave bifurcations near the critical manifold.

5.1 Setting up the Problem

We consider the two-speed one-dimensional system of Euler equations, i.e.,

$$\begin{cases} \partial_t \rho + \partial_x(\rho U) = 0, \\ \partial_t(\rho U) + \partial_x(\rho U^2 + 2P) = 0, \\ \partial_t(c_0 \rho u_2) + \partial_x(c_0 \rho u_2^2 + P) = 0, \end{cases} \tag{5.1}$$

where $c_0 = \text{const}$ is the concentration of impurities, ρ is the mixture density, U is the mean speed of mixture, and u_2 is the speed of impurities. In that follows, we use the following notation

$$V = (\rho,\ U,\ u_2)^T$$

and discuss the Riemann problem for the system (5.1), i.e., the Cauchy problem with the initial conditions

$$V|_{t=0} = \begin{cases} V_-, & x < 0 \\ V_+, & x > 0. \end{cases}$$

V.V. Palin (✉) · E.V. Radkevich
Lomonosov Moscow State University, GSP-1, Leninskie Gory, Moscow 119991, Russian Federation
e-mail: grey_stranger84@mail.ru

E.V. Radkevich
e-mail: evrad07@gmail.com

© Springer International Publishing Switzerland 2015

V.A. Sadovnichiy and M.Z. Zgurovsky (eds.), *Continuous and Distributed Systems II*, Studies in Systems, Decision and Control 30,
DOI 10.1007/978-3-319-19075-4_5

Viscosity regularization for (5.1) have the following form:

$$\begin{cases} \partial_t \rho + \partial_x(\rho U) = \varepsilon \partial_x^2 \rho, \\ \partial_t(\rho U) + \partial_x(\rho U^2 + 2P) = \varepsilon \partial_x^2 U, \\ \partial_t(c_0 \rho u_2) + \partial_x(c_0 \rho u_2^2 + P) = \varepsilon \partial_x^2 u_2. \end{cases} \tag{5.2}$$

5.2 Algebraical Deduction

We rewrite the system (5.1) in a Cauchy's form:

$$\begin{cases} \partial_t \rho + U \partial_x \rho + \rho \partial_x U = 0, \\ \partial_t U + \frac{2P'}{\rho} \partial_x \rho + U \partial_x U = 0, \\ \partial_t u_2 + \frac{c_0 u_2 (u_2 - U) + P'}{c_0 \rho} \partial_x \rho - u_2 \partial_x U + 2u_2 \partial_x u_2 = 0, \end{cases}$$

that can be written in the matrix notation as

$$\partial_t V + \mathscr{A} \partial_x V = \varepsilon \partial_x^2 V, \tag{5.3}$$

where

$$\mathscr{A} = \begin{pmatrix} U & \rho & 0 \\ \frac{2P'}{\rho} & U & 0 \\ \frac{c_0 u_2 (u_2 - U) + P'}{c_0 \rho} & -u_2 & 2u_2 \end{pmatrix}.$$

the eigenvalues of this matrix

$$\lambda_\pm = U \pm \sqrt{2P'}, \ \lambda_3 = 2u_2.$$

If $\rho > 0$, we have violation of the hyperbolicity conditions on the manifolds

$$\Sigma_\pm = \{(\rho, U, u_2) | 2u_2 = U \pm \sqrt{2P'}\}.$$

5.2.1 *Critical Manifold Σ_+: Condition of Jordanity*

We consider some neighborhood of the manifold Σ_+ (the similar reasoning for Σ_-). Let us denote

$$\mathscr{A}_\pm = \mathscr{A} - \lambda_\pm E.$$

Then on the critical manifold Σ_+, we have

$$\mathscr{A}_{\pm}|_{\Sigma_+} = \begin{pmatrix} \mp\sqrt{2P'} & \rho & 0 \\ \frac{2P'}{\rho} & \mp\sqrt{2P'} & 0 \\ \frac{c_0 u_2(\sqrt{2P'}-u_2)+P'}{c_0\rho} & -u_2 & \sqrt{2P'}\mp\sqrt{2P'} \end{pmatrix}.$$

Immediately, we obtain for matrix $\mathscr{A}_+|_{\Sigma_+}$ that if **jordanity condition**

$$c_0 u_2^2 - P' \neq 0, \tag{5.4}$$

holds, then the rank of $\mathscr{A}_+|_{\Sigma_+}$ is equal to 2, i.e., λ_+ corresponds 2×2 Jordan box. Corresponding eigenvector $\nu = (0, 0, 1)^T$ we call critical vector.

5.3 The Form of One- and Two-Front Solutions of Regularized System

We will search the solutions of (5.2) in the form of traveling waves. In the case of one-front solution, we set

$$\rho = \rho(\frac{x - x^*(t)}{\varepsilon}), \ U = U(\frac{x - x^*(t)}{\varepsilon}),$$
$$u_2 = u_2(\frac{x - x^*(t)}{\varepsilon}). \tag{5.5}$$

In the case of two-front solution, we set

$$\rho = \rho(\frac{x - x^*(t)}{\varepsilon}), \ U = U(\frac{x - x^*(t)}{\varepsilon}),$$
$$u_2 = a(\frac{x - x^*(t)}{\varepsilon}) + b(\frac{x - x_1(t)}{\varepsilon}). \tag{5.6}$$

Also we denote

$$\dot{x}^*(t) = \omega, \ \dot{x}_1(t) = \omega_1.$$

5.3.1 *Stabilization Conditions: The Choice of Parameters of Problem*

We suppose that for one-front solutions the **stabilization condition** holds:

$$\begin{aligned}\rho(\pm\infty) = \rho_\pm,\ U(\pm\infty) = U_\pm,\ u_2(\pm\infty) = u_{2,\pm},\\ \dot\rho(\pm\infty) = 0,\ \dot U(\pm\infty) = 0,\ \dot u_2(\pm\infty) = 0.\end{aligned} \tag{5.7}$$

For two-front solutions, the stabilization condition takes the form

$$\begin{aligned}\rho(\pm\infty) = \rho_\pm,\ U(\pm\infty) = U_\pm,\ a(-\infty) = u_{2,-},\\ b(-\infty) = 0,\ (a+b)(+\infty) = u_{2,+}\end{aligned} \tag{5.8}$$

Moreover, we assume that

$$0 \le \omega_1 < \omega. \tag{5.9}$$

and the pressure $P(\rho)$ satisfies the equation of the state

$$P(\rho) = p_0\rho^\gamma, \tag{5.10}$$

where $p_0 > 0$, $\gamma > 1$ are fixed constants.

5.3.2 ODE System for One-Front Equation: Rankine–Hugoniot Conditions

After the substitution of the one-front solution from (5.5) to (5.2), we obtain the ODE system

$$\begin{cases} -\omega\dot\rho + (\rho U)^{\dot{}} = \ddot\rho, \\ -\omega(\rho U)^{\dot{}} + (\rho U^2 + 2P)^{\dot{}} = \ddot U, \\ -\omega(c_0\rho u_2)^{\dot{}} + (c_0\rho u_2^2 + P)^{\dot{}} = \ddot u_2. \end{cases} \tag{5.11}$$

Integrating this system over the interval $(-\infty; +\infty)$, we obtain, by the stabilization conditions (5.7), the Rankine–Hugoniot conditions for one-front case:

$$\begin{cases} -\omega[\rho] + [\rho U] = 0, \\ -\omega[\rho U] + [\rho U^2 + 2P] = 0, \\ -\omega[c_0\rho u_2] + [c_0\rho u_2^2 + P] = 0. \end{cases} \tag{5.12}$$

5.3.3 The ODE System for Two-Front Solution

Let us notice that the regularized system (5.2) admits a factorization: first two equations of this system do not contain unknown function u_2. After the substitution of the explicit form of two-front solution to (5.2) the ode system, corresponding to the first two equations of (5.2) have the form

$$\begin{cases} -\omega\dot{\rho} + (\rho U)^{\dot{}} = \ddot{\rho}, \\ -\omega(\rho U)^{\dot{}} + (\rho U^2 + 2P)^{\dot{}} = \ddot{U}. \end{cases} \tag{5.13}$$

Let us consider the ode which corresponds the last equation of (5.2). After the substitution of two-front solution (5.6), we have

$$-\omega c_0(\dot{\rho}(a+b) + \rho\dot{a}) - \omega_1 c_0\rho\dot{b} + (c_0\rho(a+b)^2 + P)^{\dot{}} = (a+b)^{\ddot{}}.$$

Through the regrouping, this equation obtain the form

$$\begin{aligned} &-\omega c_0(\rho a)^{\dot{}} - \omega_1 c_0(\rho b)^{\dot{}} \\ &+(\omega_1 - \omega)c_0\dot{\rho}b + (c_0\rho(a+b)^2 + P)^{\dot{}} = (a+b)^{\ddot{}}. \end{aligned} \tag{5.14}$$

5.3.4 Rankine–Hugoniot Conditions

Similarly, the construction of Rankine–Hugoniot conditions for one-front solution we obtain:

$$\begin{cases} -\omega[\rho] + [\rho U] = 0, \\ -\omega[\rho U] + [\rho U^2 + 2P] = 0, \\ -\omega[\rho u_2] + (\omega - \omega_1)[b]\rho_- + [\rho u_2^2 + \frac{P}{c_0}] = 0. \end{cases} \tag{5.15}$$

Let us note that in the last equation we use the following equality

$$\int_{-\infty}^{+\infty} \dot{\rho}(\tau)b(\tau)d\tau = [b]\rho_-,$$

which follows from condition (5.9).

5.3.5 Lax's Condition : One-Front Solution

For one-front case, we move on from the second-order system to the first-order system:

$$\begin{cases} \dot{\rho} = \tilde{\rho}, \\ \dot{U} = \tilde{U}, \\ \dot{u}_2 = \tilde{u_2}, \\ \dot{\tilde{\rho}} = -\omega\tilde{\rho} + (\rho U)^{\dot{}}, \\ \dot{\tilde{U}} = -\omega(\rho U)^{\dot{}} + (\rho U^2 + 2P)^{\dot{}}, \\ \dot{\tilde{u_2}} = -\omega c_0(\rho u_2)^{\dot{}} + (c_0\rho u_2^2 + P)^{\dot{}}. \end{cases} \tag{5.16}$$

Under the Rankine–Hugoniot conditions (5.12), this system always has two singular points:

$$\rho = \rho_{\pm}, \; U = U_{\pm}, \; u_2 = u_{2,\pm}, \; \tilde{\rho} = 0, \; \tilde{U} = 0, \; \widetilde{u_2} = 0.$$

Let us assume that for the linearized system (5.16), the counts of stable and unstable eigenvalues correspondingly in a neighborhood of "minuses" and "pluses" were accorded. Linearization in the neighborhood of both points have the equal structure. For brevity, we designate

$$\tilde{V} = (\tilde{\rho}, \; \tilde{U}, \; \widetilde{u_2})^T.$$

Writing the linearized system in the neighborhood of "minus", we obtain:

$$\begin{cases} \dot{V} = \tilde{V}, \\ \dot{\tilde{V}} = \mathscr{F}^- \tilde{V}, \end{cases} \tag{5.17}$$

where

$$\mathscr{F}^- = \begin{pmatrix} -\omega + U_- & \rho_- & 0 \\ -\omega U_- + U_-^2 + 2P'_- & -\omega\rho_- + 2\rho_- U_- & 0 \\ -\omega c_0 u_{2,-} + c_0 u_{2,-}^2 + P'_- & 0 & -\omega c_0 \rho_- + 2\rho_- u_{2,-} c_0 \end{pmatrix}.$$

The characteristic polynomial for this system is

$$\det \begin{pmatrix} -\lambda E & E \\ 0 & \mathscr{F}^- - \lambda E \end{pmatrix} = -\lambda^3 \det(\mathscr{F}^- - \lambda E).$$

From the form of polynomial, we immediately deduce that three eigenvalues of the linearized matrix of system (5.16) are equal to zero in the neighborhood of both singular points. Hence, the Lax's condition takes the form: there exist positive eigenvalue of matrix $\mathscr{F}^-$ which corresponds to negative eigenvalue of matrix $\mathscr{F}^+$. We can formulate the necessary condition in jumps terms:

$$[\rho] < 0, [U] < 0, [u_2] < 0. \tag{5.18}$$

5.3.6 Lax's Condition for Two-Front Solution: Condition of Monotonicity

For the case of two-front solution, we demand that Lax condition for system (5.13) holds.

Let us notice that first two equations in the ode system of traveling waves a similar both in one- and two-front cases. Moreover, from Lax condition the existence of traveling waves for ρ, U arise. Let us assume that condition

$$\dot{\rho} < 0 \tag{5.19}$$

holds. It is easy to see that in case $|[\rho]| \ll 1$ condition of monotonicity holds.

5.4 Shock Waves in a Small Neighborhood of Σ_+: Condition for Speed ω

Let us denote

$$V_- = (\rho_-, U_-, u_{2,-})^T, \; V_+ = (\rho_+, U_+, u_{2,+})^T.$$

As in Lax's notation, we set $V_+ = V_+(q)$, $V_+(0) = V_-$. Suppose that $V_- \in \Sigma_+$. By differentiation first two of Rankine–Hugoniot conditions by parameter q, notting by stroke the derivative with respect to q, we derive:

$$\begin{cases} -\omega'[\rho] - \omega\rho'_+ + (\rho_+ U_+)' = 0, \\ -\omega'[\rho U] - \omega(\rho_+ U_+)' + (\rho_+ U_+^2 + 2P_+)' = 0. \end{cases} \tag{5.20}$$

Hence, substituting $q = 0$ after the equivalent transformations, we obtain

$$\begin{pmatrix} U_- & \rho_- \\ \frac{2P'_-}{\rho_-} & U_- \end{pmatrix} \begin{pmatrix} \rho'_+|_{q=0} \\ U'_+|_{q=0} \end{pmatrix} = \omega|_{q=0} \begin{pmatrix} \rho'_+|_{q=0} \\ U'_+|_{q=0} \end{pmatrix}, \tag{5.21}$$

i.e., $\omega|_{q=0}$ must be the eigenvalue of the upper-left block of matrix $\mathscr{A}|_{V_-}$. So, there is two different cases: $\omega|_{q=0} = \lambda_\pm|_{q=0}$.

5.4.1 The Existence of One-Front Solution then Moving in the Noncritical Eigenvector Direction

Theorem 5.1 *Suppose that $V_- \in \Sigma_+$, $\omega|_{q=0} = \lambda_-$, easily formulated conditions (on speed and monotonicity) holds and jumps are small enough. Then there is the solution of*

$$-\,\omega(c_0\rho u_2\dot{)} + (c_0\rho u_2^2 + P\dot{)} = \ddot{u}_2, \tag{5.22}$$

which satisfy stabilization conditions (5.7).

Proof Integrating (5.22) over the interval $(-\infty, z)$ and applying the stabilization condition (5.7), we obtain

$$-\omega c_0(\rho u_2 - \rho_- u_{2,-}) + c_0\rho u_2^2 + P - c_0\rho_- u_{2,-}^2 - P_- = \dot{u}_2. \tag{5.23}$$

Level 0 isoclines for this ODE satisfy the following equation

$$-\omega c_0(\rho u_2 - \rho_- u_{2,-}) + c_0\rho u_2^2 + P - c_0\rho_- u_{2,-}^2 - P_- = 0.$$

Calculating the discriminant, we derive that

$$\begin{aligned} D &= c_0\rho(\omega^2 c_0\rho + 4c_0\rho_- u_{2,-}^2 - 4\omega c_0\rho_- u_{2,-} + 4(P_- - P)) = \\ &= c_0\rho(c_0\rho_-(\omega - 2u_{2,-})^2 + 4(P_- - P) - \omega^2 c_0(\rho_- - \rho)) > 0 \end{aligned}$$

then $|[\rho]| \ll 1$ is small enough, since

$$\omega - 2u_{2,-} \neq 0, \;\; D(-\infty) = c_0^2\rho_-^2(\omega - 2u_{2,-})^2 > 0.$$

From this calculations it follows that there exist two curves which are level 0 isoclines for (5.23):

$$\widetilde{u}_2^{\pm} = \frac{1}{2c_0\rho}(\omega c_0\rho \pm \sqrt{D}).$$

Since $V_- \in \Sigma_+$, the eigenvalue $\lambda_- = U_- - \sqrt{2P'_-} < \lambda_+ = 2u_{2,-}$. So, for the lower isocline the inequality

$$\widetilde{u}_2^-(-\infty) = u_{2,-}$$

holds, which leads to the existence of stabilizing solution of (5.22).

5.4.2 One-Front Solution as the Traveling Wave for $\omega = \lambda_+|_{q=0}$

Theorem 5.2 *Suppose that $V_- \in \Sigma_+$, $\omega|_{q=0} = \lambda_+$, conditions for parameters, formulated easily, holds and jumps are small enough. Then*
(1.) If inequality

$$c_0\omega^2 \leq 4P'_-, \tag{5.24}$$

holds, then there exist the solution (5.22) satisfying stabilization conditions (5.7).
(2.) If inequality

$$c_0\omega^2 > 4\frac{[P]}{[\rho]}, \tag{5.25}$$

holds there is no one solution of (5.22), satisfying stabilization conditions (5.7).

Proof Integrating (5.22) over the interval $(-\infty, z)$ and substituting $2u_{2,-} = \omega$ we obtain, by applying the stabilization conditions (5.7) that

$$-\omega c_0 \rho u_2 + c_0 \rho u_2^2 + P + \frac{1}{4} c_0 \omega^2 \rho_- - P_- = \dot{u}_2. \tag{5.26}$$

Level 0 isoclines for this equation are the solutions of algebraic equation

$$-\omega c_0 \rho u_2 + c_0 \rho u_2^2 + P + \frac{1}{4} c_0 \omega^2 \rho_- - P_- = 0. \tag{5.27}$$

Calculating the discriminant, we have

$$D = c_0 \rho (\omega^2 c_0 (\rho - \rho_-) - 4(P - P_-)) = c_0 \rho D_2(z).$$

It is easy to see that

$$D_2(-\infty) = 0,$$

$$\dot{D}_2(z) = (\omega^2 c_0 - 4P'(\rho))\dot{\rho}. \tag{5.28}$$

Suppose that inequality (5.24) holds. Then, from the virtue of inequality $\dot{\rho} < 0$ from (5.28), we derive that $D_2(z) > 0$ for all $z \in \mathbb{R}$. Hence, there exists the stabilizing solution of ODE (5.26), close to level 0 lower isocline.

Now we assume that inequality

$$c_0 \omega^2 > 4 \frac{[P]}{[\rho]}$$

holds. Then

$$D_2(+\infty) = \omega^2 c_0 [\rho] - 4[P - P_-] < 0,$$

so, in the neighborhood of $z - +\infty$ the right-hand part of ODE (5.26) is strictly positive. From (5.25) it follows that $\dot{u}_2(+\infty) > 0$ which comes to collision with identity $\dot{u}_2(+\infty) = 0$ which holds for any stabilizing solution $u_2(z)$.

5.5 The Existence of Two-Front Solutions as $\omega = \lambda_+|_{q=0}$

Theorem 5.3 *Suppose that $V_- \in \Sigma_+$, $\omega|_{q=0} = \lambda_+$, conditions for parameters, formulated easily, holds and jumps are small enough. Then there is some $\omega_* > 0$ such that the two-front solution of (5.14) satisfied the conditions (5.8) exists for any $0 < \omega_1 < \omega_*$.*

Proof Let us show that there exist the bounded smooth function $b(z)$ which is satisfy identity $b(-\infty) = 0$ and ODE

$$-\omega_1 c_0(\rho b)^{\dot{}} + (c_0\rho b^2 + P)^{\dot{}} = \ddot{b}. \tag{5.29}$$

Integrating this equation over the interval $(-\infty, z)$ and applying stabilization conditions, we obtain

$$\dot{b} = c_0\rho b^2 - \omega_1 c_0\rho b + P - P_-. \tag{5.30}$$

Level 0 isoclines for this equation have the form:

$$c_0\rho b^2 - \omega_1 c_0\rho b + P - P_- = 0,$$

and discriminant

$$D = \omega_1^2 c_0^2 \rho^2 - 4c_0\rho(P - P_-) \geq \omega_1^2 c_0^2 \rho^2 > 0$$

by condition of sign of the jumps (5.18). From this the existence of bounded solution of (5.30) follows. This solution is close to lower level 0 isocline as $|z| \to \infty$.

Now we assume that $b(z)$ is the solution of (5.29) constructed before. Then from the Eq. (5.14), we obtain the following problem for unknown $a(z)$:

$$(\omega_1 - \omega)c_0\dot{\rho}b - \omega c_0(\rho a)^{\dot{}} + (c_0\rho(a^2 + 2ab))^{\dot{}} = \ddot{a}. \tag{5.31}$$

Integrating this equation over the interval $(-\infty, z)$ using equality $a(-\infty) = \frac{1}{2}\omega$ and condition of stabilization (5.8), we obtain

$$\begin{aligned} \dot{a} &= c_0\rho a^2 + c_0\rho(2b - \omega)a + \\ &+ \frac{1}{4}\omega^2 c_0\rho_- + (\omega_1 - \omega)c_0 I(z), \end{aligned} \tag{5.32}$$

where $I(z) = \int\limits_{-\infty}^{z} \dot{\rho} b d\tau$. Level 0 isoclines are

$$\rho a^2 + \rho(2b - \omega)a + \frac{1}{4}\omega^2\rho_- + (\omega_1 - \omega)I(z) = 0,$$

discriminant

$$D = \rho((2b - \omega)^2\rho - \rho_-\omega^2 + 4(\omega - \omega_1)I(z)) = \rho D_1.$$

Let us notice that $D_1(-\infty) = 0$. We need

Lemma 5.1 *Suppose that $\rho(z)$ is the bounded monotone solution of (5.13), satisfied condition of (5.19) and $b(z)$ is the bounded solution of problem (5.30) constructed before. Then there exist a constant $K(\omega_1) > 0$ such that the inequality*

$$\frac{\dot{\rho}}{\dot{b}} \leq K(\omega_1) \tag{5.33}$$

holds for all z.

Suppose that there is $\omega_* > 0$ such that for all $0 < \omega_1 < \omega_*$

$$\omega_1 + \frac{1}{4\rho_-}(\omega^2 - 4\frac{[P]}{c_0\rho_+})K(\omega_1) \leq \omega. \tag{5.34}$$

Then

$$4(\omega - \omega_1) \geq \frac{1}{\rho}(\omega^2 - 4\frac{[P]}{c_0\rho_+})K(\omega_1) \geq \frac{1}{\rho}(\omega^2 - 4\frac{[P]}{c_0\rho_+})\frac{\dot{\rho}}{\dot{b}},$$

this inequality is equivalent to

$$\frac{\dot{\rho}}{\rho} + \frac{4(\omega - \omega_1)\dot{b}}{\omega^2 - 4\frac{[P]}{c_0\rho_+}} \geq 0. \tag{5.35}$$

Using the inequality

$$\frac{4(\omega_1 - \omega)\dot{b}}{4b^2 - 4\omega_1 b + \omega^2} \geq \frac{4(\omega_1 - \omega)\dot{b}}{4b_+^2 - 4\omega_1 b_+ + \omega^2} = \frac{4(\omega_1 - \omega)\dot{b}}{\omega^2 - 4\frac{[P]}{c_0\rho_+}}$$

we obtain

$$\frac{\dot{\rho}}{\rho} + \frac{4(\omega_1 - \omega)\dot{b}}{4b^2 - 4\omega_1 b + \omega^2} \geq 0,$$

or, in the equivalent form

$$\dot{\rho}(4b^2 - 4\omega_1 b + \omega^2) + 4(\omega_1 - \omega)\rho\dot{b} \geq 0.$$

Let us notice that

$$\dot{D}_1 = \dot{\rho}(4b^2 - 4\omega_1 b + \omega^2) + \rho(4b^2 - 4\omega_1 b + \omega^2\dot{)} + 4(\omega_1 - \omega)\rho\dot{b}$$

and the inequality

$$(4b^2 - 4\omega_1 b + \omega^2\dot{)} = ((2b - \omega_1)^2 + \omega^2 - \omega_1^2\dot{)} = 4(2b - \omega_1)\dot{b} \geq 0$$

holds. Thus, $\dot{D}_1 \geq 0$.

Hence, in previously formulated conditions there are two level 0 isoclines for the equation (5.32) for all $z \in \mathbb{R}$, which started from the similar value $\frac{1}{2}\omega$ as $z = -\infty$. So the stabilizing as $|z| \to \infty$ solution of (5.31), exists.

5.5.1 Different Forms of Two-Front Solutions

Let us notice that the stabilizing as $|z| \to \infty$ solution of (5.31), is close to lower of level 0 isoclines for (5.31)

$$\tilde{a}^- = \frac{\omega}{2} - b - \frac{\sqrt{D}}{2\rho}$$

for $z \gg 1$ which is big enough, where $D = \rho((2b-\omega)^2\rho - \rho_-\omega^2 + 4(\omega-\omega_1)I(z))$. Besides, the solution $b(z)$ of problem (5.29) is monotonically decreasing. Thus, there are three cases for $u_2(z)$ curve:
1. Monotone decrease of $a(z)$ then $u_2(z)$ is the sum of two monotonically decreasing functions, i.e., "shelf".
2. Monotone increase of $a(z)$ then $u_2(z)$ is the sum of two functions with different monotonicity, i.e., "humped kink".
3. The $a(+\infty) = 0$ situation, i.e., "lagged wave".

For the monotone increase of $a(z)$, i.e., for "humped kink" case it is enough that the inequality

$$\tilde{a}^-(+\infty) > u_{2,-} = \frac{1}{2}\omega$$

holds, i.e., the inequality fulfilled:

$$-b_+ - \frac{\sqrt{D_+}}{2\rho_+} > 0.$$

Let us notice that $a_+ + b_+ = u_{2,+}$, so, the sufficient condition of "humped kink" -shaped solution can be written in the form

$$[u_2] > b_+. \tag{5.36}$$

Similarly the sufficient condition of "shelf" existence can be described by

$$[u_2] < b_+. \tag{5.37}$$

5.5.2 *The Evident Appearance of b_+: The Conditions for Two-Front Solutions Type in the Terms of Front Speed*

The value of b_+ can be obtained as the solution of algebraic equation

$$\frac{1}{2}\omega - \frac{\sqrt{D_+}}{2\rho_+} = u_{2,+}.$$

Writing D_+ in the evident form and using equality

$$I(+\infty) = [\rho]b_+,$$

we derive

$$b_+ = \frac{(\omega\rho_- + \omega_1[\rho]) - \sqrt{(\omega\rho_- + \omega_1[\rho])^2 - \rho_+(\omega^2[\rho] - 4\rho_+[u_2]^2)}}{2\rho_+}. \tag{5.38}$$

Substituting obtained value to inequality (5.36), we have

$$(\omega\rho_- + \omega_1[\rho]) - \sqrt{(\omega\rho_- + \omega_1[\rho])^2 - \rho_+(\omega^2[\rho] - 4\rho_+[u_2]^2)} < 2\rho_+[u_2],$$

or, in the equivalent form

$$\omega\rho_- + \omega_1[\rho] - 2\rho_+[u_2] < \sqrt{(\omega\rho_- + \omega_1[\rho])^2 - \rho_+(\omega^2[\rho] - 4\rho_+[u_2]^2)}.$$

After squaring and combination of similar summands, we obtain

$$\omega^2[\rho] - 4\omega\rho_-[u_2] < 4\omega_1[\rho][u_2]. \tag{5.39}$$

This inequality is a sufficient condition for "humped kink" form of solution.

Similarly, we can formulate a sufficient condition for the "shelf"-formed two-front solutions:

$$\omega^2[\rho] - 4\omega\rho_-[u_2] > 4\omega_1[\rho][u_2]. \tag{5.40}$$

5.5.3 *The Conditions for Two-Front Solutions of Humped Kink Type in the ω Terms*

If we make previously formulated conditions stronger, we can dismiss the restrictions for ω_1 and formulate the conditions for type of two front solutions in ω terms.

Let us notice that the right-hand part of (5.39) is positive. Hence, we can formulate the sufficient condition of "humped kink" shape of solution in the form:

$$\omega > 4\frac{[u_2]}{[\rho]}\rho_-. \tag{5.41}$$

5.5.4 *The Conditions for Two-Front Solutions of Shelf Type in the ω Terms*

Let us notice that, under assumptions $0 \le \omega_1 < \omega$, to satisfy the inequality (5.40), it is enough that

$$\omega[\rho] - 4\rho_-[u_2] > 4[\rho][u_2], \tag{5.42}$$

which is equivalent to

$$\omega[\rho] > 4[u_2](\rho_- + [\rho]).$$

Dividing the last inequality by $[\rho] < 0$, we obtain the sufficient condition for the existence of shelf-type solution, similar to (5.41):

$$\omega < 4\frac{[u_2]}{[\rho]}\rho_- + 4[u_2]. \tag{5.43}$$

5.6 Bifurcation of Rarefaction Waves on the Critical Manifold Σ_+

Let us study the behavior of rarefaction wave as $V_- \in \Sigma_+$. Let us notice that, according the classic theory, the rarefaction waves are continuous solutions depended on $\frac{x}{t}$ in a traditional definition–are sticking into one on the manifold Σ_+. Indeed, if $V = V(\frac{x}{t})$ is a rarefaction wave, then it satisfies the ODE system

$$\begin{cases} -\frac{x}{t^2}\dot{\rho} + \frac{1}{t}(\rho U\dot{)} = 0, \\ -\frac{x}{t^2}(\rho U\dot{)} + \frac{1}{t}(\rho U^2 + 2P\dot{)} = 0, \\ -\frac{x}{t^2}(\rho u_2\dot{)} + \frac{1}{t}(\rho u_2^2 + \frac{1}{c_0}P\dot{)} = 0, \end{cases} \tag{5.44}$$

thus, the vector $\dot{V}$ must be the eigenvector of $\mathscr{A}$ as $q \to 0$.

5.6.1 The Two-Front Analogue of the Rarefaction Wave: rs-Type

Let us prove that the system (5.1) have the generalized (in D' terms) solution

$$\rho = \rho(\frac{x}{t}),\ U = U(\frac{x}{t}),\ u_2 = a(\frac{x}{t}) + \beta\theta(x - \omega t), \tag{5.45}$$

where β is a scalar and $\rho(\cdot)$, $U(\cdot)$, $a(\cdot)$ are the continuous functions. Solutions of (5.45) kind are called the rs-type solutions, because the main part of them is the rarefaction wave. Let us demand the reverse for inequality (5.18) conditions for jump's signs:

$$[\rho] > 0,\ [U] > 0. \tag{5.46}$$

5.6.2 The Proof of the Existence of rs-Type Solution

Substituting (5.45) into the first two equations of (5.1), we derive

$$\begin{cases} -\frac{x}{t^2}\dot{\rho} + \frac{1}{t}(\rho U\dot{)} = 0, \\ -\frac{x}{t^2}(\rho U\dot{)} + \frac{1}{t}(\rho U^2 + 2P\dot{)} = 0. \end{cases} \tag{5.47}$$

Thus, the functions $\rho(\cdot)$ and $U(\cdot)$ are the solutions of traveling wave problem for single-speed system of Euler equations. Existence of this solutions follows from (5.46). Substituting the expression for u_2 in the third equation of (5.1) and writing the continuous part separately, we have

$$-\frac{x}{t^2}(\dot{\rho}a + \rho\dot{a}) + \frac{1}{t}(\rho a^2 + \frac{1}{c_0}P(\rho)\dot{)} = 0. \tag{5.48}$$

It is easy to see that $(\rho, U, a)^T$ is the solution of the system (5.44). In that follows for brevity $z = \frac{x}{t}$. Writing the coefficient $\theta(x - \omega t)$ for the discontinuous summand, we obtain

$$\beta(-z\dot{\rho} + \beta\dot{\rho} + 2\rho\dot{a}) = 0,\ z > \omega. \tag{5.49}$$

And the delta-function coefficient gives

$$-\omega\beta\rho(\omega) + \beta^2\rho(\omega) + 2\rho(\omega)a(\omega)\beta = 0. \tag{5.50}$$

For the sake of rs-type solution and classic rarefaction wave non-comparing it is necessary that $\beta \neq 0$. From this condition, inequality $\rho > 0$ and equalities (5.49) and (5.50) we obtain that

$$\begin{cases} (\beta - z)\dot{\rho} + 2\rho\dot{a} = 0, \ z > \omega, \\ -\omega + \beta + 2a(\omega) = 0. \end{cases} \tag{5.51}$$

Let us notice that the first of equations of (5.51) automatically holds for $\dot{\rho} = 0$, $\dot{a} = 0$, and $z > \omega$, i.e., if

$$\omega > U_+ + 2\sqrt{P'_+}. \tag{5.52}$$

wherein, it is clear that $\beta = \omega - 2a(\omega)$. So, if conditions (5.46) and relation (5.52) hold, there exist the rs-type solution of (5.1).

5.6.3 The Two-Front Analogue of Rarefaction Wave: sr-Type

Similarly to the previous case, we can prove that under the conditions for sign of jumps

$$[\rho] < 0, \ [U] < 0, \ [u_2] > 0 \tag{5.53}$$

there exist the (5.1) solution of following type

$$\begin{aligned} \rho = \rho_- + [\rho]\theta(x - \omega t), \ U = U_- + [U]\theta(x - \omega t), \\ u_2 = u_{2,-} + a(\frac{x}{t}) + \beta\theta(x - \omega t), \end{aligned} \tag{5.54}$$

which we call the sr-type solution.

References

1. Lax, P.D.: Hyperbolic Partial Differential Equations. Courant institute of Mathematical Sciences, New York (2006)
2. Evans, L.C.: Partial Differential Equations. Graduate Studies Mathematical. American Mathematical Society, Providence (1998)
3. Yakovlev, N.N., Lukashev, E.A., Palin, V.V., Radkevich, E.V.: Nonclassical regularization of the multicomponent Euler system. J. Math. Sci. **196**(3), 322–345 (2014)

4. Yakovlev, N.N., Lukashev, E.A., Palin, V.V., Radkevich, E.V.: On non-viscosity solutions of the multicomponent Euler system. Contemp. Math. Fund. Dir. **53**, 133–154 (2014)
5. Radkevich, E.V.: On the nature of bifurcations of one-front solutions of the truncated Euler system. J. Math. Sci. **196**(3), 388–404 (2014)
6. Palin, V.V., Radkevich, E.V.: On the nature of bifurcations of the Riemann problem solutions for the truncated Euler system. Differ. Equ. (2014)
7. Rykov, Y.G., Lysov, V.G., Feodoritova, O.B.: The emergence of non-classic shocks in a flow model of two-component, two-velocity medium. Keldysh Institute preprints **74**, 20 (2012)
8. Lukashev, E.A., Radkevich, E.V.: Solidification and structuresation of instability zones. Appl. Math. **1**, 159–178 (2010)
9. Lukashev, E.A., Radkevich, E.V., Yakovlev, N.N.: Structuresation of instability zone and cristallization. Trudy Seminara I.G. Petrovskogo **28**, 229–264 (2011)

Part II
Non-autonomous and Stochastic Dynamical Systems

Chapter 6
Dynamics of Nonautonomous Chemostat Models

Tomás Caraballo, Xiaoying Han, Peter E. Kloeden and Alain Rapaport

Abstract Chemostat models have a long history in the biological sciences as well as in biomathematics. Hitherto most investigations have focused on autonomous systems, that is, with constant parameters, inputs, and outputs. In many realistic situations these quantities can vary in time, either deterministically (e.g., periodically) or randomly. They are then nonautonomous dynamical systems for which the usual concepts of autonomous systems do not apply or are too restrictive. The newly developing theory of nonautonomous dynamical systems provides the necessary concepts, in particular that of a nonautonomous pullback attractor. These will be used here to analyze the dynamical behavior of nonautonomous chemostat models with or without wall growth, time-dependent delays, variable inputs and outputs. The possibility of overyielding in nonautonomous chemostats will also be discussed.

T. Caraballo (✉)
Dpto. Ecuaciones Diferenciales y Análisis Numérico, Universidad de Sevilla,
Apdo. de Correos 1160, 41080 Sevilla, Spain
e-mail: caraball@us.es

X. Han
Department of Mathematics and Statistics, Auburn University, 221 Parker Hall,
Auburn 36849, USA
e-mail: xzh0003@auburn.edu

P.E. Kloeden
School of Mathematics and Statistics, Huazhong University of Science & Technology,
Wuhan 430074, China
e-mail: kloeden@mathematik.uni-kl.de

P.E. Kloeden
Felix-Klein-Zentrum Für Mathematik, TU Kaiserslautern, 67663 Kaiserslautern, Germany

A. Rapaport
UMR INRA/SupAgro MISTEA, 2 Place Viala, 34060 Montpellier, France
e-mail: rapaport@supagro.inra.fr

A. Rapaport
MODEMIC Team, INRIA, rte des Lucioles, 06 902 Sophia-Antipolis, France

© Springer International Publishing Switzerland 2015

V.A. Sadovnichiy and M.Z. Zgurovsky (eds.), *Continuous and Distributed Systems II*, Studies in Systems, Decision and Control 30,
DOI 10.1007/978-3-319-19075-4_6

6.1 Introduction

Traditional models of the chemostat assume fixed availability of the nutrient and its supply rate, as well as fast flow rates to avoid the tendency of microorganisms to attach to container walls. However, these assumptions become unrealistic when the availability of a nutrient depends on the nutrient consumption rate and input nutrient concentration and when the flow rate is not fast enough. On the other hand, the appearance of delay terms in chemostat models [1, 2] can be fully justified since the future behavior of a dynamical system does not in general only depend on the present but also on its history. Sometimes only a short piece of history provides the relevant influence (bounded or finite delay), while in other cases it is the whole history that has to be taken into account (unbounded or infinite delay). In this article we will discuss chemostat models with a variable nutrient supplying rate and a variable input nutrient concentration, along with time-variable delays and wall growth.

Denote by $x(t)$ the concentration of the growth-limiting nutrient and by $y(t)$ the concentration of the microorganism at any time t. When wall attachment is taken into account (see e.g., [3–7]), we can regard the consumer population $y(t)$ as an aggregate of two categories of populations, one in the growth medium, denoted by $y_1(t)$, and the other on the walls of the container, denoted by $y_2(t)$. Suppose that the nutrient is equally available to both of the categories, so it can be assumed that both categories consume the same amount of nutrient and at the same rate. Let D be the rate at which the nutrient is supplied and also the rate at which the contents of the growth medium are removed, and I be the input nutrient concentration which describes the quantity of nutrient available with the system at any time. Assume that D and I vary continuously in time (e.g., periodically [8] or randomly) in bounded positive intervals $D(t) \in [d_m, d_M]$ and $I(t) \in [i_m, i_M]$, respectively, for all $t \in \mathbb{R}$. In addition, let $\tau_1(t)$ and $\tau_2(t)$ be the time delay into material recycling and in the growth response of the consumer species, respectively. The consideration of variable inputs, variable delays, and wall growth result in the following system of nonautonomous delay differential equations:

$$\frac{dx(t)}{dt} = D(t)[I(t) - x(t)] - aU(x(t))[y_1(t) + y_2(t)] + b\gamma y_1(t - \tau_1(t)), \quad (6.1)$$

$$\frac{dy_1(t)}{dt} = -[\gamma + D(t)]y_1(t) + cU(x(t - \tau_2(t)))y_1(t) - r_1 y_1(t) + r_2 y_2(t), \quad (6.2)$$

$$\frac{dy_2(t)}{dt} = -\gamma y_2(t) + cU(x(t - \tau_2(t)))y_2(t) + r_1 y_1(t) - r_2 y_2(t), \quad (6.3)$$

where $a > 0$ is the maximal consumption rate of the nutrient and also the maximum specific growth rate of microorganisms, c with $0 < c \leq a$ is the growth rate coefficient of the consumer species, γ is the collective death rate of microorganisms, $b \in (0, 1)$ is the fraction of dead biomass that is recycled, r_1 and r_2 are the rates at which the species stick onto and shear off from the walls respectively, and U

is the uptake function describing how the nutrient is consumed by the species and satisfying:

(1) $U(0) = 0$ and $U(x) > 0$ for all $x > 0$;
(2) $\lim_{x\to\infty} U(x) = L < \infty$;
(3) U is continuously differentiable;
(4) U is monotonically increasing.

In this article, when concrete computations are sought, we choose the uptake function to have the Michaelis-Menten or Holling type II form, given by

$$U(x) = \frac{x}{\lambda + x}, \tag{6.4}$$

where $\lambda > 0$ is the half-saturation constant. The results in all but the last section are collected from the papers [9–11].

6.2 Preliminaries on Nonautonomous Dynamical Systems

Given a real number $h \geq 0$, denote by $C_h := C([-h, 0], \mathbb{R}^n)$ the Banach space of continuous functions mapping the interval $[-h, 0]$ into $\mathbb{R}^n$ equipped with the usual supremum norm

$$\|\phi\|_{C_h} = \sup_{\theta\in[-h,0]} |\phi(\theta)|.$$

Note that $C_h \cong \mathbb{R}^n$ when $h = 0$.

Consider the functional differential equation

$$\dot{z}(t) = f(t, z_t) \tag{6.5}$$

where $f : \mathbb{R} \times C_h \to \mathbb{R}^n$ is continuous and maps bounded sets into bounded sets and $z_t(\cdot) \in C_h$ is given by

$$z_t(\theta) = z(t + \theta), \quad \theta \in [-h, 0],$$

for any given continuous function $z(\cdot) : \mathbb{R} \to \mathbb{R}^n$ and $t \in \mathbb{R}$. Note that Eq. (6.5) is a general formulation and includes ordinary differential equations ($h = 0$)

$$\dot{z}(t) = f(t, z(t)),$$

in which case the state space C_h reduces to $\mathbb{R}^n$.

Assume that an initial function $\psi \in C_h$ prescribed at the initial time $t_0 \in \mathbb{R}$ is associated with (6.5) to form an initial value problem. The solution of this initial value problem for which an existence and uniqueness theorem holds then defines a solution map, $Z(t, t_0)\colon \psi \mapsto z_t(\cdot; t_0, \psi) \in C_h$ for $t \geq t_0$, which is, in fact, a process (also called a two-parameter semigroup) satisfying

- $Z(t, t_0) : C_h \to C_h$ is a continuous map for all $t \geq t_0$;
- $Z(t_0, t_0) = Id_{C_h}$, the identity on C_h, for all $t_0 \in \mathbb{R}$;
- $Z(t, t_0) = Z(t, s)Z(s, t_0)$ for $t \geq s \geq t_0$.

Definition 6.1 Let Z be a process on a complete metric space X. A family $\mathscr{A} = \{A(t)\}_{t\in\mathbb{R}}$ of compact subsets of X is called a pullback attractor for Z if it is

- invariant: $Z(t, t_0)A(t_0) = A(t)$ for all $t \geq t_0$;
- pullback attracting: for any nonempty bounded subset D of X

$$\text{dist}_X\{Z(t, t - t_0)D, A(t)\} \to 0 \quad \text{as} \quad t_0 \to \infty \quad (\text{for each } t \in \mathbb{R})$$

where dist_X denotes the Hausdorff semi-distance.

Pullback attraction uses information about the dynamical system from the past in contrast with the usual forward convergence with $t \to \infty$ for fixed t_0 which uses information about the future.

Definition 6.2 A family $\{B(t)\}_{t\in\mathbb{R}}$ of nonempty subsets of X is said to be pullback absorbing with respect to a process Z if for each $t \in \mathbb{R}$, and every nonempty bounded subset D of X, there exists $T_D(t) > 0$ such that

$$Z(t, t - \sigma)D \subseteq B(t), \quad \text{for all } \sigma \geq T_D(t).$$

The following result (see [12]) shows that the existence of a family of compact absorbing sets implies the existence of a pullback attractor.

Theorem 6.1 *Let $Z(t, t_0)$ be a process on a complete metric space X. If there exists a family $\{B(t)\}_{t\in\mathbb{R}}$ of compact absorbing sets, then there exists a pullback attractor $\mathscr{A} = \{A(t)\}_{t\in\mathbb{R}}$ such that $A(t) \subset B(t)$ for all $t \in \mathbb{R}$. Furthermore,*

$$A(t) = \overline{\bigcup_{\substack{D\subset X\\ \text{bounded}}} \Lambda_D(t)} \quad \text{where} \quad \Lambda_D(t) = \bigcap_{T\geq 0} \overline{\bigcup_{t_0\geq T} Z(t, t - t_0)D}.$$

For the general case of (6.5) being a delay differential equation ($h \neq 0$), the next sufficient condition ensures the existence of a pullback attractor.

Theorem 6.2 *([13, Theorem 4.1]) Suppose that $Z(t, t_0)$ maps bounded sets of C_h into bounded sets of C_h, and there exists a family $\{B(t)\}_{t\in\mathbb{R}}$ of bounded absorbing sets for Z in C_h. Then there exists a pullback attractor $\mathscr{A}$ for Eq.* (6.5).

For the particular case of (6.5) being an ordinary differential equation ($h = 0$), the following theorem ensures the existence of an attractor in both the forward and pullback senses that consist of singleton sets, i.e., a single entire solution.

Theorem 6.3 *([12, 14]) Suppose that a process Z on $\mathbb{R}^n$ is uniform strictly contracting on a positively invariant pullback absorbing family $\{B(t)\}_{t\in\mathbb{R}}$ of nonempty compact subsets of $\mathbb{R}^n$, i.e., for each $R>0$, there exist positive constants K and α such that*

$$|Z(t,t_0)x_0 - Z(t,t_0)y_0|^2 \le Ke^{-\alpha(t-t_0)}\cdot|x_0-y_0|^2, \quad \forall t\ge t_0, \quad x_0, y_0\in\mathbb{B}(0,R),$$

where $\mathbb{B}(0,R)$ is the closed ball in $\mathbb{R}^n$ centered at the origin with radius $R>0$. Then the process Z has a unique global forward and pullback attractor $\mathscr{A}=\{A(t): t\in\mathbb{R}\}$ with component sets consisting of singleton sets, i.e., $A(t)=\{\xi^*(t)\}$ for each $t\in\mathbb{R}$, where ξ^* is an entire solution of the process.

6.3 Properties of Solutions

The existence and uniqueness of solutions to (6.1)–(6.3) with initial conditions

$$x(t)=\psi_1(t-t_0),\ y_1(t)=\psi_{21}(t-t_0),\ y_2(t)=\psi_{22}(t-t_0), \quad \forall t\in[t_0-h,t_0] \tag{6.6}$$

follow immediately from the continuity of the input functions $D(t)$ and $I(t)$ and the assumptions on the uptake function U. Therefore, we have the unique solution $z(\cdot;t_0,\psi)$ of (6.1)–(6.3) such that $z_{t_0}(\cdot;t_0,\psi)=\psi$, i.e.,

$$z_{t_0}(\theta;t_0,\psi) := z(t_0+\theta;t_0,\psi)=\psi(\theta) \quad \text{for} \quad \theta\in[-h,0].$$

Consequently, we can construct a nonautonomous dynamical system or process $Z(t,t_0): C_h\to C_h$ in the phase space C_h defined for any $t\ge t_0$ as

$$Z(t,t_0)\phi = z_t(\cdot;t_0,\phi), \quad \phi\in C_h.$$

The positiveness and boundedness of solutions are stated in the following theorems:

Theorem 6.4 *For any nonnegative continuous initial condition* (6.6) *on* $[t_0-h,t_0]$, *the solutions to* (6.1)–(6.3) *are nonnegative.*

Proof We will show that if a solution starts in the octant $\mathbb{R}^3_+=\{(x,y_1,y_2): x\ge 0,\ y_1\ge 0,\ y_2\ge 0\}$, then it remains there forever. In fact, by continuity, each solution has to take value 0 before it reaches a negative value. With $x=0$ and $y_1\ge 0$, $y_2\ge 0$, Eq. (6.1) reduces to

$$x'(t) = D(t)I(t) + b\gamma y_1(t-\tau_1(t)),$$

and thus $x(t)$ is strictly increasing at $x=0$. With $y_1=0$ and $x\ge 0$, $y_2\ge 0$, the reduced ODE for $y_1(t)$ is

$$y_1'(t) = r_2 y_2 \geq 0,$$

hence $y_1(t)$ is nondecreasing at $y_1 = 0$. Similarly, y_2 is nondecreasing at $y_2 = 0$. Therefore, $(x(t), y_1(t), y_2(t)) \in \mathbb{R}^3_+$ for any t. $\square$

Theorem 6.5 *Assume that* $D : \mathbb{R} \to [d_m, d_M]$ *where* $0 < d_m < d_M < \infty$, *and* $I : \mathbb{R} \to [i_m, i_M]$ *where* $0 < i_m < i_M < \infty$ *are continuous. In addition assume that* $\tau_1'(t) \leq M_1 < 1$ *for all* $t \in \mathbb{R}$. *Then solutions to* (6.1)–(6.3) *are bounded for any bounded initial conditions provided that*

$$\mu := \min\{\delta, \gamma - c\} > 0 \quad \text{where} \quad \delta := d_m - \frac{M_1}{1 - M_1}\gamma - c. \tag{6.7}$$

Proof Define over $\mathbb{R} \times C_h$ the functional $v(\cdot, \cdot, \cdot, \cdot)$ as

$$v(t, \phi_1, \phi_{21}, \phi_{22}) := \phi_1(0) + b\phi_{21}(0) + b\phi_{22}(0) + \frac{b\gamma}{1 - M_1}\int_{-\tau_1(t)}^{0} \phi_{21}(s)ds. \tag{6.8}$$

Given a solution $z(\cdot) = (x(\cdot), y_1(\cdot), y_2(\cdot))$ of (6.1)–(6.3) corresponding to an initial datum $(\psi_1, \psi_{21}, \psi_{22}) \in C_h$, define the function $\nu(t) := \nu(t, z_t)$ for $t \in \mathbb{R}$. After a change of variable in the integral in (6.8) we obtain

$$\nu(t) = x(t) + by_1(t) + by_2(t) + \frac{b\gamma}{1 - M_1}\int_{t-\tau_1(t)}^{t} y_1(s)ds.$$

Then the time derivative of $\nu(t)$ along solutions to (6.1)–(6.3) is

$$\begin{aligned}
\frac{d\nu(t)}{dt} &= D(t)I(t) - D(t)x(t) - aU(x(t))(y_1(t) + y_2(t)) + b\gamma y_1(t - \tau_1(t)) \\
&\quad - b[\gamma + D(t)]y_1(t) - b\gamma y_2(t) + bcU(x(t - \tau_2(t))(y_1(t) + y_2(t)) \\
&\quad + \frac{b\gamma}{1 - M_1}(y_1(t) - (1 - \tau_1'(t))y(t - \tau_1(t))).
\end{aligned}$$

Since $\tau_1'(t) \leq M_1 < 1$, we have $-\frac{1}{1-M_1}(1 - \tau_1'(t)) \leq -1$. Also using the facts that $U(x) \leq 1$ for $x \geq 0$, $d_m \leq D(t) \leq d_M$ and $i_m \leq I(t) \leq i_M$ for any t, we have

$$\begin{aligned}
\frac{d\nu(t)}{dt} &\leq d_M i_M - d_m x(t) - b(\gamma + d_m)y_1(t) - b\gamma y_2(t) + bc(y_1(t) + y_2(t)) + \frac{b\gamma}{1 - M_1}y_1(t) \\
&\leq d_M i_M - d_m x(t) - b\left(\gamma + d_m - c - \frac{b\gamma}{1 - M_1}\right)y_1(t) - b(\gamma - c)y_2(t) \\
&\leq d_M i_M - d_m x(t) - b\delta y_1(t) - b(\gamma - c)y_2(t).
\end{aligned}$$

where δ is as defined in (6.7). Now define the region

$$\Omega := \{(x, y_1, y_2) \in \mathbb{R}^3_+ : d_m x + b\delta y_1 + b(\gamma - c)y_2 \leq d_M i_M\}.$$

If a trajectory starts at time t_0 from a point in $\mathbb{R}^3_+ \backslash \Omega$, then the functional $\nu(\cdot, \cdot, \cdot, \cdot)$ along a trajectory starting from this point would be decreasing for all times $t \geq t_0$ such that $(x(t), y_1(t), y_2(t)) \in \mathbb{R}^3_+ \backslash \Omega$. Therefore,

$$
\begin{aligned}
\nu(t, x_t, (y_1)_t, (y_2)_t) &\leq \nu(t, x_{t_0}, (y_1)_{t_0}, (y_2)_{t_0}) \\
&\leq x(t_0) + by_1(t_0) + by_2(t_0) + \frac{b\gamma}{1 - M_1} \int_{t_0 - \tau_1(t)}^{t_0} y_1(s)ds \\
&\leq |\psi_1| + b\left(1 + \frac{\gamma h}{1 - M_1}\right)|\psi_{21}| + b|\psi_{22}|,
\end{aligned}
$$

which implies that

$$
\begin{aligned}
&\|(x(t), y_1(t), y_2(t))\| := x(t) + y_1(t) + y_2(t) \\
&\leq \frac{1}{b}\nu(t, x_t, (y_1)_t, (y_2)_t) \leq \frac{1}{b}|\psi_1| + \left(1 + \frac{\gamma h}{1 - M_1}\right)|\psi_{21}| + |\psi_{22}|. \quad (6.9)
\end{aligned}
$$

If a trajectory starts from or enters the region Ω at $t_1 \geq t_0$ and stays in Ω forever, then by the definition of Ω we have that for any time $t \geq t_0$, $d_m x(t) + b\delta y_1(t) + b(\gamma - c)y_2(t) \leq d_M i_M$, which implies that

$$
\|(x(t), y_1(t), y_2(t))\| \leq \frac{d_m}{b\mu}x(t) + \frac{\delta}{\mu}y_1(t) + \frac{\gamma - c}{\mu}y_2(t) \leq \frac{d_M i_M}{b\mu}. \qquad (6.10)
$$

If a trajectory starts from, enters or reenters the region Ω at times $t_{2i-1} \geq t_0$ and exits at time t_{2i}, $(i = 1, 2, \ldots)$, then (6.9) holds for all times (t_{2i}, t_{2i+1}) and (6.10) holds for all times (t_{2i-1}, t_{2i}).

To summarize, for any $t > t_0$, we have

$$
\begin{aligned}
\|z_t\| &= \|(x_t, y_{1t}, y_{2t})\| = x(t + \theta) + y_1(t + \theta) + y_2(t + \theta) \\
&\leq \max\left\{\frac{|\psi_1|}{b} + \left(1 + \frac{\gamma h}{1 - M_1}\right)|\psi_{21}| + |\psi_{22}|, \frac{d_M i_M}{b\mu}\right\}.
\end{aligned}
$$

Therefore, given any $(\psi_1, \psi_{21}, \psi_{22}) \in C_h$ with $|\psi_1| + |\psi_{21}| + |\psi_{22}| \leq r$, we have $z_t = (x_t, y_{1t}, y_{2t}) \in \mathbb{B}_{C_h}(0, \tilde{r})$ for $t \geq t_0$, where

$$
\tilde{r} := \max\left\{\frac{r}{b}, r\left(1 + \frac{\gamma h}{1 - M_1}\right), \frac{d_M i_M}{b\mu}\right\}.
$$

□

In the next section we will discuss the existence of nonautonomous attractors for different variations of system (6.1)–(6.3). Geometric details of the attractors are provided for some special cases.

6.4 Pullback Attractors for Nonautonomous Chemostat Models

In this section we discuss the existence and properties of the pullback attractors for the chemostat system (6.1)–(6.3). In particular, we will study the system with wall growth and variable delays, wall growth and variable inputs, and the special case with no wall growth.

6.4.1 *Chemostats with Wall Growth, Variable Delays, and Fixed Inputs*

When $D(t) = D$, $I(t) = I$, $\tau_1(t) \neq 0$ and $\tau_2(t) \neq 0$, the existence of a pullback absorbing set can be proved by using the Razumikhin technique, which uses a Lyapunov function rather than a functional. The reader can find an interesting motivation for the Razumikhin technique in the book by Hale and Lunel [15, pp. 151]. More precisely, our result is a consequence of the uniformly ultimately boundedness of the solutions according to Theorem 4.3 on pp. 159 in [15]. To make the result more accessible to the reader, we first recall the following notation.

Given a continuous function $V : \mathbb{R} \times \mathbb{R}^n \to \mathbb{R}$ and an initial function $\phi \in C_h$, the (upper Dini) derivative of V along the solutions of (6.5) is defined to be

$$\dot{V}(t, \phi(0)) = \limsup_{\epsilon \to 0^+} \frac{1}{\epsilon}[V(t + \epsilon, z(t + \epsilon; t, \phi) - V(t, \phi(0))]. \tag{6.11}$$

Theorem 6.6 *Assume that* $D(t) = D$, $I(t) = I$, *and* $\tau_1'(t) \leq M_1 < 1$. *Then the nonautonomous dynamical system generated by* (6.1)–(6.3) *possesses a pullback attractor in* C_h *provided that*

$$\min\left\{D - \frac{M_1}{1 - M_1}\gamma - c, \gamma - c\right\} > 0 \quad \textit{and} \quad \min\{D, \gamma - c\} > b\gamma.$$

Proof Since we are interested in only nonnegative solutions, consider the function

$$V(t, x, y_1, y_2) := x + y_1 + y_2 = \|(x, y_1, y_2)\|.$$

Given any initial value $\phi \in C_h$ we consider the solution $z(\cdot; t, \phi) := (x(\cdot), y_1(\cdot), y_2(\cdot))$ of (6.1)–(6.3) passing through (t, ϕ) and we will check the assumptions in Theorem 4.3 from [15]. Observe that when V is differentiable, the upper Dini derivative coincides with the derivative of the function V along the solutions of the problem (6.5). However, the Lyapunov function will not always be differentiable, but only continuous. Hence we can write $x(t) = \phi_1(0)$, $y_1(t) = \phi_{21}(0)$ and $y_2(t) = \phi_{22}(0)$ at time t.

By letting $u(s) = s/2$ and $v(s) = 2s$, we have

$$u(\|(x, y_1, y_2)\|) \leq V(t, x, y_1, y_2) \leq v(\|(x, y_1, y_2)\|).$$

The time derivative of V along the solution of (6.1)–(6.3) through (t, ϕ) satisfies

$$\begin{aligned}\dot{V}(t, \phi(0)) &= DI - D\phi_1(t) - (\gamma + D)\phi_{21}(0) - \gamma\phi_{22}(0) + b\gamma\phi_{21}(-\tau_1(t)) \\ &\quad - [aU(\phi_1(0)) - cU(\phi_1(-\tau_2(t)))](\phi_{21}(0) + \phi_{22}(0)) \\ &\leq DI - D\phi_1(0) - (\gamma + D - c)\phi_{21}(0) - (\gamma - c)\phi_{22}(0) + b\gamma\phi_{21}(-\tau_1(t)).\end{aligned}$$

For any $q > 1$, define $p(s) = qs$. Provided that $V(t+\theta, \phi(\theta)) < p(V(t, \phi(0)))$ for $\theta \in [-h, 0]$, we have

$$\phi_{21}(-\tau_1(t)) < q(\phi_1(0) + \phi_{21}(0) + \phi_{22}(0)).$$

Consequently,

$$\begin{aligned}\dot{V}(t, \phi(0)) &\leq DI - (D - b\gamma q)\phi_1(0) - (\gamma + D - c - b\gamma q)\phi_{21}(0) - (\gamma - c - b\gamma q)\phi_{22}(0) \\ &\leq DI - G_q[\phi_1(0) + \phi_{21}(0) + \phi_{22}(0)] = DI - G_q\|\phi(0)\|,\end{aligned}$$

where$G_q = \min\{D, \gamma - c\} - b\gamma q$.

Fix $q = 1 + \epsilon$, then $G_q > 0$ when ϵ is small enough and $\min\{D, \gamma - c\} > b\gamma$. Letting

$$w(s) = \begin{cases} 0, & s \leq DI/G_q, \\ \frac{1}{2}(G_q s - DI), & s > DI/G_q, \end{cases}$$

we have $\dot{V}(t, \phi(0)) \leq -w(\|\phi(0)\|)$ for any $\|\phi(0)\| \geq 0$. It follows immediately from Theorem 4.3 on pp. 159 in [15] that the solutions to (6.1)–(6.3) are uniformly ultimately bounded, i.e., there exists $\beta > 0$ such that for any $\alpha > 0$, there is a constant $T_\alpha > 0$, which is independent of t, such that

$$\|z(t; t_0, \phi)\| \leq \beta, \quad \forall t \geq t_0 + T_\alpha, \quad \forall t_0 \in \mathbb{R}, \ \phi \in C_h, \ \|\phi\|_{C_h} \leq \alpha.$$

This implies that the absorbing sets exist, in both the pullback and forward senses. The existence of a nonautonomous attractor then follows immediately from Theorems 6.2 and 6.5. □

6.4.2 Chemostat with Wall Growth, Variable Inputs, and No Delays

For the special case with no delays, $\tau_1(t) = \tau_2(t) = 0$ the system (6.1)–(6.3) consists of ordinary differential equations. In addition to the existence of nonautonomous attractors we will be able to obtain more geometric details of the attractor. To this end, we make the following change of variables:

$$\alpha(t) = \frac{y_1(t)}{y_1(t) + y_2(t)}, \qquad z(t) = y_1(t) + y_2(t). \tag{6.12}$$

Assuming that $U(x) = \dfrac{x}{\lambda + x}$, system (6.1)–(6.3) then attains the form

$$x'(t) = D(t)[I(t) - x(t)] - \frac{ax(t)}{\lambda + x(t)}z(t) + b\gamma\alpha(t)z(t), \tag{6.13}$$

$$z'(t) = -\gamma z(t) - D(t)\alpha(t)z(t) + \frac{cx(t)}{\lambda + x(t)}z(t), \tag{6.14}$$

$$\alpha'(t) = -D(t)\alpha(t)(1 - \alpha(t)) - r_1\alpha(t) + r_2(1 - \alpha(t)). \tag{6.15}$$

Observe that $\alpha(t)$ satisfies the Riccati equation (6.15) and is not coupled with $x(t)$ and $z(t)$. For any positive y_1 and y_2 we have $0 < \alpha(t) < 1$ for all t. Note that $\alpha'|_{\alpha=0} = r_2 > 0$ and $\alpha'|_{\alpha=1} = -r_1 < 0$, so the interval (0, 1) is positively invariant. This is the biologically relevant region.

When $D(t) = D$ is a constant, there is a unique asymptotically stable steady state $\alpha^* \in (0, 1)$ given by

$$\alpha^* := \frac{D + r_1 + r_2 - \sqrt{(D + r_1 + r_2)^2 - 4Dr_2}}{2D}. \tag{6.16}$$

Hence when $t \to \infty$, replacing $\alpha(t)$ by α^* in Eqs. (6.13) and (6.14) we have

$$\frac{dx(t)}{dt} = D(I(t) - x(t)) - \frac{ax(t)}{\lambda + x(t)}z(t) + b\gamma\alpha^* z(t) \tag{6.17}$$

$$\frac{dz(t)}{dt} = -\gamma z(t) - D\alpha^* z(t) + \frac{cx(t)}{\lambda + x(t)}z(t). \tag{6.18}$$

More details of the long-term dynamics of the solutions to (6.17) and (6.18) are established in the following theorem:

Theorem 6.7 *Assume that $D(t) = D$ for all $t \in \mathbb{R}$, and $I : \mathbb{R} \to [i_m, i_M]$ with $0 < i_m < i_M < \infty$ is continuous, $a \geq c$, $b \in (0, 1)$ and $\gamma > 0$. Then system* (6.17) *and* (6.18) *has a pullback attractor $\mathbb{A} = \{A(t) : t \in \mathbb{R}\}$ inside the nonnegative quadrant. Moreover,*

(i) the entire solution $(w^*(t), 0)$ is asymptotically stable (in the usual forwards sense) in $\mathbb{R}^2_+$, where

$$w^*(t) = De^{-Dt}\int_{-\infty}^{t} I(s)e^{Ds}ds,$$

and the pullback attractor $\mathbb{A}$ *has a singleton component subset* $A(t) = \{(w^*(t), 0)\}$ *for all* $t \in \mathbb{R}$, *provided* $\gamma + D\alpha^* > c$;

(ii) the pullback attractor $\mathbb{A}$ *also contains points strictly inside the positive quadrant in addition to the set* $\{(w^*(t), 0)\}$, *provided*

$$\gamma + D\alpha^* < \frac{cDi_M}{\lambda(a - c + \gamma - b\gamma\alpha^* + D) + Di_M} := \varphi_D(i_M). \tag{6.19}$$

Proof (i) When $\gamma + D\alpha^* > c$,

$$\frac{dz(t)}{dt} = -\left(\gamma + D\alpha^* - \frac{cx(t)}{\lambda + x(t)}\right) z(t) \le 0,$$

which implies that $z(t)$ decreases to 0 as $t \to \infty$ for any $z(t_0) \ge 0$. Consequently, $x(t)$ satisfies $\frac{dx(t)}{dt} = D(I(t) - x(t))$ and has a nontrivial nonautonomous equilibrium

$$x(t) = x(t_0)e^{-D(t-t_0)} + De^{-Dt} \int_{t_0}^{t} I(s)e^{Ds} ds,$$

which converges to $w^*(t)$ as $t \to \infty$ or $t_0 \to -\infty$.

(ii) Let $u(t) := x(t) + z(t)$, then

$$u'(t) = D(I(t) - x(t)) + \frac{(c-a)x(t)}{\lambda + x(t)} z(t) + b\gamma\alpha^* z(t) - \gamma z(t) - D(t)\alpha^* z(t).$$

On the one hand,

$$\begin{aligned} u'(t) &\le D(I(t) - x(t)) - \left(\gamma - b\gamma\alpha^* + D\alpha^*\right) z(t) \\ &< DI(t) - Dx(t) - D\alpha^* z(t) \le Di_M - D\alpha^* u(t). \end{aligned}$$

On the other hand,

$$\begin{aligned} u'(t) &\ge D(I(t) - x(t)) - \left(a - c + \gamma + D\alpha^* - b\gamma\alpha^*\right) z(t) \\ &\ge DI(t) - Dx(t) - \left(a - c + \gamma - b\gamma\beta^* + D\right) z(t) \\ &> Di_m - \left(a - c + \gamma - b\gamma\beta^* + D\right) u(t). \end{aligned}$$

Therefore, we have the upper and lower bounds for $u(t)$ as

$$l := \frac{Di_M}{a - c + \gamma - b\gamma\alpha^* + D} < u(t) < \frac{i_M}{\alpha^*}. \tag{6.20}$$

For $\varepsilon > 0$ small, define T_ε to be the trapezoid

$$T_\varepsilon := \{(x, z) \in \mathbb{R}_+^2 : x \ge \varepsilon, \ \ z \ge \varepsilon, \ \ \frac{Di_M}{a - c + \gamma - b\gamma\alpha^* + D} \le x + z \le \frac{i_M}{\alpha^*}\},$$

then T_ε is absorbing. In addition, we have the following inequalities satisfied on the boundaries of T_ε:

$$\begin{aligned} x'(t)\big|_{x=\varepsilon} &= D(I(t)-\varepsilon) + (b\gamma\alpha^* - \frac{a\varepsilon}{\lambda+\varepsilon})z(t) > 0, \\ z'(t)\big|_{z=\varepsilon} &> \left(-\gamma + D\alpha^* + \frac{c(l-\varepsilon)}{\lambda+l-\varepsilon}\right)\varepsilon > 0, \\ (x(t)+z(t))'\big|_{x+z=i_M/\alpha^*} &< 0, \quad (x(t)+z(t))'\big|_{x+z=l} > 0. \end{aligned}$$

Hence T_ε is invariant and this implies that there exists a pullback attractor $\mathbb{A} = \{A(t) : t \in \mathbb{R}\}$ in T_ε. □

When $I(t) = I$ is fixed and $D(t) \in [d_m, d_M]$ varies continuously in time, a pullback attractor of the form $\mathscr{A}_\alpha = \{A_\alpha(t) : t \in \mathbb{R}\}$ in the unit interval (0, 1) exists, since the unit interval is positively invariant (see e.g., [12]), and its component subsets are given by

$$A_\alpha(t) = \bigcap_{t_0<t} \alpha\left(t, t_0, [0,1]\right), \qquad \forall t \in \mathbb{R}.$$

These component subsets have the form $A_\alpha(t) = \left[\alpha_l^*(t), \alpha_u^*(t)\right]$, where $\alpha_l^*(t)$ and $\alpha_u^*(t)$ are entire bounded solutions of the Riccati equation. Differential inequalities can be used to obtain bounds on these entire solutions as

$$A_\alpha(t) = \left[\alpha_l^*(t), \alpha_u^*(t)\right] \subset \left[\frac{r_2}{r_1+r_2+d_M}, \frac{r_2}{r_1+r_2}\right] := [\underline{\alpha}, \overline{\alpha}].$$

To investigate the case where the pullback attractor consists of a single entire solution, we need to find conditions under which $\alpha_l^*(t) \equiv \alpha_u^*(t)$ for any $t \in \mathbb{R}$. To this end, let $\Delta_\alpha(t) = \alpha_u^*(t) - \alpha_l^*(t)$. Then

$$\begin{aligned} \Delta_\alpha'(t) &= D(t)\left(\alpha_u^*(t) + \alpha_l^*(t)\right)\Delta_\alpha(t) - \left(D(t) + r_1 + r_2\right)\Delta_\alpha(t) \\ &\le d_M \cdot 2\alpha_u^*(t)\Delta_\alpha(t) - (d_m + r_1 + r_2)\Delta_\alpha(t) \\ &\le \left(\frac{2d_M r_2}{r_1+r_2} - d_m - r_1 - r_2\right)\Delta_\alpha(t). \end{aligned}$$

Thus, when $2d_M r_2 < d_m(r_1+r_2) + (r_1+r_2)^2$, we have

$$0 \le \Delta_\alpha(t) \le e^{\left(\frac{2d_M r_2}{r_1+r_2} - d_m - r_1 - r_2\right)(t-t_0)}\Delta_\alpha(t_0) \to 0 \quad \text{as } t \to \infty \quad \text{or} \quad t_0 \to -\infty.$$

Since $d_m < d_M$, this holds, e.g., if $d_M(r_2 - r_1) < (r_1+r_2)^2$. It essentially puts a restriction on the width of the interval in which $D(t)$ can take its values, unless $r_1 > r_2$. Note that $\alpha^*(t)$ is also asymptotically stable in the forward sense in this case.

Therefore for t (or $-t_0$) sufficiently large, $x(t)$ and $z(t)$ components of the system (6.13)–(6.15) satisfy

$$x'(t) = D(t)(I - x(t)) - \frac{ax(t)}{\lambda + x(t)} z(t) + b\gamma\alpha^*(t)z(t), \tag{6.21}$$

$$z'(t) = -\gamma z(t) - D(t)\alpha^*(t)z(t) + \frac{cx(t)}{\lambda + x(t)} z(t). \tag{6.22}$$

The following theorem is proved in [9].

Theorem 6.8 *Assume that* $I(t) = I$ *and* $D : \mathbb{R} \to [d_m, d_M]$ *with* $0 < d_m < d_M < \infty$ *is continuous,* $a \geq c$, $b \in (0, 1)$ *and* $\gamma > 0$. *Then system* (6.21) *and* (6.22) *has a pullback attractor* $\mathbb{A} = \{A(t) : t \in \mathbb{R}\}$ *inside the nonnegative quadrant. Moreover,*

(i) the axial steady state solution $(I, 0)$ *is asymptotically stable in the nonnegative quadrant and the pullback attractor* $\mathbb{A}$ *has a singleton component subset* $A(t) = \{(I, 0)\}$ *for all* $t \in \mathbb{R}$, *provided* $\gamma + d_m\underline{\alpha} > c$;

(ii) the pullback attractor $\mathbb{A}$ *also contains points strictly inside the positive quadrant in addition to the point* $\{(I, 0)\}$, *provided*

$$\gamma + d_M\overline{\alpha} < \frac{cd_m I}{\lambda(a - c + \gamma + d_M - b\gamma\overline{\alpha}) + d_m I} : \varphi_I(d_m). \tag{6.23}$$

6.4.3 Chemostat with No Wall Growth or Delays

In the special case where $\tau_1(t) = \tau_2(t) = 0$ and the wall growth is neglected (see e.g., [16]), system (6.1)–(6.3) reduces to the system of ODEs

$$\frac{dx(t)}{dt} = D(t)[I(t) - x(t)] - \frac{ax(t)}{\lambda + x(t)} y(t), \tag{6.24}$$

$$\frac{dy(t)}{dt} = -D(t)y(t) + \frac{ax(t)}{\lambda + x(t)} y(t). \tag{6.25}$$

We are able to obtain more details of the attractor for this special case, as stated in the following theorem, which is also proved in [9].

Theorem 6.9 *Assume that* $I(t) = I$ *fixed and* $D : \mathbb{R} \to [d_m, d_M]$ *with* $0 < d_m < d_M < \infty$ *is continuous. Then the nonautonomous dynamical system generated by the system of ODEs* (6.24) *and* (6.25) *has a pullback attractor* $\mathscr{A} = \{A(t) : t \in \mathbb{R}\}$ *in* $\mathbb{R}^2_+$. *Moreover,*

(i) when $a < d_m$, *the axial steady state solution* $(I, 0)$ *is asymptotically stable in the nonnegative quadrant and the pullback attractor* $\mathscr{A}$ *has a singleton component subset* $A(t) = \{(I, 0)\}$ *for all* $t \in \mathbb{R}$;

(ii) when $a > (1 + \lambda/I)d_M$, *the pullback attractor* $\mathscr{A}$ *also contains points strictly inside the positive quadrant in addition to the point* $\{(I, 0)\}$;

(iii) *when* $d_m < a < \frac{d_m(\lambda d_m + d_M I)^2}{(\lambda d_m + d_M I)^2 - \lambda I d_m^2}$, *the pullback attractor* $\mathscr{A}$ *consists of the axial point* $\{(I, 0)\}$ *and a single entire solution* ξ^* *that is uniformly bounded away from the axes as well as heteroclinic entire solutions between them, i.e., its component subsets are*

$$A(t) = \left\{(x, y) \in \mathbb{R}_+^2 \, : \, x + y = I; \xi^*(t) \le x \le I\right\} \quad \textit{for} \quad t \in \mathbb{R}.$$

Assume that $D(t) = D$ *fixed and* $I : \mathbb{R} \to [i_m, i_M]$ *with* $0 < i_m < i_M < \infty$ *is continuous. Then the nonautonomous dynamical system generated by the system of ODEs* (6.24) *and* (6.25) *has a pullback attractor* $\mathscr{A} = \{A(t) : t \in \mathbb{R}\}$ *in* $\mathbb{R}_+^2$. *Moreover,*

(i) *when* $D > a$, *the entire solution* $(x^*(t), y^*(t)) = (w^*(t), 0)$ *is asymptotically stable in* $\mathbb{R}_+^2$ *and the pullback attractor has singleton component sets* $A(t) = \{(w^*(t), 0)\}$ *for every* $t \in \mathbb{R}$;
(ii) *when* $ai_m > D(\lambda + i_M)$, *the pullback attractor has nontrivial component sets that include* $(w^*(t), 0)$ *and strictly positive points;*
(iii) *when* $D < a$ *and* $a\left(\lambda^2 + \lambda(2i_M - i_m) + i_M^2\right) < D(\lambda + i_M)^2$, *the pullback attractor contains a nontrivial entire solution that attracts all other strictly positive entire solutions.*

6.5 Random Chemostat Models

It practice input and output parameters may vary slightly in a random manner, taking values in bounded intervals about an ideal or mean value. The system (6.1)–(6.3) without delays then becomes a system of pathwise random ordinary differential equations (RODEs):

$$x'(t) = D_t(\omega)\,(I_t(\omega) - x(t)) - a\frac{x(t)}{m + x(t)}\,(y_1(t) + y_2(t)) + b\gamma y_1(t), \tag{6.26}$$

$$y_1'(t) = -\,(\gamma + D_t(\omega))\,y_1(t) + c\frac{x(t)}{m + x(t)}y_1(t) - r_1 y_1(t) + r_2 y_2(t), \tag{6.27}$$

$$y_2'(t) = -\gamma y_2(t) + c\frac{x(t)}{m + x(t)}y_2(t) + r_1 y_1(t) - r_2 y_2(t), \tag{6.28}$$

where the inputs are perturbed by real noise, i.e., D_t and I_t are continuous and essentially bounded with values

$$D_t(\omega) \in d\cdot[1-\varepsilon_D, 1+\varepsilon_D], \quad I_t(\omega) \in i\cdot[1-\varepsilon_I, 1+\varepsilon_I], \quad d > 0,\, i > 0,\, \varepsilon_D,\, \varepsilon_I < 1.$$

Bounded noise can be modeled in various ways. For example, given a stochastic process Z_t such as an Ornstein-Uhlenbeck process, D or I could be the stochastic process defined by [17]

$$\zeta(Z_t(\omega)) := \zeta_0\Big(1 - 2\varepsilon\frac{Z_t(\omega)}{1 + Z_t(\omega)^2}\Big), \tag{6.29}$$

where ζ_0 and ε are positive constants with $\varepsilon \in (0, 1]$. This takes values in the interval $\zeta_0[1 - \varepsilon, 1 + \varepsilon]$ and tends to peak around the points $\zeta_0(1 \pm \varepsilon)$, so is suitable for a noisy switching scenario. Another possibility is the stochastic process

$$\eta(Z_t(\omega)) := \eta_0\Big(1 - \frac{2\varepsilon}{\pi}\arctan Z_t(\omega)\Big), \tag{6.30}$$

where η_0 and ε are positive constants with $\varepsilon \in (0, 1]$, which takes values in the interval $\eta_0[1 - \varepsilon, 1 + \varepsilon]$ and is centered on η_0.

In the theory of random dynamical systems the driving noise process is represented abstractly by a canonical driving system $\theta_t(\omega)$ on the sample space Ω, and the system is analyzed in a pathwise fashion. The solutions to the system of RODEs (6.26)–(6.28) generate a cocycle mapping, and the nonautonomous system has a skew-product like structure with the noise process acting as a measure theoretical rather than topological autonomous dynamical system (see [10, 12, 18] for more details). A random attractor is a pullback attractor for this system and consists of random subsets, reducing to a single stochastic process when the random sets are singleton sets. Counterparts of the deterministic results above (without delay) are given in [10]. Convergence to a random attractor is pathwise in the pullback sense. Forward convergence also holds, but in the weaker sense of in probability due to the possibility of large deviations. Random delays could also be considered as in e.g., [19], but this has not yet been done in the chemostat context.

6.6 Overyield in Nonautonomous Chemostats

For a given amount of nutrient that is fed in a chemostat during a given period of time T, one can compare the biomass production over the time period, depending on the way the amount of nutrient is distributed over the time period. We say that there exists a biomass *overyielding* when a time varying input produces more biomass than a constant input. To illustrate the effect of overyielding in nonautonomous chemostats, we consider the chemostat model with wall growth, variable inputs, and non-delays as in Sect. 6.4.2.

When $D(t) = D$ is constant an $I(\cdot)$ a nonconstant T-periodic function with

$$\frac{1}{T}\int_t^{t+T} I(s)ds = \bar{I}\ ,$$

a periodic solution of system (6.17) and (6.18) has to fulfill the equations

$$0 = D(\bar{I} - \bar{x}) - a\frac{1}{T}\int_t^{t+T} U(x(s))z(s)ds + b\gamma\alpha^*\bar{z}\,, \tag{6.31}$$

$$0 = -(\gamma + D\alpha^*)\bar{z} + c\frac{1}{T}\int_t^{t+T} U(x(s))z(s)ds\,, \tag{6.32}$$

where $\bar{x}$, $\bar{z}$ denote the average values of the variables $x(\cdot)$, $z(\cdot)$ over the period T. Combining equations (6.31) and (6.32), one obtains the relation

$$D(\bar{I} - \bar{x}) = \left[\frac{a(\gamma + D\alpha^*)}{c} - b\gamma\alpha^*\right]\bar{z}\,. \tag{6.33}$$

One can also write from Eq. (6.18)

$$0 = \frac{1}{T}\int_t^{t+T}\frac{z'(s)}{z(s)}ds = -(\gamma + D\alpha^*) + c\frac{1}{T}\int_t^{t+T} U(x(s))ds\,.$$

As the function $U(\cdot)$ is concave and increasing, one deduces the inequality $\bar{x} > x^*$, where x^* stands for the steady state of the variable $x(\cdot)$ with the constant input $I(t) = \bar{I}$. Similarly, x^* satisfies the equality $cU(x^*) = \gamma + D\alpha^*$. One can then compare the corresponding biomass variables, with the help of Eq. (6.33), and obtain:

$$\left[b\gamma\alpha^* - \frac{a(\gamma + D\alpha^*)}{c}\right](\bar{z} - z^\star) > 0\,.$$

We conclude that an overyielding occurs when the condition

$$bc\gamma\alpha^* > a(\gamma + D\alpha^*) \tag{6.34}$$

is fulfilled. One can see that the nutrient recycling of the dead biomass ($b\gamma \neq 0$) is essential to obtain an overyielding.

Consider now the chemostat model without wall, $I(\cdot) = I$ constant and $D(\cdot)$ a nonconstant T-periodic function with

$$\frac{1}{T}\int_t^{t+T} D(s)ds = \bar{D}\,.$$

From Eqs. (6.24) and (6.25) a periodic solution has to fulfill

$$I = x(t) + y(t) \tag{6.35}$$

$$0 = \frac{1}{T}\int_t^{t+T}\frac{y'(s)}{y(s)}ds = -\bar{D} + a\frac{1}{T}\int_t^{t+T} U(x(s))ds \tag{6.36}$$

From Eq. (6.36), one obtains, as before, the inequality $\bar{x} > x^*$ and thus $\bar{y} < y^*$. Consequently, overyielding never occurs.

For the chemostat model with a wall and periodic $D(\cdot)$, we have not been able to prove if an overyielding is possible, although numerical simulations tend to show that it is not.

Remark 6.1 For more general time varying inputs (i.e., not necessarily periodic), one can also study the influence of the variations of the inputs on the characteristics of the pullback attractor. Indeed Theorems 6.7, 6.8 and 6.9 provide precise conditions for which the pullback attractor is larger than the single *washout* trajectory $\{(w^*(\cdot), 0)\}$ (i.e., absence of biomass). When enlarging the input set $[i_m, i_M]$ or $[d_m, D_M]$ allows the pullback attractor to be larger than the single washout, one can consider that a biomass survival (and thus an overyielding) could occur.

- Statements (ii) in Theorem 6.9 (chemostat with no wall) show that enlarging the input sets does not help the dynamics to avoid the washout.
- In statements (ii) of Theorems 6.8 and 6.9 (chemostat with wall), one can check that the functions $\varphi_D(\cdot)$, $\varphi_I(\cdot)$ as defined in (6.19) and (6.23), respectively, could be increasing or decreasing depending on the values of the parameters. Therefore, enlarging the input set could be beneficial for the biomass survival, which is different from the no wall case.

Acknowledgments This work has been partially supported by the Spanish Ministerio de Economía y Competitividad project MTM2011-22411 and the Consejería de Innovación, Ciencia y Empresa (Junta de Andalucía) under grant 2010/FQM314 and Proyecto de Excelencia P12-FQM-1492.

References

1. Beretta, E., Takeuchi, Y.: Qualitative properties of chemostat equations with time delays: boundedness, local and global asymptotic stability. Differ. Equ. Dyn. Syst. **2**, 19–40 (1994)
2. Beretta, E., Takeuchi, Y.: Qualitative properties of chemostat equations with time delays II. Differ. Equ. Dyn. Syst. **2**, 263–288 (1994)
3. Ballyk, M., Jones, D., Smith, H.: The biofilm model of Freter: a review. In: Magal, P., Ruan, S. (eds.) Structured Population Models in Biology and Epidemiology, pp. 265–302. Springer, Berlin (2008)
4. Jones, D., Kojouharov, H., Le, D., Smith, H.L.: The Freter model: a simple model of biofilm formation. J. Math. Biol. **47**, 137–152 (2003)
5. Pilyugin, S.S., Waltman, P.: The simple chemostat with wall growth. SIAM J. Appl. Math. **59**, 1552–1572 (1999)
6. Sree Hari Rao, V., Raja Sekhara Rao, P.: Dynamic Models and Control of Biological Systems. Springer, Heidelberg (2009)
7. Topiwala, H., Hamer, G.: Effect of wall growth in steady state continuous culture. Biotech. Bioeng. **13**, 919–922 (1971)
8. Butler, G.J., Hsu, S.B., Waltman, P.: A mathematical model of the chemostat with periodic washout rate. SIAM J. Appl. Math. **45**, 435–449 (1985)
9. Caraballo, T., Han, X., Kloeden, P.E.: Chemostats with time-dependent inputs and wall growth. Appl. Math. Inf. Sci. (to appear)

10. Caraballo, T., Han, X., Kloeden, P. E.: Chemostats with random inputs and wall growth. Math. Methods Appl. Sci. (to appear). doi:10.1002/mma.3437
11. Caraballo, T., Han X., Kloeden, P. E.: Non-autonomous chemostats with variable delays. SIAM J. Math. Anal. (to appear). doi:10.1137/14099930X
12. Kloeden, P.E., Rasmussen, M.: Nonautonomous Dynamical Systems. American Mathematical Society, Providence (2011)
13. Caraballo, T., Langa, J.A., Robinson, J.C.: Attractors for differential equations with variable delays. J. Math. Anal. Appl. **260**(2), 421–438 (2001)
14. Kloeden, P.E., Lorenz, T.: Pullback incremental stability. Nonauton. Random Dyn. Sys. 53–60 (2013). doi:10.2478/msds-2013-0004
15. Hale, J.K., Verduyn Lunel, S.M.: Introduction to Functional-Differential Equations. Applied Mathematical Sciences, vol. 99. Springer, New York (1993)
16. Smith, H.L., Waltman, P.: The Theory of the Chemostat: Dynamics of Microbial Competition. Cambridge University Press, Cambridge (1995)
17. Asai, Y., Kloeden, P.E.: Numerical schemes for random ODEs via stochastic differential equations. Commun. Appl. Anal. **17**(3 and 4), 521–528 (2013)
18. Arnold, L.: Random Dynamical Systems. Springer, Berlin (1998)
19. Caraballo, T., Kloeden, P.E., Real, J.: Discretization of asymptotically stable stationary solutions of delay differential equations with a random stationary delay. J. Dyn. Differ. Equ. **18**(4), 863–880 (2006)

Chapter 7
Asymptotic Dynamics of Stochastic Lattice Differential Equations: A Review

Xiaoying Han

Abstract This is an expository article on asymptotic dynamics of stochastic lattice differential equations. In particular, we investigate the long-term behavior of stochastic lattice differential equations, by using the concept of global random pullback attractor in the framework of random dynamical systems. General results on the existence of global compact random attractors are first provided for general random dynamical systems in weighted spaces of infinite sequences. They are then used to study the existence of global pullback random attractors for various types of stochastic lattice dynamical systems with white noise.

7.1 Introduction

Lattice differential equations (LDEs) are infinite systems of coupled ordinary differential equations or difference equations indexed by points in a lattice. They can be cast as infinite dimensional dynamical systems that arise as spatial discretization of continuum models. But more importantly, LDEs arise naturally in modeling systems with intrinsic discrete structure, such as solidification of alloys, interactions on a single strand of DNA, cellular neural networks, propagation of pulses in myelinated axons, waves in lattice gases, dispersal in patchy media or environments, and many other examples in chemical reaction, pattern recognition, image processing, etc. [6–8, 16, 17, 20, 30–32, 34, 37, 39]. Lattice differential equations (LDEs) have been studied extensively over the past two decades, for their interesting mathematical properties and plethora of applications. We refer to [2, 12, 13, 36] and references therein for traveling waves solutions; [12, 14, 15, 40] and references therein for chaotic properties of solutions; [1, 5, 21, 23, 42, 43, 47–49] and references therein for the existence and properties of global attractors.

X. Han (✉)
Department of Mathematics and Statistics, Auburn University,
221 Parker Hall, Auburn, AL 36849, USA
e-mail: xzh0003@auburn.edu

© Springer International Publishing Switzerland 2015

V.A. Sadovnichiy and M.Z. Zgurovsky (eds.), *Continuous and Distributed Systems II*, Studies in Systems, Decision and Control 30,
DOI 10.1007/978-3-319-19075-4_7

However, random spatiotemporal forcing, referred to as noise, exists in most realistic dynamical systems. This can be randomness of fluctuations either in parameters of an individual system or in equations involving variations from one individual to another. Usually, there are two sources of noise: external noise that affects the systems but are not explicitly described in the model, and internal noise due to the stochastic nature of the dynamical systems. Therefore, the consideration of noise in a dynamical system is not only to compensate for the defects in deterministic models, but more importantly to represent inherent stochastic structures of the system. Stochastic lattice differential equations (SLDEs) then become desirable tools to describe the essential dynamics of such systems with discrete structures and uncertainties taken into honest account.

In contrast to the integrated results for deterministic LDEs, studies on SLDEs have only started, mainly due to the challenge of analyzing the stochasticity and nonlinearility within the systems. During the past two decades, the theory of random dynamical systems (RDSs) has made substantial progress in describing the asymptotic behavior of systems with random forcing. In particular, the study of random attractor was initiated by Ruelle to capture the essential dynamics with possible extremely wide fluctuations [38]. The theory of random attractor in the pullback sense was further developed by Crauel, Debussche, Flandoli, Imkeller, and Schmalfuss among others [18, 19, 22, 29]. It is worth mentioning that while a stochastic system is observed by the pullback approach, in which the system runs from a distant point in the past until the present time, the geometric structures related to the stochastic dynamics emerge naturally.

In this paper, we provide a review of recent results on the existence of global random pullback attractors for various type of SLDEs, based on the theory of random dynamical systems. Note that regular spaces of infinite sequences may exclude important solutions such as traveling waves and whose components are just bounded; we consider a more inclusive weighted space of infinite sequences. The rest of the article is organized as follows. In Sect. 7.2, we provide an introduction to the pullback approach and necessary preliminaries from the theory of RDSs; in Sect. 7.3, we set up the RDS framework on the weighted space of infinite sequences and illustrate how to prove the existence of random attractors for SLDEs by three model systems with multiplicative or additive noise, and a summary is given in Sect. 7.4.

7.2 Preliminaries on Random Dynamical Systems

Since this is an expository work on the studies of random pullback attractors by the RDS theory, in this section we will explain the concept of pullback approach, followed by the definition for an RDS and a random pullback attractor, then general results on the existence of random pullback attractors for RDSs. To understand this rather novel set of concepts, we start from the concept of pullback attractor in the context of a nonautonomous dynamical system [33].

Consider an initial value problem for a nonautonomous ordinary differential equation:

$$\dot{x}(t) = f(t, x), \quad x(t_0) = x_0. \tag{7.1}$$

For solution mapping $\phi(t, t_0, x_0)$ of (7.1) when an existence and uniqueness theorem holds, we can formulate it as a process if it satisfies

(a) the initial value property: $\phi(t_0, t_0, x_0) = x_0$;
(b) the two-parameter semigroup evolution property:

$$\phi(t_2, t_0, x_0) = \phi\left(t_2, t_1, \phi(t_1, t_0, x_0)\right), \quad t_0 \leq t_1 \leq t_2;$$

(c) the continuity property: $(t, t_0, x_0) \mapsto \phi(t, t_0, x_0)$ is continuous on the state space.

Note that the solution usually depends on both the actual time t and the initial time t_0 rather than just on the elapsed time $t - t_0$ as in an autonomous system. Therefore, the asymptotic behavior when $t_0 \to -\infty$ and t fixed may be different from the one obtained in the forward sense with t_0 fixed and $t \to \infty$.

Roughly, a pullback attractor for the process generated by (7.1) satisfying properties (a)–(c) is a family of limiting objects $A(t)$ which exist in actual time t rather than asymptotically in the future, and attracts some subsets of initial data taken in the *past*. In rigorous terms, a family of nonempty compact subsets $\mathscr{A} := \{A(t) : t \in \mathbb{R}\}$ in a complete metric phase space X is called a pullback attractor if each $A(t)$ is

- invariance with respect to the process ϕ:

$$\phi(t, t_0, A(t)) = A(t) \quad \text{for all} \quad t \geq t_0;$$

- pullback attracting:

$$\lim_{t_0 \to -\infty} \operatorname{dist}_X(\phi(t, t_0, D(t_0), A(t)) = 0, \quad \text{for all} \quad D \in \mathscr{D} := \{D(t) : t \in \mathbb{R}\},$$

where dist_X denotes the Hausdorff semi-distance between two subsets in X.

When the time-dependent forcing in a dynamical system is random, the pullback attractor becomes a random pullback attractor which needs further clarification by the concept of random dynamical systems.

In general, to analyze systems with randomness, we need to have a reasonable description of their random aspects. These aspects may change over time, and thus the noise has to be modeled as a time-dependent stochastic process with certain known properties. Represented mathematically, such a stochastic process starts with a probability space $(\Omega, \mathscr{F}, \mathbb{P})$, where $\mathscr{F}$ is the $\sigma-$algebra of measurable subsets of Ω (called "events") and $\mathbb{P}$ is the probability measure. To connect the state ω in the probability space Ω at time 0 with its state after a time of t elapses, we parametrize noise by time, i.e., parameterize the probability space by time. This is done by defining a flow $\theta = \{\theta_t\}_{t \in \mathbb{R}}$ on Ω with each θ_t being a mapping $\theta_t : \Omega \to \Omega$ that satisfies

(1) $\theta_0 = \mathrm{Id}_\Omega$,
(2) $\theta_s \circ \theta_t = \theta_{s+t}$ for all $s, t \in \mathbb{R}$,
(3) the mapping $(t, \omega) \mapsto \theta_t\omega$ is measurable and
(4) the probability measure $\mathbb{P}$ is preserved by θ_t, i.e., $\theta_t\mathbb{P} = \mathbb{P}$.

This set-up establishes a time-dependent family θ that tracks the noise, and $(\Omega, \mathscr{F}, \mathbb{P}, \theta)$ is called a *driving dynamical system* (DDS) [3].

One typical example of such families is given by the class of Wiener processes which has been widely used in SDEs. In fact, let $\Omega = \{\omega \in C_0(\mathbb{R})\}$ be the collection of continuous paths $\omega(t)$ on $\mathbb{R}$ such that $\omega(0) = 0$, $\mathscr{F}$ be the associated Borel $\sigma-$field and $\mathbb{P}$ be the Wiener measure. Let the operator θ_t be the Wiener shift: $\theta_t\omega(\cdot) = \omega(\cdot + t) - \omega(t)$ for $\omega \in \Omega$ and $t \in \mathbb{R}$, then it preserves the Wiener measure.

Let $(X, \|\cdot\|_X)$ be a Hilbert space and a stochastic system can be interpreted as moving on the product space $\Omega \times X$, along a path $\theta_t\omega$ in Ω. For the above approach of modeling noise to be reasonable, the system must behave self-consistently along such a path θ_t, i.e., when ω is shifted by θ in time t to the point $\theta_t\omega$ on the base space Ω, the system moves the point in $\{\omega\} \times X$ over ω to a point in $\{\theta_t\omega\} \times X$ over $\theta_t\omega$. This can be described by the *cocycle property* of its stochastic dynamics [3].

Definition 7.1 A *cocycle* is a measurable mapping $\Phi : \mathbb{R}^+ \times \Omega \times X \to X$, $(t, \omega, x) \mapsto \Phi(t, \omega)x$ with $\Phi(0, \omega) = Id_X$ and $\Phi(t + s, \omega) = \Phi(t, \theta_s\omega) \circ \Phi(s, \omega)$ for all $s, t \geq 0$.

It appears that the skew product $(\omega, x) \mapsto (\theta_t\omega, \Phi(t, \omega)x)$ is a dynamical system on the space $\Omega \times X$. This gives the definition of a *random dynamical system* (RDS), that a random dynamical system is a driving dynamical system equipped with the cocycle property [3].

With the concept of RDS, we can extend the notion of pullback attractor to the random context. Assume that $\mathscr{A}(\omega)$ of X is a compact random set, i.e., the mapping $\omega \mapsto \mathrm{dist}_X(x, \mathscr{A}(\omega))$ is measurable for any $x \in X$ and $\mathscr{A}(\omega)$ is compact for all $\omega \in \Omega$. Furthermore, if $\omega \mapsto \mathscr{A}(\omega)$ satisfies

- invariance with respective to Φ:

$$\Phi(t, \omega)\mathscr{A}(\omega) = \mathscr{A}(\theta_t\omega), \quad \text{for each} \quad t \geq 0, \quad a.e.\, \omega \in \Omega;$$

- pullback attracting property:

$$\lim_{t\to\infty} \mathrm{dist}_X(\Phi(t, \theta_{-t}\omega)D(\theta_{-t}\omega), \mathscr{A}(\omega)) = 0, \quad \forall D \in \mathscr{D}(X),$$

where $\mathrm{d}(X)$ denotes the set of all tempered random sets of X.[1] Then $\mathscr{A}(\omega)$ is called a *global random pullback* d *attractor* for the RDS $\{\Phi(t, \omega)\}_{t\geq 0, \omega\in\Omega}$. When the content

[1]A random set $D(\omega)$ is said to be tempered with respect to $(\theta_t)_{t\in\mathbb{R}}$ if for a.e. $\omega \in \Omega$, $\lim_{t\to\infty} e^{-\gamma t} \sup_{x\in D(\theta_{-t}\omega)} \|x\|_X = 0$ for all $\gamma > 0$. A random variable $\omega \mapsto r(\omega) \in \mathbb{R}$ is said to be tempered with respect to $(\theta_t)_{t\in\mathbb{R}}$ if for a.e. $\omega \in \Omega$, $\lim_{t\to+\infty} e^{-\gamma t} \sup_{t\in\mathbb{R}} |r(\theta_{-t}\omega)| = 0$ for all $\gamma > 0$.

is clear, the word "pullback" can be omitted. Note that in general the collection d is the domain of attraction of $\mathscr{A}$. Other examples of d include the collection of all finite deterministic subsets of X (for which the random d attractor is called a point attractor), and the collection of all compact deterministic subsets of X (for which the random d attractor is called a set attractor) [9].

The next proposition provides an abstract result on the existence of a global random attractor.

Proposition 7.1 ([4, 22]) *Let $B(\omega) \in \mathscr{D}(X)$ be a random closed absorbing set for the continuous RDS $\{\Phi(t,\omega)\}_{t\geq 0,\omega\in\Omega}$, and satisfy the asymptotic compactness property, i.e., for a.e. $\omega \in \Omega$, each sequence $x_n \in \Phi(t_n, \theta_{-t_n}\omega)B(\theta_{-t_n}\omega)$ with $t_n \to +\infty$ has a convergent subsequence in X. Then the RDS $\{\Phi(t,\omega)\}_{t\geq 0,\omega\in\Omega}$ has a unique global random $\mathscr{D}$ attractor*

$$\mathscr{A}(\omega) = \bigcap_{\tau \geq T_B(\omega)} \overline{\bigcup_{t\geq\tau} \Phi(t, \theta_{-t}\omega)B(\theta_{-t}\omega)}.$$

In the next section, we will provide a literature review for recent work on random attractors for SLDEs based on Proposition 7.1.

7.3 Random Attractors for Stochastic Differential Equations

The existence of global random attractors for first-order SLDEs with additive white noise and multiplicative white noise in the regular space of infinite sequences was first studied by Bates et al. [4] and Caraballo and Lu [9], respectively. Later, these results were generalized to first-order SLDEs with random coupling among nodes in the weighted space of infinite sequences by Han et al. [27]. Other works on the random attractors for first-order SLDES could be found in [10, 11, 28, 35, 41, 45] and references therein.

Next we provide the mathematical setting of weighted spaces of infinite sequences and the sufficient conditions for the existence of global random attractors for general RDSs in weighted spaces of infinite sequences.

7.3.1 Mathematical Settings

We first introduce a weighted space of infinite sequences. Let $p \geq 1$ be a real number, and ρ be a positive function from $\mathbb{Z}$ to $(0, M_0] \subset \mathbb{R}^+$, where M_0 is a positive constant. For any $i \in \mathbb{Z}$, define $\rho_i = \rho(i)$, and

$$l^p_\rho = \left\{ u = (u_i)_{i\in\mathbb{Z}} : \sum_{i\in\mathbb{Z}} \rho_i |u_i|^p < \infty, u_i \in \mathbb{R} \right\} \tag{7.2}$$

with the norm

$$\|u\|_{\rho,p} = \left(\sum_{i\in\mathbb{Z}} \rho_i |u_i|^p \right)^{\frac{1}{p}} \quad \text{for} \quad u = (u_i)_{i\in\mathbb{Z}} \in l^p_\rho.$$

Then l^p_ρ is a Banach space with the norm $\|\cdot\|_{\rho,p}$. Moreover, l^p_ρ is separable since $\cup_{N\geq 1} l_N$ is a countable dense subset of l^p_ρ, where

$$l_N = \{(u_i)_{i\in\mathbb{Z}} : u_i \in \mathbb{Q} \quad \text{for} \quad i \in \mathbb{Z} \text{ and } u_i = 0 \quad \text{for} \quad |i| > N\}.$$

It is clear that $\cup_{N\geq 1} l_N$ is a countable subset of l^p_ρ. For any given element $u = (u_i)_{i\in\mathbb{Z}} \in l^p_\rho$ and for any $\varepsilon > 0$, there exists a positive integer $I(\varepsilon) \in \mathbb{N}$ such that $\sum_{|i|>I(\varepsilon)} \rho_i |u_i|^p < \varepsilon^p/2$. Choose $\bar{u} = (\bar{u}_i)_{i\in\mathbb{Z}}$ such that $\bar{u}_i \in \mathbb{Q}$ for $|i| \leq I(\varepsilon)$, $\bar{u}_i = 0$ for $|i| > I(\varepsilon)$, and

$$\sum_{|i|\leq I(\varepsilon)} \rho_i |u_i - \bar{u}_i|^p < \varepsilon^p/2.$$

Then $\bar{u} \in \cup_{N\geq 1} l_N$ and

$$\|u - \bar{u}\|_{\rho,p} < \varepsilon.$$

This implies that $\cup_{N\geq 1} l_N$ is dense in l^p_ρ and hence l^p_ρ is separable. In particular, l^2_ρ is a separable Hilbert space with the inner product $(u, v)_\rho = \sum_{i\in\mathbb{Z}} \rho_i u_i v_i$ and norm $||u||^2_{\rho,2} = (u, u)_\rho = \sum_{i\in\mathbb{Z}} \rho_i |u_i|^2$ for $u = (u_i)_{i\in\mathbb{Z}}$, $v = (v_i)_{i\in\mathbb{Z}} \in l^2_\rho$. We write $\|\cdot\|_{\rho,2}$ as $\|\cdot\|_\rho$ and write $\|\cdot\|_\rho$ as $\|\cdot\|$ if $\rho(i) \equiv 1$. Note that if $\rho(i) \equiv 1$, then l^2_ρ is the standard space $l^2 = \{u = (u_i)_{i\in\mathbb{Z}} : \sum_{i\in\mathbb{Z}} |u_i|^2 < \infty, u_i \in \mathbb{R}\}$ with the inner product $(\cdot, \cdot)$ and norm $\|\cdot\|$.

Let $(\Omega, \mathscr{F}, \mathbb{P}, (\theta_t)_{t\in\mathbb{R}})$ be a metric dynamical system on a probability space $(\Omega, \mathscr{F}, \mathbb{P})$. Let $\{\Phi(t, \omega)\}_{t\geq 0,\omega\in\Omega}$ be a continuous RDS over $(\Omega, \mathscr{F}, \mathbb{P}, (\theta_t)_{t\in\mathbb{R}})$ with state space l^p_ρ. The following definition is required to ensure the sequence has a "light" tail.

Definition 7.2 An RDS $\{\Phi(t, \omega)\}_{t\geq 0,\omega\in\Omega}$ is said to be *random asymptotically null* in $\mathscr{D}(l^p_\rho)$ if for a.e. $\omega \in \Omega$, any $B(\omega) \in \mathscr{D}(l^p_\rho)$, and any $\varepsilon > 0$, there exist $T(\varepsilon, \omega, B(\omega)) > 0$ and $I(\varepsilon, \omega, B(\omega)) \in \mathbb{N}$ such that

$$\left(\sum_{|i|>I(\varepsilon,\omega,B(\omega))} \rho_i |(\Phi(t, \theta_{-t}\omega)u(\theta_{-t}\omega))_i|^p \right)^{1/p} \leq \varepsilon$$

holds for every $t \geq T(\varepsilon, \omega, B(\omega))$ and $u(\omega) \in B(\omega)$.

The next proposition provides some sufficient conditions for the existence of global random attractors for general RDSs in weighted spaces of infinite sequences, for which the proof can be found in [27].

Proposition 7.2 *Suppose that*

(a) *there exists a bounded closed random absorbing set $B_0(\omega) \in \mathscr{D}(l_\rho^p)$ such that for a.e. $\omega \in \Omega$ and any $B(\omega) \in \mathscr{D}(l_\rho^p)$, there exists $T_B(\omega) > 0$ yielding*

$$\Phi(t, \theta_{-t}\omega)B(\theta_{-t}\omega) \subset B_0(\omega) \quad \text{for all} \quad t \geq T_B(\omega);$$

(b) *$\{\Phi(t, \omega)\}_{t \geq 0, \omega \in \Omega}$ is random asymptotically null on $B_0(\omega)$.*

Then the RDS $\{\Phi(t, \omega)\}_{t \geq 0, \omega \in \Omega}$ possesses a unique global random $\mathscr{D}$ attractor $\mathscr{A}(\omega)$ given by

$$\mathscr{A}(\omega) = \bigcap_{\tau \geq T_{B_0}(\omega)} \overline{\bigcup_{t \geq \tau} \Phi(t, \theta_{-t}\omega)B_0(\theta_{-t}\omega)}$$

7.3.2 Selected Results for First-Order SLDEs

To illustrate the procedure of proving the existence of random pullback attractors for SLDEs, we consider the following model systems with multiplicative and additive noise, respectively:

$$du_i = (-\lambda u_i - f_i(u_i) + g_i + \sum_{j=-q}^{q} \eta_{i,j}(\theta_t\omega)u_{i+j})dt + u_i \circ dw(t); \quad (7.3)$$

$$du_i = (-\lambda u_i - f_i(u_i) + g_i + \sum_{j=-q}^{q} \eta_{i,j}(\theta_t\omega)u_{i+j})dt + \beta_i dw_i(t), \quad (7.4)$$

where λ is a positive constant; for $i \in \mathbb{Z}$, where $\mathbb{Z}$ denotes the integer set, u_i, g_i, $\beta_i \in \mathbb{R}$; $f_i \in C^1(\mathbb{R})$ satisfies proper dissipative conditions; $w(t)$, $w_i(t)$ are Brownian motions; $\eta_{i,-q}(\omega), \ldots, \eta_{i,0}(\omega), \ldots, \eta_{i,+q}(\omega)$, $q \in \mathbb{N}$, are random variables; $(\theta_t)_{t\in\mathbb{R}}$ is a metric dynamical system on proper probability space; and $\circ$ denotes the Stratonovich sense of the stochastic term.

We choose a positive weight function $\rho : \mathbb{Z} \to \mathbb{R}$ to satisfy

$$\textbf{(P0)} \quad 0 < \rho(i) \leq M_0, \rho(i) \leq c\rho(i \pm 1), \ \forall i \in \mathbb{Z}, \quad M_0 > 0, \ c > 0.$$

Observe that $l^2 \subset l_\rho^2$ and l^2 is dense in l_ρ^2. Particularly, if $\sum_{i\in\mathbb{Z}} \rho(i) < \infty$, for example, $\rho(i) \sim \frac{1}{i^2}$, then l_ρ^2 contains any infinite sequences with bounded components and that $l^2 \subset l^\infty \subset l_\rho^2$ (see [46]).

Before fitting the problems into the RDS framework, we need to show that Eqs. (7.3) and (7.4) each generates a continuous RDS. To this end, we first transform the stochastic equations (7.3) and (7.4) into equations with random coefficients, but with no explicit appearance of white noise. Define

$$\Omega_0 = \{\omega \in C(\mathbb{R}, \mathbb{R}) : \omega(0) = 0\} = C_0(\mathbb{R}); \tag{7.5}$$
$$\Omega_1 = \{\omega \in C(\mathbb{R}, l^2) : \omega(0) = 0\}. \tag{7.6}$$

Let $\mathscr{F}_0$ and $\mathscr{F}_1$ be the Borel σ-algebra on Ω_0 and Ω_1, respectively, generated by the compact open topology [4]; $\mathbb{P}_0$ and $\mathbb{P}_1$ be the corresponding Wiener measure on $\mathscr{F}_0$ and $\mathscr{F}_1$, respectively. Let θ_t be the Wiener shift defined by

$$\theta_t\omega(\cdot) = \omega(\cdot + t) - \omega(t), \quad t \in \mathbb{R}, \quad \omega \in \Omega_j, \ j = 0, 1,$$

and note that

$$W(t) = W(t, \omega) = \sum_{i\in\mathbb{Z}} \beta_i w_i(t) e^i \quad \text{with} \quad \beta = (\beta_i)_{i\in\mathbb{Z}} \in l^2 \tag{7.7}$$

is a Brownian motion on $(\Omega_1, \mathscr{F}_1, \mathbb{P}_1)$, where the infinite sequence e^i $(i \in \mathbb{Z})$ denotes the element having 1 at position i and 0 for all other components. Then Eqs. (7.3) and (7.4) can be written into the following equations, respectively:

$$du = (-\lambda u - f(u) + g + A(\theta_t\omega)u)dt + u \circ dw(t), \quad \omega \in \Omega_0, \tag{7.8}$$
$$du = (-\lambda u - f(u) + g + A(\theta_t\omega)u)dt + dW(t), \quad \omega \in \Omega_1, \tag{7.9}$$

where $u = (u_i)_{i\in\mathbb{Z}}$, $f(u) = (f_i(u_i))_{i\in\mathbb{Z}}$, $g = (g_i)_{i\in\mathbb{Z}}$, $A(\omega)$ is a linear operator on l^2_ρ and $(A(\omega)u)_i = \sum_{j=-q}^{q} \eta_{i,j}(\omega)u_{i+j}$.

To transform (7.8) and (7.9) into random differential equations, we introduce the Ornstein–Uhlenbeck (O–U) processes

$$\delta(\theta_t\omega) = -\int_{-\infty}^{0} e^s \theta_t\omega(s)\mathrm{d}s, \quad t \in \mathbb{R}, \quad \omega \in \Omega_0, \tag{7.10}$$

$$\tilde{\delta}(\theta_t\omega) = -\lambda \int_{-\infty}^{0} e^{\lambda s} \theta_t\omega(s)\mathrm{d}s, \quad t \in \mathbb{R}, \quad \omega \in \Omega_1, \tag{7.11}$$

which solves the Ornstein–Uhlenbeck equations

$$\mathrm{d}\delta + \delta \mathrm{d}t = \mathrm{d}w(t), \quad \mathrm{d}\tilde{\delta} + \lambda\tilde{\delta}\mathrm{d}t = \mathrm{d}W(t),$$

respectively. The following lemma provides crucial estimates on the O–U processes which have been widely used for general stochastic differential equations.

Lemma 7.1 *[3, 4, 9] There exists a θ_t -invariant set $\tilde{\Omega}_1 \in \mathscr{F}_1$ of Ω_1 of full $\mathbb{P}_1$ measure such that for $\omega \in \tilde{\Omega}_1$,*

(i) the random variable $\|\tilde{\delta}(\omega)\|$ is tempered;

(ii) the mapping $(t,\omega) \to \tilde{\delta}(\theta_t\omega) = -\lambda \int_{-\infty}^{0} e^{\lambda s}\omega(t+s)ds + W(t) \in l^2$ is a stationary solution of Ornstein–Uhlenbeck equation in l^2;

(iii)

$$\lim_{t\to\pm\infty} \frac{\|\tilde{\delta}(\theta_t\omega)\|}{t} = \lim_{t\to\pm\infty} \frac{1}{t}\int_0^t \tilde{\delta}(\theta_s\omega)ds = 0. \tag{7.12}$$

Note that an estimate for δ in (7.10) can be easily obtained as an analog to Lemma 7.1, with $(\Omega_1, \mathscr{F}_1, \mathbb{P}_1)$ to $(\Omega_0, \mathscr{F}_0, \mathbb{P}_0)$, l^2 to $\mathbb{R}$ and $\lambda = 1$.

Next we perform the following change of variables

$$v(t,\omega) = e^{-\delta(\theta_t\omega)}u(t,\omega) := \Pi^{-1}(\theta_t\omega)u(t,\omega), \quad \omega \in \Omega_0, \tag{7.13}$$

$$\tilde{v}(t,\omega) = u(t,\omega) - \tilde{\delta}(\theta_t\omega), \quad \omega \in \Omega_1. \tag{7.14}$$

Stochastic equations (7.8) and (7.9) then become the following random differential equations with no white noise, respectively:

$$\frac{dv}{dt} = -\lambda v + A(\theta_t\omega)v + \delta(\theta_t\omega)v - e^{-\delta(\theta_t\omega)}f(e^{\delta(\theta_t\omega)}v) + e^{-\delta(\theta_t\omega)}g. \tag{7.15}$$

$$\frac{d\tilde{v}}{dt} = -\lambda\tilde{v} + A(\theta_t\omega)\tilde{v} - f(\tilde{v} + \tilde{\delta}(\theta_t\omega)) + A(\theta_t\omega)\tilde{\delta}(\theta_t\omega) + g. \tag{7.16}$$

The following standing assumptions are required for the results in the sequel, where $\mathbf{E}(\cdot)$ denotes the expectation of a random variable.

(H1) $g = (g_i)_{i\in\mathbb{Z}} \in l^2_\rho$.

(H2) Let $\eta(\omega) = \sup\{|\eta_{i,-q}(\omega)|, \ldots, |\eta_{i,0}(\omega)|, \ldots, |\eta_{i,+q}(\omega)| : i \in \mathbb{Z}\} \geq 0$, for all $q \in \mathbb{N}$, $\eta(\theta_t\omega)$ $(< \infty)$ belongs to $L^1_{\text{loc}}(\mathbb{R})$ with respect to $t \in \mathbb{R}$ for each $\omega \in \Omega_0$ or Ω_1, $\mathbf{E}(\eta) = \lim_{t\to\pm\infty} \frac{1}{t}\int_0^t \eta(\theta_t\omega)ds < \infty$; and $\eta(\omega)$ is tempered.

(H3) $\lambda > \tilde{q}\mathbf{E}|\eta(\omega)| = \lim_{t\to\pm\infty} \frac{1}{t}\int_0^t (q + \sum_{k=0}^q c_0^k)\eta(\theta_t\omega)ds$, where $\tilde{q} = q + \sum_{k=0}^q c_0^k$.

(H4) *There exists a function $R(r) \in C(\mathbb{R}^+, \mathbb{R}^+)$ such that* $\sup_{i\in\mathbb{Z}} \max_{s\in[-r,r]} |f_i'(s)| \leq R(r), \forall r \in \mathbb{R}^+$.

(H5) $f_i \in C^1(\mathbb{R},\mathbb{R})$, $f_i(0) = 0, sf_i(s) \geq -b_i^2, b = (b_i)_{i\in\mathbb{Z}} \in l^2_\rho$, *and there exists a constant $a \geq 0$ such that $f_i'(s) \geq -a, \forall s \in \mathbb{R}, i \in \mathbb{Z}$.*

(H6) $f_i \in C^1(\mathbb{R},\mathbb{R})$ satisfies $sf_i(s) \geq \mu s^{2(p+1)} - d_i^2$, $|f_i(s)| \leq d_f|s|(|s|^{2p} + 1)$, $\forall s \in \mathbb{R}$, $i \in \mathbb{Z}$, *where μ, d_i, d_f are positive constants, $p \in \mathbb{N}$, and* $d = (d_i)_{i\in\mathbb{Z}} \in l^2$.

The following theorem on existence and uniqueness of solutions was proved in [27].

Theorem 7.1 *(1) Let* $T > 0$, $\omega \in \Omega_0$, *and (P0), (H1)–(H2), (H4)–(H5) hold. Then*

(a) for any $\omega \in \Omega_0$ *and any initial data* $v_0 \in l^2_\rho$, *Eq.* (7.15) *admits a unique mild solution* $v(\cdot; \omega, v_0, g) \in C([0, T), l^2_\rho)$ *with* $v(0; \omega, v_0, g) = v_0$, $v(t; \omega, v_0, g)$ *being continuous in* $v_0, g \in l^2_\rho$, *and* $v(t; \omega, v_0, g) \in l^2$ *if* $v_0, g \in l^2$.

(b) Equation (7.15) *generates a continuous random dynamical system* $\{\psi(t, \omega)\}_{t\geq 0, \omega\in\Omega_0}$ *over* $(\Omega_0, \mathscr{F}_0, \mathbb{P}_0, (\theta_t)_{t\in\mathbb{R}})$ *with state space* l^2_ρ:

$$\psi(t, \omega)v_0 := v(t; \omega, v_0) \quad \text{for} \quad v_0 \in l^2_\rho,\ t \geq 0,\ \omega \in \Omega_0.$$

Moreover,

$$\Phi(t, \omega)u_0 := \Pi(\theta_t\omega)\psi(t, \omega)\Pi^{-1}(\omega)u_0 \quad \text{for} \quad u_0 \in l^2_\rho,\ t \geq 0,\ \omega \in \Omega_0$$

defines a continuous RDS $\{\Phi(t, \omega)\}_{t\geq 0, \omega\in\Omega_0}$ *over* $(\Omega_0, \mathscr{F}_0, \mathbb{P}_0, (\theta_t)_{t\in\mathbb{R}})$ *associated with* (7.8).

(2) Let $T > 0$, $\omega \in \Omega_1$ *and (P0), (H1)–(H2), (H4), (H6) hold. Then*

(a) for every $\omega \in \Omega_1$ *and any initial data* $\tilde{v}_0 \in l^2_\rho$, *Eq.* (7.16) *admits a unique mild solution* $\tilde{v}(\cdot; \omega, \tilde{v}_0, g) \in C([0, T), l^2_\rho)$ *with* $\tilde{v}(0; \omega, \tilde{v}_0, g) = \tilde{v}_0$, $\tilde{v}(t; \omega, v_0, g)$ *being continuous in* $v_0, g \in l^2_\rho$, *and* $\tilde{v}(t; \omega, v_0, g) \in l^2$ *if* $v_0, g \in l^2$.

(b) Equation (7.16) *generates a continuous RDS* $\{\tilde{\psi}(t, \omega)\}_{t\geq 0, \omega\in\Omega_1}$ *over* $(\Omega_1, \mathscr{F}_1, \mathbb{P}_1, (\theta_t)_{t\in\mathbb{R}})$ *with state space* l^2_ρ :

$$\tilde{\psi}(t, \omega)\tilde{v}_0 := \tilde{v}(t; \omega, \tilde{v}_0) \quad \text{for} \quad \tilde{v}_0 \in l^2_\rho, t \geq 0, \omega \in \Omega_1.$$

Moreover,

$$\tilde{\Phi}(t, \omega)u_0 := \tilde{\psi}(t, \omega)(u_0 - \tilde{\delta}(\omega)) + \tilde{\delta}(\theta_t\omega) \quad \text{for} \quad u_0 \in l^2_\rho, t \geq 0, \omega \in \Omega_1$$

defines a continuous RDS $\{\tilde{\Phi}(t, \omega)\}_{t\geq 0, \omega\in\Omega_1}$ *over* $(\Omega_1, \mathscr{F}_1, \mathbb{P}_1, (\theta_t)_{t\in\mathbb{R}})$ *associated with* (7.9).

After sophisticated arguments on the existence of tempered random bounded absorbing sets for the RDS associated with (7.8) and (7.9), as well as the asymptotic nullness of solutions to (7.8) and (7.9), the following result on existence of global random attractors are obtained.

Theorem 7.2 *(1) If* $\omega \in \Omega_0$ *and (P0), (H1)–(H5) hold, then the RDS* $\{\Phi(t, \omega)\}_{t\geq 0, \omega\in\Omega_0}$ *generated by Eq.* (7.8) *possesses a unique global random* *attractor given by*

$$\mathscr{A}_{0\rho}(\omega) = \bigcap_{\tau \geq T_{B_{0\rho}}(\omega)} \overline{\bigcup_{t \geq \tau} \Phi(t, \theta_{-t}\omega) B_{0\rho}(\theta_{-t}\omega)} \subset l^2_\rho.$$

(2) *If $\omega \in \Omega_1$ and (P0), (H1)–(H4), (H6) hold, then the RDS $\{\tilde{\Phi}(t,\omega)\}_{t \geq 0, \omega \in \Omega}$ generated by Eq. (7.9) possesses a unique global random attractor given by*

$$\mathscr{A}_{1\rho}(\omega) = \bigcap_{\tau \geq T_{B_{1\rho}}(\omega)} \overline{\bigcup_{t \geq \tau} \tilde{\Phi}(t, \theta_{-t}\omega) B_{1\rho}(\theta_{-t}\omega)} \in l^2_\rho.$$

7.3.3 A Brief on Second-Order SLDEs

The existence of global attractors for a second-order SLDEs with additive white noise was first investigated by Wang et al. [44]. The stochastic sine-Gordon lattice systems were studied by Zhao and Zhou [45] and Han [24] in regular and weighted spaces, respectively. Later, Han studied the second-order SLDEs with multiplicative noise and additive noise in weighted spaces and on $\mathbb{Z}^k$ lattice [25, 26]. As the concepts and methodology used are similar to the first-order systems, here we only elaborate the different parts. To be more inclusive for the contents, we consider the $\mathbb{Z}^k$ ($k \in \mathbb{N}$) lattice instead of $\mathbb{Z}$. For illustration purpose, we choose the following system as a model system:

$$\ddot{u}_i + \lambda \dot{u}_i + (Au)_i + \alpha_i u_i + f_i(u_i) = g_i + \beta_i \dot{w}_i, \quad i = (i_1, i_2, \ldots, i_k) \in \mathbb{Z}^k, \ t > 0, \tag{7.17}$$

with initial conditions

$$u_i(0) - u_{0i}, \quad \dot{u}_i(0) - u_{1i}$$

where u_i, $\dot{u}_i$, $\ddot{u}_i$, u_{0i}, u_{1i}, g_i, $\beta_i \in \mathbb{R}$, $\lambda > 0$, $\alpha_i > 0$, $f_i \in C(\mathbb{R}, \mathbb{R})$ for $i \in \mathbb{Z}^k$, $\{w_i(t) : i \in \mathbb{Z}^k\}$ are independent two-sided Brownian motions, and A is a bounded linear operator with the form $A = \sum_{j=1}^k A_j$, of which each A_j ($j = 1, 2, \ldots, k$) is a nonnegative self-adjoint linear operator on the space of infinite sequences with decomposition $A_j = D_j^* D_j = D_j D_j^*$ where D_j is defined by $(D_j u)_i = \sum_{l=-q}^{q} d_{j,l}\, u_{i_{j,l}}$ where $i_{j,l} \doteq (i_1, \ldots, i_{j-1}, i_j + l, i_{j+1}, \ldots, i_k)$, $\forall u = (u_i)_{i \in \mathbb{Z}^k}$, $j = 1, 2, \ldots, k$, with $q \in \mathbb{N}$, $|d_{j,l}| \leq d$ for $1 \leq j \leq k$, $-q \leq l \leq q$, and D_j^* is the adjoint of D_j.

7.3.4 Mathematical Setting

Consider the same metric dynamical system $(\Omega_1, \mathscr{F}_1, \mathbb{P}_1, (\theta_t)_{t\in\mathbb{R}})$ as defined in Sect. 7.2. Let $W(t)$ be defined as in (7.7), and let $\dot{u} = v$, Eq. (7.17) is equivalent to

$$\begin{cases} \dot{u} = v, \\ \dot{v} = (-\lambda v - \alpha u - f(u) + g - Au) + \dot{W}(t), \\ u(0) = u_0, \quad v(0) = u_1. \end{cases} \tag{7.18}$$

where $u = (u_i)_{i\in\mathbb{Z}^k}$, $\dot{u} = (\dot{u_i})_{i\in\mathbb{Z}^k}$, $\ddot{u} = (\ddot{u_i})_{i\in\mathbb{Z}^k}$, $u_0 = (u_{0i})_{i\in\mathbb{Z}^k}$, $u_1 = (u_{1i})_{i\in\mathbb{Z}^k}$, $Au = (Au_i)_{i\in\mathbb{Z}^k}$, $\alpha u = (\alpha_i u_i)_{i\in\mathbb{Z}^k}$, $(f(u) = f_i(u_i))_{i\in\mathbb{Z}^k}$ and $g = (g_i)_{i\in\mathbb{Z}^k}$.

Let $\tilde{\delta}(\theta_t\omega)$ be the OU process defined as in (7.11) and let

$$\tilde{v}(t,\omega) = v + \xi u - \tilde{\delta}(\theta_t\omega),$$

then Eq. (7.18) can be transformed into the following random differential equations

$$\begin{cases} \dot{u} = \tilde{v} - \xi u + \tilde{\delta}(\theta_t\omega), \\ \dot{\tilde{v}} = (-A - \alpha + \xi(\lambda - \xi))u - (\lambda - \xi)\tilde{v} - f(u) + \xi\tilde{\delta}(\theta_t\omega) + g, \\ u(0) = u_0, \quad \tilde{v}(0) = u_1 + \xi u_0 - \tilde{\delta}(\omega). \end{cases} \tag{7.19}$$

The weight function $\rho : \mathbb{Z}^k \to \mathbb{R}^+$ is chosen to satisfy the following conditions which are slightly different from **(P0)** in Sect. 7.2:

(P1) there exist positive constants M_0, a, b such that

$$0 < \rho(i) \le M_0, \quad \rho(i_{j,\pm 1}) \le a \cdot \rho(i), \quad \left|\rho(i_{j,\pm 1}) - \rho(i)\right| \le b \cdot \rho(i), \quad \forall\, i \in \mathbb{Z}^k.$$

In addition, we make the following standing assumptions on functions f_i and α_i, and g_i, $i \in \mathbb{Z}^k$:

(A1) *there exist positive real numbers M_i $(i \in \mathbb{Z}^k)$ and L such that*

$$|f_i(0)| \le M_i, \quad |f_i(s_1) - f_i(s_2)| \le L|s_1 - s_2|, \quad \forall s_1, s_2 \in \mathbb{R},\ i \in \mathbb{Z}^k;$$

(A2) $M = (M_i)_{i\in\mathbb{Z}^k}$, $g = (g_i)_{i\in\mathbb{Z}^k} \in l^2_\rho$;

(A3) *there exist two positive constants* $m_1, m_2 > 0$ *such that*

$$0 < m_1 \le \alpha_i \le m_2 < +\infty, \quad \forall i \in \mathbb{Z}^k.$$

For any $u = (u_i)_{i\in\mathbb{Z}^k}$ and $v = (v_i)_{i\in\mathbb{Z}^k} \in l^2_\rho$, define a bilinear operate via

$$\langle u, v\rangle_{\alpha,\rho} = \sum_{j=1}^{k} \left(D_j u, D_j v\right)_\rho + \sum_{i\in\mathbb{Z}^k} \rho_i \alpha_i u_i v_i, \tag{7.20}$$

then $\langle \cdot, \cdot \rangle_{\alpha,\rho}$ is an inner product on l^2_ρ. It is obvious that for any $u = (u_i)_{i \in \mathbb{Z}^k}$,

$$m_1 \|u\|^2_\rho \le \|u\|^2_{\alpha,\rho} \le \left(k(2q+1)^2 d^2 a^q + m_2\right) M_0 \|u\|^2$$

where $\|\cdot\|$ is the regular l^2−norm, thus $\|\cdot\|_{\alpha,\rho}$ is equivalent to $\|\cdot\|_\rho$.

Denote $l^2_{\alpha,\rho} = \left(l^2_\rho, \langle \cdot \rangle_{\alpha,\rho}, \|\cdot\|_{\alpha,\rho}\right)$, and define $E = l^2_{\alpha,\rho} \times l^2_\rho$ with the inner product

$$\langle z_1, z_2 \rangle_E = \left\langle u^{(1)}, u^{(2)} \right\rangle_{\alpha,\rho} + \left(v^{(1)}, v^{(2)}\right)_\rho,$$

$$\forall z_j = \begin{pmatrix} u^{(j)} \\ v^{(j)} \end{pmatrix} = \begin{pmatrix} u_i^{(j)} \\ v_i^{(j)} \end{pmatrix}_{i \in \mathbb{Z}^k} \in E,\ j = 1, 2,$$

and norm

$$\|z\|^2_E = \langle z, z \rangle_E = \|u\|^2_{\alpha,\rho} + \|v\|^2_\rho, \quad \forall\, z = \begin{pmatrix} u \\ v \end{pmatrix} \in E.$$

Then E is a separable Hilbert space.

7.3.5 Existence of Random Attractors

Following a similar procedure to Sect. 7.2, we now brief the main results obtained for system (7.18) and (7.19), for which the proofs can be found in [26].

- Existence and uniqueness of solutions: let $T > 0$ and assume (A1)–(A3) hold. Then

  (i) for any $\omega \in \Omega$ and any initial data $\psi_0 \in E$, Eq. (7.19) admits a unique solution $\psi(\cdot, \omega, \psi_0) \in C^1([0, T), E)$.

  (ii) system (7.19) generates a continuous RDS $\{\psi(t, \omega, \cdot)\}_{t \ge 0, \omega \in \Omega}$ with state space E, over $(\Omega, \mathscr{F}, \mathbb{P}, (\theta_t)_{t \in \mathbb{R}})$. Moreover,

  $$\Phi(t, \omega, \Phi_0) = \begin{pmatrix} u \\ v \end{pmatrix} = \begin{pmatrix} u \\ \tilde{v} - \xi u + \tilde{\delta}(\theta_t \omega) \end{pmatrix} \quad \text{for} \quad \Phi_0 \in E, \omega \in \Omega \tag{7.21}$$

  defines a continuous RDS $\{\Phi(t, \omega, \cdot)\}_{t \ge 0, \omega \in \Omega}$ over $(\Omega, \mathscr{F}, \mathbb{P}, (\theta_t)_{t \in \mathbb{R}})$ associated with (7.18).

- Existence of tempered random bounded absorbing set for $\{\Phi(t, \omega, \cdot)\}_{t \ge 0, \omega \in \Omega}$: let

$$\xi = \frac{m_1 \lambda}{2\lambda^2 + 3m_1}, \quad K = \frac{\xi}{2} - \frac{8L^2}{\lambda m_1} > 0.$$

Then there exists a closed random bounded absorbing set $B_0(\omega) \in \mathscr{D}(E)$ of $\{\Phi(t, w, \cdot)\}_{t\geq 0, \omega\in\Omega}$ such that for any $B \in \mathscr{D}(E)$ and $\omega \in \Omega$, there exists $T_B(\omega) > 0$ yielding $\Phi(t, \theta_{-t}\omega, B(\theta_{-t}\omega)) \subset B_0(\omega), \forall t \geq T_B(\omega)$. In particular, there exists $T_{B_0}(\omega) > 0$ yielding $\Phi(t, \theta_{-t}\omega, B_0(\theta_{-t}\omega)) \subset B_0(\omega), \forall t \geq T_{B_0}(\omega)$.

- Existence of global random attractor: if $K > 0$, then the RDS $\{\Phi(t, \omega, \cdot)\}_{t\geq 0, \omega\in\Omega}$ generated by (7.18) is random asymptotically null on $B_0(\omega)$, and possesses a unique global random attractor $\mathscr{A}(\omega)$ given by

$$\mathscr{A}(\omega) = \bigcap_{\tau\geq T_{B_0}(\omega)} \overline{\bigcup_{t\geq\tau} \Phi(t, \theta_{-t}\omega) B_0(\theta_{-t}\omega)}.$$

It is worth mentioning that the "asymptotically null" here is slightly different from which was defined in Sect. 7.2. An RDS $\{\Phi(t, \omega, \cdot)\}_{t\geq 0, \omega\in\Omega}$ is said to be *random asymptotically null* in $\mathscr{D}(E)$, if for a.e. $\omega \in \Omega$, any $D(\omega) \in \mathscr{D}(E)$, and any $\varepsilon > 0$, there exist $T(\varepsilon, \omega, D(\omega)) > 0$ and $I(\varepsilon, \omega, D(\omega)) \in \mathbb{N}$ such that

$$\left(\sum_{\|i\|>I(\varepsilon,\omega,D(\omega))} |(\Phi(t, \theta_{-t}\omega, z(\theta_{-t}\omega))_i|_E^2 \right)^{1/2} \leq \varepsilon, \quad where \quad \|i\| \doteq \max_{j=1,\dots,k} \{|i_j|\}$$

for all $t \geq T(\varepsilon, \omega, D(\omega))$ and $z(\omega) = \begin{pmatrix} u(\omega) \\ v(\omega) \end{pmatrix} \in D(\omega)$.

7.4 Closing Remarks

This article is aimed to have an explanatory review on recent studies of global random attractors for stochastic lattice differential equations with white noise, by using the theory of random dynamical systems. We introduce the concept of random pullback attractors and use it to study the long-term behavior for different type of SLDEs. In particular, we choose two first-order SLDEs on lattice $\mathbb{Z}$ with multiplicative and additive noise, respectively, and one second-order SLDEs on lattice $\mathbb{Z}^k$, as model systems to illustrate the underlying methodology. All systems are considered in the weighted space of infinite sequences, which is more inclusive than the regular space of infinite sequences but with more challenging computations.

For a particular SLDE system with additive or with multiplicative noise, different Ornstein–Uhlenbeck processes can be utilized to obtain random differential equations with no explicit appearance of white noise. The theory of RDS can then be used to prove the existence of global random attractors. These are usually done by three steps: first, showing the existence and uniqueness of solutions and that the system generates a continuous RDS, second finding a tempered random bounded absorbing set, and third proving the asymptotical nullness of the RDS. It is worth mentioning that although the underlying framework and strategy are uniform, the associated technical calculations are rather different for different systems. Up to date there are many results on the existence of random attractors for SLDEs. However, more details of the random attractors in addition to the existence, such as finite dimensionality and geometric descriptions, still needs further investigation.

Acknowledgments This work has been partially supported by FEDER and the Spanish Ministerio de Economía y Competitividad project MTM2011-22411.

References

1. Abdallah, A.Y.: Uniform exponential attractors for first order non-autonomous lattice dynamical systems. J. Differ. Equ. **251**(6), 1489–1504 (2011)
2. Afraimovich, V.S., Nekorkin, V.I.: Chaos of traveling waves in a discrete chain of diffusively coupled maps. Int. J. Bifur. Chaos **4**, 631–637 (2004)
3. Arnold, L.: Random Dynamical Systems. Springer, Berlin (1998)
4. Bates, P.W., Lisei, H., Lu, K.: Attractors for stochastic lattice dynamical systems. Stoch. Dyn. **6**, 1–21 (2006)
5. Bates, P.W., Lu, K., Wang, B.: Attractors for lattice dynamical systems. Int. J. Bifur. Chaos **11**, 143–153 (2001)
6. Bell, J.: Some threshold results for models of myelinated nerves. Math. Biosci. **54**(3–4), 181–190 (1981)
7. Bell, J., Cosner, C.: Threshold behavior and propagation for nonlinear differential-difference systems motivated by modeling myelinated axons. Q. Appl. Math. **42**(1), 1–14 (1984)
8. Cahn, J.W.: Theory of Crystal Growth and Interface Motion in Crystalline Materials, pp. 51–52. Wiley, New York (2013)
9. Caraballo, T., Lu, K.: Attractors for stochastic lattice dynamical systems with a multiplicative noise. Front. Math. China **3**(3), 317–335 (2008)
10. Caraballo, T., Morillas, F., Valero, J.: Random attractors for stochastic lattice systems with non-Lipschitz nonlinearity. J. Differ. Equ. Appl. **17**(2), 161–184 (2011)
11. Caraballo, T., Morillas, F., Valero, J.: Attractors of stochastic lattice dynamical systems with a multiplicative noise and non-Lipschitz nonlinearities. J. Differ. Equ. **253**(2), 667–693 (2012)
12. Chow, S.-N., Mallet-Paret, J.: Pattern formulation and spatial chaos in lattice dynamical systems. I. IEEE Trans. Circuits Syst. **42**, 746–751 (1995)
13. Chow, S.-N., Mallet-Paret, J., Shen, W.: Traveling waves in lattice dynamical systems. J. Differ. Equ. **149**, 248–291 (1998)
14. Chow, S.-N., Mallet-Paret, J., Van Vleck, E.S.: Pattern formation and spatial chaos in spatially discrete evolution equations. Random Comput. Dyn. **4**, 109–178 (1996)
15. Chow, S.-N., Shen, W.: Dynamics in a discrete Nagumo equation: spatial topological chaos. SIAM J. Appl. Math. **55**, 1764–1781 (1995)
16. Chua, L.O., Yang, L.: Cellular neural networks: applications. IEEE Trans. Circuits Syst. **35**(10), 1273–1290 (1988)
17. Chua, L.O., Yang, L.: Cellular neural networks: theory. IEEE Trans. Circuits Syst. **35**(10), 1257–1272 (1988)
18. Crauel, H., Debussche, A., Flandoli, F.: Random attractors. J. Dyn. Differ. Equ. **9**(2), 307–341 (1997)
19. Crauel, H., Flandoli, F.: Attractors for random dynamical systems. Probab. Theory Relat. Fields **100**, 365–393 (1994)
20. Erneux, T., Nicolis, G.: Propagating waves in discrete bistable reaction-diffusion systems. Phys. D **67**(1–3), 237–244 (1993)
21. Fan, X., Yang, H.: Exponential attractor and its fractal dimension for a second order lattice dynamical system. J. Math. Anal. Appl. **367**(2), 350–359 (2010)

22. Flandol, F.I., Schmalfuss, B.: Random attractors for the 3d stochastic navier-stokes equation with multiplicative white noise. Stoch. Stoch. Rep. **59**, 21–45 (1996)
23. Han, X.: Exponential attractors for lattice dynamical systems in weighted spaces. Discret. Contin. Dyn. Syst. **31**(2), 445–467 (2011)
24. Han, X.: Random attractors for stochastic sine-Gordon lattice systems with multiplicative white noise. J. Math. Anal. Appl. **376**(2), 481–493 (2011)
25. Han, X.: Random attractors for second order stochastic lattice dynamical systems with multiplicative noise in weighted spaces. Stoch. Dyn. **12**(3), 1150024 (2012)
26. Han, X.: Asymptotic behaviors for second order stochastic lattice dynamical systems on $\mathbb{Z}^k$ in weighted spaces. J. Math. Anal. Appl. **397**(1), 242–254 (2013)
27. Han, X., Shen, W., Zhou, S.: Random attractors for stochastic lattice dynamical systems in weighted spaces. J. Differ. Equ. **250**(3), 1235–1266 (2011)
28. Huang, J.: The random attractor of stochastic FitzHugh-Nagumo equations in an infinite lattice with white noises. Phys. D **233**(2), 83–94 (2007)
29. Imkeller, P., Schmalfuss, B.: The conjugacy of stochastic and random differential equations and the existence of global attractors. J. Dynam. Differ. Equ. **13**(2), 215–249 (2001)
30. Kapval, R.: Discrete models for chemically reacting systems. J. Math. Chem. **6**, 113–163 (1991)
31. Keener, J.P.: Propagation and its failure in the discrete Nagumo equation. In Ordinary and partial differential equations (Dundee, 1986), vol. 157 of Pitman Research Notes in Mathematics, pp. 95–112. Longman Sci. Tech., Harlow (1987)
32. Keener, J.P.: The effects of discrete gap junction coupling on propagation in myocardium. J. Theor. Biol. **148**, 49–82 (1991)
33. Kloeden, P.E., Rasmussen, M.: Nonautonomous dynamical systems. Mathematical Surveys and Monographs, vol. 176. American Mathematical Society, Providence (2011)
34. Laplante, J.P., Erneux, T.: Propagating failure in arrays of coupled bistable chemical reactors. J. Phys. Chem. **96**, 4931–4934 (1992)
35. Lv, Y., Sun, J.H.: Dynamical behavior for stochastic lattice systems. Chaos Solitons Fractals **27**, 1080–1090 (2006)
36. Mallet-Paret, J.: The global structure of traveling waves in spatially discrete dynamical systems. J. Dyn. Differ. Equ. **11**(1), 49–127 (1999)
37. Rashevsky, N.: Mathematical Biophysics: Physico-mathematical Foundations of Biology, vol. 2, 3rd revised edn. Dover Publications Inc., New York (1960)
38. Ruelle, D.: Characteristic exponents for a viscous fluid subjected to time dependent forces. Commun. Math. Phys. **93**(3), 285–300 (1984)
39. Scott, A.C.: Analysis of a myelinated nerve model. Bull. Math. Biophys. **26**, 247–254 (1964)
40. Shen, W.: Lifted lattices, hyperbolic structures, and topological disorders in coupled map lattices. SIAM J. Appl. Math. **56**, 1379–1399 (1996)
41. Van Vleck, E., Wang, B.: Attractors for lattice FitzHugh-Nagumo systems. Phys. D **212**(3–4), 317–336 (2005)
42. Wang, B.: Dynamics of systems on infinite lattices. J. Differ. Equ. **221**(1), 224–245 (2006)
43. Wang, B.: Asymptotic behavior of non-autonomous lattice systems. J. Math. Anal. Appl. **331**(1), 121–136 (2007)
44. Wang, X., Li, X., Xu, D.: Random attractors for second-order stochastic lattice dynamical systems. Nonlinear Anal. Theory Methods Appl. **72**(1), 483–494 (2010)
45. Zhao, C., Zhou, S.: Sufficient conditions for the existence of global random attractors for stochastic lattice dynamical systems and applications. J. Math. Anal. Appl. **354**(1), 78–95 (2009)
46. Zhao, X., Zhou, S.: Kernel sections for processes and nonautonomous lattice systems. Discret. Contin. Dyn. Syst. Ser. B **9**(3–4), 763–785 (2008)
47. Zhou, S.: Attractors for first order dissipative lattice dynamical systems. Phys. D: Nonlinear Phenom. **178**(12), 51–61 (2003)
48. Zhou, S.: Attractors and approximations for lattice dynamical systems. J. Differ. Equ. **200**(2), 342–368 (2004)
49. Zhou, S., Han, X.: Pullback exponential attractors for non-autonomous lattice systems. J. Dyn. Differ. Equ. **24**(3), 601–631 (2012)

Chapter 8
A Generalized Cahn-Hilliard Equation with Logarithmic Potentials

Alain Miranville

Abstract Our aim in this paper is to study the well-posedness for a generalized Cahn-Hilliard equation with a proliferation term and singular potentials. We also prove the existence of the global attractor.

Keywords Generalized cahn-hilliard equation · Proliferation term · Logarithmic potential · Well-posedness · Global attractor

8.1 Introduction

The Cahn-Hilliard equation plays an important role in materials science and describes phase separation processes. This can be observed, e.g., when a binary alloy is cooled down sufficiently. One then observes a partial nucleation (i.e., the apparition of nucleides in the material) or a total nucleation, the so-called spinodal decomposition: the material quickly becomes inhomogeneous, forming a fine-grained structure in which each of the two components appears more or less alternatively. In a second stage, which is called coarsening and occurs at a slower time scale, these microstructures coarsen. Such phenomena play an essential role in the mechanical properties of the material, e.g., strength. We refer the reader to, e.g., [5, 6, 9, 13, 17, 18, 20, 21, 27, 28] for more details.

It is interesting to note that the Cahn-Hilliard equation, or some of its variants, is also relevant in other contexts, in which phase separation and coarsening/clustering processes can be observed or come into play. We can mention, for instance, population dynamics (see [11]), bacterial films (see [16]), thin films (see [30, 33]), image processing and inpainting (see [1, 2, 4, 7, 8, 12]) and even the rings of Saturn (see [34]) and the clustering of mussels (see [19]).

In [15], the authors proposed the following equation:

A. Miranville (✉)
Laboratoire de Mathématiques et Applications, UMR CNRS 7348, SP2MI, Université de Poitiers, 86962 Chasseneuil Futuroscope Cedex, France
e-mail: alain.miranville@math.univ-poitiers.fr

© Springer International Publishing Switzerland 2015
V.A. Sadovnichiy and M.Z. Zgurovsky (eds.), *Continuous and Distributed Systems II*, Studies in Systems, Decision and Control 30,
DOI 10.1007/978-3-319-19075-4_8

$$\frac{\partial u}{\partial t} - \frac{\partial^2}{\partial x^2}[\ln(1-q)\frac{\partial^2 u}{\partial x^2} + F'(u)] + \alpha u(u-1) = 0 \tag{8.1}$$

to model cells which move, proliferate and interact via diffusion and cell-cell adhesion in wound healing and tumor growth (in two space dimensions, one also models the clustering of malignant tumor cells). Here, u is the local density of cells, $\alpha > 0$ is the proliferation rate and F is the local free energy (or potential). Furthermore,

$$q = 1 - \exp(-\frac{J}{k_B T}), \tag{8.2}$$

where J corresponds to the interatomic interactions (e.g., the coupling strength in the Ising model), k_B is the Boltzmann's constant and T is the absolute temperature, assumed constant. Finally, F is a double-well potential of the form

$$F(s) = \frac{1}{4}a(s - \frac{1}{2})^4 + \frac{1}{2}b(s - \frac{1}{2})^2, \tag{8.3}$$

where a and b are taken in the form

$$a = (\frac{q - q_{\mathrm{cr}}}{1 - 16\frac{(1-q)^2}{q^4}})^{\frac{1}{4}} \frac{c(q)}{q_{\mathrm{cr}}^{\frac{1}{4}}}, \quad b = -\frac{q - q_{\mathrm{cr}}}{|q - q_{\mathrm{cr}}|^{\frac{3}{4}}} \frac{c(q)}{q_{\mathrm{cr}}^{\frac{1}{4}}}, \tag{8.4}$$

$q_{\mathrm{cr}} > 0$ and $q > q_{\mathrm{cr}}$ (this corresponds to the unstable region; note however that, as noted in [15], even in the case $q < q_{\mathrm{cr}}$, when the proliferation term is turned on, the initially homogeneous state can become inhomogeneous, leading also to phase separation and clustering). Furthermore, the positive constant $c(q)$ is such that

$$\lim_{q \to 0} a = 0, \ \lim_{q \to 0} b = \frac{1}{4}, \tag{8.5}$$

hence

$$\lim_{q \to 0} c(q) = \frac{1}{4}. \tag{8.6}$$

We proved in [23] the existence and uniqueness of solutions to (8.1), endowed with Dirichlet boundary conditions (actually, there, we considered a more general equation and also proved the existence of finite-dimensional attractors). The case of Neumann boundary conditions was addressed in [10]. There, it was proved that, in that case, the situation is completely different, in the sense that one can have blow up in finite time. Now, for the biologically relevant solutions, i.e., those which remain in the interval [0, 1], one does not have blow up in finite time. Unfortunately, as it is actually the case for the original Cahn-Hilliard equation (see, e.g., [31]), there exist solutions which do not stay in the (biologically) relevant interval (and, as shown in numerical simulations, blow up).

A natural way, in the Cahn-Hilliard context, to force the solutions to remain in the relevant interval, is to consider, instead of a regular potential as in (8.3), a logarithmic one; note indeed that, in the Cahn-Hilliard equation, regular (polynomial) potentials are approximations of thermodynamically relevant logarithmic potentials which follow from a mean field model.

Our aim in this paper is to consider such logarithmic potentials for (8.1) and prove the existence and uniqueness of solutions, as well as the existence of the global attractor.

We consider here Dirichlet boundary conditions only. The case of Neumann boundary conditions is more delicate. Indeed, in that case, one has to estimate the spatial average of the order parameter, $\langle u\rangle := \frac{1}{\mathrm{Vol}(\Omega)}\int_\Omega u\,dx$. For the original Cahn-Hilliard equation, estimating such a quantity is straightforward, as one has the conservation of mass, i.e., of $\langle u\rangle$. However, for (8.1), setting $u = \langle u\rangle + v$, we have (see [10])

$$\frac{d\langle u\rangle}{dt} + \alpha\langle u\rangle(\langle u\rangle - 1) = -\langle v\rangle^2.$$

In particular, this shows that $\langle u\rangle$ can blow up in finite time and one essential difficulty is to prove that also $\langle u\rangle$ remains in the biologically relevant interval $[0, 1]$. This will be addressed elsewhere.

8.2 Setting of the Problem

As is usual in the mathematical analysis of the Cahn-Hilliard equation, we rescale the order parameter (setting $v = 2u - 1$ and calling again u the new order parameter), so that the new relevant interval is $(-1, 1)$; note that the nonlinear term $\alpha u(u-1)$ in (8.1) becomes $\frac{\alpha}{4}(u+1)(u-1)$ in the new equation.

We then consider the following more general initial and boundary value problem in a bounded and regular domain $\Omega \subset \mathbf{R}^N$, $N = 1, 2$ or 3, with boundary Γ:

$$\frac{\partial u}{\partial t} + \Delta^2 u - \Delta f(u) + g(u) = 0, \tag{8.7}$$

$$u = \Delta u = 0 \text{ on } \Gamma, \tag{8.8}$$

$$u|_{t=0} = u_0. \tag{8.9}$$

As far as the nonlinear terms f and g are concerned, we make the following assumptions:

$$f = F', \text{ where } F(s) = \frac{\lambda_1}{2}(1-s^2) + \frac{\lambda_2}{2}((1-s)\ln\frac{1-s}{2} + (1+s)\ln\frac{1+s}{2}), \tag{8.10}$$

$$0 < \lambda_2 < \lambda_1,\ s \in (-1, 1),$$

$$\text{hence } f(s) = -\lambda_1 s + \frac{\lambda_2}{2} \ln \frac{1+s}{1-s},\ s \in (-1, 1), \text{ and}$$

$$f' \geq -\lambda_1 \tag{8.11}$$

(note that F is bounded and $f(0) = 0$);

$$g \text{ is of class } \mathscr{C}^1,\ |g(s)| \leq c_1(|s|^p + 1),\ |g'(s)| \leq c_2(|s|^p + 1), \tag{8.12}$$

$$p \in \mathrm{N},\ p \geq 2,\ c_1,\ c_2 \geq 0,\ s \in (0, 1).$$

Remark 8.1 The cases $p = 0$ and $p = 1$ can also be treated (in particular, when $g(s) = \alpha s$, $\alpha > 0$, we recover the Cahn-Hilliard-Oono equation proposed in [29] (see also [22]) in order to model long-ranged (i.e., nonlocal) interactions). Furthermore, we can more generally consider a function g depending also on x, i.e., $g = g(x, s)$, allowing, in particular, to consider the fidelity term proposed in [1, 2] (see also [4, 8]).

We write

$$F(s) = \frac{\lambda_1}{2}(1 - s^2) + F_1(s) \tag{8.13}$$

and define, following [14], $F_{1,\varepsilon}$, $\varepsilon > 0$, as follows:

$$F_{1,\varepsilon}^{(2p)}(s) = \begin{cases} F_1^{(2p)}(1-\varepsilon),\ s \geq 1-\varepsilon, \\ F_1^{(2p)}(s),\ -1+\varepsilon \leq s \leq 1-\varepsilon, \\ F_1^{(2p)}(-1+\varepsilon),\ s \leq -1+\varepsilon, \end{cases} \tag{8.14}$$

$$F_{1,\varepsilon}^{(i)}(0) = F_1^{(i)}(0),\ i = 0,\ \cdots,\ 2p-1. \tag{8.15}$$

In particular, $F_{1,\varepsilon}$ satisfies, for ε small enough (see [14]),

$$F_{1,\varepsilon}(s) \geq c_3 s^{2p} - c_4,\ c_3 > 0,\ c_4 \geq 0,\ s \in \mathbf{R}, \tag{8.16}$$

where c_3 and c_4 are independent of ε, and, setting $F_\varepsilon(s) = \frac{\lambda_1}{2}(1 - s^2) + F_{1,\varepsilon}(s)$, $f_\varepsilon = F_\varepsilon'$ and $f_{1,\varepsilon} = F_{1,\varepsilon}'$, there holds

$$f_{1,\varepsilon}' \geq 0,\ f_\varepsilon' \geq -\lambda_1. \tag{8.17}$$

Furthermore, there holds, owing to (8.15)–(8.17) and for ε small enough,

$$f_\varepsilon(s)s = -\lambda_1 s^2 + f_{1,\varepsilon}(s)s \tag{8.18}$$

$$\geq -\lambda_1 s^2 + F_{1,\varepsilon}(s)$$

$$\geq \frac{1}{2} F_{1,\varepsilon}(s) - c_5, \ c_5 \geq 0, \ s \in \mathbf{R},$$

where the constant c_5 is independent of ε.

We finally consider the following approximated problems:

$$\frac{\partial u^\varepsilon}{\partial t} + \Delta^2 u^\varepsilon - \Delta f_\varepsilon(u^\varepsilon) + g(u^\varepsilon) = 0, \tag{8.19}$$

$$u^\varepsilon = \Delta u^\varepsilon = 0 \text{ on } \Gamma, \tag{8.20}$$

$$u^\varepsilon|_{t=0} = u_0. \tag{8.21}$$

We denote by $((\cdot, \cdot))$ the usual L^2-scalar product, with associated norm $\|\cdot\|$. We further set $\|\cdot\|_{-1} = \|(-\Delta)^{-\frac{1}{2}} \cdot \|$, where $(-\Delta)^{-1}$ denotes the inverse minus Laplace operator associated with Dirichlet boundary conditions; this norm is equivalent to the usual H^{-1}-norm, where $H^{-1}(\Omega)$ is the topological dual of $H_0^1(\Omega)$. More generally, $\|\cdot\|_X$ denotes the norm on the Banach space X.

Throughout the paper, the same letters c and c' (and, sometimes, c'') denote (generally positive) constants which may vary from line to line and which are independent of ε.

8.3 A Priori Estimates

We rewrite (8.19)–(8.20) in the (formally) equivalent form

$$(-\Delta)^{-1} \frac{\partial u^\varepsilon}{\partial t} - \Delta u^\varepsilon + f_\varepsilon(u^\varepsilon) + (-\Delta)^{-1} g(u^\varepsilon) = 0, \tag{8.22}$$

$$u^\varepsilon = 0 \text{ on } \Gamma. \tag{8.23}$$

Multiplying (8.22) by u^ε, we have

$$\frac{1}{2} \frac{d}{dt} \|u^\varepsilon\|_{-1}^2 + \|\nabla u^\varepsilon\|^2 + ((f_\varepsilon(u^\varepsilon), u^\varepsilon)) + (((-\Delta)^{-1} g(u^\varepsilon), u^\varepsilon)) = 0. \tag{8.24}$$

Noting that, owing to (8.12) and (8.16),

$$|(((-\Delta)^{-1} g(u^\varepsilon), u^\varepsilon))| = |((g(u^\varepsilon), (-\Delta)^{-1} u^\varepsilon))| \tag{8.25}$$

$$\leq \|g(u^\varepsilon)\|_{L^1(\Omega)} \|(-\Delta)^{-1} u^\varepsilon\|_{L^\infty(\Omega)} \leq c(\|u^\varepsilon\|_{L^p(\Omega)}^{p+1} + 1) \leq \frac{1}{4} \int_\Omega F_{1,\varepsilon}(u^\varepsilon)\, dx + c',$$

where we have used the continuous embedding $H^2(\Omega) \subset L^\infty(\Omega)$, it follows from (8.18) and (8.24)–(8.25) that

$$\frac{d}{dt}\|u^\varepsilon\|_{-1}^2 + c(\|u^\varepsilon\|_{H^1(\Omega)}^2 + \int_\Omega F_{1,\varepsilon}(u^\varepsilon)\,dx) \le c',\ c > 0. \tag{8.26}$$

We then multiply (8.19) by u^ε and obtain, in view of (8.17),

$$\frac{1}{2}\frac{d}{dt}\|u^\varepsilon\|^2 + \|\Delta u^\varepsilon\|^2 + ((g(u^\varepsilon), u^\varepsilon)) \le \lambda_1\|\nabla u^\varepsilon\|^2. \tag{8.27}$$

Noting that, owing again to (8.12) and (8.16),

$$|((g(u^\varepsilon), u^\varepsilon))| \le c(\|u^\varepsilon\|_{L^{p+1}(\Omega)}^{p+1} + 1) \le c'\int_\Omega F_{1,\varepsilon}(u^\varepsilon)\,dx + c'', \tag{8.28}$$

we find

$$\frac{d}{dt}\|u^\varepsilon\|^2 + c\|u^\varepsilon\|_{H^2(\Omega)}^2 \le c'(\|u^\varepsilon\|_{H^1(\Omega)}^2 + \int_\Omega F_{1,\varepsilon}(u^\varepsilon)\,dx) + c'',\ c > 0. \tag{8.29}$$

Summing finally (8.26) and (8.29) multiplied by δ_1, where $\delta_1 > 0$ is chosen small enough, we have

$$\frac{dE_{1,\varepsilon}}{dt} + c(\|u^\varepsilon\|_{H^2(\Omega)}^2 + \int_\Omega F_{1,\varepsilon}(u^\varepsilon)\,dx) \le c',\ c > 0, \tag{8.30}$$

where

$$E_{1,\varepsilon} = \|u^\varepsilon\|_{-1}^2 + \delta_1\|u^\varepsilon\|^2.$$

Next, we multiply (8.22) by $\frac{\partial u^\varepsilon}{\partial t}$ and obtain

$$\frac{1}{2}\frac{d}{dt}\|\nabla u^\varepsilon\|^2 + \frac{d}{dt}\int_\Omega F_\varepsilon(u^\varepsilon)\,dx + \|\frac{\partial u^\varepsilon}{\partial t}\|_{-1}^2 + ((g(u^\varepsilon), (-\Delta)^{-1}\frac{\partial u^\varepsilon}{\partial t})) = 0. \tag{8.31}$$

Employing once more (8.16), we find

$$|((g(u^\varepsilon), (-\Delta)^{-1}\frac{\partial u^\varepsilon}{\partial t}))| \le \frac{1}{2}\|\frac{\partial u^\varepsilon}{\partial t}\|_{-1}^2 + c\int_\Omega F_{1,\varepsilon}(u^\varepsilon)\,dx + c' \tag{8.32}$$

and (8.31) yields

$$\frac{d}{dt}(\|\nabla u^\varepsilon\|^2 + 2\int_\Omega F_\varepsilon(u^\varepsilon)\,dx) + \|\frac{\partial u^\varepsilon}{\partial t}\|_{-1}^2 \le c\int_\Omega F_{1,\varepsilon}(u^\varepsilon)\,dx + c'. \tag{8.33}$$

Summing (8.30) and (8.33) multiplied by δ_2, where $\delta_2 > 0$ is chosen small enough, we have

$$\frac{dE_{2,\varepsilon}}{dt} + c(E_{2,\varepsilon} + \|u^\varepsilon\|^2_{H^2(\Omega)} + \|\frac{\partial u^\varepsilon}{\partial t}\|^2_{-1}) \leq c',\ c > 0, \tag{8.34}$$

where

$$E_{2,\varepsilon} = E_{1,\varepsilon} + \delta_2(\|\nabla u^\varepsilon\|^2 + 2\int_\Omega F_\varepsilon(u^\varepsilon)\,dx)$$

satisfies

$$E_{2,\varepsilon} \geq c(\|u^\varepsilon\|^2_{H^1(\Omega)} + \|u^\varepsilon\|^{2p}_{L^{2p}(\Omega)}) - c',\ c > 0,\ c' \geq 0. \tag{8.35}$$

Differentiating (8.22) with respect to time, we obtain

$$(-\Delta)^{-1}\frac{\partial}{\partial t}\frac{\partial u^\varepsilon}{\partial t} - \Delta\frac{\partial u^\varepsilon}{\partial t} + f_\varepsilon'(u^\varepsilon)\frac{\partial u^\varepsilon}{\partial t} + (-\Delta)^{-1}(g'(u^\varepsilon)\frac{\partial u^\varepsilon}{\partial t}) = 0, \tag{8.36}$$

$$\frac{\partial u^\varepsilon}{\partial t} = 0 \text{ on } \Gamma. \tag{8.37}$$

Multiplying then (8.36) by $t\frac{\partial u^\varepsilon}{\partial t}$, we find, owing to (8.17),

$$\frac{1}{2}\frac{d}{dt}(t\|\frac{\partial u^\varepsilon}{\partial t}\|^2_{-1}) + t\|\nabla\frac{\partial u^\varepsilon}{\partial t}\|^2 + t((g'(u^\varepsilon)\frac{\partial u^\varepsilon}{\partial t}, (-\Delta)^{-1}\frac{\partial u^\varepsilon}{\partial t})) \tag{8.38}$$
$$\leq \lambda_1 t\|\frac{\partial u^\varepsilon}{\partial t}\|^2 + \frac{1}{2}\|\frac{\partial u^\varepsilon}{\partial t}\|^2_{-1}.$$

Noting that

$$|((g'(u^\varepsilon)\frac{\partial u^\varepsilon}{\partial t}, (-\Delta)^{-1}\frac{\partial u^\varepsilon}{\partial t}))| \leq c\int_\Omega (|u^\varepsilon|^p + 1)|\frac{\partial u^\varepsilon}{\partial t}||(-\Delta)^{-1}\frac{\partial u^\varepsilon}{\partial t}|\,dx \tag{8.39}$$
$$\leq c(\|u^\varepsilon\|^p_{L^{2p}(\Omega)} + 1)\|\frac{\partial u^\varepsilon}{\partial t}\|^2$$
$$\leq c(\|u^\varepsilon\|^p_{L^{2p}(\Omega)} + 1)\|\nabla\frac{\partial u^\varepsilon}{\partial t}\|\|\frac{\partial u^\varepsilon}{\partial t}\|_{-1},$$

where we have used the continuous embedding $H^2(\Omega) \subset L^\infty(\Omega)$ and the interpolation inequality

$$\|v\|^2 \leq c\|v\|_{H^1(\Omega)}\|v\|_{-1},\ v \in H_0^1(\Omega), \tag{8.40}$$

we deduce from (8.38) to (8.40) that

$$\frac{d}{dt}(t\|\frac{\partial u^\varepsilon}{\partial t}\|^2_{-1}) + t\|\frac{\partial u^\varepsilon}{\partial t}\|^2_{H^1(\Omega)} \leq c(t+1)(\|u^\varepsilon\|^{2p}_{L^{2p}(\Omega)} + 1)\|\frac{\partial u^\varepsilon}{\partial t}\|^2_{-1}. \tag{8.41}$$

Multiplying also (8.36) by $\frac{\partial u^\varepsilon}{\partial t}$, we have, proceeding as above,

$$\frac{d}{dt}\|\frac{\partial u^\varepsilon}{\partial t}\|_{-1}^2 + \|\frac{\partial u^\varepsilon}{\partial t}\|_{H^1(\Omega)}^2 \le c(\|u^\varepsilon\|_{L^{2p}(\Omega)}^{2p} + 1)\|\frac{\partial u^\varepsilon}{\partial t}\|_{-1}^2. \tag{8.42}$$

We now rewrite (8.22) as an elliptic equation, for $t > 0$ fixed,

$$-\Delta u^\varepsilon + f_\varepsilon(u^\varepsilon) = -(-\Delta)^{-1}\frac{\partial u^\varepsilon}{\partial t} - (-\Delta)^{-1}g(u^\varepsilon), \tag{8.43}$$

$$u^\varepsilon = 0 \text{ on } \Gamma. \tag{8.44}$$

Multiplying (8.43) by $-\Delta u^\varepsilon$, we obtain, owing to (8.17) and proceeding as above (employing, in particular, (8.16)),

$$\|u^\varepsilon\|_{H^2(\Omega)}^2 \le c(E_{2,\varepsilon} + \|\frac{\partial u^\varepsilon}{\partial t}\|_{-1}^2). \tag{8.45}$$

We finally multiply (8.22) by $f_\varepsilon(u^\varepsilon)$ and find, owing to (8.16)–(8.17),

$$\|f_\varepsilon(u^\varepsilon)\|^2 \le c(E_{2,\varepsilon} + \|\frac{\partial u^\varepsilon}{\partial t}\|_{-1}^2). \tag{8.46}$$

8.4 The Continuous Semigroup

We have the

Theorem 8.1 *We assume that $u_0 \in \Phi := \{v \in H_0^1(\Omega),\ \|u_0\|_{L^\infty(\Omega)} < 1\}$. Then, (8.7)–(8.9) possesses a unique solution u such that*

$$u \in \mathscr{C}([0,T]; H^{-1}(\Omega)) \cap L^\infty(0,T; H_0^1(\Omega)) \cap L^2(0,T; H^2(\Omega)) \cap L^\infty(\tau,T; H^2(\Omega)),$$

$$\frac{\partial u}{\partial t} \in L^2(0,T; H^{-1}(\Omega)) \cap L^2(\tau,T; H_0^1(\Omega)) \cap L^\infty(\tau,T; H^{-1}(\Omega))$$

and

$$f(u) \in L^2((0,T)\times\Omega) \cap L^\infty(\tau,T; L^2(\Omega)),$$

$\forall 0 < \tau < T$. Furthermore, $-1 < u(t,x) < 1$ a.e. (t,x).

Proof **(a) Uniqueness**. Let u^1 and u^2 be two solutions to (8.7)–(8.9) with initial data u_0^1 and u_0^2, respectively. Then, we have, setting $u = u^1 - u^2$ and $u_0 = u_0^1 - u_0^2$,

$$(-\Delta)^{-1}\frac{\partial u}{\partial t} - \Delta u + f(u^1) - f(u^2) + (-\Delta)^{-1}(g(u^1) - g(u^2)) = 0, \tag{8.47}$$

$$u = 0 \text{ on } \Gamma, \tag{8.48}$$

$$u|_{t=0} = u_0. \tag{8.49}$$

Multiplying (8.47) by u, we obtain, owing to (8.11),

$$\frac{1}{2}\frac{d}{dt}\|u\|_{-1}^2 + \|\nabla u\|^2 \leq \lambda_1 \|u\|^2 - ((g(u^1) - g(u^2), (-\Delta)^{-1}u)). \tag{8.50}$$

Noting that

$$|((g(u^1) - g(u^2), (-\Delta)^{-1}u))| \leq c\|u\|^2, \tag{8.51}$$

it follows from (8.50) to (8.51) and the interpolation inequality (8.40) that

$$\frac{d}{dt}\|u\|_{-1}^2 + c\|u\|_{H^1(\Omega)}^2 \leq c'\|u\|_{-1}^2, \quad c > 0, \tag{8.52}$$

and Gronwall's lemma yields the continuous dependence with respect to the initial data (in the H^{-1}-norm), as well as the uniqueness.

(b) Existence. We consider the solution u^ε to the approximated problem (8.19)–(8.21) (the proof of existence, uniqueness and regularity of u^ε can be found in [23]). Then, it follows from the a priori estimates derived in the previous section and (8.16) that, up to a subsequence, u^ε converges to a limit function u such that

$$u^\varepsilon \to u \text{ in } L^\infty(0,T;H_0^1(\Omega)) \text{ and } L^\infty(0,T;L^{2p}(\Omega)) \text{ weak} - \star \text{ and in } L^2(0,T;H^2(\Omega)) \text{ weak},$$

$$u^\varepsilon \to u \text{ a.e. } (t,x).$$

The only difficulty here is to pass to the limit in the nonlinear term $f_\varepsilon(u^\varepsilon)$ (say, within a proper variational formulation).

First, it follows from (8.34) and (8.46) that $f_\varepsilon(u^\varepsilon)$ is bounded, independently of ε, in $L^1((0,T)\times\Omega)$. Then, it follows from the explicit expression of f_ε that

$$\text{meas}(E_{\varepsilon,\varepsilon'}) \leq \varphi(\varepsilon), \quad 0 < \varepsilon' \leq \varepsilon,$$

where

$$E_{\varepsilon,\varepsilon'} = \{(t,x) \in (0,T)\times\Omega, \ |u^{\varepsilon'}(t,x)| > 1-\varepsilon\}$$

and

$$\varphi(s) = \frac{c}{|f(1-s)|}.$$

Here, the constant c is independent of ε and ε'. Note indeed that, owing to the explicit expression of f_ε, there holds

$$\int_0^T \int_\Omega |f_{\varepsilon'}(u^{\varepsilon'})|\,dx\,dt \geq \int_{E_{\varepsilon,\varepsilon'}} |f_{\varepsilon'}(u^{\varepsilon'})|\,dx\,dt \geq c'\text{meas}(E_{\varepsilon,\varepsilon'})|f(1-\varepsilon)|, \tag{8.53}$$

where the constant c' is independent of ε and ε' (recall also that $f_\varepsilon(u^\varepsilon)$ is bounded, independently of ε, in $L^1((0,T)\times\Omega)$ and note that $f(-s) = -f(s)$). We can pass to the limit $\varepsilon' \to 0$ (employing Fatou's lemma, see (8.53)) and then $\varepsilon \to 0$ (noting that $\lim_{s\to 0}\varphi(s) = 0$) to find

$$\text{meas}\{(t,x) \in (0,T)\times\Omega,\ |u(t,x)| \geq 1\} = 0,$$

so that

$$-1 < u(t,x) < 1 \text{ a.e. } (t,x).$$

Next, it follows from the above almost everywhere convergence of u^ε to u (and also from the explicit expression of f_ε) that

$$f_\varepsilon(u^\varepsilon) \to f(u) \text{ a.e. } (t,x) \in (0,T)\times\Omega. \tag{8.54}$$

Finally, it follows from (8.34) and (8.46), which yield that $f_\varepsilon(u^\varepsilon)$ is bounded, independently of ε, in $L^2((0,T)\times\Omega)$, and from (8.54) that $f_\varepsilon(u^\varepsilon) \to f(u)$ in $L^2((0,T)\times\Omega)$ weak, which finishes the proof of the passage to the limit (the additional regularity results follow from the a priori estimates derived in the previous section).

Remark 8.2 If we further assume that $u_0 \in H^3(\Omega)$, then it follows from (8.42) that $\frac{\partial u}{\partial t} \in L^\infty(0,T;H^{-1}(\Omega)) \cap L^2(0,T;H_0^1(\Omega))$, $\forall T > 0$.

It follows from Theorem 8.1 that we can define the continuous (for the H^{-1}-norm) semigroup

$$S(t) : \Phi \to \Phi,\ u_0 \mapsto u(t),\ t \geq 0.$$

Furthermore, it follows from (8.34) and Gronwall's lemma that $S(t)$ is dissipative on Φ, i.e., it possesses a bounded absorbing set $\mathscr{B}_0 \subset \Phi$ (in the sense that, $\forall B \subset \Phi$ bounded (for the H^1-norm), $\exists t_0 = t_0(B)$ such that $t \geq t_0$ implies $S(t)B \subset \mathscr{B}_0$).

Actually, it follows from the H^{-1}-continuity that we can extend, in a unique way and by continuity, $S(t)$ to the closure of Φ in the H^{-1}-topology, namely, to the space

$$L := \{v \in H^{-1}(\Omega),\ -1 \leq v(x) \leq 1 \text{ a.e.}\}.$$

We then deduce from (8.26) that $S(t)$ is dissipative in L. The above results also show an instantaneous regularization and mixing of the pure states, i.e., $S(t)L \subset \Phi$, as soon as $t > 0$. Finally, it follows from (8.34), (8.42), (8.45) and the uniform Gronwall's lemma (see, e.g., [32]) that there exists $\mathscr{B}_1 \subset H^2(\Omega)$ and $t_1 > 0$ such that $t \geq t_1$ implies $S(t)\mathscr{B}_0 \subset \mathscr{B}_1$. We thus deduce (see, e.g., [3, 25, 32]) the

Theorem 8.2 *The semigroup $S(t)$ possesses the global attactor $\mathscr{A}$ in L which is compact in L and bounded in $H^2(\Omega)$.*

Remark 8.3 Having the existence of the global attractor, we can prove, adapting the strategies devised in [24, 26], that it has finite fractal dimension.

Remark 8.4 We recall that the global attractor $\mathscr{A}$ is the smallest (for the inclusion) compact set of the phase space which is invariant by the flow (i.e., $S(t)\mathscr{A} = \mathscr{A}$, $\forall t \geq 0$) and attracts all bounded sets of initial data as time goes to infinity; it thus appears as a suitable object in view of the study of the asymptotic behavior of the system. Furthermore, the finite-dimensionality means, roughly speaking, that, even though the initial phase space is infinite-dimensional, the reduced dynamics is, in some proper sense, finite-dimensional and can be described by a finite number of parameters. We refer the reader to [3, 25, 32] for more details and discussions on this.

References

1. Bertozzi, A., Esedoglu, S., Gillette, A.: Analysis of a two-scale Cahn-Hilliard model for binary image inpainting. Multiscale Model. Simul. **6**, 913–936 (2007)
2. Bertozzi, A., Esedoglu, S., Gillette, A.: Inpainting of binary images using the Cahn-Hilliard equation. IEEE Trans. Image Process. **16**, 285–291 (2007)
3. Babin, A.V., Vishik, M.I.: Attractors of Evolution Equations. Amsterdam, New York (1992)
4. Burger, M., He, L., Schönlieb, C.: Cahn-Hilliard inpainting and a generalization for grayvalue images. SIAM J. Imaging Sci. **3**, 1129–1167 (2009)
5. Cahn, J.W.: On spinodal decomposition. Acta Metall. **9**, 795–801 (1961)
6. Cahn, J.W., Hilliard, J.E.: Free energy of a nonuniform system I. Interfacial free energy. J. Chem. Phys. **28**, 258–267 (1958)
7. Chalupeckí, V.: Numerical studies of Cahn-Hilliard equations and applications in image processing. In: Proceedings of Czech-Japanese Seminar in Applied Mathematics 2004, Czech Technical University in Prague, 4–7 August 2004
8. Cherfils, L., Fakih, H., Miranville, A.: Finite-dimensional attractors for the Bertozzi Esedoglu-Gillette-Cahn-Hilliard equation in image inpainting. Inv. Prob. Imaging. **9**, 105 125 (2015)
9. Cherfils, L., Miranville, A., Zelik, S.: The Cahn-Hilliard equation with logarithmic potentials. Milan J. Math. **79**, 561–596 (2011)
10. Cherfils, L., Miranville, A., Zelik, S.: On a generalized Cahn-Hilliard equation with biological applications. Discret. Contin. Dyn. Syst. B **19**, 2013–2026 (2014)
11. Cohen, D., Murray, J.M.: A generalized diffusion model for growth and dispersion in a population. J. Math. Biol. **12**, 237–248 (1981)
12. Dolcetta, I.C., Vita, S.F.: Area-preserving curve-shortening flows: from phase separation to image processing. Interfaces Free Bound. **4**, 325–343 (2002)
13. Elliott, C.M.: The Cahn-Hilliard model for the kinetics of phase separation. In: Mathematical Models for Phase Change Problems, Rodrigues, J.F. (ed.), International Series of Numerical Mathematics, vol. 88. Birkhäuser, Basel (1989)
14. Frigeri, S., Grasselli, M.: Nonlocal Cahn-Hilliard-Navier-Stokes systems with singular potentials. Dyn. PDE **9**, 273–304 (2012)
15. Khain, E., Sander, L.M.: A generalized Cahn-Hilliard equation for biological applications. Phys. Rev. E **77**, 051129 (2008)
16. Klapper, I., Dockery, J.: Role of cohesion in the material description of biofilms. Phys. Rev. E **74**, 0319021 (2006)

17. Kohn, R.V., Otto, F.: Upper bounds for coarsening rates. Commun. Math. Phys. **229**, 375–395 (2002)
18. Langer, J.S.: Theory of spinodal decomposition in alloys. Ann. Phys. **65**, 53–86 (1975)
19. Liu, Q.-X., Doelman, A., Rottschäfer, V., de Jager, M., Herman, P.M.J., Rietkerk, M., van de Koppel, J.: Phase separation explains a new class of self-organized spatial patterns in ecological systems. In: Proceedings of the National Academy of Sciences. http://www.pnas.org/cgi/doi/10.1073/pnas.1222339110 (2013)
20. Maier-Paape, S., Wanner, T.: Spinodal decomposition for the Cahn-Hilliard equation in higher dimensions. Part I: Probability and wavelength estimate. Commun. Math. Phys. **195**, 435–464 (1998)
21. Maier-Paape, S., Wanner, T.: Spinodal decomposition for the Cahn-Hilliard equation in higher dimensions: nonlinear dynamics. Arch. Ration. Mech. Anal. **151**, 187–219 (2000)
22. Miranville, A.: Asymptotic behavior of the Cahn-Hilliard-Oono equation. J. Appl. Anal. Comput. **1**, 523–536 (2011)
23. Miranville, A.: Asymptotic behavior of a generalized Cahn-Hilliard equation with a proliferation term. Appl. Anal. **92**, 1308–1321 (2013)
24. Miranville, A., Zelik, S.: Robust exponential attractors for Cahn-Hilliard type equations with singular potentials. Math. Methods Appl. Sci. **27**, 545–582 (2004)
25. Miranville, A., Zelik, S.: Attractors for dissipative partial differential equations in bounded and unbounded domains. In: Dafermos, C.M., Pokorny, M. (eds.) Handbook of Differential Equations, Evolutionary Partial Differential Equations, vol. 4, pp. 103–200. Elsevier, Amsterdam (2008)
26. Miranville, A., Zelik, S.: The Cahn-Hilliard equation with singular potentials and dynamic boundary conditions. Discret. Contin. Dyn. Syst. **28**, 275–310 (2010)
27. Novick-Cohen, A.: The Cahn-Hilliard equation: mathematical and modeling perspectives. Adv. Math. Sci. Appl. **8**, 965–985 (1998)
28. Novick-Cohen, A.: The Cahn-Hilliard equation. In: Dafermos, C.M., Pokorny, M. (eds.) Handbook of Differential Equations, Evolutionary Partial Differential Equations, vol. 4, pp. 201–228. Elsevier, Amsterdam (2008)
29. Oono, Y., Puri, S.: Computationally efficient modeling of ordering of quenched phases. Phys. Rev. Lett. **58**, 836–839 (1987)
30. Oron, A., Davis, S.H., Bankoff, S.G.: Long-scale evolution of thin liquid films. Rev. Mod. Phys. **69**, 931–980 (1997)
31. Pierre, M.: Habilitation thesis, Université de Poitiers (1997)
32. Temam, R.: Infinite-Dimensional Dynamical Systems in Mechanics and Physics, vol. 68. Springer, New York (1997)
33. Thiele, U., Knobloch, E.: Thin liquid films on a slightly inclined heated plate. Phys. D **190**, 213–248 (2004)
34. Tremaine, S.: On the origin of irregular structure in Saturn's rings. Astron. J. **125**, 894–901 (2003)

Chapter 9
Attractors for Multivalued Processes with Weak Continuity Properties

Piotr Kalita and Grzegorz Łukaszewicz

Abstract A method is proposed to deal with some multivalued processes with weak continuity properties. An application to a nonautonomous contact problem for the Navier–Stokes flow with nonmonotone multivalued frictional boundary condition is presented.

9.1 Introduction

Various notions have been proposed in recent years to study the asymptotic behavior of nonautonomous infinite dimensional dynamical systems governed by PDEs. These notions include uniform attractors, pullback attractors, and skew-product flows [2, 7, 13–16, 28]. In this article, we focus on the idea of pullback attractors; in particular, we use the theory of pullback $\mathscr{D}$-attractors, which allows to consider the general case of unbounded in the past forcing term present in the underlying PDE. We prove an abstract theorem on the existence of a pullback $\mathscr{D}$-attractor and then use it to study the problem in fluid dynamics with multivalued frictional contact conditions which yield the solution nonuniqueness.

We are interested in the evolution problems for which the solutions for a given initial conditions are not unique. To deal with such problems, the theory of m-semiflows and their global attractors was introduced in [19, 20] and later extended to nonautonomous case [6]. Results on the existence of pullback $\mathscr{D}$-attractors for the problems without uniqueness of solutions were obtained in [1, 5, 18]. The abstract result on the existence of a pullback $\mathscr{D}$-attractor shown in the present paper is based on these works; however, we show some extension to the results presented there:

P. Kalita (✉)
Faculty of Mathematics and Computer Science, Jagiellonian University,
ul. Łojasiewicza 6, 30-348 Krakow, Poland
e-mail: piotr.kalita@uj.edu.pl

G. Łukaszewicz
Faculty of Mathematics, Informatics and Mechanics, University of Warsaw,
ul. Banacha 2, 02-957 Warsaw, Poland
e-mail: glukasz@mimuw.edu.pl

© Springer International Publishing Switzerland 2015

V.A. Sadovnichiy and M.Z. Zgurovsky (eds.), *Continuous and Distributed Systems II*, Studies in Systems, Decision and Control 30,
DOI 10.1007/978-3-319-19075-4_9

our alternative proof of pullback $\mathscr{D}$-attractor existence is based on the notion of ω-$\mathscr{D}$-limit compactness and seems particularly transparent, we weaken the upper-semicontinuity assumption from [1, 5, 18] replacing it by the graph closedness, and we additionally study not only necessary but also the necessary and sufficient conditions for attractor existence.

Finally, we introduce the so-called condition (*NW*), earlier studied for autonomous case in [29] and recently in [10], and for nonautonomous case in [27] in context of uniform attractors. This condition, as it turns out in the second part of the present work, is easy and natural to verify for particular problems of mathematical physics governed by PDEs.

As an application of the abstract theory of attractors, we study a two-dimensional incompressible Navier–Stokes flow with a general form of nonmonotone frictional boundary conditions. Such conditions, considered for example in [4, 26] represent the frictional contact between the fluid and the wall, where the friction force depends in a nonmonotone and even discontinuous way on the slip rate. In this way, the present work is a follow-up of the works [12, 16]. Here we consider a much more general form of the frictional condition then in these works, but in contrast to [12, 16] we are able to obtain only the attractor existence, while, for example the question of its fractal dimension for the case without uniqueness remains an open problem. As we exemplify, for the studied problem the requirements for the abstract theorem on the attractor existence, and in particular the condition (*NW*), are natural to check.

In Sect. 9.2 we present abstract results on the existence of a pullback $\mathscr{D}$-attractor while in Sect. 9.3 we present a problem coming from contact mechanics, for which we show the existence of a pullback $\mathscr{D}$-attractor by means of proposed abstract framework.

9.2 Abstract Theory of Pullback $\mathscr{D}$-Attractors for Multivalued Processes

In this section, we recall basic definitions of the theory of pullback $\mathscr{D}$-attractors for multivalued processes, prove a theorem which gives some useful criteria of existence of pullback $\mathscr{D}$-attractors for such processes, and introduce the notion of "condition (NW)".

Let (H, ρ) be a complete metric space, and $P(H)$ be the family of all nonempty subsets of H. By $\text{dist}_H(A, B)$, we denote the Hausdorff semidistance between the sets $A, B \subset H$ defined as

$$\text{dist}_H(A, B) = \sup_{x \in A} \inf_{y \in B} \rho(x, y).$$

If the set $A \subset H$ is bounded, then we define its Kuratowski measure of noncompactness $\kappa(A)$ as

$$\kappa(A) = \{\inf \delta > 0 \mid A \text{ has finite open cover of sets of diameter less then } \delta\}.$$

For the properties of κ, see for example [8, 10]. We denote $\mathbb{R}_d = \{(t, \tau) \in \mathbb{R}^2 : t \geq \tau\}$. Subsets of $\mathbb{R} \times H$ will be called *nonautonomous sets*. For a nonautonomous set $\mathbb{D} \subset \mathbb{R} \times H$, we identify $\mathbb{D}$ with the family of its fibers, i.e.,

$$\mathbb{D} = \{D(t) \mid t \in \mathbb{R}\}, \quad \text{where} \quad x \in D(t) \Leftrightarrow (x, t) \in \mathbb{D}.$$

A family $\mathscr{D}$ of nonautonomous sets such that for every $\mathbb{D} \in \mathscr{D}$ and all $t \in \mathbb{R}$ the fiber $D(t)$ is nonempty (i.e., $D(t) \in P(H)$) is called an *attraction universe* over H.

Definition 9.1 The map $U : \mathbb{R}_d \times H \to P(H)$ is called a multivalued process (m-process) if:

(1) $U(\tau, \tau, z) = z$ for all $z \in H$.
(2) $U(t, \tau, z) \subset U(t, s, U(s, \tau, z))$ for all $z \in H$ and all $t \geq s \geq \tau$.

If in (2), in place of inclusion we have the equality $U(t, \tau, z) = U(t, s, U(s, \tau, z))$, then the m-process is said to be *strict*.

Definition 9.2 Let $\mathscr{D}$ be an attraction universe over H. The m-process U in H is pullback ω-$\mathscr{D}$-limit compact if for every $\mathbb{D} = \{D(\tau) \mid \tau \in \mathbb{R}\} \in \mathscr{D}$ and $t \in \mathbb{R}$ we have

$$\kappa\left(\bigcup_{\tau \leq s} U(t, \tau, D(\tau))\right) \to 0 \quad \text{as} \quad s \to -\infty.$$

Definition 9.3 Let $\mathscr{D}$ be an attraction universe over H. The m-process U in H is pullback $\mathscr{D}$-asymptotically compact if for any $t \in \mathbb{R}$, every $\mathbb{D} \in \mathscr{D}$, and all sequences $\tau_n \to -\infty$ and $\xi_n \in U(t, \tau_n, D(\tau_n))$, there exists a subsequence $\{\xi_{n_k}\}$ such that $\xi_{n_k} \to \xi$ in H for some $\xi \in H$.

Lemma 9.1 *Let $\mathscr{D}$ be an attraction universe over H. The m-process U in H is ω-$\mathscr{D}$-limit compact if and only if it is pullback $\mathscr{D}$-asymptotically compact.*

Proof The proof is similar to that for multivalued semiflows provided in [10]. □

Definition 9.4 Let $\mathscr{D}$ be an attraction universe over H. We say that an m-process U has a pullback $\mathscr{D}$-absorbing nonautonomous set $\mathbb{B} = \{B(t) \mid t \in \mathbb{R}\} \in \mathscr{D}$ if for every $t \in \mathbb{R}$ and $\mathbb{D} \in \mathscr{D}$ there exists $\tau_0 \leq t$ depending on t and $\mathbb{D}$ such that for $\tau \leq \tau_0$,

$$U(t, \tau, D(\tau)) \subset B(t).$$

Definition 9.5 Let $\mathscr{D}$ be an attraction universe over H and let U be an m-process on H. The nonautonomous set $\mathbb{A} = \{A(t) \mid t \in \mathbb{R}\} \subset \mathbb{R} \times H$ is called a pullback $\mathscr{D}$-attractor for U if:

(1) $A(t)$ is compact in H and nonempty for every $t \in \mathbb{R}$.

(2) $A(t) \subset U(t,\tau,A(\tau))$ for every $t \in \mathbb{R}$ and $\tau \le t$ ($\mathbb{A}$ is negatively semi-invariant).
(3) $\mathbb{A}$ is pullback $\mathscr{D}$-attracting, that is for every $t \in \mathbb{R}$ and $\mathbb{D} \in \mathscr{D}$,

$$\mathrm{dist}_H(U(t,\tau,D(\tau)),A(t)) \to 0 \quad \text{as} \quad \tau \to -\infty.$$

(4) If $\mathbb{C}$ is a nonautonomous set such that $C(t)$ are closed for $t \in \mathbb{R}$ and $\mathbb{C}$ is pullback $\mathscr{D}$-attracting then $A(t) \subset C(t)$ for every $t \in \mathbb{R}$ (minimality property).

Remark 9.1 Condition (4) in the above definition implies that pullback $\mathscr{D}$-attractor is always defined uniquely.

Remark 9.2 If we assume that, by definition, $\mathbb{A} \in \mathscr{D}$ then condition (4) (minimality) follows from conditions (1)–(3). In such a case, by Remark 9.1, conditions (1)–(3) suffice to guarantee the uniqueness of $\mathbb{A}$.

The minimality property can be proved as follows. Suppose that there exists a nonautonomous set $\mathbb{C}$ such that $C(t)$ are closed for all $t \in \mathbb{R}$, $\mathbb{C}$ is attracting, and for some $t \in \mathbb{R}$, $C(t) \subset A(t)$ with proper inclusion. As $\mathbb{A} \in \mathscr{D}$, we have $\mathrm{dist}_H(U(t,\tau,A(\tau)),C(t)) \to 0$ for $\tau \to -\infty$. But $A(t) \subset U(t,\tau,A(\tau))$, and we have $\mathrm{dist}_H(A(t),C(t)) \le \mathrm{dist}_H(U(t,\tau,A(t)),C(t)) \to 0$ as $\tau \to -\infty$, whence $A(t) \subset C(t)$, a contradiction.

Definition 9.6 The m-process U on H is *closed* if for all $\tau \le t$ the graph of the multivalued mapping $x \to U(t,\tau,x)$ is a closed set in the product topology on $H \times H$.

Now we formulate a theorem about the existence of pullback $\mathscr{D}$-attractors in complete metric spaces. A similar result is proved in [1, 5, 18] (see also [19, 20] for the case of m-semiflows and [6, 28] for the case the universe of bounded and constant in time sets is considered).

Theorem 9.1 *Let H be a complete metric space and let the m-process U on H be closed. Assume that*

(i) U has an pullback $\mathscr{D}$-absorbing nonautonomous set $\mathbb{B} \in \mathscr{D}$.
(ii) U is pullback ω-$\mathscr{D}$-limit compact (equivalently, U is pullback $\mathscr{D}$-asymptotically compact).

Then there exists a pullback $\mathscr{D}$-attractor for U.

Proof The proof is similar to that of Theorem 2.1 in [10].
Step 1. We define a candidate set for a pullback $\mathscr{D}$-attractor by setting

$$A(t) = \bigcap_{s \le t} \overline{\bigcup_{\tau \le s} U(t,\tau,B(\tau))}, \tag{9.2.1}$$

where $\mathbb{B} = \{B(\tau) \mid t \in \mathbb{R}\}$ is a nonautonomous pullback $\mathscr{D}$-absorbing set.
Step 2. $A(t)$ is nonempty and compact. We have, by ω-$\mathscr{D}$-limit compactness of U,

$$\kappa\left(\overline{\bigcup_{\tau\le s} U(t,\tau,B(\tau))}\right) \to 0 \quad \text{as} \quad s \to -\infty. \tag{9.2.2}$$

Since the sets $X_s = \overline{\bigcup_{\tau\le s} U(t,\tau,B(\tau))}$ are nonempty, bounded and closed in H, and the family of sets $\{X_s \mid s \le t\}$ is not increasing as $s \to -\infty$, we can apply a well known property of κ (cf. eg. [8]) to get the claims.

Step 3. Attraction property. Since for every $t \in \mathbb{R}$, every nonautonomous set $\mathbb{D} \in \mathscr{D}$, and every $r \le t$ it is $U(t,\tau,D(\tau)) \subset U(t,r,U(r,\tau,D(\tau))) \subset U(t,r,B(r))$ for all $\tau \le \tau_0$, for some $\tau_0(\mathbb{D}, r) \le r$, it suffices to prove that $\text{dist}_H(U(t,\tau,B(\tau)), A(t)) \to 0$ as $\tau \to -\infty$. Assume, to the contrary, that there exist $\tau_n \to -\infty$ and $\xi_n \in U(t,\tau_n,B(\tau_n))$ such that $\inf_{y\in A(t)} \rho(\xi_n, y) \ge \varepsilon > 0$. By the pullback $\mathscr{D}$-asymptotic compactness property, there exists a convergent subsequence, $\xi_\mu \to \xi$ in H, and by the definition of the pullback $\mathscr{D}$-attractor in (9.2.1), $\xi \in A(t)$, which gives a contradiction.

Step 4. We prove that $A(t) \subset U(t,\tau,A(\tau))$ for all $\tau \le t$. Let $x \in A(t)$ and $\tau \le t$. We shall prove that $x \in U(t,\tau,p)$ for some $p \in A(\tau)$. Since $x \in A(t)$ defined in (9.2.1), there exist $t_n \to -\infty$ and $\xi_n \in U(t,t_n,B(t_n))$ such that $\xi_n \to x$ in H. We have,

$$\xi_n \in U(t,t_n,B(t_n)) \subset U(t,\tau,U(\tau,t_n,B(t_n)))$$

for large n, whence there exist $z_n \in B(t_n)$ such that $\xi_n \in U(t,\tau,U(\tau,t_n,z_n))$, and $p_n \in U(\tau,t_n,z_n)$ such that $\xi_n \in U(t,\tau,p_n)$. By the pullback $\mathscr{D}$-asymptotic compactness it follows that for a subsequence of $\{p_n\}$ we have $p_\mu \to p$ in H. From (9.2.1) it follows that $p \in A(\tau)$. Since $\xi_\mu \to x$ and $p_\mu \to p$ with $\xi_\mu \in U(t,\tau,p_\mu)$ from the fact that U is closed it follows that $x \in U(t,\tau,p)$. The proof is complete.

Step 5. To prove the minimality property, assume that there exists a pullback $\mathscr{D}$-attracting nonautonomous set $\mathbb{C}$ such that $C(t)$ are closed. Then due to the fact that $\mathbb{B} \in \mathscr{D}$ we have $\text{dist}_H(U(t,\tau,B(\tau)), C(t)) \to 0$ as $\tau \to -\infty$. Let $x \in A(t)$. By (9.2.1) there exist sequences $\tau_n \to -\infty$ and $\xi_n \in U(t,\tau_n,B(\tau_n))$ such that $\xi_n \to x$, whereas, as $C(t)$ is closed, we have $x \in C(t)$ and the proof is complete. □

Remark 9.3 The above theorem still holds if we assume that, in the definition of pullback $\mathscr{D}$-asymptotic compactness, any set $\mathbb{D} \in \mathscr{D}$ is replaced with the pullback $\mathscr{D}$-absorbing set $\mathbb{B}$. In such a case, this set does not have to belong to the universe $\mathscr{D}$ (see [1, 5, 18]).

Remark 9.4 The existence of the pullback $\mathscr{D}$-attractor implies condition (ii) in the above theorem, that is the ω-$\mathscr{D}$-limit compactness of U. To prove it we follow the argument provided e.g. in [29]. Denote by $\mathscr{O}_\varepsilon(A(t))$ the closed ε-neighborhood of $A(t)$. Since the set $A(t)$ is compact, then $\kappa(A(t)) = 0$, and $\kappa(\mathscr{O}_\varepsilon(A(t))) \le 2\varepsilon$. From the attraction property of $\mathbb{A}$, for any $\mathbb{D} \in \mathscr{D}$ we have $\text{dist}_H(U(t,\tau,D(\tau)), A(t)) \le 2\varepsilon$ for all $\tau \le \tau_0$ for some $\tau_0(\mathbb{D}, t)$. Thus

$$\kappa\left(\bigcup_{\tau\le\tau_0} U(t,\tau,D(\tau))\right) \le 2\varepsilon,$$

which implies (ii).

Remark 9.5 In general, it is not true that the existence of a pullback $\mathscr{D}$-attractor for U implies (i). From the existence of a pullback $\mathscr{D}$-attractor $\mathbb{A}$ for U, it does not follow the existence of a pullback $\mathscr{D}$-absorbing nonautonomous set in $\mathscr{D}$.

Indeed, consider the equation $u'(t) = 0$ in $\mathbb{R}$. Then $U(t, \tau) = S(t - \tau) = id$. Let the universe $\mathscr{D}$ consist of only one nonautonomous set $\mathbb{D}$, given by $D(\tau) = \{\sin \tau\}$. Then the attractor will be given as $A(t) = A = [-1, 1]$ for $t \in \mathbb{R}$. There is no choice for a family of $\mathscr{D}$-absorbing set but $B(\tau) = D(\tau)$. But $U(t, \tau, D(\tau)) = D(\tau) \neq B(t)$ as $\sin \tau \neq \sin t$. Thus, in this case there does not exist a pullback $\mathscr{D}$-absorbing nonautonomous set in $\mathscr{D}$.

It is natural to ask what assumptions on the universe $\mathscr{D}$ are required to guarantee that existence of pullback $\mathscr{D}$-attractor implies (i). We make the following assumptions.
$(H\mathscr{D}1)$ The universe $\mathscr{D}$ is *inclusion closed*, that is if $\mathbb{D} = \{D(t) \mid t \in \mathbb{R}\} \in \mathscr{D}$ and $\mathbb{D}' = \{D'(t) \mid t \in \mathbb{R}\}$ is such that $D'(t) \subset \overline{D(t)}$ and $D'(t) \in P(H)$ for all $t \in \mathbb{R}$, then $\mathbb{D}' \in \mathscr{D}$.
$(H\mathscr{D}2)$ If $\mathbb{D} = \{D(t) \mid t \in \mathbb{R}\} \in \mathscr{D}$, then for some $\varepsilon > 0$ the nonautonomous set $\{\mathscr{O}_\varepsilon(D(t)) \mid t \in \mathbb{R}\}$ also belongs to $\mathscr{D}$.

Corollary 9.1 *Let the universe $\mathscr{D}$ satisfy $(H\mathscr{D}1)$ and let the m-process U satisfy the assumptions of Theorem 9.1. Then we have $\mathbb{A} \in \mathscr{D}$. If we, moreover, assume that U is strict then $\mathbb{A}$ is invariant, i.e., $U(t, \tau, A(\tau)) = A(t)$ for all $t \in \mathbb{R}$ and $\tau \leq t$.*

Proof The proof follows the lines of the proofs of 6. in Theorem 3 in [18] and (5) (6) in Theorem 3.10 in [1]. Since $\mathbb{B}$ is pullback $\mathscr{D}$-absorbing we have $\mathrm{dist}_H(U(t, \tau, D(\tau)), \overline{B(t)}) = 0$ for all $\mathbb{D} \in \mathscr{D}$ and $\tau \leq t_0(\mathbb{D}, t)$. Hence as $\tau \to -\infty$, we have $\mathrm{dist}_H(U(t, \tau, D(\tau)), \overline{B(t)}) \to 0$ and $\{\overline{B(t)} \mid t \in \mathbb{R}\}$ is pullback $\mathscr{D}$-attracting. Then from the attractor minimality we have $A(t) \subset \overline{B(t)}$ and the assertion follows from $(H\mathscr{D}1)$.
For the proof of positive invariance let $t \geq \tau$ be fixed and let $r \leq 0$. We have

$$U(t, \tau, A(\tau)) \subset U(t, \tau, U(\tau, \tau + r, A(\tau + r))) = U(t, \tau + r, A(\tau + r)).$$

Since $\mathbb{A} \in \mathscr{D}$, it must pullback attract itself, and

$$\lim_{r \to -\infty} \mathrm{dist}_H(U(t, \tau + r, A(\tau + r)), A(t)) = 0.$$

Hence

$$\lim_{r \to -\infty} \mathrm{dist}_H(U(t, \tau, A(\tau)), A(t)) = 0,$$

and $\mathrm{dist}_H(U(t, \tau, A(\tau)), A(t)) = 0$, whence, by closedness of $A(t)$, $U(t, \tau, A(\tau)) \subset A(t)$ and the assertion follows.

□

Corollary 9.2 *Let the universe $\mathscr{D}$ satisfy $(H\mathscr{D}2)$ and let the m-process U have the pullback $\mathscr{D}$-attractor $\mathbb{A} \in \mathscr{D}$. Then U has a pullback $\mathscr{D}$-absorbing nonautonomous set $\mathbb{B} \in \mathscr{D}$.*

Proof Define $B(t) = \mathscr{O}_\varepsilon(A(t))$, where ε is given in $(H\mathscr{D}2)$. Then $\mathbb{B} = \{B(t)\,|\,t \in \mathbb{R}\} \in \mathscr{D}$. Suppose, for contradiction, that $B(t)$ is not pullback $\mathscr{D}$-absorbing. This means that there exist sequences $t_n \to -\infty$ and $\xi_n \in U(t, t_n, D(t_n))$ such that $\xi_n \notin B(t)$ for large n. From Remark 9.4 it follows that U is pullback $\mathscr{D}$-asymptotically compact, and, for a subsequence, $\xi_\mu \to \xi$. As $\mathbb{A}$ is pullback $\mathscr{D}$-attracting and $A(t)$ is closed, we have $\xi \in A(t)$, a contradiction with $\xi_n \notin \mathscr{O}_\varepsilon(A(t))$. □

Summarizing, we have a following theorem which is a consequence of Theorem 9.1, Remark 9.4, and Corollaries 9.1 and 9.2.

Theorem 9.2 *Let $\mathscr{D}$ be a universe of nonautonomous sets satisfying $(H\mathscr{D}1)$ and $(H\mathscr{D}2)$ and let U be a closed m-process. Then U has a pullback $\mathscr{D}$-attractor $\mathbb{A} \in \mathscr{D}$ if and only if U is pullback $\mathscr{D}$-asymptotically compact and has a pullback $\mathscr{D}$-absorbing nonautonomous set $\mathbb{B} \in \mathscr{D}$.*

In the sequel of this section, we confine ourselves to Banach spaces where we will consider both strong and weak topologies. We introduce condition (*NW*), "norm-to-weak", that generalizes to the nonautonomous multivalued case the norm-to-weak continuity assumed in [29] for semigroups (see Definition 3.4 in [29]) and in [10] for multivalued semiflows (see Definition 2.11 in [10]). A similar condition is introduced for the nonautonomous multivalued case in [27] (see condition (3) in Definition 2.6 in [27]), where only the strict case is considered and, instead of a subsequence, whole sequence is assumed to converge weakly.

Definition 9.7 The m-process U on a Banach space H satisfies condition (*NW*) if for every $t \in \mathbb{R}$ and $\tau \leq t$, from $x_n \to x$ in H and $\xi_n \in U(t, \tau, x_n)$ it follows that there exists a subsequence $\{\xi_{k_n}\}$, such that $\xi_{k_n} \to \xi$ weakly in H with $\xi \in U(t, \tau, x)$.

Lemma 9.2 *If a multivalued process U on a Banach space H satisfies condition (NW) then it is closed.*

Proof An elementary proof follows directly from the definitions. □

From Theorem 9.1 together with Lemma 9.2 we have the following

Corollary 9.3 *If the m-process U on a Banach space H has a pullback $\mathscr{D}$-absorbing nonautonomous set $\mathbb{B} \in \mathscr{D}$, is pullback $\mathscr{D}$-asymptotically compact (equivalently, ω-$\mathscr{D}$-limit compact), and satisfies condition (NW) then there exists a pullback $\mathscr{D}$-attractor for U.*

The above corollary provides useful criteria of existence of pullback $\mathscr{D}$-attractors for multivalued processes. In the next section, we apply them to study the time asymptotic behavior of solutions of a problem originating from contact mechanics.

9.3 Application

Problem formulation. The flow of an incompressible fluid in a two-dimensional domain Ω is described by the equation of motion

$$u'(t) - \eta\Delta u(t) + (u(t)\cdot\nabla)u(t) + \nabla p(t) = f(t) \quad \text{in} \quad \Omega\times(t_0,\infty) \tag{9.3.1}$$

and the incompressibility condition

$$\operatorname{div} u(t) = 0 \quad \text{in} \quad \Omega\times(t_0,\infty), \tag{9.3.2}$$

where the unknowns are the velocity $u : \Omega\times(t_0,\infty)\to\mathbb{R}^2$ and pressure $p : \Omega\times(t_0,\infty)\to\mathbb{R}$, $\eta>0$ is the viscosity coefficient and $f : \Omega\times(t_0,\infty)\to\mathbb{R}^2$ is the volume mass force density. To define the domain Ω of the flow, let Ω_∞ be the infinite channel

$$\Omega_\infty = \{x=(x_1,x_2)\in\mathbb{R}^2 \mid x_2\in(0,h(x_1))\},$$

where $h:\mathbb{R}\to\mathbb{R}$ is a positive function (we assume that $h(x_1)\ge h_0>0$ for all $x_1\in\mathbb{R}$), smooth, and L-periodic in x_1. Then we set

$$\Omega = \{x=(x_1,x_2)\in\mathbb{R}^2 \mid x_1\in(0,L),\quad x_2\in(0,h(x_1))\},$$

with the boundary $\partial\Omega = \overline{\Gamma}_D\cup\overline{\Gamma}_C\cup\overline{\Gamma}_L$, where $\Gamma_D=\{(x_1,h(x_1))\mid x_1\in(0,L)\}$, $\Gamma_C=(0,L)\times\{0\}$, and $\Gamma_L=\{0,L\}\times(0,h(0))$ and are the top, bottom and lateral parts of $\partial\Omega$, respectively. We will use the notation $e_1=(1,0)$ and $e_2=(0,1)$ for the canonic basis of $\mathbb{R}^2$. Note, that on Γ_C the outer normal unit vector is given by $\nu=-e_2$.

We are interested in the solutions of (9.3.1)–(9.3.2) such that $u(0,x_2,t) = u(L,x_2,t)$, $\frac{\partial u_2(0,x_2,t)}{\partial x_1} = \frac{\partial u_2(L,x_2,t)}{\partial x_1}$, and $p(0,x_2,t)=p(L,x_2,t)$ for $x_2\in[0,h(0)]$ and $t\in\mathbb{R}^+$. The first condition represents the L-periodicity of velocities, while the latter two ones, the L-periodicity of normal stresses in the space of divergence-free functions. Moreover, we assume that

$$u(t)=0 \quad \text{on} \quad \Gamma_D\times(t_0,\infty). \tag{9.3.3}$$

On the contact boundary Γ_C, we decompose the velocity into the normal component $u_\nu = u\cdot\nu$, where ν is the unit outward normal vector, and the tangential one $u_\tau = u\cdot e_1$. Note, that since the domain Ω is two dimensional it is possible to consider the tangential components as scalars, for the sake of the ease of notation. Likewise, we decompose the stress on the boundary Γ_C into its normal component $\sigma_\nu=\sigma\nu\cdot\nu$ and the tangential one $\sigma_\tau=\sigma\nu\cdot e_1$. The stress tensor is related to the velocity and pressure through the linear constitutive law $\sigma_{ij}=-p\delta_{ij}+\eta(u_{i,j}+u_{j,i})$.

We assume that there is no flux across Γ_C and hence we have

$$u_\nu(t) = 0 \text{ at } \Gamma_C \times (t_0, \infty), \tag{9.3.4}$$

and that the tangential component of the velocity u_τ on Γ_C is in the following relation with the tangential stresses σ_τ,

$$-\sigma_\tau(t) \in \partial j(t, u_\tau(t)) \quad \text{at} \quad \Gamma_C \times (t_0, \infty). \tag{9.3.5}$$

In above formula $j : \Gamma_C \times \mathbb{R} \times \mathbb{R} \to \mathbb{R}$ is a potential which is locally Lipschitz and not necessarily convex with respect to the last variable, and ∂ is the subdifferential in the sense of Clarke (see eg. [21]). Note that the function j is assumed to depend on both space and time variables, however, sometimes the dependence on the space variable $x \in \Gamma_C$ is skipped for the ease of notation. In the end, we have the initial condition

$$u(x, t_0) = u_0(x) \quad \text{in} \quad \Omega. \tag{9.3.6}$$

The relation (9.3.5) is the generalized version of the Tresca friction law. We assume that the friction force is related by the multivalued and nonmonotone law with the tangential velocity. Our setup is more general than in the previous works [12, 16]. In contract to [12], we do not assume any monotonicity-type conditions on the potential j and hence the solutions to the formulated problem can be nonunique here. Similar laws were used for example in [4, 26] for static problems in elasticity, see Fig. 4.6 in [4] and Fig. 2(a) in [26] for examples of particular nonmonotone friction laws that satisfy the assumptions $H(j)$ listed below and hence they can be used in our study. Complex nonmonotone behavior of the friction force density is related to the presence of asperities on the surface [23]. Similar friction law is also used for example in the study of the motion of tectonic plates, see [9, 22, 25] and the references therein.

Weak formulation and existence of solutions. We introduce the variational formulation of the problem, and, for the convenience of the readers, we describe the relations between the classical and the weak formulations.

We begin with some basic definitions. Let

$$\begin{aligned}\widetilde{V} = \{u \in C^\infty(\Omega)^2 \mid \operatorname{div} u = 0 \text{ in } \Omega, \quad & u \text{ is L-periodic in } x_1, \\ u = 0 \text{ at } \Gamma_D, \quad & u \cdot \nu = 0 \text{ at } \Gamma_C\}\end{aligned}$$

and

$$V = \text{closure of } \widetilde{V} \text{ in } H^1(\Omega)^2, \qquad H = \text{closure of } \widetilde{V} \text{ in } L^2(\Omega)^2.$$

We define scalar products in H and V, respectively, by

$$(u, w) = \int_\Omega u(x) \cdot w(x)\, dx \quad \text{and} \quad (\nabla u, \nabla w) = \int_\Omega \nabla u(x) : \nabla w(x)\, dx$$

and their associated norms by

$$\|u\|_H = (u, u)^{\frac{1}{2}} \quad \text{and} \quad \|u\| = (\nabla u, \nabla u)^{\frac{1}{2}}.$$

The duality in $V \times V^*$ is denoted as $\langle \cdot, \cdot \rangle$. Moreover, let, for u, w, and z in V

$$a(u, w) = (\nabla u, \nabla w) \quad \text{and} \quad b(u, z, w) = ((u \cdot \nabla) z, w).$$

We define a linear operator $A : V \to V^*$ as $\langle Au, w \rangle = a(u, w)$, and a nonlinear operator $\langle Bu, w \rangle = b(u, u, w)$. For $w \in V$, we have the Poincaré inequality $\lambda_1 \|w\|_H^2 \leq \|w\|^2$, where λ_1 is the first eigenvalue of the Stokes operator A in V.

The linear and continuous trace mapping leading from V to $L^2(\Gamma_C)^2$ is denoted by γ, we will use the same symbol to denote the Nemytskii trace operator on the spaces of time dependent functions. The norm of trace operator will be denoted by $\|\gamma\| \equiv \|\gamma\|_{\mathscr{L}(V; L^2(\Gamma_C)^2)}$.

If $u \in L^2(\Gamma_C)$, we will denote by $S^2_{\partial j(t,u)}$ the set of all L^2 selections of $\partial j(t, u)$, i.e., all functions $\xi \in L^2(\Gamma_C)$ such that $\xi(x) \in \partial j(x, t, u(x))$ for a.e. $x \in \Gamma_C$.

The variational formulation of the problem is as follows.

Problem (P) Given $u_0 \in H$, find $u : (t_0, \infty) \to H$ such that for all $T > t_0$,

$$u \in C([t_0, T]; H) \cap L^2(t_0, T; V), \qquad \textit{with} \quad u' \in L^2(t_0, T; V^*)$$

and

$$\langle u'(t) + \eta A u(t) + Bu(t), z \rangle + (\xi(t), z_\tau)_{L^2(\Gamma_C)} = \langle f(t), z \rangle, \tag{9.3.7}$$
$$-\xi(t) \in S^2_{\partial j(t, u_\tau(t))},$$

for a.e. $t \geq t_0$ and for all $z \in V$.

The assumptions on problem data are the following.

H_0: $u_0 \in H$, $f \in L^2_{loc}(\mathbb{R}; H)$,
$H(j)$: $j : \Gamma_C \times \mathbb{R} \times \mathbb{R} \to \mathbb{R}$ is a function such that

(i) $j(x, t, \cdot)$ is locally Lipschitz for a.e. $(x, t) \in \Gamma_C \times \mathbb{R}$,
(ii) $j(\cdot, \cdot, s)$ is measurable for all $s \in \mathbb{R}$,
(iii) ∂j satisfies the growth condition $|\xi| \leq a + b|s|$ for all $s \in \mathbb{R}$ and $\xi \in \partial j(x, t, s)$ with $a, b > 0$ a.e. $(x, t) \in \Gamma_C \times \mathbb{R}$,
(iv) ∂j satisfies the dissipativity condition $\xi s \geq c - d|s|^2$ for all $s \in \mathbb{R}$ and $\xi \in \partial j(x, t, s)$ with $c \in \mathbb{R}$ and $d \in \left(0, \frac{\eta}{\|\gamma\|^2}\right)$ a.e. $(x, t) \in \Gamma_C \times \mathbb{R}$.

$H(f)$: for certain $t \in \mathbb{R}$ (and thus for all $t \in \mathbb{R}$) we have

$$\int_{-\infty}^{t} e^{(\eta - d\|\gamma\|^2)\lambda_1 s} \|f(s)\|_H^2 \, ds < \infty.$$

Note that by $H(j)(iii)$ it follows that for $u \in L^2(\Gamma_C)$ the set $S^2_{\partial j(t,u)}$ is in fact the set of all measurable selections of $\partial j(t, u)$.

We have the following relations between classical and weak formulations.

Proposition 9.1 *Every classical solution of* (9.3.1)–(9.3.6) *is also a solution of Problem* ***(P)****. On the other hand, every solution of Problem* ***(P)*** *which is smooth enough is also a classical solution of* (9.3.1)–(9.3.6).

Proof Let u be a classical solution of (9.3.1)–(9.3.6). As it is (by assumption) sufficiently regular, we have to check only (9.3.7). Remark first that (9.3.1) can be written as

$$u'(t) - \operatorname{Div} \sigma(u(t), p(t)) + (u(t) \cdot \nabla)u(t) = f(t). \tag{9.3.8}$$

Let $z \in V$. Multiplying (9.3.8) by z, integrating by parts and using the Green formula, we obtain

$$\begin{aligned} \int_\Omega u'(t) \cdot z \, dx + \int_\Omega \sigma_{ij}(u(t), p(t)) z_{i,j} \, dx + b(u(t), u(t), z) \\ = \int_{\partial\Omega} \sigma_{ij}(u(t), p(t)) \nu_j z_i \, d\Gamma + \int_\Omega f(t) \cdot z \, dx \end{aligned} \tag{9.3.9}$$

for $t \in (t_0, \infty)$. As $u(t)$ and z are in V, after some calculations we obtain

$$\int_\Omega \sigma_{ij}(u(t), p(t)) z_{i,j} dx = \eta a(u(t), z). \tag{9.3.10}$$

Taking into account the boundary conditions, we get

$$\int_{\partial\Omega} \sigma_{ij}(u(t), p(t)) \nu_j z_i \, d\Gamma = \int_{\Gamma_C} \sigma_\tau(u(t), p(t)) z_\tau \, d\Gamma + \int_{\Gamma_C} \sigma_\nu(u(t), p(t)) z_\nu \, d\Gamma$$

and the last integral equals zero as z satisfies (9.3.4) a.e. on Γ_C. Thus, (9.3.7) holds. Conversely, suppose that u is a sufficiently smooth solution to Problem **(P)**. We have immediately (9.3.2)–(9.3.4) and (9.3.6).

Now, let z be in the space $(H^1_0(div, \Omega))^2 = \{z \in V \mid z = 0 \text{ on } \partial\Omega\}$. We take such z in (9.3.7) to get

$$\langle u'(t) - \eta\Delta u(t) + (u(t) \cdot \nabla)u(t) - f(t), z\rangle = 0 \qquad \forall z \in (H^1_0(div, \Omega))^2.$$

Thus, there exists a distribution $p(t)$ on Ω such that

$$u'(t) - \eta\Delta u(t) + (u(t) \cdot \nabla)u(t) - f(t) = \nabla p(t) \quad \text{in } \Omega \times (t_0, \infty), \tag{9.3.11}$$

so that (9.3.1) holds.

Now, we shall derive the subdifferential boundary condition (9.3.5) from the weak formulation. Take $z \in V$. We have

$$\int_{\Omega} \sigma_{ij}(u(t), p(t)) z_{i,j} dx = -\int_{\Omega} \operatorname{Div} \sigma(u(t), p(t)) \cdot z dx + \int_{\partial\Omega} \sigma \nu \cdot z \, d\Gamma. \tag{9.3.12}$$

Since $z \in V$ we have $z_\nu = 0$ a.e. on Γ_C and hence $\sigma \nu \cdot z = \sigma_\tau z_\tau$. Applying (9.3.10) and (9.3.12) to (9.3.7) we get

$$\int_{\Omega} (u'(t) - \operatorname{Div} \sigma(u(t), p(t)) + (u(t) \cdot \nabla) u(t) - f(t)) \cdot z \, dx$$
$$+ (\xi(t), z_\tau)_{L^2(\Gamma_C)} + \int_{\partial\Omega} \sigma_\tau z_\tau \, d\Gamma = 0.$$

By (9.3.11) we have (9.3.8) and so the first integral on the left-hand side vanishes. Thus we obtain the subdifferential boundary condition. □

Theorem 9.3 *Assuming H_0, and $H(j)(i) - (iii)$, Problem* **(P)** *has a solution.*

The proof uses the standard approach by the mollification of the nonsmooth term and the Galerkin method. Since it is, on one hand, technical and quite involved, and, on the other hand the methodology is the same as in the proof of Theorem 3.1 in [11] and Theorem 12.11 in [12], with the necessary modifications for the nonautonomous terms, we skip the proof here.

Multivalued process and its attractor. We associate with Problem **(P)** the multifunction $U : \mathbb{R}_d \times H \to P(H)$, where $U(t, t_0, u_0)$ is the set of states attainable at time t from the initial condition u_0 taken at time t_0. Observe that U is a strict m-process.

Lemma 9.3 *Assume H_0, $H(j)(i) - (iv)$. If $u \in L^2_{loc}(t_0, +\infty; V)$ with $u' \in L^2_{loc}(t_0, +\infty; V^\star)$ solves Problem* **(P)***, then*

$$\frac{d}{dt}\|u(t)\|_H^2 + (\eta - d\|\gamma\|^2)\|u(t)\|^2 \le C_1(1 + \|f(t)\|_H^2), \tag{9.3.13}$$
$$\|u'(t)\|_{V^*} \le C_2(1 + \|f(t)\|_H) + C_3(1 + \|u(t)\|_H)\|u(t)\|, \tag{9.3.14}$$

for a.e. $t \in (t_0, +\infty)$ with $C_1, C_2, C_3 > 0$ independent on t_0, u_0, t.

Proof We take $z = u(t)$ in (9.3.7) and make use of

$$b(w, w, w) = 0 \quad \text{for} \quad w \in V. \tag{9.3.15}$$

to obtain

$$\frac{1}{2}\frac{d}{dt}\|u(t)\|_H^2 + \eta\|u(t)\|^2 + (\xi(t), u_\tau(t))_{L^2(\Gamma_C)} = (f(t), u(t))_H,$$

where $\xi(t) \in S^2_{\partial j(t,u_\tau(t))}$ for a.e. $t \geq t_0$. From $H(j)(iv)$ we obtain with arbitrary $\varepsilon > 0$

$$\frac{1}{2}\frac{d}{dt}\|u(t)\|_H^2 + \eta\|u(t)\|^2 \leq \frac{\varepsilon}{2}\|u(t)\|^2 + \frac{1}{2\varepsilon}\|f(t)\|_{V^*}^2 + d\|u_\tau(t)\|^2_{L^2(\Gamma_C)} - cL.$$

Taking $\varepsilon = \eta - d\|\gamma\|^2$, we obtain with a constant $C > 0$

$$\frac{1}{2}\frac{d}{dt}\|u(t)\|_H^2 + \frac{\eta - d\|\gamma\|^2}{2}\|u(t)\|^2 \leq C(1 + \|f(t)\|_{V^*}^2),$$

which proves (9.3.13), as $H \subset V^*$ is a continuous embedding.

To show (9.3.14) let us observe that for $z \in V$ and a.e. $t \in \mathbb{R}^+$, we have

$$\begin{aligned}\langle u'(t), z\rangle \leq \|A\|_{\mathscr{L}(V;V^*)}\|u(t)\|\|z\| + \|f(t)\|_{V^*}\|z\| \\ + \|\xi(t)\|_{L^2(\Gamma_C)}\|\gamma\|\|z\| + c\|u(t)\|_H\|u(t)\|\|z\|,\end{aligned}$$

with $\xi(t) \in S^2_{\partial j(u_\tau(t))}$, where we used that fact, that, in view of the Ladyzhenskaya inequality (cf. [16]), for $w, z \in V$ we have

$$|b(w, w, z)| \leq c\|w\|_H\|w\|\|z\|. \tag{9.3.16}$$

Now (9.3.14) follows directly from the growth condition $H(j)(iii)$ and trace inequality. □

Denote $\sigma = (\eta - d\|\gamma\|^2)\lambda_1$ and define $R(t)$ by

$$R(t)^2 := \frac{C_1}{\sigma} + C_1\int_{-\infty}^t e^{\sigma s}\|f(s)\|_H^2\, ds + 1 \tag{9.3.17}$$

for $t \in \mathbb{R}$. By $H(f)$ the quantity $R(t)$ is well defined and finite. We denote by $\mathscr{R}_\sigma$ the family of all functions $r : \mathbb{R} \to (0, \infty)$ such that

$$\lim_{t\to-\infty} e^{\sigma t} r^2(t) = 0,$$

and by $\mathscr{D}_\sigma$ we denote the family of all nonautonomous sets $\mathbb{D} = \{D(t) \mid t \in \mathbb{R}\}$ such that $\mathbb{D} \in \mathscr{D}_\sigma$ if and only if $D(t) \in P(H)$ for all $t \in \mathbb{R}$ and there exists $r_\mathbb{D} \in \mathscr{R}_\sigma$ such that for all $t \in \mathbb{R}$ and for all $w \in D(t)$ we have $\|w\|_H \leq r_\mathbb{D}(t)$.

Lemma 9.4 *Assume H_0, $H(j)(i) - (iv)$, and $H(f)$. The nonautonomous set $\mathbb{B} = \{B(t) \mid t \in \mathbb{R}\}$ defined as $B(t) = \{v \in H \mid \|v\|_H \leq R(t)\}$ is $\mathscr{D}_\sigma$-pullback absorbing and $\mathbb{B} \in \mathscr{R}_\delta$.*

Proof Obviously, $R \in \mathscr{R}_\sigma$ and hence $\mathbb{B} \in \mathscr{D}_\sigma$. Take $t \in \mathbb{R}$ and $\mathbb{D} \in \mathscr{D}_\sigma$. There exists $r_\mathbb{D} \in \mathscr{R}_\sigma$ with $\|w\|_H \leq r_\mathbb{D}(s)$ for all $w \in D(s)$ and $s \in \mathbb{R}$. We can choose $t_0 = t_0(\mathbb{D}, t)$ small enough such that $r_\mathbb{D}(\tau)^2 e^{\sigma\tau} \leq e^{\sigma t}$ for all $\tau \leq t_0$. We will consider a solution u for Problem **(P)** with the initial time $\tau \leq t_0$ and the initial condition $u_0 \in D(\tau)$. From (9.3.13) by the Poincaré inequality, we obtain

$$\frac{d}{dt}\|u(s)\|_H^2 + (\eta - d\|\gamma\|^2)\lambda_1\|u(s)\|_H^2 \leq C_1(1 + \|f(s)\|_H^2),$$

for a.e. $s \in (\tau, t)$. From the Gronwall Lemma, we get

$$\|u(t)\|_H^2 \leq \|u(\tau)\|_H^2 e^{-\sigma(t-\tau)} + \frac{C_1}{\sigma} + C_1 e^{-\sigma t}\int_\tau^t e^{\sigma s}\|f(s)\|_H^2\, ds. \qquad (9.3.18)$$

Hence

$$\|u(t)\|_H^2 \leq \|u(\tau)\|_H^2 e^{-\sigma(t-\tau)} + R^2(t) - 1.$$

Since $u(\tau) \in D(\tau)$, then

$$\|u(t)\|_H^2 \leq r_{\mathbb{D}}(\tau)^2 e^{\sigma\tau} e^{-\sigma t} - 1 + R^2(t).$$

Due to the bound $r_{\mathbb{D}}(\tau)^2 e^{\sigma\tau} \leq e^{\sigma t}$, we have

$$\|u(t)\|_H^2 \leq R^2(t),$$

whence the assertion follows. □

We pass to the next lemma which states that the m-process U is $\mathscr{D}_\sigma$-pullback asymptotically compact.

Lemma 9.5 *Assume H_0, $H(j)(i) - (iv)$, and $H(f)$. The m-process U is $\mathscr{D}_\sigma$-asymptotically compact.*

Proof The proof uses the technique of Proposition 7.4 in [3] (alternatively it is also possible to obtain the same assertion by the method of [24]). Let $\{D(t) \mid t \in \mathbb{R}\} = \mathbb{D} \in \mathscr{D}_\sigma$ and let $u_0^n \in D(t_0^n)$ with $t_0^n \to -\infty$. Moreover, $z_n \in U(t, t_0^n, u_0^n)$. We must show that the sequence $\{z_n\}$ is relatively compact in H. There exists a sequence of functions $u_n \in L^2(t_0^n, t+1; V) \cap C([t_0^n, t+1]; H)$ with $u_n' \in L^2(t_0^n, t+1; V^*)$, solutions of Problem **(P)**, such that $u_n(t_0^n) = u_0^n$ and $u_n(t) = z_n$. From Lemma 9.4 it follows that there exists N_0 such that for all natural $n \geq N_0$, we have $\|u_n(t-1)\|_H \leq R(t-1)$. From now on, we will consider the sequence $\{u_n\}_{n=N_0}^{\infty}$. The functions ξ corresponding to u_n will be denoted by $\xi_n(t) \in S^2_{\partial j(t, u_{n\tau}(t))}$. Note that the restrictions of u_n, ξ_n to the interval $[t-1, t+1]$ solve **(P)** on this interval, with the initial conditions $u_n(t-1)$ taken at $t-1$. Using (9.3.13) it follows that the sequence u_n is uniformly bounded in $L^2(t-1, t+1; V) \cap C([t-1, t+1]; H)$, and, by (9.3.14) it follows that u_n' is uniformly bounded in $L^2(t-1, t+1; V^*)$. The growth condition $H(j)(iii)$ implies that ξ_n are bounded in $L^2(t-1, t+1; L^2(\Gamma_C))$. These bounds are sufficient to extract a subsequence, not renumbered, such that for certain $u \in L^2(t-1, t+1; V)$ with $u' \in (t-1, t+1; V^*)$ and $\xi \in L^2(t-1, t+1; L^2(\Gamma_C))$ the following convergences hold

$$u_n \to u \text{ weakly in } L^2(t-1,t+1;\, V), \tag{9.3.19}$$

$$u_n' \to u' \text{ weakly in } L^2(t-1,t+1;\, V^*), \tag{9.3.20}$$

$$u_n \to u \text{ strongly in } L^2(t-1,t+1;\, H), \tag{9.3.21}$$

$$u_n(s) \to u(s) \text{ weakly in } H \text{ for all } s \in [t-1,t+1], \tag{9.3.22}$$

$$u_{n\tau} \to u_\tau \text{ strongly in } L^2(t-1,t+1;\, L^2(\Gamma_C)), \tag{9.3.23}$$

$$\xi_n \to \xi \text{ weakly in } L^2(t-1,t+1;\, L^2(\Gamma_C)),. \tag{9.3.24}$$

In view of (9.3.22) it is sufficient to show that $\|u_n(t)\|_H \to \|u(t)\|_H$, whence, as H is a Hilbert space, it follows that $u_n(t) \to u(t)$ strongly in H and the assertion will be proved. Note that by (9.3.21), for another subsequence, also not renumbered, the strong convergence $u_n(s) \to u(s)$ holds in H for a.e. $s \in (t-1, t+1)$. Taking the test function $u_n(t)$ in (9.3.7) written for u_n, where the corresponding ξ is denoted by ξ_n, and integrating from $t-1$ to $s \in [t-1, t+1]$, we get

$$\begin{aligned}
&\frac{1}{2}\|u_n(s)\|_H^2 + \eta \int_{t-1}^{s} a(u_n(r), u_n(r))\, dr + \int_{t-1}^{s} (\xi_n(r), u_{n\tau}(r))\, dr \\
=&\frac{1}{2}\|u_n(t-1)\|_H^2 + \int_{t-1}^{s} (f(r), u_n(r))_H\, dr.
\end{aligned} \tag{9.3.25}$$

We define the functions $V_n : [t-1, t+1] \to \mathbb{R}$ as

$$V_n(s) = \frac{1}{2}\|u_n(t-1)\|_H^2 - \eta \int_{t-1}^{s} a(u_n(r), u_n(r))\, dr.$$

Note that the coercivity of a implies that the functions V_n are nonincreasing. By the energy Eq. (9.3.25), we have

$$V_n(s) = \frac{1}{2}\|u_n(s)\|_H^2 + \int_{t-1}^{s} (\xi_n(r), u_{n\tau}(r))\, dr - \int_{t-1}^{s} (f(r), u_n(r))_H\, dr.$$

From (9.3.21), (9.3.23), and (9.3.24) it follows that

$$\lim_{n\to\infty} V_n(s) = \frac{1}{2}\|u(s)\|_H^2 + \int_{t-1}^{s} (\xi(r), u_\tau(r))\, dr - \int_{t-1}^{s} (f(r), u(r))_H\, dr := V(s),$$

for a.e. $s \in (t-1, t+1)$. We can choose sequences $p_k \nearrow t$ and $q_k \searrow t$ such that $V_n(p_k) \to V(p_k)$ and $V_n(q_k) \to V(q_k)$ as $n \to \infty$ for all $k \in \mathbb{N}$. We have

$$V_n(p_k) \geq V_n(t) \geq V_n(q_k),$$

and hence

$$V(p_k) = \lim_{n\to\infty} V_n(p_k) \geq \limsup_{n\to\infty} V_n(t) \geq \liminf_{n\to\infty} V_n(t) \geq \lim_{n\to\infty} V_n(q_k) = V(q_k).$$

We pass with k to infinity and use the fact that V, by definition, is continuous. We get

$$V(t) \geq \limsup_{n\to\infty} V_n(t) \geq \liminf_{n\to\infty} V_n(t) \geq V(t),$$

whence $V_n(t) \to V(t)$ as $n \to \infty$. But we have

$$\int_{t-1}^{t} (\xi_n(r), u_{n\tau}(r))\, dr - \int_{t-1}^{t} (f(r), u_n(r))_H\, dr \longrightarrow \int_{t-1}^{t} (\xi(r), u_\tau(r))\, dr - \int_{t-1}^{t} (f(r), u(r))_H\, dr,$$

which gives us the convergence $\|u_n(t)\|_H \to \|u(t)\|_H$, and the proof is complete. □

Lemma 9.6 *If for every integer n, u_n solves Problem **(P)** with initial condition u_n^0 taken at t_0, and $u_n^0 \to u_0$ strongly in H then for any $t > t_0$, for a subsequence, $u_n(t) \to u(t)$ weakly in H, where u is a solution of Problem **(P)** with initial condition u_0 taken at t_0. In consequence, the m-process U on H satisfies condition (NW).*

Proof The proof is standard since the a priori estimates of Lemma 9.3 provide enough convergence to pass to the limit. Indeed, the sequence u_n is bounded in $L^2(t_0, t; V)$ with u_n' bounded in $L^2(t_0, t; V^*)$. Hence, for a subsequence, $u_n \to u$ weakly in $L^2(t_0, t; V)$ with $u_n' \to u'$ weakly in $L^2(t_0, t; V^*)$. In consequence, for all $s \in [0, t]$ we have $u_n(s) \to u(s)$ weakly in H, which means that $u(t_0) = u_0$ and $u_n(t) \to u(t)$ weakly in H. We must show that u solves Problem **(P)** on (t_0, t). We only discuss passing to the limit in the multivalued term since for other terms it is standard. Let $\xi_n \in L^2(t_0, t; L^2(\Gamma_C))$ be such that $\xi_n(s) \in S^2_{\partial j(s, u_{n\tau}(s))}$ a.e. $s \in (t_0, t)$ and (9.3.7) holds. From the growth condition $H(j)(iii)$ it follows that ξ_n is bounded in $L^2(t_0, t; L^2(\Gamma_C))$ and thus, for a subsequence, we have $\xi_n \to \xi$ weakly in $L^2(t_0, t; L^2(\Gamma_C))$ and weakly in $L^2(\Gamma_C \times (t_0, t))$. To conclude the proof, we need to show that $\xi(s) \in S^2_{\partial j(s, u_\tau(s))}$ for a.e. $s \in (t_0, t)$. From the compactness of the Nemytskii trace operator it follows that $u_{n\tau} \to u_\tau$ strongly in $L^2(t_0, t; L^2(\Gamma_C))$ and, in consequence, for a subsequence, $u_{n\tau}(x, s) \to u_\tau(x, s)$ in $L^2(\Gamma_C \times (t_0, t))$ for a.e. $(x, s) \in \Gamma_C \times (t_0, t)$, with $|u_{n\tau}(x, s)| \leq h(x, s)$ for certain $h \in L^2(\Gamma_C \times (t_0, t))$. By the growth condition $H(j)(iii)$ it follows that $|\xi_n(x, s)| \leq a + bh(x, s)$ a.e. $(x, s) \in \Gamma_C \times (t_0, t)$. We are in a position to use Proposition 3.16 in [21] whence it follows that

$$\xi(x, s) \in \overline{conv}(K - \limsup_{n\to\infty}\{\xi_n(x, s)\}),$$

for a.e. $(x, s) \in \Gamma_C \times (t_0, t)$ where $K - \limsup_{n\to\infty}$ stands for Kuratowski–Painlevé upper limit of sets (in this case, singletons) defined by

$$K - \limsup_{n\to\infty} A_n = \{x \in \mathbb{R} \mid x = \lim_{k\to\infty} x_{n_k}, x_{n_k} \in A_{n_k}, n_1 < n_2 < \ldots < n_k < \ldots\}.$$

Since $\xi_n(x, s) \in \partial j(x, s, u_{n\tau}(x, s))$ a.e. $(x, s) \in \Gamma_C \times (t_0, t)$, and the graph of the multivalued mapping $\lambda \to \partial j(x, t, \lambda)$ is closed, we must have

$$\overline{conv}(K - \limsup_{n\to\infty}\{\xi_n(x, s)\}) \subset \overline{conv}(\partial j(x, s, u_\tau(x, s))) = \partial j(x, s, u_\tau(x, s)).$$

Thus $\xi(x, s) \in \partial j(x, s, u_\tau(x, s))$ a.e. $(x, s) \in \Gamma_C \times (t_0, t)$, and the assertion follows. □

From Lemmata 9.4, 9.5 and 9.6 it follows that all assumptions of Corollary 9.3 hold and hence we have shown the following Theorem.

Theorem 9.4 *The m-process U on H associated with Problem **(P)** has a $\mathscr{D}_\sigma$-pullback attractor. This attractor is, moreover, invariant and it belongs to $\mathscr{D}_\sigma$.*

In the above theorem, the attractor invariance and the fact that it belongs to the universe $\mathscr{D}_\sigma$ follow from Corollary 9.1 as the m-process is strict and the attraction universe $\mathscr{D}_\sigma$ satisfies $(H\mathscr{D}1)$.

Acknowledgments Work of P.K. was supported by the Marie Curie International Research Staff Exchange Scheme Fellowship within the 7th European Community Framework Programme under Grant Agreement No. 295118, by the International Project co-financed by Polish Ministry of Science and Higher Education under grant no. W111/7.PR/2012, and by Polish National Science Center under grant no. DEC-2012/06/A/ST1/00262. We wish to thank the referee for useful comments.

References

1. Anguiano, M.: Attractors for nonlinear and non-autonomous parabolic PDEs in unbounded domains. Ph.D. thesis, Universidad de Sevilla (2011)
2. Balibrea, F., Caraballo, T., Kloeden, P.E., Valero, J.: Recent developments in dynamical systems: three perspectives. Int. J. Bifurc. Chaos **20**, 2591–2636 (2010)
3. Ball, J.M.: Continuity properties and global attractors of generalized semiflows and the Navier-Stokes equations. Nonlinear Sci. **7**, 475–502 (1997). Erratum, ibid 8, p. 233 (1998). Corrected version appears in “Mechanics: From Theory to Computation”, pp. 447–474. Springer (2000)
4. Baniotopoulos, C.C., Haslinger, J., Morávková, Z.: Mathematical modeling of delamination and nonmonotone friction problems by hemivariational inequalities. Appl. Math. **50**, 1–25 (2005)
5. Caraballo, T., Kloeden, P.: Non-autonomous attractors for integro-differential evolution equations. Discret. Contin. Dyn. Syst.-Ser. S **2**, 17–36 (2009)
6. Caraballo, T., Langa, J.A., Melnik, V.S., Valero, J.: Pullback attractors of nonautonomous and stochastic multivalued dynamical systems. Set-Valued Anal. **11**, 153–201 (2003)
7. Carvalho, A.N., Langa, J.A., Robinson, J.C.: Attractors for Infinite-Dimensional Non-Autonomous Dynamical Systems. Springer, New York (2012)

8. Hale, J.K.: Asymptotic Behavior of Dissipative Systems. AMS, Providence (1988)
9. Ionescu, I.R., Nguyen, Q.L.: Dynamic contact problems with slip dependent friction in viscoelasticity. Int. J. Appl. Math. Comput. **12**, 71–80 (2002)
10. Kalita, P., Łukaszewicz, G.: Global attractors for multivalued semiflows with weak continuity properties. Nonlinear Anal.-Theory **101**, 124–143 (2014)
11. Kalita, P., Łukaszewicz, G.: Attractors for Navier-Stokes flows with multivalued and nonmonotone subdifferential boundary conditions. Nonlinear Anal.-Real **19**, 75–88 (2014)
12. Kalita, P., Łukaszewicz, G.: Theory, Numerical Analysis, and Applications. In: Han, W., Migórski, S., Sofonea, M. (eds.) On large time asymptotics for two classes of contact problems, to appear in Advances in Variational and Hemivariational Inequalities. Springer, New York (2015)
13. Kapustyan, O.V., Kasyanov, P.O., Valero, J.: Pullback attractors for some class of extremal solutions of 3D Navier-Stokes system. J. Math. Anal. Appl. **373**, 535–547 (2011)
14. Kapustyan, O.V., Kasyanov, P.O., Valero, J., Zgurovsky, M.Z.: Structure of uniform global attractor for general non-autonomous reaction-diffusion system. In: Zgurovsky, M.Z., Sadovnichiy, V.A. (eds.) Continuous and Distributed Systems, pp. 163–180. Springer (2014)
15. Li, Y., Zhong, C.: Pullback attractors for the norm-to-weak continuous process and application to the nonautonomous reaction-diffusion equations. Appl. Math. Comput. **190**, 1020–1029 (2007)
16. Łukaszewicz, G.: On the existence of an exponential attractor for a planar shear flow with the Tresca friction condition. Nonlinear Anal.-Real **14**, 1585–1600 (2013)
17. Ma, Q.F., Wang, S.H., Zhong, C.K.: Necessary and sufficient conditions for the existence of global attractors for semigroups and applications. Indiana Univ. Math. J. **51**(6), 1541–1559 (2002)
18. Marín-Rubio, P., Real, J.: Pullback attractors for 2d-Navier-Stokes equations with delays in continuous and sub-linear operators. Discrete Contin. Dyn. Syst.-A **26**, 989–1006 (2010)
19. Melnik, V.S., Valero, J.: On attractors of multivalued semiflows and differential inclusions. Set-Valued Anal. **6**, 83–111 (1998)
20. Melnik, V.S., Valero, J.: Addendum to on attractors of multivalued semiflows and differential inclusions [Set-Valued Anal. **6**, 83–111 (1998)]. Set-Valued Anal. **16**, 507–509 (2008)
21. Migórski, S., Ochal, A., Sofonea, M.: Nonlinear Inclusions and Hemivariational Inequalities. Models and Analysis of Contact Problems. Springer, New York (2013)
22. Perrin, G.J., Rice, J.R., Zheng, G.: Self-healing slip pulse on a frictional surface. J. Mech. Phys. Solids **43**, 1461–1495 (1995)
23. Persson, B.N.J.: Sliding Friction. Physical Principles and Applications. Springer, New York (2000)
24. Rosa, R.: The global attractor for the 2d Navier-Stokes flow on some unbounded domains. Nonlinear Anal.-Theory **32**, 71–85 (1998)
25. Scholz, C.H.: The Mechanics of Earthquakes and Faulting. Cambridge University Press, Cambridge (1990)
26. Šestak, I., Jovanović, B.S.: Approximation of thermoelasticity contact problem with nonmonotone friction. Appl. Math. Mech. **31**, 77–86 (2010)
27. Wang, Y., Zhou, S.: Kernel sections and uniform attractors of multivalued semiprocesses. J. Differ. Equ. **232**, 573–622 (2007)
28. Zgurovsky, M.Z., Kasyanov, P.O., Kapustyan, O.V., Valero, J., Zadoianchuk, N.V.: Evolution inclusions and Variation Inequalities for Earth Data Processing III. Springer, Berlin (2012)
29. Zhong, C.K., Yang, M.H., Sun, C.Y.: The existence of global attractors for the norm-to-weak continuous semigroup and application to the nonlinear reaction-diffusion equations. J. Differ. Equ. **223**, 367–399 (2006)

Chapter 10
Lévy–Areas of Ornstein–Uhlenbeck Processes in Hilbert–Spaces

María J. Garrido-Atienza, Kening Lu and Björn Schmalfuß

Abstract In this paper, we investigate the existence and some useful properties of the Lévy-areas of Ornstein-Uhlenbeck processes associated to Hilbert space-valued fractional Brownian motions with Hurst parameter $H \in (1/3, 1/2]$. We prove that this stochastic area has a Hölder-continuous version with sufficiently large Hölder-exponent and that can be approximated by smooth areas. In addition, we prove the stationarity of this area.

10.1 Introduction

During the last decades, new techniques have been developed generalizing the well-known Ito- or Stratonovich-integration. For fundamental publications in this area, see for instance Lyons an Qian [18] for the so-called *Rough-Paths* theory and Zähle [22] for the *Fractional Calculus*[1] theory. In particular, these techniques allow to have stochastic integrators which are more general than the Brownian motion. A candidate for such an integrator is for instance the fractional Brownian motion. This stochastic process in general does not have the semi-martingale property, which would allow to define the stochastic integral as a limit in probability, if the integrator satisfies special measurability and integrability conditions, see e.g., Karatzas

[1] The name fractional calculus theory goes back to D. Nualart.

M.J. Garrido-Atienza (✉)
Dpto. Ecuaciones Diferenciales Y Análisis Numérico, Universidad de Sevilla,
Apdo. de Correos 1160, 41080 Sevilla, Spain
e-mail: mgarrido@us.es

K. Lu
346 TMCB Brigham Young University, Provo, UT 84602, USA
e-mail: klu@math.byu.edu

B. Schmalfuß
Institut Für Stochastik, Friedrich Schiller Universität Jena, Ernst Abbe Platz 2,
77043 Jena, Germany
e-mail: bjoern.schmalfuss@uni-jena.de

© Springer International Publishing Switzerland 2015

V.A. Sadovnichiy and M.Z. Zgurovsky (eds.), *Continuous and Distributed Systems II*, Studies in Systems, Decision and Control 30,
DOI 10.1007/978-3-319-19075-4_10

and Shreve [14]. Another advantage of these new techniques is to treat stochastic integrals in a pathwise way. In particular, for any sufficient regular integrand and integrator with more general properties than the bounded variation property, an integral can be defined. This kind of integrals goes back to Young [21] which allows to consider integrals for Hölder-continuous integrands and integrators fulfilling special conditions with respect to the Hölder-exponents. In practice, this integral cannot be used to replace the well-known Ito-integral by a Young-integral if the integrator is a Brownian motion and the integrand has a Hölder-exponent less than or equal to $1/2$.

Our reason for dealing with this type of new integral is to take advantage of the pathwise property in order to further introduce *Random Dynamical Systems* (RDS) for infinite dimensional differential equations, namely, for stochastic evolution equations and stochastic partial differential equations as well. The random driver of this kind of equation in general will be a trace-class fractional Brownian motion with Hurst parameter $H \in (1/3, 1/2]$. To introduce an RDS, one needs, at first, a model for a noise which is called a *metric dynamical system*: Consider the quadruple $(\Omega, \mathscr{F}, \mathbb{P}, \theta)$ where $(\Omega, \mathscr{F}, \mathbb{P})$ is a probability space and θ is a *measurable flow* on Ω:

$$\theta : (\mathbb{R} \times \Omega, \mathscr{B}(\mathbb{R}) \otimes \mathscr{F}) \to (\Omega, \mathscr{F})$$
$$\theta_t \circ \theta_\tau = \theta_t \theta_\tau = \theta_{t+\tau}, \qquad \theta_0 = \mathrm{id}_\Omega.$$

An Hilbert (or topological) space RDS is a measurable mapping

$$\phi : (\mathbb{R}^+ \times \Omega \times V, \mathscr{B}(\mathbb{R}^+) \otimes \mathscr{F} \otimes \mathscr{B}(V)) \to (V, \mathscr{B}(V))$$

satisfying the cocycle property

$$\phi(t + \tau, \omega, x) = \phi(t, \theta_\tau \omega, \phi(\tau, \omega, x)), \qquad \phi(0, \omega, x) = x,$$

for all $t,\ \tau \in \mathbb{R}^+$, $x \in V$ and *for all* $\omega \in \Omega$ or at least for all ω of a θ-invariant set $\tilde{\Omega} \in \mathscr{F}$ of full $\mathbb{P}$-measure which is independent of $x,\ t,\ \tau$. This is different to the fact that *a property holds almost surely* for any $x \in V$, which is often used in Stochastic Analysis, since exceptional sets depending on $t,\ \tau$ and x are in most cases forbidden when dealing with the cocycle property.

An example of a metric dynamical system is, for instance, $(C_0, \mathscr{B}(C_0), \mathbb{P}_H, \theta)$, where the probability space $(C_0, \mathscr{B}(C_0), \mathbb{P}_H)$ is the canonical probability space such that C_0 is the space of continuous functions on $\mathbb{R}$ with values in a separable Hilbert–space V that are zero at zero, $\mathscr{B}(C_0)$ is its Borel-σ-algebra, $\mathbb{P}_H$ is the distribution on $\mathscr{B}(C_0)$ of a trace-class fractional Brownian motion ω with Hurst parameter $H \in (0, 1)$, and $\theta : (\mathbb{R} \times C_0, \mathscr{B}(\mathbb{R}) \otimes \mathscr{B}(C_0)) \to (C_0, \mathscr{B}(C_0))$ is given by

$$\theta_\tau(t) = \omega(t + \tau) - \omega(\tau), \quad \text{for } t, \tau \in \mathbb{R}, \tag{10.1}$$

see Arnold [1] or Maslowski and Schmalfuß [19].

It is known that under typical conditions on the coefficients an ordinary Ito-equation generates an RDS, see Arnold [1]. The main instrument to obtain this property is the Kolmogorov-theorem on the existence of a Hölder-continuous version of a random field. Such a random field is derived from an Ito-equation where the parameters of this field are the time and the nonrandom initial condition. The so-called perfection technique then can be used to conclude the existence of a version of the ordinary Ito-equation which defines an RDS. However, there is no appropriate version of the Kolmogorov-theorem for infinite dimensional random fields that could be applied to show that solutions of stochastic evolution equations generate an RDS; and therefore this property is rather an open problem, although there are partial results for particular cases, see, e.g., the recent papers [3, 5, 7, 9, 10].

Consider a stochastic evolution equation

$$du = Audt + G(u)d\omega$$

where A is the generator of an analytic stable semigroup S on the separable Hilbert space V and ω is a β-Hölder-continuous fractional Brownian motion in V with Hurst parameter $H \in (1/3, 1/2]$, so that $\beta \in (1/3, 1/2)$. This equation has the mild interpretation

$$u(t) = S(t)u_0 + \int_0^t S(t-r)G(u(r))d\omega(r) \tag{10.2}$$

where u_0 is a nonrandom initial condition in V. The integral in the above equation has to be interpreted in a fractional sense. For a good understanding of that integral, we refer to [11, 12].

For the following, let $\bar{\Delta}_{a,b} \subset \mathbb{R}^2$ be the set of the pairs (s,t) such that $-\infty < a \le s \le t \le b < \infty$. Let $V \otimes V$ be the Hilbert tensor space of V with tensor product $\otimes_V$. We consider now functions

$$\bar{\Delta}_{0,T} \ni (s,t) \mapsto (u \otimes \omega)(s,t).$$

The reason to introduce these elements is to interpret the integral of (10.2) in a fractional sense. We need to consider the tensor product of a possible solution u and a noise path ω. In particular, for a smooth ω this tensor product is given by

$$\begin{aligned}(u\otimes\omega)(s,t) = &\int_s^t (S(\xi - s) - \mathrm{id})u(s) \otimes_V \omega'(\xi)d\xi \\ &+ \int_s^t \int_s^\xi S(\xi - r)G(u(r))\omega'(r)dr \otimes_V \omega'(\xi)d\xi,\end{aligned} \tag{10.3}$$

and, exchanging the order of integration, the last integral of (10.3) can be written as[2]

[2] D_1 means the first derivative w.r.t. the first variable of $(\omega \otimes_S \omega)(\cdot, \cdot)$.

$$-\int_s^t G(u(r))D_1(\omega \otimes_S \omega)(r,t)dr$$

where $(\omega \otimes_S \omega)(s,t)$ is defined by

$$\begin{aligned} L_2(V,\hat{V}) \ni E \mapsto E(\omega \otimes_S \omega)(s,t) &= \int_s^t \int_s^\xi S(\xi - r)E\omega'(r) \otimes_V \omega'(\xi)drd\xi \\ &= \int_s^t \int_r^t S(\xi - r)E\omega'(r) \otimes_V \omega'(\xi)d\xi dr \end{aligned} \tag{10.4}$$

for $(s,t) \in \bar{\Delta}_{0,T}$. Here $\hat{V}$ is another Hilbert space to be determined later and E is an element of the Hilbert–Schmidt space $L_2(V,\hat{V})$. Let $C_{2\beta}(\bar{\Delta}_{a,b}, L_2(L_2(V,\hat{V}), V \otimes V))$ the space of 2β-Hölder-continuous fields on $\bar{\Delta}_{a,b}$ with values in $L_2(L_2(V,\hat{V}), V \otimes V)$ with norm

$$\|v\|_{2\beta} = \|v\|_{2\beta,a,b} = \sup_{(s,t)\in\bar{\Delta}_{a,b}} \frac{\|v(s,t)\|_{L_2(L_2(V,\hat{V}),V\otimes V)}}{|t-s|^{2\beta}} < \infty. \tag{10.5}$$

We know that $\omega \in C_\beta([0,T];V)$ for every $\beta < H$. We also denote ω^n a piecewise linear (continuous) approximation of ω with respect to an equidistant partition of length $2^{-n}T := \delta$ such that $\omega^n(t) = \omega(t)$ for the partition points t. For these ω^n, we can define $(\omega^n \otimes_S \omega^n)$ by the right-hand side of (10.4).

The main purpose of this paper is to prove the following result, a property which is needed to establish the existence of solutions to (10.2). For a detailed description on the construction of solutions to (10.2), we refer the reader to the paper [11].

Theorem 10.1 *Let $(\omega^n)_{n\in\mathbb{N}}$ be the sequence of piecewise linear approximations of some ω introduced above, such that $((\omega^n \otimes_S \omega^n))_{n\in\mathbb{N}}$ is defined by* (10.4). *Then for any $\beta < H$, the sequence $((\omega^n, (\omega^n \otimes_S \omega^n)))_{n\in\mathbb{N}}$ converges to $(\omega, (\omega \otimes_S \omega))$ in $C_\beta([0,T];V) \times C_{2\beta}(\bar{\Delta}_{0,T}; L_2(L_2(V,\hat{V}), V \otimes V))$ on a set of full measure. In particular $(\omega \otimes_S \omega)$ is continuous.*

In the following, we consider only the statements of this theorem with respect to $(\omega \otimes_S \omega)$. The convergence properties of $(\omega^n)_{n\in\mathbb{N}}$ follow in a similar and simpler manner and we omit here.

The object $(\omega \otimes_S \omega)$ will be called *Lévy area of the Ornstein-Uhlembeck process.* Since a priori it is not clear whether $(\omega \otimes_S \omega)$ is well-defined, in what follows we are going to give an appropriate meaning to this term by presenting two proofs of the above theorem. The first one is related to β-Hölder-continuous paths covering the case of a fractional Brownian motion for an appropriate Hilbert space $\hat{V}$. The second proof deals with the case of a Brownian motion for a more general space $\hat{V}$.

As we have said, A is the generator of the analytic semigroup S on the separable Hilbert space V. We also suppose that $-A$ is positive and symmetric such that its inverse is compact. Then $-A$ has a positive point spectrum $(\lambda_i)_{i\in\mathbb{N}}$ of finite multiplicity such that $\lim_{i\to\infty}\lambda_i = \infty$. We denote the associated eigenelements by $(e_i)_{i\in\mathbb{N}}$ which form a complete orthonormal system in V. Furthermore, $-A$ generates for $\kappa \in \mathbb{R}$ the separable Hilbert spaces

$$D((-A)^\kappa) =: V_\kappa = \{u = \sum_i \hat{u}_i e_i : \|u\|^2_{V_\kappa} = \sum_i |\hat{u}_i|^2 \lambda_i^{2\kappa} < \infty\},$$

with $V = V_0$.

10.2 The Construction of $(\omega \otimes_S \omega)$ for a Fractional Brownian motion

Let us begin this section by introducing a V-valued fractional Brownian motion. For brevity, we suppose that ω can be presented by

$$\omega(t) = \sum_i q_i^{\frac{1}{2}} \omega_i(t) e_i$$

where $(\omega_i)_{i\in\mathbb{N}}$ is a sequence of one-dimensional independent β-Hölder-continuous standard fractional Brownian motions for any $\beta < H$, such that $\sum_i q_i < \infty$. Then ω can be interpreted as a β-Hölder-continuous fractional Brownian motion in V with Hurst parameter $H \in (0, 1)$. The covariance is given by the operator Q which is a diagonal operator in our standard basis with diagonal elements q_i. Let us also denote the piecewise linear approximations by ω^n, ω_i^n with respect to the equidistant partition $\{t_i^n\}$ of $[0, T]$ of length $2^{-n}T = \delta$.

Throughout this section, we assume that ω has a Hurst parameter $H \in (1/3, 1/2]$. Since ω^n is smooth, we can define $(\omega^n \otimes_S \omega^n)(s, t)$ as a Bochner-integral with respect to the Lebesgue-measure:

$$E(\omega^n \otimes_S \omega^n)(s, t) = \int_s^t \int_s^\xi S(\xi - r) E d\omega^n(r) \otimes_V d\omega^n(\xi)$$

for $E \in L_2(V, \hat{V})$. By an integration by parts argument, this integral can be rewritten as

$$\int_s^t E(\omega^n(\xi) - \omega^n(s)) \otimes_V d\omega^n(\xi) + \int_s^t A \int_s^\xi S(\xi - r) E(\omega^n(r) - \omega^n(s)) dr \otimes_V d\omega^n(\xi).$$

This motivates to interpret $(\omega \otimes_S \omega)(s, t)$ as

$$\int_s^t E(\omega(\xi) - \omega(s)) \otimes_V d\omega(\xi) + \int_s^t A \int_s^\xi S(\xi - r) E(\omega(r) - \omega(s)) dr \otimes_V d\omega(\xi) \tag{10.6}$$

where we have to give an appropriate meaning to both integrals. We start with the first one which is abbreviated in the following by $E(\omega \otimes \omega)(s,t)$.

Lemma 10.1 *Let* $\hat{V} = V_\kappa$ *for* $\kappa > 0$ *such that* $\sum_i \lambda_i^{-2\kappa} < \infty$. *The sequence* $((\omega^n \otimes \omega^n))_{n\in\mathbb{N}}$, *which elements can be represented component-wise by*

$$q_j^{\frac{1}{2}} q_k^{\frac{1}{2}} \int_s^t (\omega_j^n(\xi) - \omega_j^n(s)) d\omega_k^n(\xi)$$

converges on a set of full measure in $C_{2\beta}(\bar{\Delta}_{0,T}; L_2(L_2(V, V_\kappa), V \otimes V))$ *to* $(\omega \otimes \omega)$.

Proof Consider the orthonormal basis $(E_{ij})_{i,j\in\mathbb{N}}$ of $L_2(V, V_\kappa)$ given by

$$E_{ij} e_k = \begin{cases} 0 & : j \neq k \\ \frac{e_i}{\lambda_i^\kappa} & : j = k, \end{cases}$$

and $(e_l \otimes_V e_k)_{l,k\in\mathbb{N}}$, the orthonormal basis of $V \otimes V$. First note that, for $(s,t) \in \bar{\Delta}_{0,T}$,

$$\begin{aligned} \|(\omega^n \otimes \omega^n)(s,t)\|^2_{L_2(L_2(V,V_\kappa),V\otimes V)} &= \sum_{i,j} \sum_{l,k} (E_{ij}(\omega^n \otimes \omega^n)(s,t), e_l \otimes_V e_k)^2_{V\otimes V} \\ &\leq \sum_i \lambda_i^{-2\kappa} \sum_{j,k} q_j q_k \left(\int_s^t (\omega_j^n(\xi) - \omega_j^n(s)) d\omega_k^n(\xi) \right)^2, \end{aligned}$$

and hence we have to study the behavior of the last sum. Let us denote

$$A^n_{j,k}(s,t) := \int_s^t (\omega_j(\xi) - \omega_j(s)) d\omega_k(\xi) - \int_s^t (\omega_j^n(\xi) - \omega_j^n(s)) d\omega_k^n(\xi).$$

By symmetry, we can assume $j \leq k$. In fact we assume that $j < k$ since the case $j = k$ is easier, see a comment at the end of the proof.[3]

To estimate the continuous element $A^n_{j,k}(s,t)$, we apply the Lemma 3.7 in Deya et al. [6], which claims that for $p \geq 1$ there exists $K_{\beta,p}$ such that

$$\|A^n_{j,k}\|_{2\beta} \leq K_{\beta,p}(R^{j,k}_{n,p} + \|\omega_j - \omega_j^n\|_\beta \|\omega_k\|_\beta + \|\omega_k - \omega_k^n\|_\beta \|\omega_j^n\|_\beta), \tag{10.7}$$

where

$$R^{j,k}_{n,p} := \left(\int_0^T \int_0^T \frac{|A^n_{j,k}(s,t)|^{2p}}{|t-s|^{4\beta p+2}} ds dt \right)^{1/(2p)}.$$

[3] In [6], $A^n_{j,k}(s,t)$ are considered over the square $[0,T]^2$. However, since they are symmetric w.r.t. the diagonal of $[0,T]$ it is sufficient to consider these elements over $\bar{\Delta}_{0,T}$.

In particular, from the proof of Lemma 3.7 in [6] we know that

$$\mathbb{E}(R_{n,p}^{j,k})^{2p} \leq cn^{-4p(H-\beta')} < \infty,$$

for $\beta < \beta' < H$, being β' close enough to H and p large enough. Indeed, let us take p large enough such that $4p(H-\beta') > 1$, and thus

$$\mathbb{P}\left(\sum_{j,k} q_j q_k (R_{p,n}^{j,k})^2 > o(n)^2\right) \leq \frac{(\mathrm{tr}_V Q)^{2(p-1)}}{o(n)^{2p}} \sum_{j,k} q_j q_k \mathbb{E}(R_{n,p}^{j,k})^{2p} \leq \frac{cn^{-4p(H-\beta')}}{o(n)^{2p}}.$$

For an appropriate sequence $(o(n))_{n\in\mathbb{N}}$ with limit zero, the right-hand side has a finite sum. Then by the Borel-Cantelli-lemma, $(\sum_{j,k} q_j q_k (R_{p,n}^{j,k})^2)_{n\in\mathbb{N}}$ tends to zero almost surely. In a similar manner, we obtain the convergence of the last terms in (10.7). It suffices to take into account that, for $\beta < \beta' < H$,

$$\|\omega_j - \omega_j^n\|_\beta \leq G_{\beta'}(j,\omega) n^{\beta-\beta'}, \ \|\omega_j\|_\beta \leq G_{\beta'}(j,\omega), \ \|\omega_j^n\|_\beta \leq G_{\beta'}(j,\omega) \tag{10.8}$$

where $G_{\beta'}(j,\omega) \geq \|\omega_j\|_{\beta'}$ and $G_{\beta'}(j,\omega) \in L_p(\Omega)$ for any $p \in \mathbb{N}$ are iid random variables, see Kunita [16] Theorem 1.4.1. We then have

$$\mathbb{P}\left(\sum_{j,k} q_j q_k \|\omega_j^n - \omega_j\|_\beta^2 \|\omega_k\|_\beta^2 > o(n)^2\right)$$
$$\leq \frac{(\mathrm{tr}_V Q)^{2(p-1)}}{o(n)^{2p}} \sum_{j,k} q_j q_k (\mathbb{E}G_{\beta'}(j,\omega)^{4p})^{\frac{1}{2}} (\mathbb{E}G_{\beta'}(k,\omega)^{4p})^{\frac{1}{2}} n^{2p(\beta-\beta')} \leq \frac{cn^{2p(\beta-\beta')}}{o(n)^{2p}}.$$

For p chosen sufficiently large and an appropriate zero-sequence $(o(n))_{n\in\mathbb{N}}$, we obtain the almost sure convergence of $(\sum_{j,k} q_j q_k \|\omega_j^n - \omega_j\|_\beta^2 \|\omega_k\|_\beta^2)_{n\in\mathbb{N}}$. Similarly we can treat the last term of (10.7), that is, $(\sum_{j,k} q_j q_k \|\omega_j^n - \omega_j\|_\beta^2 \|\omega_k^n\|_\beta^2)_{n\in\mathbb{N}}$. Finally,

$$A_{j,j}^n(s,t) = \frac{1}{2}(\omega_j(t) - \omega_j(s))^2 - \frac{1}{2}(\omega_j^n(t) - \omega_j^n(s))^2$$

and thanks to (10.8) $\|A_{j,j}^n\|_{2\beta} \leq G_{\beta'}(j,\omega)^2 n^{2(\beta-\beta')}$, which completes the proof.

Lemma 10.2 *Suppose that there exists γ such that $\gamma + \beta > 1$ and*

$$\sum_i \lambda_i^{2\gamma - 2\kappa} < \infty.$$

Then the mapping

$$((s,t),E)\in\bar{\Delta}_{0,T}\times L_2(V,V_\kappa)\mapsto\int_s^t\left(A\int_s^\xi S(\xi-r)E(\omega(r)-\omega(s))dr\right)\otimes_V d\omega(\xi)$$

is in $C_{2\beta}(\bar{\Delta}_{0,T};L_2(L_2(V,V_\kappa),V\otimes V))$.

Proof Thanks to Pazy [20] Theorem 4.3.5 (iii),

$$\begin{aligned}\left|A\int_s^\xi S(\xi-r)E_{ij}(\omega(r)-\omega(s))dr\right|&=\frac{q_j^{\frac12}}{\lambda_i^\kappa}\left|\int_s^\xi AS(\xi-r)e_i(\omega_j(r)-\omega_j(s))dr\right|\\&\le c\frac{q_j^{\frac12}}{\lambda_i^\kappa}\|\omega_j\|_\beta(\xi-s)^\beta,\end{aligned}\tag{10.9}$$

and applying Bensoussan and Frehse [4] Corollary 2.1 we also have

$$\begin{aligned}&\left|A\int_s^\xi S(\xi-r)E_{ij}(\omega(r)-\omega(s))dr-A\int_s^{\xi'}S(\xi'-r)E_{ij}(\omega(r)-\omega(s))dr\right|\\&\le\frac{q_j^{\frac12}}{\lambda_i^{\kappa-\gamma}}\left|\int_s^\xi(-A)^{1-\gamma}S(\xi-r)e_i(\omega_j(r)-\omega_j(s))dr\right.\\&\qquad\left.-\int_s^{\xi'}(-A)^{1-\gamma}S(\xi'-r)e_i(\omega_j(r)-\omega_j(s))dr\right|\le c\frac{q_j^{\frac12}}{\lambda_i^{\kappa-\gamma}}\|\omega_j\|_\beta|\xi-\xi'|^\gamma.\end{aligned}\tag{10.10}$$

Therefore, for an $\alpha<\gamma$ such that $\beta>1-\alpha$, we can define the integral

$$\begin{aligned}&\int_s^t\left(A\int_s^\xi S(\xi-r)e_i(\omega_j(r)-\omega_j(s))dr\right)d\omega_k(\xi)\\&=(-1)^\alpha\int_s^t D_{s+}^\alpha\left(A\int_s^\cdot S(\cdot-r)e_i(\omega_j(r)-\omega_j(s))dr\right)[\xi]D_{t-}^{1-\alpha}(\omega_k)_{t-}[\xi]d\xi.\end{aligned}$$

For the definition of the so-called fractional derivatives D_{s+}^α and $D_{t-}^{1-\alpha}$ and the definition of the stochastic integral in terms of these expressions, we refer to [22]. Note that as a consequence of (10.9) and (10.10), and because $\gamma>\beta$,

$$\left|D_{s+}^\alpha\left(A\int_s^\cdot S(\cdot-r)e_i(\omega_j(r)-\omega_j(s))dr\right)[\xi]\right|\le c\|\omega_j\|_\beta(\xi-s)^{\beta-\alpha}.$$

Since $\beta+\alpha>1$, $|D_{t-}^{1-\alpha}(\omega_k)_{t-}[\xi]|\le c\|\omega_k\|_\beta(t-\xi)^{\alpha+\beta-1}$, and then

$$\left(\sum_{i,j,k}\left(\int_s^t\left(\frac{q_j^{\frac12}}{\lambda_i^\kappa}\int_s^\xi AS(\xi-r)e_i(\omega_j(r)-\omega_j(s))dr\right)\otimes_V q_k^{\frac12}d\omega_k(\xi)\right)^2\right)^{\frac12}$$
$$\le\left(\sum_i\lambda_i^{2\gamma-2\kappa}\right)^{\frac12}\left(\sum_{j,k}q_jq_k\|\omega_j\|_\beta^2\|\omega_k\|_\beta^2\right)^{\frac12}\int_s^t(\xi-s)^{\beta-\alpha}(t-\xi)^{\alpha+\beta-1}d\xi,$$

see [11] to find the estimate of the integral in terms of the norms of the fractional derivatives. Now it suffices to take into account that the last integral can be estimated by $c(t-s)^{2\beta}$, which follows from the definition of the Beta function.

Remark 10.1 Replacing ω by $\omega^n-\omega$ in the above proof, we obtain

$$\lim_{n\to\infty}\int_s^t A\int_s^\xi S(\xi-r)E(\omega^n(r)-\omega^n(s))dr\otimes_V d\omega^n(\xi)$$
$$=\int_s^t A\int_s^\xi S(\xi-r)E(\omega(r)-\omega(s))dr\otimes_V d\omega(\xi)$$

in $C_{2\beta}(\bar{\Delta}_{0,T};L_2(L_2(V,V_\kappa),V\otimes V))$. Indeed, in the proof of Lemma 10.1 we have shown

$$\lim_{n\to\infty}\sum_{i,j}q_iq_j(\|\omega_i-\omega_i^n\|_\beta^2\|\omega_j\|_\beta^2+\|\omega_j-\omega_j^n\|_\beta^2\|\omega_i^n\|_\beta^2)=0\quad\text{a.s.}$$

In view of Lemma 10.1 and Remark 10.1, we conclude with the proof of Theorem 10.1.

Corollary 10.1 *The mapping $\bar{\Delta}_{0,T}\ni(s,t)\to(\omega\otimes_S\omega)(s,t)$ is continuous.*

Indeed, from the above estimate we have the convergence

$$\lim_{n\to\infty}\sup_{(s,t)\in\bar{\Delta}_{0,T}}\|(\omega\otimes_S\omega)(s,t)-(\omega^n\otimes_S\omega^n)(s,t)\|_{L_2(L_2(V,V_\kappa),V\otimes V)}=0$$

and straightforwardly

$$\bar{\Delta}_{0,T}\ni(s,t)\to(\omega^n\otimes_S\omega^n)(s,t)$$

is continuous.

10.3 The Construction of $(\omega \otimes_S \omega)$ for a Brownian motion

We consider ω to be a trace-class Brownian motion with a positive symmetric trace-class covariance operator Q. Therefore, in that case ω is a trace-class fractional Brownian motion with Hurst parameter $H = 1/2$. We want to formulate a weaker condition for the existence of $(\omega \otimes_S \omega)$ than the assumption of Lemma 10.2. Again we will take $\hat{V} = V_\kappa$ and will determine the conditions that κ must satisfy.

Now let us split $(\omega \otimes_S \omega)$ into its components. Taking into account that $e^{-\lambda_i t}$ gives the decomposition of $S(t)$ with respect to the base $(e_i)_{i\in\mathbb{N}}$, we characterize $(\omega \otimes_S \omega)$ for $l = j$ by

$$
\begin{aligned}
(E_{ij}(\omega \otimes_S \omega)(s,t), e_l \otimes_V e_k)_{V\otimes V} &= \frac{q_j^{\frac{1}{2}} q_k^{\frac{1}{2}}}{\lambda_i^\kappa} \int_s^t \int_s^\xi e^{-\lambda_i(\xi-r)} d\omega_j(r) \circ d\omega_k(\xi) \\
= \frac{q_j^{\frac{1}{2}} q_k^{\frac{1}{2}}}{\lambda_i^\kappa} \Bigg(\int_s^t (\omega_j(\xi) - \omega_j(s)) \circ d\omega_k(\xi) \\
- \lambda_i \int_s^t \int_s^\xi e^{-\lambda_i(\xi-r)} (\omega_j(r) - \omega_j(s)) dr d\omega_k(\xi) \Bigg) \\
= \frac{q_j^{\frac{1}{2}} q_k^{\frac{1}{2}}}{\lambda_i^\kappa} \int_s^t \int_s^\xi e^{-\lambda_i(\xi-r)} d\omega_j(r) d\omega_k(\xi) - \frac{1}{2}\delta_{jk} \frac{q_j^{\frac{1}{2}} q_k^{\frac{1}{2}}}{\lambda_i^\kappa}(t-s) \qquad (10.11)
\end{aligned}
$$

and for $l \neq i$ by 0. Here $\circ$ means Stratonovich-integration where for the inner stochastic integration Ito- and Stratonovich-integrals are the same. On the right-hand side we have Ito-integration, and the last term there is the Ito-correction which only appears for $j = k$, expressed by the Kronnecker-symbol δ_{jk}. Note that the second line in (10.11) is obtained by stochastic integration by parts. Let us abbreviate

$$
\int_s^t \int_s^\xi e^{-\lambda_i(\xi-r)} d\omega_j(r) \circ d\omega_k(\xi) =: (\omega \otimes_S \omega)_{ijk}.
$$

In this section, ω^n, ω_i^n are piecewise linear approximations of the Brownian motions ω, ω_i with respect to the equidistant partition $\{t_i^n\}$ of $[0, T]$ of length $2^{-n}T = \delta$.

We now deal with the convergence of $(\omega^n \otimes_S \omega^n)_{ijk}$ to $(\omega \otimes_S \omega)_{ijk}$. At first we present some lemmata which will be needed for this purpose.

As a preparatory result for the following, we need (see Karatzas and Shreve [14], Exercise 3.25, p. 163):

Lemma 10.3 *Let ω denote an one-dimensional Brownian motion and x be a measurable, adapted process satisfying*

$$
\mathbb{E} \int_0^T |x(\xi)|^{2p} d\xi < \infty
$$

for some real numbers $T > 0$ and $p \geq 1$, then

$$\mathbb{E}\left(\int_0^T x(\xi)d\omega(\xi)\right)^{2p} \leq (p(2p-1))^p T^{p-1}\mathbb{E}\int_0^T |x(\xi)|^{2p}d\xi.$$

Lemma 10.4 *For any $p \in \mathbb{N}$ there exists a $c_p > 0$ such that for any $M \in \mathbb{N}$*

$$\sum_{k_1+\cdots+k_M=p} \frac{(2p)!}{(2k_1)!(2k_2)!\cdots(2k_M)!} \leq c_p M^p.$$

Let $(x_i)_{i=1,\cdots,M}$ be a sequence of independent random variables in L_{2p} where the odd moments are zero. If in addition $\mathbb{E}x_i^{2p} \leq c$, $\mathbb{E}x_i^0 = 1$ for $x_i \not\equiv 0$ then

$$\mathbb{E}\left(\sum_{i=1}^M x_i\right)^{2p} \leq c_p c M^p,$$

assuming that all terms containing at least one odd power disappear.

Proof We have

$$\sum_{k_1+\cdots+k_M=p} 1 \leq M^p.$$

On the other hand

$$\sup_{k_1+\cdots+k_M=p} \frac{(2p)!}{(2k_1)!(2k_2)!\cdots(2k_M)!} \leq c_p \leq (2p)!.$$

Hence a bound for this expression can be chosen independently of M.

To conclude the proof it suffices to apply the multinomial theorem, which reduces to the following situation since the terms containing odd powers are neglected:

$$\begin{aligned}
\mathbb{E}\left(\sum_{i=1}^M x_i\right)^{2p} &= \sum_{a_1+\cdots+a_M=2p} \binom{2p}{a_1,\cdots,a_M} \mathbb{E}x_1^{a_1}\cdots\mathbb{E}x_M^{a_M} \\
&= \sum_{k_1+\cdots+k_M=p} \binom{2p}{2k_1,\cdots,2k_M} \mathbb{E}x_1^{2k_1}\cdots\mathbb{E}x_M^{2k_M} \\
&\leq \sum_{k_1+\cdots+k_M=p} \binom{2p}{2k_1,\cdots,2k_M} \prod_{m=1,\cdots,M,k_m>0} (\mathbb{E}x_m^{2p})^{\frac{2k_m}{2p}} \\
&\leq c_p c M^p.
\end{aligned}$$

If some of the $k_m = 0$ then $\mathbb{E}x_m^0 = 1$ and the corresponding terms would be removed from the above product.

Lemma 10.5 *For every $T > 0$, $p \in \mathbb{N}$ and a sufficiently small $\epsilon > 0$ there exists $c_p > 0$ such that for $(s,t) \in \bar{\Delta}_{0,T}$, $i,j,k,n \in \mathbb{N}$*

$$\mathbb{E}\left|(\omega \otimes_S \omega)_{ijk} - (\omega^n \otimes_S \omega^n)_{ijk}\right|^{2p} \le c_p \delta^\epsilon \lambda_i (t-s)^{2p-\epsilon}. \tag{10.12}$$

Proof We here only study the case $j = k$ since the case $j \neq k$ can be studied similarly, see also Friz and Hairer [8] p. 33f. For the following, we assume that $\lambda_1 \ge 1$ without lost of generality. For a partition interval $[t_{m-1}^n, t_m^n)$, we use the notation

$$\Delta_k^\delta(m) = \omega_k(t_m^n) - \omega_k(t_{m-1}^n)$$

and when $s < t \in [t_{m-1}^n, t_m^n)$ we will also use the notation

$$\Delta_k^{t-s}(m) = \omega_k(t) - \omega_k(s).$$

We divide the proof in several cases:

(i) We consider at first the case that $s, t \in [t_{m-1}^n, t_m^n]$. In that situation, the difference of double integrals we want to estimate is given for any i by

$$\begin{aligned}&\frac{\Delta_k^\delta(m)^2}{\delta^2\lambda_i^2}(e^{-\lambda_i(t-s)} + \lambda_i(t-s) - 1) - \frac{\Delta_k^{t-s}(m)^2}{2}\\ &+ \lambda_i \int_s^t \int_s^\xi e^{-\lambda_i(\xi - r)}(\omega_k(r) - \omega_k(s))dr d\omega_k(\xi)\end{aligned} \tag{10.13}$$

which follows by the second part of (10.11), and where the first expression corresponds to the integral with respect to the piecewise linear approximated Brownian motions. The first two expressions of (10.13) can be estimated by

$$\begin{aligned}&\left|\frac{\Delta_k^\delta(m)^2}{\delta^2\lambda_i^2}(e^{-\lambda_i(t-s)} + \lambda_i(t-s) - 1) - \frac{1}{2}\Delta_k^{t-s}(m)^2\right|\\ &\le \Delta_k^{t-s}(m)^2\left|\frac{1}{2} - \frac{e^{-\lambda_i(t-s)} + \lambda_i(t-s) - 1}{(t-s)^2\lambda_i^2}\right|\\ &+ \left|\frac{\Delta_k^\delta(m)^2(t-s)^2}{\delta^2} - \Delta_k^{t-s}(m)^2\right|\frac{e^{-\lambda_i(t-s)} + \lambda_i(t-s) - 1}{(t-s)^2\lambda_i^2}.\end{aligned} \tag{10.14}$$

Note that the function

$$\mathbb{R}^+ \ni x \mapsto \frac{1}{2} - \frac{e^{-x} + x - 1}{x^2} = O(x) \quad \text{for } x \to 0^+$$

defined by zero at zero, is increasing and bounded by $1/2$, having derivatives bounded by 1 for $x > 0$ and

$$\mathbb{R}^+ \ni x \mapsto \frac{e^{-x} + x - 1}{x^2} \in [0, \frac{1}{2}]$$

given by $1/2$ at zero. Hence

$$\left(\frac{1}{2} - \frac{e^{-\lambda_i(t-s)} + \lambda_i(t-s) - 1}{(t-s)^2\lambda_i^2}\right)^{2p} \leq \frac{1}{2} - \frac{e^{-\lambda_i(t-s)} + \lambda_i(t-s) - 1}{(t-s)^2\lambda_i^2} \leq \lambda_i(t-s).$$

Since $t - s < \delta$, the $2p$-moment of the first expression on the right-hand side of (10.14) can be estimated by $\tilde{c}_p^1(t-s)^{2p}\lambda_i\delta$ and the second one by $\hat{c}_p^1(t-s)^{2p-\epsilon}\delta^\epsilon$ for a sufficiently small ϵ and for appropriate constants $\tilde{c}_p^1$, $\hat{c}_p^1$. Considering the last expression of (10.13), by Hölder's inequality for the inner integral we obtain

$$\begin{aligned}
&\mathbb{E}\left(\int_s^\xi e^{-\lambda_i(\xi-r)}(\omega_k(r) - \omega_k(s))dr\right)^{2p} \\
&\leq \left(\int_s^\xi e^{\frac{-2p}{2p-1}\lambda_i(\xi-r)}dr\right)^{2p-1} \int_s^\xi \mathbb{E}(\omega_k(r) - \omega_k(s))^{2p}dr \\
&\leq \bar{c}_p^1\left(\frac{2p-1}{2p}\right)^{2p-1}\frac{1}{\lambda_i^{2p-1}}(1 - e^{\frac{-2p}{2p-1}\lambda_i(\xi-s)})^{2p-1}(\xi - s)^{p+1}.
\end{aligned}$$

Applying Lemma 10.3 we get

$$\mathbb{E}\left(\lambda_i\int_s^t\int_s^\xi e^{-\lambda_i(\xi-r)}(\omega_k(r) - \omega_k(s))drd\omega_k(\xi)\right)^{2p} \leq \tilde{c}_p^1(t-s)^{2p}\delta\lambda_i.$$

Note that all the constants depending on p that have appeared can be chosen independently of i, k.

(ii) If s, t are in two neighbored intervals, say, for instance, $t_{m-1}^n \leq s < t_m^n \leq t \leq t_{m+1}^n$, we can estimate the expression in (10.12) in a similar way than before, just dividing the region of integration into two triangles $\bar{\Delta}_{s,t_m^n}$, $\bar{\Delta}_{t_m^n,t}$ and a rectangle $[s, t_m^n] \times [t_m^n, t]$. The integrals with respect to the triangles can be estimated as in the step (i) while the estimates with respect to the rectangle are considered below.

(iii) Let us now assume in general $t - s > \delta$. In addition suppose that $0 \leq t_{m_0-1}^n \leq s < t_{m_0}^n < t_{m_1}^n < t \leq t_{m_1+1}^n \leq T$, we begin to consider the integrals, denoted by $I_m^{\Delta,\delta,i,k,k}$, of (10.12) for $j = k$ and over any of the triangles $\bar{\Delta}_{t_m^n,t_{m+1}^n}$ along the hypothenuse of $\bar{\Delta}_{0,T}$. Then following the step (i) of the proof, since in the same triangle $\bar{\Delta}_{t_m^n,t_{m+1}^n}$ the distance is just given by δ, the corresponding estimate of (10.12) is bounded by $\tilde{c}_p^1\lambda_i\delta^{2p+1}$ (the second term of (10.14) cancels out in this case).

To estimate the $2p$-moment of the sum of all these integrals, we apply Hölder's inequality. Since the number of these triangles is of order $(t-s)/\delta$, we get

$$\mathbb{E}\Big(\sum_m I_m^{\Delta,\delta,i,k,k}\Big)^{2p} = O\left(\left(\frac{t-s}{\delta}\right)^{2p-1}\frac{t-s}{\delta}\right)\delta^{2p+1}\lambda_i \quad \text{for } \delta \to 0^+.$$

Hence taking also the triangles $\bar{\Delta}_{s,t_{m_0}^n}$, $\bar{\Delta}_{t_{m_1}^n,t}$ into account there exists an c_p such that for any δ and $t_{m_0}-s \le \delta$ and $t-t_{m_1} \le \delta$,

$$\mathbb{E}\Big(\sum_m I_m^{\Delta,\delta,i,k,k}\Big)^{2p} \le c_p\lambda_i(t-s)^{2p-\epsilon}\delta^{\epsilon}.$$

(iv) Consider now the integrals over one of the rectangles $[t_{m'-1}^n, t_{m'}^n] \times [t_m^n, t_{m+1}^n]$ such that $m \ge m'$. First we note that

$$\begin{aligned}
I_{m'-1,m}^{R,\delta,i,k,k} &:= \int_{m\delta}^{(m+1)\delta}\int_{(m'-1)\delta}^{m'\delta} e^{-\lambda_i(\xi-r)}d\omega_k(r)d\omega_k(\xi)\\
&\quad - \int_{m\delta}^{(m+1)\delta}\int_{(m'-1)\delta}^{m'\delta} e^{-\lambda_i(\xi-r)}d\omega_k^n(r)d\omega_k^n(\xi)\\
&=_{distr}\, e^{-\lambda_i\delta(m-m')}\int_{m\delta}^{(m+1)\delta}\int_{(m-1)\delta}^{m\delta}\left(e^{-\lambda_i(\xi-r)} - \frac{(1-e^{-\lambda_i\delta})^2}{\delta^2\lambda_i^2}\right)d\omega_k(r)d\omega_k(\xi)\\
&= e^{-\lambda_i\delta(m-m')}I_{m-1,m}^{R,\delta,i,k,k}
\end{aligned}$$

where the factor in front of the integral is less than 1. Considering the $2p$-moment of these integrals, thanks to Lemma 10.3 and Hölder's inequality, we have

$$\begin{aligned}
\mathbb{E}(I_{m-1,m}^{R,\delta,i,k,k})^{2p} &\le c_p\delta\lambda_i\frac{1}{(\delta\lambda_i)}\delta^{2p-2}\int_{m\delta}^{(m+1)\delta}\int_{(m-1)\delta}^{m\delta}\left(e^{-\lambda_i(\xi-r)} - \frac{(1-e^{-\lambda_i\delta})^2}{\delta^2\lambda_i^2}\right)^{2p} drd\xi\\
&\le c_p\delta\lambda_i\delta^{2p}\int_0^1\int_{-1}^0\left(e^{-\lambda_i\delta(\xi-r)} - \frac{(1-e^{-\lambda_i\delta})^2}{\delta^2\lambda_i^2}\right)^{2p}/(\delta\lambda_i)drd\xi\\
&\le c_p\delta\lambda_i\delta^{2p}\int_0^1\int_{-1}^0 \sup_{x>0,y\in[0,2]}\left(e^{-xy} - \frac{(1-e^{-x})^2}{x^2}\right)^{2p}/xdrd\xi.
\end{aligned}$$

The above supremum is finite. Hence it is easily seen that this integrand is bounded independently of y for $x \to +\infty$. Consider now the integral in (10.12) over the union of squares inside $\bar{\Delta}_{0,T}$. Expanding

$$\mathbb{E}\left(\sum_{m'\le m} I_{m',m}^{R,\delta,i,k,k}\right)^{2p}$$

we can cancel all terms containing an odd power by the independence of all these integrals, such that the number of terms in this multinomial is of order $((t-s)/\delta)^{2p}$ by Lemma 10.4. Hence

$$\mathbb{E}\left(\sum_{m'\le m} I_{m-1,m}^{R,\delta,i,k,k}\right)^{2p} = O\left(\frac{(t-s)^{2p}}{\delta^{2p}}\right)\lambda_i\delta^{2p+1} = \lambda_i O((t-s)^{2p})\delta, \quad \delta\to 0^+$$

for $\delta \to 0^+$. Therefore, for any $p,\ i,\ k$ there exists a c_p such that

$$\mathbb{E}\left(\sum_{m'\le m} I_{m-1,m}^{R,\delta,i,k,k}\right)^{2p} \le c_p\lambda_i\delta(t-s)^{2p}.$$

(v) In a similar manner, we can consider the rectangles along the cathetus of $\bar{\Delta}_{0,T}$. We omit the calculations.

Lemma 10.6 *Suppose that for a $\nu > 0$ and $p \in \mathbb{N}$ we have*

$$\sum_i \lambda_i^{\frac{-\nu p}{p-1}} < \infty \quad \textit{and} \quad \sum_i \lambda_i^{p(\nu-2\kappa)+1} < \infty.$$

Then there exists a $c_p > 0$ such that for all $(s,t) \in \bar{\Delta}_{0,T}$

$$\mathbb{E}\|(\omega\otimes_S\omega)(s,t)\|^{2p}_{L_2(L_2(V,V_\kappa),V\otimes V)} \le c_p(t-s)^{2p}.$$

In addition the random field $(\omega\otimes_S\omega)$ has a continuous version.

Proof In the proof, c_p denotes a constant that can vary from one line to other. First of all,

$$\mathbb{E}(\omega\otimes_S\omega)^{2p}_{ijk} \le c_p(t-s)^{2p} + c_p\mathbb{E}\left(\int_s^t\int_s^\xi e^{-\lambda_i(\xi-r)}d\omega_j(r)d\omega_k(\xi)\right)^{2p}.$$

To estimate this last expectation, note that

$$\mathbb{E}\left(\int_s^\xi e^{-\lambda_i(\xi-r)}d\omega_j(r)\right)^{2p} \le c_p\left(\mathbb{E}\left(\int_s^\xi e^{-\lambda_i(\xi-r)}d\omega_j(r)\right)^2\right)^p \le c_p(\xi-s)^p,$$

due to the fact that the above integral is a Gauß-variable. Applying Lemma 10.3, we obtain that $\mathbb{E}(\omega \otimes_S \omega)_{ijk}^{2p} \le c_p(t-s)^{2p}$. Now, by Hölder's inequality we have

$$\mathbb{E}\|(\omega \otimes_S \omega)(s,t)\|^{2p}_{L_2(L_2(V,V_\kappa),V\otimes V))} = \mathbb{E}\Big(\sum_{i,j,k} q_j q_k \lambda_i^{-2\kappa}(\omega \otimes_S \omega)^2_{ijk}\Big)^p$$

$$\le \Big(\sum_{ijk} q_j q_k \lambda_i^{\frac{-\nu p}{p-1}}\Big)^{p-1} \sum_{ijk} q_j q_k \lambda_i^{p\nu - 2p\kappa}\mathbb{E}(\omega \otimes_S \omega)^{2p}_{ijk}$$

$$\le \Big(\sum_{ijk} q_j q_k \lambda_i^{\frac{-\nu p}{p-1}}\Big)^{p-1} \sum_{ijk} q_j q_k \lambda_i^{p\nu - 2p\kappa} c_p(t-s)^{2p}.$$

Next we would like to sketch that the random field $(\omega \otimes_S \omega)$ has a continuous version in $L_2(L_2(V,V_\kappa), V\otimes V)$. Let us first consider the Ito-version of the Hilbert space-valued stochastic integral $(\omega \otimes_S \omega)(0,\cdot)$ given by

$$t \mapsto E(\omega \otimes_S \omega)(0,t) + \frac{1}{2}\sum_i (e_i, Ee_i)(e_i \otimes_V e_i)t. \tag{10.15}$$

To see that such an Ito-Integral makes sense, we consider at first a predictable stochastic process $G : [0,T] \mapsto V$. We define

$$w \mapsto (G(\xi)\otimes)w = G(\xi)\otimes_V w \in V\otimes V, \quad w \in V.$$

If $G\otimes$ satisfies the condition

$$\mathbb{E}\Big(\int_0^t (G(\xi)\otimes)d\omega(\xi)\Big)^2 = \int_0^t \sum_i q_i \mathbb{E}\|(G(\xi)\otimes)e_i\|^2_{V\otimes V}d\xi$$

$$= \mathbb{E}\int_0^t \sum_{ijk} q_i(e_j \otimes_V e_k, G(\xi)\otimes_V e_i)^2_{V\otimes V}d\xi$$

$$= \sum_{ij}\int_0^t q_i \mathbb{E}(e_j, G(\xi))^2 d\xi = \sum_i q_i \int_0^T \mathbb{E}\|G(\xi)\|^2 d\xi < \infty$$

then the Ito-integral

$$\int_0^t (G(\xi)\otimes)d\omega(\xi) \in V\otimes V$$

is well-defined. We apply this formula to define an $L_2(L_2(V,V_\kappa), V\otimes V)$-valued Ito-integral. In particular we set

$$G(\xi) = G_{ij}(\xi) = \int_0^\xi S(\xi - r)E_{ij}d\omega(r)$$

where $(E_{ij})_{i,j\in\mathbb{N}}$ is the complete orthonormal system of $L_2(V, V_\kappa)$. Then $(\omega \otimes_S \omega)(s,t)$ is well defined on $L_2(L_2(V, V_\kappa), V \otimes V)$.

Since additivity holds almost surely for the Ito-integrals, the following equality follows

$$\cdot(\omega \otimes_S \omega)(s,t) = \cdot(\omega \otimes_S \omega)(0,t) - \cdot(\omega \otimes_S \omega)(0,s) \\ - (-1)^{-\alpha} \int_s^t \int_0^s S(\xi - r) \cdot d\omega(r) \otimes_V d\omega(\xi) \tag{10.16}$$

almost surely for $(s,t) \in \bar{\Delta}_{0,T}$. Note that $t \mapsto (\omega \otimes_S \omega)(0,\cdot)$ is continuous on a set of full measure. Furthermore, the continuity of

$$\bar{\Delta}_{0,T} \ni (s,t) \mapsto \int_s^t \int_0^s S(\xi - r) \cdot d\omega(r) \otimes_V d\omega(\xi)$$

follows by [11] by a fractional calculus argument based on fractional integrals. By (10.16) $(\omega \otimes_S \omega)(s,t)$ is continuous on $\bar{\Delta}_{0,T}$ on a set of full measure. Outside this set of full measure, we set $(\omega \otimes_S \omega) \equiv 0$. Hence $(\omega \otimes_S \omega)$ is continuous on $\bar{\Delta}_{0,T}$.

Lemma 10.7 *Suppose that the conditions for* ν, p *of the last lemma hold. Then*

$$\mathbb{E}\|(\omega \otimes_S \omega)(s,t) - (\omega^n \otimes_S \omega^n)(s,t)\|^{2p}_{L_2(L_2(V,V_\kappa),V\otimes V)} \le c_p \delta^\epsilon (t-s)^{2p-\epsilon}.$$

Proof Having Lemma 10.5 in mind we obtain

$$\begin{aligned}
&\mathbb{E}\|(\omega \otimes_S \omega)(s,t) - (\omega^n \otimes_S \omega^n)(s,t)\|^{2p}_{L_2(L_2(V,V_\kappa),V\otimes V)} \\
&= \mathbb{E}\Big(\sum_{i,j,k} q_j q_k \lambda_i^{-2\kappa}((\omega \otimes_S \omega)_{ijk} - (\omega^n \otimes_S \omega^n)_{ijk})^2\Big)^p \\
&\le \Big(\sum_{ijk} q_j q_k \lambda_i^{\frac{-p\nu}{p-1}}\Big)^{p-1} \sum_{ijk} q_j q_k \lambda_i^{p\nu\ \ 2p\kappa} \mathbb{E}((\omega \otimes_S \omega)_{ijk} - (\omega^n \otimes_S \omega^n)_{ijk})^{2p} \\
&\le c\Big(\sum_{ijk} q_j q_k \lambda_i^{\frac{-p\nu}{p-1}}\Big)^{p-1} \sum_{ijk} q_j q_k \lambda_i^{p(\nu-2\kappa)} \delta^\epsilon \lambda_i (t-s)^{2p-\epsilon}.
\end{aligned}$$

Note that

$$\int_\tau^t \int_s^\tau S(\xi - r) E d\omega(r) \otimes_V d\omega(\xi) = \int_\tau^t S(\xi - \tau) \int_s^\tau S(\tau - r) E d\omega(r) \otimes_V d\omega(\xi),$$

and this is the reason to define the following operators:

$$e \in V \mapsto \omega_S(s,\tau)e := (-1)^{-\alpha}\int_s^\tau (S(\xi - s)e)\otimes_V d\omega(\xi)$$

$$E \in L_2(V, V_\kappa) \mapsto S_\omega(\tau, t)E := \int_\tau^t S(t-r)Ed\omega(r). \tag{10.17}$$

We refer to Garrido-Atienza et al. [11] to check that these operators are well-defined and for its additional properties.

Now we can formulate the main result of this section:

Lemma 10.8 *Suppose that ω is a trace-class canonical Brownian motion. Then Theorem 10.1 holds.*

Proof First we show that $(\omega \otimes_S \omega)$ exists in the sense of Theorem 10.1. We apply the Garsia-Rodemich-Rumsey lemma for Deya et al. [6], Lemma 3.4, which is possible since $(\omega \otimes_S \omega)$ is continuous as we have already explained. By Lemma 10.6 we obtain that

$$\mathbb{E}\int_0^T\int_0^t \frac{\|(\omega\otimes_S\omega)(s,t)\|^{2p}_{L_2(L_2(V,V_\kappa),V\otimes V)}}{|t-s|^{4\beta p+2}}dsdt \le c_p\int_0^T\int_0^t \frac{|t-s|^{2p}}{|t-s|^{4\beta p+2}}dsdt < \infty$$

when p is sufficiently large since $\beta < 1/2$. This shows that there exists a set $\Omega' \subset \Omega$ of $\mathbb{P}$-measure one on which $(\omega \otimes_S \omega) \in C_{2\beta}(\bar{\Delta}_{0,T}; L_2(L_2(V, V_\kappa), V \otimes V))$.

To see the convergence property of the theorem, we note that from Lemma 10.7

$$\begin{aligned}\mathbb{E}&\int_0^T\int_0^t \frac{\|(\omega\otimes_S\omega)(s,t)-(\omega^n\otimes_S\omega^n)(s,t)\|^{2p}_{L_2(L_2(V,V_\kappa),V\otimes V))}}{|t-s|^{4\beta p+2}}dsdt\\ &\le c_p\delta^\epsilon\int_0^T\int_0^t\frac{|t-s|^{2p-\epsilon}}{|t-s|^{4\beta p+2}}dsdt \le \delta^\epsilon c\end{aligned}$$

for p sufficiently large such that $p(2-4\beta) > 1+\epsilon$. Recall that $\delta = \delta(n) = 2^{-n}T$ and take a sequence $(o(n))_{n\in\mathbb{N}}$ tending to zero for $n \to \infty$ such that $\sum_n \delta(n)^\epsilon/o(n) < \infty$. Then by the Chebyshev-lemma

$$\begin{aligned}\sum_n\mathbb{P}&\left(\int_0^T\int_0^t \frac{\|(\omega\otimes_S\omega)(s,t)-(\omega^n\otimes_S\omega^n)(s,t)\|^{2p}_{L_2(L_2(V,V_\kappa),V\otimes V)}}{|t-s|^{4\beta p+2}}dsdt > o(n)\right)\\ &\le \sum_n\frac{1}{o(n)}\mathbb{E}\int_0^T\int_0^t\frac{\|(\omega\otimes_S\omega)(s,t)-(\omega^n\otimes_S\omega^n)(s,t)\|^{2p}_{L_2(L_2(V,V_\kappa),V\otimes V)}}{|t-s|^{4\beta p+2}}dsdt\\ &\le c\sum_n\frac{\delta(n)^\epsilon}{o(n)} < \infty.\end{aligned}$$

Hence by the Borel-Cantelli-lemma we obtain the convergence of the integrals inside the above probability to zero for $n \to \infty$ with probability 1. In addition, we have

$$\lim_{n\to\infty} \sup_{0<s<r<t<T} \frac{\|\omega_S(r,t)S_{(\omega^n-\omega)}(s,r) - (\omega^n-\omega)_S(r,t)S_{\omega^n}(s,r)\|_{L_2(L_2(V,V_\kappa),V\otimes V)}}{|r-s|^\beta|t-r|^\beta} = 0, \tag{10.18}$$

see [11]. Now by Lemma 3.4 of [6] we get

$$\begin{aligned}
&\|(\omega\otimes_S\omega) - (\omega^n\otimes_S\omega^n)\|_{C_{2\beta}(\bar{\Delta}_{0,T},L_2(L_2(V,V_\kappa),V\otimes V))}\\
&\quad\le c\left(\int_0^T\int_0^t \frac{\|(\omega\otimes_S\omega)(s,t) - (\omega^n\otimes_S\omega^n)(s,t)\|^{2p}_{L_2(L_2(V,V_\kappa),V\otimes V)}}{|t-s|^{4\beta p+2}}dsdt\right)^{1/2p}\\
&\quad + c\sup_{0<s<r<t<T} \frac{\|\omega_S(r,t)S_{(\omega^n-\omega)}(s,r) - (\omega^n-\omega)_S(r,t)S_{\omega^n}(s,r)\|_{L_2(L_2(V,V_\kappa),V\otimes V)}}{|r-s|^\beta|t-r|^\beta}.
\end{aligned}$$

Hence we have the convergence conclusion of Theorem 10.1 on a set Ω_0 of full $\mathbb{P}$-measure.

Remark 10.2 If $-A$ is given by the Laplace-operator with homogenous Dirichlet- or Neumann-boundary conditions on a bounded smooth domain in $\mathbb{R}^d$ such that $\lambda_i \sim i^{2/d}$ and $d = 1$, the conditions of Lemma 10.7 are fulfilled when choosing $\nu = 1/2$, $\kappa > 1/4$ and p sufficiently large.

10.4 Additional Properties of $(\omega \otimes_S \omega)$

In this section, we study additional properties of $(\omega \otimes_S \omega)$, which are valid for the two considered constructions.

Let us denote by Δ the set of pairs $(s,t) \in \mathbb{R}^2$ such that $s \le t$. We consider the set $C(\Delta, L_2(L_2(V,V_\kappa), V\otimes V))$ equipped with a countable family of semi-norms for $m \in \mathbb{N}$

$$\sup_{(s,t)\in\bar{\Delta}_{-m,m}} \|F(s,t)\|_{L_2(L_2(V,V_\kappa),V\otimes V)},$$

for $F \in C(\Delta, L_2(L_2(V,V_\kappa), V\otimes V))$. This gives us a Polish-space.

In the previous sections, we found that we have $(\omega\otimes_S\omega) \in C(\bar{\Delta}_{0,T}, L_2(L_2(V,V_\kappa), V\otimes V))$ on a set of probability one. Since our noise is defined on $\mathbb{R}$, we obtain that for any $m \in \mathbb{N}$

$$(\omega\otimes_S\omega) \in C(\bar{\Delta}_{-m,m}, L_2(L_2(V,V_\kappa), V\otimes V))$$

and hence that

$$(\omega\otimes_S\omega) \in C(\Delta, L_2(L_2(V,V_\kappa), V\otimes V))$$

on a set of probability one. On this set, we have for every $m \in \mathbb{N}$ that $(\omega \otimes_S \omega)$ is 2β Hölder-continuous in the sense of (10.5) on $\bar{\Delta}_{-m,m}$. In addition, for every $m \in \mathbb{N}$

$$\lim_{n\to\infty} \|(\omega \otimes_S \omega) - (\omega^n \otimes_S \omega^n)\|_{2\beta,-m,m} = 0.$$

Outside this set, let us define $(\omega^n \otimes_S \omega^n) \equiv 0$ such that those relations hold for all ω.

Theorem 10.2 *There exists a set of full measure where we have the Chen-equality: On a set of probability one for* $-\infty < s \leq \tau \leq t < \infty$

$$(\omega \otimes_S \omega)(s,t) - (\omega \otimes_S \omega)(\tau,t) - (\omega \otimes_S \omega)(s,\tau) = (-1)^{-\alpha}\omega_S(r,t)S_\omega(s,r).$$

It is a simple exercise to check this Chen-equality for $(\omega^n \otimes_S \omega^n)$ which follows easily by the semigroup properties and by the additivity of the integral with respect to the integral bounds. To see this inequality for $(\omega \otimes_S \omega)$, we can apply the convergence conclusion of Theorem 10.1.

Finally, we show that $(\omega \otimes_S \omega)$ is strongly stationary. To do this, we recall the definition of strong stationarity of the metric dynamical system $(\Omega, \mathscr{F}, \mathbb{P}, \theta)$.

Definition 10.1 The stochastic process $X = (X_t)_{t\in\mathbb{R}}$ is called strongly stationary if there exists a θ-invariant set $\tilde{\Omega} \in \mathscr{F}$ of full measure such that for any $\tau \in \mathbb{R}$ and $\omega \in \tilde{\Omega}$ we have that

$$X_0(\theta_\tau \omega) = X_\tau(\omega). \tag{10.19}$$

Such a process is called *weakly stationary* if for any $\tau \in \mathbb{R}$ there exists a set $\Omega_\tau \in \mathscr{F}$ of full $\mathbb{P}$-measure such that for $\omega \in \Omega_\tau$ we have (10.19).

To establish the last result of this paper, we need to use the following theorem, see Lederer [17]. Similar results can be found in Arnold and Scheutzow [2], Imkeller and Lederer [13], and Kümmel [15].

Theorem 10.3 *Let S be a Polish-space and let* $X = (X_t)_{t\in\mathbb{R}}$ *be a stochastic process over the metric dynamical system* $(\Omega, \mathscr{F}, \mathbb{P}, \theta)$. *Suppose that X is weakly stationary. Then there exists a continuous stochastic process* $\hat{X} = (\hat{X})_{t\in\mathbb{R}}$ *such that X and* $\hat{X}$ *are indistinguishable, and* $\hat{X}$ *is strongly stationary.*

Now we apply this theorem.

Theorem 10.4 *The random fields* $(\omega \otimes_S \omega)$ *defined in Sects. 10.2 and 10.3 have a strong stationary version satisfying the convergence property of Theorem 10.1 with respect to* $\bar{\Delta}_{-m,m}$.

Proof We know that $(\omega \otimes_S \omega)$ is continuous. Since $\bar{\Delta}_{-m,m}$ is compact $(\omega \otimes_S \omega)$ restricted to this set is uniformly continuous, and hence $\tau \mapsto (\omega \otimes_S \omega)(\cdot + \tau, \cdot + \tau) \in C(\bar{\Delta}_{-m,m}, L_2(L_2(V, V_\kappa), V \otimes V))$.

In addition, shifting the integrals of the random field $(\omega \otimes_S \omega)$ we easily obtain that $\tau \mapsto (\omega \otimes_S \omega)(\cdot, \cdot)$ is weakly stationary: For every $\tau \in \mathbb{R}$ we have that almost surely

$$(\omega \otimes_S \omega)(\cdot + \tau, \cdot + \tau) = (\theta_\tau \omega \otimes_S \theta_\tau \omega)(\cdot, \cdot).$$

In particular, the weak stationarity of $(\omega \otimes \omega)$ of Sect. 10.2 follows by Deya et al. [6] Sect. 10.2, where this area is presented by a double integral of smooth processes, while for the second expression of (10.6) we use that $D_{\cdot -}^{1-\alpha}$ and θ_t commute in some sense, see [11] for the meaning of this commutation. Therefore, taking $S := C(\Delta, L_2(L_2(V, V_\kappa), V \otimes V))$ we can apply Theorem 10.3 to $\tau \mapsto X_\tau(\omega) := (\omega \otimes_S \omega)(\cdot + \tau, \cdot + \tau) \in S$, obtaining a version which is strongly stationary. However, this version is in addition indistinguishable such that we obtain the regularity, the convergence properties and the Chen-equality for this new version of $(\omega \otimes_S \omega)$.

References

1. Arnold, L.: Random Dynamical Systems. Springer, Berlin (1998)
2. Arnold, L., Scheutzow, M.: Perfect cocycles through stochastic differential equations. Probab. Theory Relat. Fields **110**, 559–588 (1998)
3. Bártek, J., Garrido-Atienza, M.J., Maslowski, B.: Stochastic porous media equation driven by fractional Brownian motion. Stoch. Dyn. **13**(4), 1350010 (2013)
4. Bensoussan, A., Frehse, J.: Local solutions for stochastic Navier stokes equations. Math. Model. Numer. Anal. **34**(2), 241–273 (2000)
5. Caraballo, T., Kloeden, P., Schmalfuß, B.: Exponentially stable stationary solutions for stochastic evolution equations and their perturbation. Appl. Math. Optim. **50**, 183–207 (2004)
6. Deya, A., Neuenkirch, A., Tindel, S.: A Milstein-type scheme without Lévy area terms for SDES driven by fractional Brownian motion. Ann. Inst. H. Poincaré Probab. Stat. **48**(2), 518–550 (2012)
7. Flandoli, F., Lisei, H.: Stationary conjugation of flows for parabolic SPDEs with multiplicative noise and some applications. Stoch. Anal. Appl. **22**, 1385–1420 (2004)
8. Friz, P., Hairer, M.: A Course on Rough Paths (2014). To appear
9. Gao, H., Garrido-Atienza, M.J., Schmalfuß, B.: Random attractors for stochastic evolution equations driven by fractional Brownian motion. SIAM J. Math. Anal. **46**(4), 2281–2309 (2014)
10. Garrido-Atienza, M.J., Lu, K., Schmalfuß, B.: Random dynamical systems for stochastic partial differential equations driven by a fractional Brownian motion. Discret. Contin. Dyn. Syst. Ser. B **14**(2), 473–493 (2010)
11. Garrido-Atienza, M.J., Lu, K., Schmalfuß, B.: Pathwise local solutions to stochastic partial differential equations driven by fractional Brownian motions with Hurst parameters in (1/3, 1/2]. To appear in Discrete and Continuous Dynamical Systems, Series B
12. Hu, Y., Nualart, D.: Rough path analysis via fractional calculus. Trans. Am. Math. Soc. **361**(5), 2689–2718 (2009)
13. Imkeller, P., Lederer, C.: The cohomology of stochastic and random differential equations, and local linearization of stochastic flows. Stoch. Dyn. **2**(2), 131–159 (2002)
14. Karatzas, I., Shreve, S.E.: Brownian Motion and Stochastic Calculus, 2nd edn. Springer, New York (1991)
15. Kümmel, K.: Dynamics of Marcus type SDEs. In preparation
16. Kunita, H.: Stochastic Flows and Stochastic Differential Equations. Cambridge University Press, Cambridge (1990)

17. Lederer, C.: Konjugation stochastischer und zufälliger stationärer Differentialgleichungen und eine Version des lokalen Satzes von Hartman-Grobman für stochastische Differentialgleichungen. Ph.D. thesis. HU Berlin (2001)
18. Lyons, T., Qian, Z.: System Control and Rough Paths. Oxford Mathematical Monographs. Oxford Science Publications. Oxford University Press, Oxford (2002)
19. Maslowski, B., Schmalfuß, B.: Random dynamical systems and stationary solutions of differential equations driven by the fractional Brownian motion. Stoch. Anal. Appl. **22**, 1577–1607 (2004)
20. Pazy, A.: Semigroups of Linear Operators and Applications to Partial Differential Equations. Springer Applied Mathematical Series. Springer, Berlin (1983)
21. Young, L.C.: An integration of Hölder type, connected with Stieltjes integration. Acta Math. **67**, 251–282 (1936)
22. Zähle, M.: Integration with respect to fractal functions and stochastic calculus. I. Probab. Theory Relat. Fields **111**(3), 333–374 (1998)

Chapter 11
Periodic and Almost Periodic Random Inertial Manifolds for Non-Autonomous Stochastic Equations

B. Wang

Abstract By the Lyapunov-Perron method, we prove the existence of random inertial manifolds for a class of equations driven simultaneously by nonautonomous deterministic and stochastic forcing. These invariant manifolds contain tempered pullback random attractors if such attractors exist. We also prove pathwise periodicity and almost periodicity of inertial manifolds when nonautonomous deterministic forcing is periodic and almost periodic in time, respectively.

11.1 Introduction

In this paper, we investigate the existence of random inertial manifold (IM) for the following nonautonomous stochastic equation on a separable Hilbert space H for $t > \tau$ with $\tau \in \mathbf{R}$:

$$\frac{du}{dt} + Au = F(u) + g(t) + \frac{dW}{dt} \quad \text{with } u(\tau) = u_\tau, \tag{11.1}$$

where $A : D(A) \subseteq H \to H$ is a symmetric positive operator with compact inverse, F is a nonlinearity, g is a time-dependent external forcing, and W is a H-valued Wiener process on a probability space $(\Omega, \mathscr{F}, P)$.

If g does not depend on time, then (11.1) is said to be an autonomous stochastic equation. The existence of IMs for autonomous random systems has been studied by experts in [2, 3, 5–8, 12] and the references therein. For such a system, an IM is a random set $\mathscr{M} = \{\mathscr{M}(\omega) : \omega \in \Omega\}$ which is invariant, exponentially attracts all solutions, and is given by the graph of a finite-dimensional Lipschitz map. In the present paper, we want to consider the nonautonomous case where g varies in time. It seems that there is no result reported in the literature on IMs when g is time dependent. As we will see in Sect. 11.3, an IM for a nonautonomous

B. Wang (✉)
Department of Mathematics, New Mexico Institute of Mining and Technology,
801 Leroy Place, Socorro, NM 87801, USA
e-mail: bwang@nmt.edu

© Springer International Publishing Switzerland 2015

V.A. Sadovnichiy and M.Z. Zgurovsky (eds.), *Continuous and Distributed Systems II*, Studies in Systems, Decision and Control 30,
DOI 10.1007/978-3-319-19075-4_11

stochastic equation is parametrized not only by $\omega \in \Omega$, but also by initial times $\tau \in \mathbf{R}$. More precisely, an IM in this case is a finite-dimensional Lipschitz manifold $\mathcal{M} = \{\mathcal{M}(\tau, \omega) : \tau \in \mathbf{R}, \omega \in \Omega\}$ that is invariant and exponentially attracts all solutions. We will first prove the existence of such IMs for Eq. (11.1) in Sect. 11.3 under the classical gap condition. We will also prove that these IMs must contain tempered pullback random attractors if such an attractor exists. In addition, we will investigate the dependence of IMs on the external forcing g. If g is almost periodic, we show the random IMs are pathwise almost periodic. We will also prove pathwise periodicity of IMs when g is periodic in time.

We remark that constructions of random IMs are quite similar to random invariant manifolds which have been investigated recently in [1, 4, 9–11, 13–15, 17]. Of course, for IMs, one must establish the exponential attracting property which is also called the asymptotic completeness of IMs. Interestingly enough, the proof of asymptotic completeness is often based on similar arguments for constructing IMs but under even more restrictive conditions.

In the next section, we review some concepts of random dynamical systems and define IMs for nonautonomous stochastic equations. Section 11.3 is devoted to the existence of IMs for nonautonomous random systems under a gap condition. In the last section, we show the periodicity and almost periodicity of IMs when g is periodic and almost periodic in time, respectively.

11.2 Notation

In this section, we introduce the concept of random IMs for nonautonomous stochastic equations. The reader is referred to [2, 3, 5–8, 12] on IMs for autonomous random systems.

Assume X is a separable Banach space with norm $\|\cdot\|$, and $(\Omega, \mathcal{F}, P, \{\theta_t\}_{t\in\mathbf{R}})$ is a metric dynamical system as in [1]. Let $D = \{D(\tau, \omega) : \tau \in \mathbf{R}, \omega \in \Omega\}$ be a family of nonempty bounded subsets of X. Then D is said to be tempered if $\lim_{t\to-\infty} e^{ct}\|D(\tau+t, \theta_t\omega)\|_X = 0$ for every $c > 0$, where $\|D\|_X = \sup_{x\in D}\|x\|$. A family $D = \{D(\tau, \omega) : \tau \in \mathbf{R}, \omega \in \Omega\}$ of nonempty closed subsets is said to be measurable if it is measurable in ω for each fixed τ. The following is the definition of cocycle for nonautonomous random equations.

Definition 11.1 Suppose $\Phi: \mathbf{R}^+ \times \mathbf{R} \times \Omega \times X \to X$ satisfies, for any $\tau \in \mathbf{R}, \omega \in \Omega$ and $t, s \in \mathbf{R}^+$,

(i) $\Phi(\cdot, \tau, \cdot, \cdot) : \mathbf{R}^+ \times \Omega \times X \to X$ is $(\mathcal{B}(\mathbf{R}^+) \times \mathcal{F} \times \mathcal{B}(X), \mathcal{B}(X))$-measurable;
(ii) $\Phi(0, \tau, \omega, \cdot)$ is the identity on X;
(iii) $\Phi(t+s, \tau, \omega, \cdot) = \Phi(t, \tau+s, \theta_s\omega, \cdot) \circ \Phi(s, \tau, \omega, \cdot)$.

Such Φ is called a cocycle in X over $(\Omega, \mathcal{F}, P, \{\theta_t\}_{t\in\mathbf{R}})$. If, in addition, $\Phi(t, \tau, \omega, \cdot) : X \to X$ is continuous, then we say Φ is a continuous cocycle in X.

Definition 11.2 A measurable family $\mathscr{M} = \{\mathscr{M}(\tau, \omega) : \tau \in \mathbf{R}, \omega \in \Omega\}$ of subsets of X is called a random inertial manifold of Φ if the following conditions (i)–(iii) are satisfied:

(i) There exists a finite-dimensional subspace X_1 such that $X = X_1 \oplus X_2$ and there is a mapping $m : \mathbf{R} \times \Omega \times X_1 \to X_2$ such that for every $\tau \in \mathbf{R}$ and $\omega \in \Omega$,

$$\mathscr{M}(\tau, \omega) = \{x + m(\tau, \omega)(x) : x \in X_1\} \tag{11.1}$$

where $m(\tau, \omega)(x)$ is measurable in $\omega \in \Omega$ and is Lipschitz continuous in $x \in X_1$;

(ii) $\mathscr{M}$ is invariant: $\Phi(t, \tau, \omega, \mathscr{M}(\tau, \omega)) = \mathscr{M}(\tau + t, \theta_t \omega)$ for all $t \in \mathbf{R}^+$, $\tau \in \mathbf{R}$ and $\omega \in \Omega$;

(iii) $\mathscr{M}$ is asymptotically complete: for every $\tau \in \mathbf{R}, \omega \in \Omega$ and $x \in X$, there exists $z = z(\tau, \omega, x) \in \mathscr{M}(\tau, \omega)$ such that for all $t \geq 0$,

$$\|\Phi(t, \tau, \omega, x) - \Phi(t, \tau, \omega, z)\| \leq c_1 e^{-c_2 t},$$

where c_1 and c_2 are positive numbers depending on τ, ω and x.

Next, we define pathwise almost periodic random inertial manifolds. Let $g : \mathbf{R} \to X$ be a continuous function. Then g is said to be almost periodic (see, e.g., [18]) if for every $\varepsilon > 0$ there exists a positive number $l = l(\varepsilon)$ such that every interval of length l contains a number t_0 for which $\|g(t + t_0) - g(t)\| < \varepsilon$ for all $t \in \mathbf{R}$. Motivated by almost periodic functions, we introduce the following concept.

Definition 11.3 Let $\mathscr{M} = \{\mathscr{M}(\tau, \omega) : \tau \in \mathbf{R}, \omega \in \Omega\}$ be a random inertial manifold of Φ given by (11.1). Then $\mathscr{M}$ is said to be pathwise almost periodic if for every $\varepsilon > 0$, there exists a positive number $l = l(\varepsilon)$ such that any interval of length l contains a number τ_0 such that

$$\sup_{x \in X_1} \|m(\tau + \tau_0, \omega)(x) - m(\tau, \omega)(x)\| \leq \varepsilon, \quad \text{for all } \tau \in \mathbf{R} \text{ and } \omega \in \Omega.$$

If there exists $T > 0$ such that $m(\tau + T, \omega)(x) = m(\tau, \omega)(x)$ for all $\tau \in \mathbf{R}$, $\omega \in \Omega$ and $x \in X_1$, then $\mathscr{M}$ is called a pathwise periodic random inertial manifold with period T.

It is evident that if $\mathscr{M}$ is a T-periodic random inertial manifold in the sense of Definition 11.3, then $\mathscr{M}(\tau + T, \omega) = \mathscr{M}(\tau, \omega)$ for all $\tau \in \mathbf{R}$ and $\omega \in \Omega$. Similarly, if $\mathscr{M}$ is almost periodic, then for every $\varepsilon > 0$, there exists a positive number $l = l(\varepsilon)$ such that any interval of length l contains a number τ_0 such that for all $\tau \in \mathbf{R}$ and $\omega \in \Omega$,

$$d^H(\mathscr{M}(\tau + \tau_0, \omega), \mathscr{M}(\tau, \omega)) \leq \varepsilon \text{ and } d^H(\mathscr{M}(\tau, \omega), \mathscr{M}(\tau + \tau_0, \omega)) \leq \varepsilon, \tag{11.2}$$

where d^H is the Hausdorff semi-distance between two subsets of X. Note that (11.2) actually indicates the Hausdorff distance of $\mathscr{M}(\tau + \tau_0, \omega)$ and $\mathscr{M}(\tau, \omega)$ is controlled by ε.

11.3 Existence of Random Inertial Manifolds

In this section, we construct random inertial manifolds for the stochastic equation (11.1) in a separable Hilbert space H with norm $\|\cdot\|$. Assume that W in (11.1) is a two-sided H-valued Wiener process with covariance operator Q of trace class on a probability space $(\Omega, \mathscr{F}, P)$ where $\Omega = \{\omega \in C(\mathbf{R}, H) : \omega(0) = 0\}$ with compact-open topology, $\mathscr{F}$ is the Borel σ-algebra and P the Wiener measure. Suppose $A : D(A) \subseteq H \to H$ is a symmetric positive operator with compact inverse. Then H has an orthonormal basis $\{e_n\}_{n=1}^{\infty}$ consisting of eigenvectors of A:

$$Ae_n = \lambda_n e_n, \quad 0 < \lambda_1 \le \lambda_2 \le \cdots \le \lambda_n \le \cdots \quad \text{with } \lambda_n \to \infty.$$

Let $F : D(A^\alpha) \to H$ be a Lipschitz nonlinearity for some $\alpha \in [0, \frac{1}{2})$; that is, there exists a positive constant L such that

$$F(0) = 0, \quad \|F(u_1) - F(u_2)\| \le L\|A^\alpha(u_1 - u_2)\| \tag{11.1}$$

for all $u_1, u_2 \in D(A^\alpha)$. This implies that for all $u \in D(A^\alpha)$,

$$\|F(u)\| \le L\|A^\alpha u\|. \tag{11.2}$$

For the nonautonomous term g in (11.1), we assume that $g \in L^2_{loc}(\mathbf{R}, D(A^\alpha))$ and

$$\int_{-\infty}^{0} e^{\lambda_1 s}\|A^\alpha g(s)\|ds < \infty, \tag{11.3}$$

where λ_1 is the first eigenvalue of A. These assumptions imply that for every $\tau \in \mathbf{R}$,

$$\int_{-\infty}^{0} e^{\lambda_1 s}\|A^\alpha g^\tau(s)\|ds < \infty, \tag{11.4}$$

where $g^\tau(\cdot) = g(\cdot + \tau)$ is the translation of g by τ.

Given an integer $n \ge 1$, let $P_n : H \to span\{e_1, \cdots, e_n\}$ be the orthogonal projection and $Q_n = I - P_n$. Then we have (see, e.g., [5]) for all $u \in H$,

$$\|A^\alpha e^{-At}P_n u\| \le \lambda_n^\alpha e^{-\lambda_n t}\|u\|, \quad t \le 0, \tag{11.5}$$

$$\|e^{-At}Q_n u\| \le e^{-\lambda_{n+1} t}\|u\|, \quad t \ge 0, \tag{11.6}$$

and

$$\|A^\alpha e^{-At}Q_n u\| \le \left(\alpha^\alpha t^{-\alpha} + \lambda_{n+1}^\alpha\right)e^{-\lambda_{n+1} t}\|u\|, \quad t > 0. \tag{11.7}$$

Note that $e^{-At}: P_n H \to P_n H$ is invertible for any $t \geq 0$, and its inverse is denoted by e^{At}. This means $e^{-At}: P_n H \to P_n H$ is defined for all $t \in \mathbf{R}$. Let $\{\theta_t\}_{t\in\mathbf{R}}$ be a group of translations on Ω given by: $\theta_t\omega(\cdot) = \omega(\cdot + t) - \omega(t)$ for all $\omega \in \Omega$ and $t \in \mathbf{R}$. Then $(\Omega, \mathscr{F}, P, \{\theta_t\}_{t\in\mathbf{R}})$ is a metric dynamical system as in [1]. Let $z: \Omega \to D(A^\alpha)$ be the unique stationary solution of

$$dz(\theta_t\omega) + Az(\theta_t\omega) = dW. \tag{11.8}$$

Then by [5], $z(\theta_t\omega)$ has a continuous version from $\mathbf{R}$ to $D(A^\alpha)$ for every fixed ω, and is tempered in the sense that for every $c > 0$ and $\omega \in \Omega$,

$$\lim_{t\to-\infty} e^{ct} \|A^\alpha z(\theta_t\omega)\| = 0. \tag{11.9}$$

In terms of (11.8), we can transfer the stochastic equation (11.1) into a pathwise random one by introducing a new variable $v(t) = u(t) - z(\theta_t\omega)$. By (11.1) and (11.8) we get

$$\frac{dv}{dt} + Av = F(v + z(\theta_t\omega)) + g(t), \quad t > \tau, \quad v(\tau) = v_\tau. \tag{11.10}$$

By (11.1) and the Banach fixed point theory as in [3], one can prove that for every $v_\tau \in D(A^\alpha)$ with $0 \leq \alpha < \frac{1}{2}$, problem (11.10) has a unique mild solution in $C([\tau, \infty), D(A^\alpha))$, which is measurable in ω and continuous in v_τ in $(D(A^\alpha))$. To indicate the dependence on all related parameters, we write the solution of (11.10) as $v(t, \tau, \omega, g, v_\tau)$ which is given by

$$v(t, \tau, \omega, g, v_\tau) = e^{-A(t-\tau)}v_\tau + \int_\tau^t e^{-A(t-s)} \left(F(v(s) + z(\theta_s\omega)) + g(s)\right) ds.$$

Let $\Psi: \mathbf{R}^+ \times \mathbf{R} \times \Omega \times D(A^\alpha) \to D(A^\alpha)$ be a mapping given by, for all $t \in \mathbf{R}^+$, $\tau \in \mathbf{R}$, $\omega \in \Omega$ and $v_\tau \in D(A^\alpha)$,

$$\Psi(t, \tau, \omega, v_\tau) = v(t + \tau, \tau, \theta_{-\tau}\omega, g, v_\tau) = v(t, 0, \omega, g^\tau, v_\tau), \tag{11.11}$$

where $g^\tau(\cdot) = g(\cdot + \tau)$ as usual. Then we find that Ψ is a continuous cocycle over $(\Omega, \mathscr{F}, P, \{\theta_t\}_{t\in\mathbf{R}})$. Note that if v is a solution of (11.10), then the process

$$u(t, \tau, \omega, g, u_\tau) = v(t, \tau, \omega, g, v_\tau) + z(\theta_t\omega) \quad \text{with} \quad u_\tau = v_\tau + z(\theta_\tau\omega) \tag{11.12}$$

is a mild solution of the stochastic equation (11.1). Based on this fact, we can define a continuous cocycle Φ for (11.1) by

$$\Phi(t, \tau, \omega, u_\tau) = u(t + \tau, \tau, \theta_{-\tau}\omega, g, u_\tau) \tag{11.13}$$

for all $t \in \mathbf{R}^+$, $\tau \in \mathbf{R}$, $\omega \in \Omega$ and $u_\tau \in D(A^\alpha)$. By (11.11)–(11.13) we have

$$\Phi(t, \tau, \omega, u_\tau) = \Psi(t, \tau, \omega, u_\tau - z(\omega)) + z(\theta_t\omega). \tag{11.14}$$

We often need to replace g by g^τ on the right-hand side of (11.10) and consider the equation

$$\frac{dv}{dt} + Av = F(v + z(\theta_t\omega)) + g^\tau(t), \quad t > r, \quad v(r) = v_r. \tag{11.15}$$

The solution of (11.15) is given by, for $t \geq r$,

$$v(t, r, \omega, g^\tau, v_r) = e^{-A(t-r)}v_r + \int_r^t e^{-A(t-s)} \left(F(v(s) + z(\theta_s\omega)) + g(s + \tau)\right) ds. \tag{11.16}$$

Note that the solution of (11.15) is defined only for forward time $t \geq r$ in general. But we need to consider backward solutions when constructing inertial manifolds. For that purpose, we introduce the following concept of solutions defined on $(-\infty, 0]$.

Definition 11.4 Given $\tau \in \mathbf{R}$ and $\omega \in \Omega$, a continuous mapping $\xi : (-\infty, 0] \to D(A^\alpha)$ is called a mild solution of (11.15) on $(-\infty, 0]$ if $v(t, r, \omega, g^\tau, \xi(r)) = \xi(t)$ for all $r \leq t \leq 0$, where $v(t, r, \omega, g^\tau, \xi(r))$ is the unique solution of (11.15) with initial value $\xi(r)$.

We now assume that there is $n \in \mathbb{N}$ such that

$$\lambda_{n+1} - \lambda_n \geq \frac{2L}{k}\left(\lambda_{n+1}^\alpha + \lambda_n^\alpha + c_\alpha(\lambda_{n+1} - \lambda_n)^\alpha\right) \quad \text{for some } k \in (0, 1), \tag{11.17}$$

where c_α is a nonnegative number given by

$$c_\alpha = \alpha^\alpha \int_0^\infty s^{-\alpha}e^{-s}ds \quad \text{if } \alpha > 0 \quad \text{and } c_\alpha = 0 \quad \text{if } \alpha = 0.$$

For convenience, we set

$$\mu = \lambda_n + \frac{2L}{k}\lambda_n^\alpha. \tag{11.18}$$

Then by (11.17) we have $\mu \in (\lambda_n, \lambda_{n+1})$. Let $\mathscr{S}$ be the Banach space defined by

$$\mathscr{S} = \{\xi \in C((-\infty, 0], D(A^\alpha)) : \sup_{t \leq 0} e^{\mu t}\|A^\alpha\xi(t)\| < \infty\}$$

with norm $\|\xi\|_{\mathscr{S}} = \sup_{t \leq 0} e^{\mu t}\|A^\alpha\xi(t)\|$. Given $\tau \in \mathbf{R}$ and $\omega \in \Omega$, denote by $\mathscr{M}(\tau, \omega) = \{\xi(0) : \xi \in \mathscr{S}$ and is a mild solution of (11.15) on $(-\infty, 0]$ by Definition 11.4$\}$

$$= \{\xi(0) : \xi \in \mathscr{S} \text{ and } v(t, r, \omega, g^\tau, \xi(r)) = \xi(t) \text{ for all } r \le t \le 0\}. \qquad (11.19)$$

We will show $\mathscr{M} = \{\mathscr{M}(\tau, \omega) : \tau \in \mathbf{R}, \omega \in \Omega\}$ is an inertial manifold of (11.15) for which we need:

Lemma 11.1 *Suppose* (11.1), (11.3) *and* (11.17)–(11.18) *hold, and* $\xi \in \mathscr{S}$. *Then* ξ *is a mild solution of* (11.15) *on* $(-\infty, 0]$ *in the sense of Definition 11.4 if and only if there exists* $x \in P_n H$ *such that for all* $t \le 0$,

$$\xi(t) = e^{-At}x - \int_t^0 e^{-A(t-s)} P_n \left(F(\xi(s) + z(\theta_s\omega)) + g(s+\tau)\right) ds$$

$$+ \int_{-\infty}^t e^{-A(t-s)} Q_n \left(F(\xi(s) + z(\theta_s\omega)) + g(s+\tau)\right) ds. \qquad (11.20)$$

Proof Given $\xi \in \mathscr{S}$ and $t \le 0$, it is evident that the first integral on $(t, 0)$ in (11.20) is well-defined. We now prove the second integral on $(-\infty, t)$ exists. By (11.4), (11.6) and $\lambda_{n+1} \ge \lambda_1$ we have

$$\int_{-\infty}^t \|A^\alpha e^{-A(t-s)} Q_n g(s+\tau)\| ds \le \int_{-\infty}^t e^{-\lambda_{n+1}(t-s)} \|A^\alpha g(s+\tau)\| ds < \infty. \qquad (11.21)$$

By (11.2) and (11.7) we have for $t \le 0$,

$$\int_{-\infty}^t \|A^\alpha e^{-A(t-s)} Q_n F(\xi(s) + z(\theta_s\omega))\| ds$$

$$\le L \int_{-\infty}^t \left(\frac{\alpha^\alpha}{(t-s)^\alpha} + \lambda_{n+1}^\alpha\right) e^{-\lambda_{n+1}(t-s)} \|A^\alpha \xi(s) + A^\alpha z(\theta_s\omega)\| ds$$

$$\le L \int_{-\infty}^t \left(\frac{\alpha^\alpha}{(t-s)^\alpha} + \lambda_{n+1}^\alpha\right) e^{-\lambda_{n+1}t} e^{(\lambda_{n+1}-\mu)s} e^{\mu s} \|A^\alpha \xi(s) + A^\alpha z(\theta_s\omega)\| ds$$

$$\le L \int_{-\infty}^t \left(\frac{\alpha^\alpha}{(t-s)^\alpha} + \lambda_{n+1}^\alpha\right) e^{-\lambda_{n+1}t} e^{(\lambda_{n+1}-\mu)s} \left(\sup_{r\le 0} e^{\mu r} \|A^\alpha \xi(r) + A^\alpha z(\theta_r\omega)\|\right) ds$$

$$\le L \sup_{r\le 0} e^{\mu r} \|A^\alpha \xi(r) + A^\alpha z(\theta_r\omega)\| \int_{-\infty}^t \left(\frac{\alpha^\alpha}{(t-s)^\alpha} + \lambda_{n+1}^\alpha\right) e^{-\lambda_{n+1}t} e^{(\lambda_{n+1}-\mu)s} ds$$

$$\le L\|\xi + z\|_{\mathscr{S}} \int_{-\infty}^t \left(\frac{\alpha^\alpha}{(t-s)^\alpha} + \lambda_{n+1}^\alpha\right) e^{-\lambda_{n+1}t} e^{(\lambda_{n+1}-\mu)s} ds < \infty, \qquad (11.22)$$

where $\|\xi + z\|_{\mathscr{S}} = \sup_{r\le 0} e^{\mu r}\|A^\alpha \xi(r) + A^\alpha z(\theta_r\omega)\|$ and the last inequality follows from $\mu < \lambda_{n+1}$ and the temperedness of z as in (11.9). By (11.21) and (11.22), the integrals in (11.20) are well-defined in $D(A^\alpha)$ for all $\xi \in \mathscr{S}$ and $t \le 0$. If $\xi \in \mathscr{S}$ is a solution of (11.15) on $(-\infty, 0]$, then by (11.16) and Definition 11.4 we get for all $r < t \le 0$,

$$\xi(t) = e^{-A(t-r)}\xi(r) + \int_r^t e^{-A(t-s)}(F(\xi(s) + z(\theta_s\omega)) + g(s+\tau))ds,$$

which implies for all $r < t \le 0$,

$$P_n\xi(t) = e^{-A(t-r)}P_n\xi(r) + \int_r^t e^{-A(t-s)}P_n(F(\xi(s) + z(\theta_s\omega)) + g(s+\tau))ds, \tag{11.23}$$

and

$$Q_n\xi(t) = e^{-A(t-r)}Q_n\xi(r) + \int_r^t e^{-A(t-s)}Q_n(F(\xi(s) + z(\theta_s\omega)) + g(s+\tau))ds. \tag{11.24}$$

We get from (11.23) for $t = 0$

$$P_n\xi(0) = e^{Ar}P_n\xi(r) + \int_r^0 e^{As}P_n(F(\xi(s) + z(\theta_s\omega)) + g(s+\tau))ds$$

and hence

$$e^{-At}P_n\xi(0) = e^{-A(t-r)}P_n\xi(r) + \int_r^0 e^{-A(t-s)}P_n(F(\xi(s) + z(\theta_s\omega)) + g(s+\tau))ds. \tag{11.25}$$

It follows from (11.23) to (11.25) that

$$P_n\xi(t) = e^{-At}P_n\xi(0) - \int_t^0 e^{-A(t-s)}P_n(F(\xi(s)+z(\theta_s\omega))+g(s+\tau))ds. \tag{11.26}$$

On the other hand, since $\mu \in (\lambda_n, \lambda_{n+1})$, by (11.6) we have for $r < t \le 0$,

$$\|e^{-A(t-r)}Q_n\xi(r)\|_{D(A^\alpha)} \le e^{-\lambda_{n+1}(t-r)}\|\xi(r)\|_{D(A^\alpha)} \le e^{-\lambda_{n+1}t}e^{(\lambda_{n+1}-\mu)r}\|\xi\|_{\mathscr{S}} \to 0 \tag{11.27}$$

as $r \to -\infty$. Taking the limit of (11.24) as $r \to -\infty$, by (11.26) and the fact $\xi(t) = P_n\xi(t) + Q_n\xi(t)$ we find that ξ satisfies (11.20) with $x = P_n\xi(0)$.

Suppose now $\xi \in \mathscr{S}$ satisfies (11.20). Then by simple calculations, one can verify that for all $r \le t \le 0$, $v(t, r, \omega, g^\tau, \xi(r)) = \xi(t)$ and hence ξ is a solution of (11.15) on $(-\infty, 0]$. □

Next, we will find all solutions $\xi \in \mathscr{S}$ of (11.15) on $(-\infty, 0]$ in order to characterize the structure of $\mathscr{M}$ given by (11.19). By Lemma 11.1, we only need to find all $\xi \in \mathscr{S}$ satisfying (11.20). Given $\tau \in \mathbf{R}$, $\omega \in \Omega$, $x \in P_n H$ and $\xi \in \mathscr{S}$, denote by

$$\mathscr{I}(\xi, x, \omega, \tau)(t) = e^{-At}x - \int_t^0 e^{-A(t-s)} P_n \left(F(\xi(s) + z(\theta_s\omega)) + g(s+\tau)\right) ds$$

$$+ \int_{-\infty}^t e^{-A(t-s)} Q_n \left(F(\xi(s) + z(\theta_s\omega)) + g(s+\tau)\right) ds. \tag{11.28}$$

Then ξ satisfies (11.20) if and only if ξ is a fixed point of $\mathscr{I}$ in $\mathscr{S}$. We first show $\mathscr{I}$ maps $\mathscr{S}$ into itself.

Lemma 11.2 *Suppose* (11.1), (11.3) *and* (11.17)–(11.18) *hold. Then for every fixed* $x \in P_n H$, $\omega \in \Omega$ *and* $\tau \in \mathbf{R}$, $\mathscr{I}(\cdot, x, \omega, \tau) : \mathscr{S} \to \mathscr{S}$ *is well-defined.*

Proof By (11.28) and (11.5)–(11.7) we have for each $\xi \in \mathscr{S}$ and $t \le 0$,

$$\|\mathscr{I}(\xi, x, \omega, \tau)(t)\|_{D(A^\alpha)} \le e^{-\lambda_n t}\|A^\alpha x\| + \int_t^0 e^{-\lambda_n(t-s)} \left(\lambda_n^\alpha \|F(\xi + z(\theta_s\omega))\| + \|A^\alpha g(s+\tau)\|\right) ds$$

$$+ \int_{-\infty}^t \left(\frac{\alpha^\alpha}{(t-s)^\alpha} + \lambda_{n+1}^\alpha\right) e^{-\lambda_{n+1}(t-s)} \|F(\xi(s) + z(\theta_s\omega))\| ds + \int_{-\infty}^t e^{-\lambda_{n+1}(t-s)} \|A^\alpha g(s+\tau)\| ds. \tag{11.29}$$

By (11.2) and (11.18) we have for all $t \le 0$,

$$e^{\mu t} \int_t^0 e^{-\lambda_n(t-s)} \left(\lambda_n^\alpha \|F(\xi(s) + z(\theta_s\omega))\| + \|A^\alpha g(s+\tau)\|\right) ds$$

$$\le L\lambda_n^\alpha \|\xi + z\|_{\mathscr{S}} \int_t^0 e^{(\mu-\lambda_n)(t-s)} ds + \int_t^0 e^{(\mu-\lambda_n)t} e^{\lambda_n s} \|A^\alpha g(s+\tau)\| ds$$

$$\le \frac{\lambda_n^\alpha L}{\mu - \lambda_n} \|\xi + z\|_{\mathscr{S}} + \int_t^0 e^{\lambda_1 s} \|A^\alpha g(s+\tau)\| ds. \tag{11.30}$$

Similarly, for all $t \le 0$ we get

$$e^{\mu t} \int_{-\infty}^t \left(\frac{\alpha^\alpha}{(t-s)^\alpha} + \lambda_{n+1}^\alpha\right) e^{-\lambda_{n+1}(t-s)} \|F(\xi(s) + z(\theta_s\omega))\| ds$$

$$\le L\|\xi + z\|_{\mathscr{S}} \int_{-\infty}^t \left(\frac{\alpha^\alpha}{(t-s)^\alpha} + \lambda_{n+1}^\alpha\right) e^{(\mu-\lambda_{n+1})(t-s)} ds$$

$$\le L(\lambda_{n+1} - \mu)^{-1} \left(\lambda_{n+1}^\alpha + c_\alpha(\lambda_{n+1} - \mu)^\alpha\right) \|\xi + z\|_{\mathscr{S}}. \tag{11.31}$$

In addition, by (11.4) and (11.18) we have for $t \le 0$,

$$e^{\mu t}\int_{-\infty}^{t} e^{-\lambda_{n+1}(t-s)}\|A^\alpha g(s+\tau)\|ds \le \int_{-\infty}^{t} e^{(\mu-\lambda_{n+1})(t-s)}e^{\mu s}\|A^\alpha g(s+\tau)\|ds$$

$$\le \int_{-\infty}^{t} e^{\lambda_1 s}\|A^\alpha g(s+\tau)\|ds < \infty. \tag{11.32}$$

Since $\lambda_{n+1} - \mu \le \lambda_{n+1} - \lambda_n$ by (11.18), it follows from (11.29) to (11.32) that

$$\|\mathscr{I}(\xi, x, \omega, \tau)\|_{\mathscr{S}} \le L\left(\frac{\lambda_n^\alpha}{\mu-\lambda_n} + \frac{\lambda_{n+1}^\alpha + c_\alpha(\lambda_{n+1}-\lambda_n)^\alpha}{\lambda_{n+1}-\mu}\right)\|\xi + z\|_{\mathscr{S}} + \|A^\alpha x\|$$

$$+\int_{-\infty}^{0} e^{\lambda_1 s}\|A^\alpha g(s+\tau)\|ds \le k\|\xi+z\|_{\mathscr{S}} + \|A^\alpha x\| + \int_{-\infty}^{0} e^{\lambda_1 s}\|A^\alpha g(s+\tau)\|ds, \tag{11.33}$$

where the last inequality follows from (11.17). Therefore, we have $\mathscr{I}(\xi, x, \omega, \tau) \in \mathscr{S}$. □

We now establish existence of fixed points of $\mathscr{I}$ in $\mathscr{S}$.

Lemma 11.3 *Suppose* (11.1), (11.3) *and* (11.17)–(11.18) *hold. Then for every* $x \in P_n H$, $\omega \in \Omega$ *and* $\tau \in \mathbf{R}$, $\mathscr{I}(\cdot, x, \omega, \tau) : \mathscr{S} \to \mathscr{S}$ *has a unique fixed point which is Lipschitz continuous in* $x \in P_n H$.

Proof Given $\xi_1, \xi_2 \in \mathscr{S}$, following the proof of Lemma 11.2, one can verify

$$\|\mathscr{I}(\xi_1, x, \omega, \tau) - \mathscr{I}(\xi_2, x, \omega, \tau)\|_{\mathscr{S}} \le k\|\xi_1 - \xi_2\|_{\mathscr{S}}. \tag{11.34}$$

Since $k \in (0, 1)$, by (11.34) we find that $\mathscr{I}(\cdot, x, \omega, \tau)$ has a unique fixed point $\xi^*(x, \omega, \tau)$ in $\mathscr{S}$. For every $t \le 0$, $\xi^*(x, \cdot, \tau)(t) : \Omega \to D(A^\alpha)$ is measurable since it is a limit of iterations of measurable functions starting at zero. In addition, by (11.33) we have

$$(1-k)\|\xi^*(x, \omega, \tau)\|_{\mathscr{S}} \le k\|z\|_{\mathscr{S}} + \|A^\alpha x\| + \int_{-\infty}^{0} e^{\lambda_1 s}\|A^\alpha g(s+\tau)\|ds, \tag{11.35}$$

where $\|z\|_{\mathscr{S}} = \sup_{t\le 0} e^{\mu t}\|A^\alpha z(\theta_t\omega)\|$. We now prove the continuity of ξ^* in $x \in P_n H$. Let $\xi_1^* = \xi^*(x_1, \omega, \tau)$ and $\xi_2^* = \xi^*(x_2, \omega, \tau)$. By (11.18), (11.28) and (11.34) we have

$$\|\xi_1^* - \xi_2^*\|_{\mathscr{S}} = \|\mathscr{I}(\xi_1^*, x_1, \omega, \tau) - \mathscr{I}(\xi_2^*, x_2, \omega, \tau)\|_{\mathscr{S}}$$

$$\leq \|\mathscr{I}(\xi_1^*, x_1, \omega, \tau) - \mathscr{I}(\xi_1^*, x_2, \omega, \tau)\|_{\mathscr{S}} + \|\mathscr{I}(\xi_1^*, x_2, \omega, \tau) - \mathscr{I}(\xi_2^*, x_2, \omega, \tau)\|_{\mathscr{S}}$$

$$\leq \sup_{t\leq 0} e^{(\mu-\lambda_n)t}\|x_1 - x_2\|_{D(A^\alpha)} + k\|\xi_1^* - \xi_2^*\|_{\mathscr{S}} \leq \|x_1 - x_2\|_{D(A^\alpha)} + k\|\xi_1^* - \xi_2^*\|_{\mathscr{S}}. \tag{11.36}$$

This implies the Lipschitz continuity of ξ^* in x. □

By the fixed point ξ^* established by Lemma 11.3, for each $\tau \in \mathbf{R}$ and $\omega \in \Omega$, we can define a mapping $m(\tau, \omega) : P_n H \to Q_n D(A^\alpha)$ by

$$m(\tau, \omega)(x) = Q_n\xi^*(x, \omega, \tau)(0) = \int_{-\infty}^{0} e^{As} Q_n \left(F(\xi^*(x, \omega, \tau)(s) + z(\theta_s\omega)) + g(s + \tau)\right) ds \tag{11.37}$$

for $x \in P_n H$. It follows from (11.36) that

$$\|m(\tau, \omega)(x_1) - m(\tau, \omega)(x_2)\|_{D(A^\alpha)} \leq \frac{1}{1-k}\|x_1 - x_2\|_{D(A^\alpha)}. \tag{11.38}$$

By (11.35) and (11.37) we get

$$\|m(\tau, \omega)(x)\|_{D(A^\alpha)} \leq \frac{k}{1-k}\|z\|_{\mathscr{S}} + \frac{1}{1-k}\|A^\alpha x\| + \frac{1}{1-k}\int_{-\infty}^{0} e^{\lambda_1 s}\|A^\alpha g(s + \tau)\| ds. \tag{11.39}$$

By (11.19), (11.37) and Lemma 11.1 we find that for all $\tau \in \mathbf{R}$ and $\omega \in \Omega$,

$$\mathscr{M}(\tau, \omega) = \{\xi^*(x, \omega, \tau)(0) : x \in P_n H\} = \{x + m(\tau, \omega)(x) : x \in P_n H\}. \tag{11.40}$$

Since $m(\tau, \omega)(x)$ is measurable in ω and continuous in x, the measurability of $\mathscr{M}(\tau, \cdot)$ follows.

Lemma 11.4 *Suppose* (11.1), (11.3) *and* (11.17)–(11.18) *hold. Then* $\mathscr{M} = \{\mathscr{M}(\tau, \omega) : \tau \in \mathbf{R}, \omega \in \Omega\}$ *given by* (11.19) *is a Lipschitz invariant manifold of* (11.15).

Proof Given $\tau \in \mathbf{R}$ and $\omega \in \Omega$, by (11.40), $\mathscr{M}(\tau, \omega)$ is the graph of $m(\tau, \omega)$ which is Lipschitz continuous by Lemma 11.3. Next, we show the invariance of $\mathscr{M}$. Given $t > 0$ and $v_0 \in \mathscr{M}(\tau, \omega)$, by (11.19) there exists $\xi \in \mathscr{S}$ such that $\xi(0) = v_0$ and ξ is a solution of (11.15) on $(-\infty, 0]$. Let

$$\widetilde{\xi}(r) = \begin{cases} v(r + t, 0, \omega, g^\tau, \xi(0)) & \text{if } -t \leq r \leq 0; \\ \xi(r + t) & \text{if } r < -t. \end{cases} \tag{11.41}$$

Note that $\widetilde{\xi} \in \mathscr{S}$ since $\xi \in \mathscr{S}$. By straightforward calculations, one can verify, for all $r_2 \leq r_1 \leq 0$,

$$v(r_1, r_2, \theta_t\omega, g^{\tau+t}, \widetilde{\xi}(r_2)) = \widetilde{\xi}(r_1), \tag{11.42}$$

where $v(r_1, r_2, \theta_t\omega, g^{\tau+t}, \widetilde{\xi}(r_2))$ is given by (11.16) with ω and g^τ replaced by $\theta_t\omega$ and $g^{\tau+t}$, respectively. By (11.19) and (11.42), we find that $\widetilde{\xi}(0) \in \mathscr{M}(\tau + t, \theta_t\omega)$, which along with (11.11) and (11.41) indicates that $\Psi(t, \tau, \omega, v_0) = v(t, 0, \omega, g^\tau, v_0) \in \mathscr{M}(\tau + t, \theta_t\omega)$ for all $v_0 \in \mathscr{M}(\tau, \omega)$, and hence

$$\Psi(t, \tau, \omega, \mathscr{M}(\tau, \omega)) \subseteq \mathscr{M}(\tau + t, \theta_t\omega), \quad \text{for all } t \geq 0. \tag{11.43}$$

On the other hand, for every $v_0 \in \mathscr{M}(\tau + t, \theta_t\omega)$, by (11.19) there exists $\xi \in \mathscr{S}$ such that $\xi(0) = v_0$ and for all $s_2 \leq s_1 \leq 0$,

$$v(s_1, s_2, \theta_t\omega, g^{\tau+t}, \xi(s_2)) = \xi(s_1). \tag{11.44}$$

Let $\widetilde{\xi} : (-\infty, 0] \to D(A^\alpha)$ be given by

$$\widetilde{\xi}(r) = \xi(r - t) \quad \text{for all } r \leq 0. \tag{11.45}$$

Then $\widetilde{\xi} \in \mathscr{S}$, and by (11.44)–(11.45), we have for $t \geq 0$ and for all $r_2 \leq r_1 \leq 0$,

$$\widetilde{\xi}(r_1) = \xi(r_1 - t) = v(r_1 - t, r_2 - t, \theta_t\omega, g^{\tau+t}, \xi(r_2 - t)) = v(r_1, r_2, \omega, g^\tau, \widetilde{\xi}(r_2)). \tag{11.46}$$

By (11.19), (11.46) and we find that $\widetilde{\xi}(0) = \xi(-t) \in \mathscr{M}(\tau, \omega)$. By (11.11) and (11.44) we get

$$\Psi(t, \tau, \omega, \xi(-t)) = v(t + \tau, \tau, \theta_{-\tau}\omega, g, \xi(-t)) = v(0, -t, \theta_t\omega, g^{\tau+t}, \xi(-t)) = \xi(0). \tag{11.47}$$

Since $\xi(-t) \in \mathscr{M}(\tau, \omega)$ and $\xi(0) = v_0$ where v_0 is an arbitrary given point in $\mathscr{M}(\tau + t, \theta_t\omega)$, by (11.47) we obtain

$$\mathscr{M}(\tau + t, \theta_t\omega) \subseteq \Psi(t, \tau, \omega, \mathscr{M}(\tau, \omega)), \quad \text{for all } t \geq 0. \tag{11.48}$$

Then the invariance of $\mathscr{M}$ follows from (11.43) to (11.48) immediately. □

The next result is concerned with the exponential attraction property of $\mathscr{M}$.

Lemma 11.5 *Suppose* (11.1), (11.3) *and* (11.17)–(11.18) *hold with* $k \in (0, \frac{1}{2})$. *Then for every* $\tau \in \mathbf{R}$, $\omega \in \Omega$ *and* $v_0 \in D(A^\alpha)$, *there exists a random variable* $v_0^*(\tau, \omega) \in \mathscr{M}(\tau, \omega)$ *such that for all* $t \geq 0$,

$$\|\Psi(t, \tau, \omega, v_0^*) - \Psi(t, \tau, \omega, v_0)\|_{D(A^\alpha)} \leq \frac{1}{1 - \delta} e^{-\mu t} \|Q_n v_0 - m(\tau, \omega)(P_n v_0)\|_{D(A^\alpha)},$$

where $\delta = k + \frac{k}{2-2k} \in (0, 1)$ *for* $0 < k < \frac{1}{2}$.

Proof We argue as in [5]. Let $\mathscr{S}^+ = \{\xi \in C([0, \infty), D(A^\alpha)) : \sup_{t\geq 0} e^{\mu t}\|A^\alpha \xi(t)\| < \infty\}$ with norm $\|\xi\|_{\mathscr{S}^+} = \sup_{t\geq 0} e^{\mu t}\|A^\alpha \xi(t)\|$. We will find $v_0^* \in \mathscr{M}(\tau, \omega)$ such that $v(t, 0, \omega, g^\tau, v_0^*) - v(t, 0, \omega, g^\tau, v_0) \in \mathscr{S}^+$. To that end, we need to solve the the equation

$$\xi(t) = e^{-At} y_0 + \int_0^t e^{-A(t-s)} Q_n \left(F(\xi(s) + z(\theta_s \omega) + v(s)) - F(v(s) + z(\theta_s \omega))\right) ds$$

$$- \int_t^\infty e^{-A(t-s)} P_n \left(F(\xi(s) + z(\theta_s \omega) + v(s)) - F(v(s) + z(\theta_s \omega))\right) ds, \quad (11.49)$$

where $v(s) = v(s, 0, \omega, g^\tau, v_0)$ and y_0 is determined by

$$y_0 = -Q_n v_0 + m(\tau, \omega)\left(P_n v_0 - \int_0^\infty e^{As} P_n(F(\xi + v + z(\theta_s \omega)) - F(v + z(\theta_s \omega)))ds\right). \quad (11.50)$$

Given $\xi \in \mathscr{S}^+$, denote the right-hand side of (11.49) by $\mathscr{I}^+(\xi)$. We will find a fixed point of $\mathscr{I}^+$ in $\mathscr{S}^+$. By (11.5)–(11.7) and (11.1) we get for $t \geq 0$ and $\xi \in \mathscr{S}^+$,

$$e^{\mu t}\|A^\alpha \mathscr{I}^+(\xi)(t)\| \leq e^{(\mu-\lambda_{n+1})t}\|A^\alpha y_0\| + Le^{\mu t}\int_t^\infty \lambda_n^\alpha e^{-\lambda_n(t-s)}\|A^\alpha \xi(s)\|ds$$

$$+Le^{\mu t}\int_0^t \left(\frac{\alpha^\alpha}{(t-s)^\alpha} + \lambda_{n+1}^\alpha\right) e^{-\lambda_{n+1}(t-s)}\|A^\alpha \xi(s)\|ds$$

$$\leq \|A^\alpha y_0\| + L\|\xi\|_{\mathscr{S}^+}\left(\frac{\lambda_n^\alpha}{\mu - \lambda_n} + \frac{\lambda_{n+1}^\alpha}{\lambda_{n+1} - \mu} + c_\alpha(\lambda_{n+1} - \mu)^{\alpha-1}\right). \quad (11.51)$$

By (11.50), (11.38), (11.5) and (11.1) we get

$$\|A^\alpha y_0\| \leq \| - Q_n v_0 + m(\tau, \omega)(P_n v_0)\|_{D(A^\alpha)}$$

$$+\|m(\tau, \omega)(P_n v_0) - m(\tau, \omega)\left(P_n v_0 - \int_0^\infty e^{As} P_n(F(\xi + v + z(\theta_s \omega)) - F(v + z(\theta_s \omega)))ds\right)\|_{D(A^\alpha)}$$

$$\leq \|Q_n v_0 - m(\tau, \omega)(P_n v_0)\|_{D(A^\alpha)} + \frac{1}{1-k}\int_0^\infty \|e^{As} P_n(F(\xi + v + z) - F(v + z))ds\|_{D(A^\alpha)}$$

$$\leq \|Q_n v_0 - m(\tau, \omega)(P_n v_0)\|_{D(A^\alpha)} + \frac{L\lambda_n^\alpha}{1-k}\int_0^\infty e^{(\lambda_n - \mu)s} e^{\mu s}\|A^\alpha \xi(s)\|ds$$

$$\leq \|Q_n v_0 - m(\tau,\omega)(P_n v_0)\|_{D(A^\alpha)} + \frac{L\lambda_n^\alpha}{(1-k)(\mu-\lambda_n)}\|\xi\|_{\mathscr{S}^+}. \tag{11.52}$$

It follows from (11.51)–(11.52) and (11.17)–(11.18) that

$$\|\mathscr{I}^+(\xi)\|_{\mathscr{S}^+} \leq \|Q_n v_0 - m(\tau,\omega)(P_n v_0)\|_{D(A^\alpha)}$$

$$+L\|\xi\|_{\mathscr{S}^+}\left(\frac{(2-k)\lambda_n^\alpha}{(1-k)(\mu-\lambda_n)} + \frac{\lambda_{n+1}^\alpha}{\lambda_{n+1}-\mu} + c_\alpha(\lambda_{n+1}-\mu)^{\alpha-1}\right)$$

$$\leq \|Q_n v_0 - m(\tau,\omega)(P_n v_0)\|_{D(A^\alpha)} + \delta\|\xi\|_{\mathscr{S}^+}, \quad \text{with } \delta = k + \frac{k}{2-2k}. \tag{11.53}$$

This shows that $\mathscr{I}^+$ maps $\mathscr{S}^+$ into itself. Note that for $\xi_1, \xi_2 \in \mathscr{S}^+$, we have

$$\|\mathscr{I}^+(\xi_1) - \mathscr{I}^+(\xi_2)\|_{D(A^\alpha)} \leq \|e^{-At}y_{0,1} - e^{-At}y_{0,2}\|_{D(A^\alpha)}$$

$$+\int_0^t \|e^{-A(t-s)}Q_n(F(\xi_1+z+v) - F(\xi_2+z+v))\|_{D(A^\alpha)}ds$$

$$+\int_t^\infty \|e^{-A(t-s)}P_n(F(\xi_1+z+v) - F(\xi_2+z+v))\|_{D(A^\alpha)}ds.$$

Following the proof of (11.53) we can get

$$\|\mathscr{I}^+(\xi_1) - \mathscr{I}^+(\xi_2)\|_{\mathscr{S}^+} \leq \delta\|\xi_1 - \xi_2\|_{\mathscr{S}^+} \tag{11.54}$$

where δ is given in (11.53) and $\delta \in (0,1)$ for $k \in (0, \frac{1}{2})$. Since $\mathscr{I}^+ : \mathscr{S}^+ \to \mathscr{S}^+$ is a contraction by (11.54), it has a unique fixed point ξ in $\mathscr{S}^+$ which satisfies (11.49)–(11.50). This fixed point is measurable since it can be obtained by a limit of iterations of measurable functions. Further, by (11.53) we obtain

$$\|\xi\|_{\mathscr{S}^+} = \|\mathscr{I}^+\|_{\mathscr{S}^+} \leq \|Q_n v_0 - m(\tau,\omega)(P_n v_0)\|_{D(A^\alpha)} + \delta\|\xi\|_{\mathscr{S}^+}$$

and hence the fixed point ξ satisfies, for all $t \geq 0$,

$$\|\xi(t)\|_{D(A^\alpha)} \leq \frac{1}{1-\delta}e^{-\mu t}\|Q_n v_0 - m(\tau,\omega)(P_n v_0)\|_{D(A^\alpha)}. \tag{11.55}$$

By (11.49) we have

$$\xi(0) = y_0 - \int_0^\infty e^{As}P_n\left(F(\xi+z+v) - F(v+z)\right)ds. \tag{11.56}$$

By (11.49) and (11.56) we get

$$\xi(t) = e^{-At}\xi(0) + \int_0^t e^{-A(t-s)} (F(\xi + z + v) - F(v + z))\, ds. \tag{11.57}$$

Note that $v(t, 0, \omega, g^\tau, v_0)$ is a solution of (11.15) with $r = 0$, which along with (11.57) implies that $v^*(t) = \xi(t) + v(t, 0, \omega, g^\tau, v_0)$ satisfies

$$v^*(t) = e^{-At}v^*(0) + \int_0^t e^{-A(t-s)}(F(v^*(s) + z(\theta_s\omega)) + g(s+\tau))ds \text{ with } v^*(0) = \xi(0) + v_0.$$

This shows that $v^*(t)$ is a solution of (11.15) with initial condition $v^*(0)$. By the uniqueness of solutions, we have $v^*(t) = v(t, 0, \omega, g^\tau, v^*(0))$. Since $\xi(t) = v(t, 0, \omega, g^\tau, v^*(0)) - v(t, 0, \omega, g^\tau, v_0)$, by (11.55) we get for all $t \geq 0$,

$$\|v(t, 0, \omega, g^\tau, v^*(0)) - v(t, 0, \omega, g^\tau, v_0)\|_{D(A^\alpha)} \leq \frac{1}{1-\delta} e^{-\mu t} \|Q_n v_0 - m(\tau, \omega)(P_n v_0)\|_{D(A^\alpha)}. \tag{11.58}$$

By (11.56) and (11.50) we get $v^*(0) = \xi(0) + v_0 = x_0 + m(\tau, \omega)(x_0) \in \mathcal{M}(\tau, \omega)$ where $x_0 = P_n v_0 - \int_0^\infty e^{As} P_n(F(v^* + z) - F(v + z))ds$, which along with (11.58) completes the proof. □

As an immediate consequence of Lemmas 11.4 and 11.5, we obtain the existence of inertial manifolds for Eq. (11.10).

Corollary 11.1 Suppose (11.1), (11.3) and (11.17)–(11.18) hold with $k \in (0, \frac{1}{2})$. Then the cocycle Ψ defined by (11.11) for Eq. (11.10) has an inertial manifold $\mathcal{M} = \{\mathcal{M}(\tau, \omega) : \tau \in \mathbf{R}, \omega \in \Omega\}$ as given by (11.19) and (11.40).

Based on Corollary 11.1, we are able to establish the existence of inertial manifolds for the nonautonomous stochastic equation (11.1).

Theorem 11.1 *Suppose* (11.1), (11.3) *and* (11.17)–(11.18) *hold with* $k \in (0, \frac{1}{2})$. *Then the cocycle* Φ *defined by* (11.13) *for Eq.* (11.1) *has an inertial manifold* $\widetilde{\mathcal{M}} = \{\widetilde{\mathcal{M}}(\tau, \omega) : \tau \in \mathbf{R}, \omega \in \Omega\}$ *which is given by, for each* $\tau \in \mathbf{R}$ *and* $\omega \in \Omega$,

$$\widetilde{\mathcal{M}}(\tau, \omega) = \{x + z(\omega) + m(\tau, \omega)(x) : x \in P_n H\} = \mathcal{M}(\tau, \omega) + z(\omega). \tag{11.59}$$

Proof Given $\tau \in \mathbf{R}$, $\omega \in \Omega$ and $x \in P_n H$, let $y = x + P_n z(\omega)$. Then we have

$$x + z(\omega) + m(\tau, \omega)(x) = y + Q_n z(\omega) + m(\tau, \omega)(y - P_n z(\omega)),$$

which along with (11.59) implies

$$\widetilde{\mathcal{M}}(\tau,\omega) = \{y + Q_n z(\omega) + m(\tau,\omega)(y - P_n z(\omega)) : y \in P_n H\} = \{y + \widetilde{m}(\tau,\omega)(y) : y \in P_n H\} \tag{11.60}$$

where

$$\widetilde{m}(\tau,\omega)(\cdot) = Q_n z(\omega) + m(\tau,\omega)(\cdot - P_n z(\omega)) \tag{11.61}$$

is a Lipschitz map from $P_n H$ to $Q_n H$. By (11.14), (11.59) and the invariance of $\mathcal{M}$ under Ψ we get

$$\Phi(t,\tau,\omega,\widetilde{\mathcal{M}}(\tau,\omega)) = \{\Phi(t,\tau,\omega,x + z(\omega) + m(\tau,\omega)(x) : x \in P_n H\}$$

$$= \{\Psi(t,\tau,\omega,x+m(\tau,\omega)(x)+z(\theta_t\omega) : x \in P_n H\} = \Psi(t,\tau,\omega,\mathcal{M}(\tau,\omega))+z(\theta_t\omega)$$

$$= \mathcal{M}(\tau+t,\theta_t\omega) + z(\theta_t\omega) = \widetilde{\mathcal{M}}(\tau+t,\theta_t\omega), \tag{11.62}$$

where the last equality follows from (11.59) again. Therefore, $\widetilde{\mathcal{M}}$ is invariant under Φ. Finally, we prove the attraction property of $\widetilde{\mathcal{M}}$. Given $u_0 \in D(A^\alpha)$, let $v_0 = u_0 - z(\omega)$. By Lemma 11.5, there exists a random variable $v_0^*(\tau,\omega) \in \mathcal{M}(\tau,\omega)$ such that for all $t \geq 0$,

$$\|\Psi(t,\tau,\omega,v_0^*) - \Psi(t,\tau,\omega,v_0)\|_{D(A^\alpha)} \leq \frac{1}{1-\delta} e^{-\mu t} \|Q_n v_0 - m(\tau,\omega)(P_n v_0)\|_{D(A^\alpha)}. \tag{11.63}$$

Let $u_0^* = v_0^* + z(\omega)$. Since $v_0^* \in \mathcal{M}(\tau,\omega)$, by (11.59) we find $u_0^* \in \widetilde{\mathcal{M}}(\tau,\omega)$. By (11.14) we get

$$\|\Phi(t,\tau,\omega,u_0^*) - \Phi(t,\tau,\omega,u_0)\|_{D(A^\alpha)} = \|\Psi(t,\tau,\omega,v_0^*) - \Psi(t,\tau,\omega,v_0)\|_{D(A^\alpha)}$$

which along with (11.63) yields

$$\|\Phi(t,\tau,\omega,u_0^*) - \Phi(t,\tau,\omega,u_0)\|_{D(A^\alpha)} \leq \frac{1}{1-\delta} e^{-\mu t} \|Q_n u_0 - \widetilde{m}(\tau,\omega)(P_n u_0)\|_{D(A^\alpha)}. \tag{11.64}$$

By (11.60), (11.62) and (11.64) we conclude the proof. □

We now establish relationships between tempered random attractors and inertial manifolds. Recall that a tempered family $\mathcal{A} = \{\mathcal{A}(\tau,\omega) : \tau \in \mathbf{R}, \omega \in \Omega\}$ of nonempty compact subsets of $D(A^\alpha)$ is called a tempered random pullback attractor of Φ if $\mathcal{A}$ is measurable, invariant, and pullback attracts all tempered family of bounded subsets of $D(A^\alpha)$ (see, e.g., [16] and the references therein). Based on this notation, the random inertial manifold constructed in Theorem 11.1 must contain tempered random attractors as subsets.

Theorem 11.2 *Suppose* (11.1), (11.3) *and* (11.17)–(11.18) *hold with* $k \in (0, \frac{1}{2})$. *If* Φ *has a tempered pullback random attractor* $\mathscr{A} = \{\mathscr{A}(\tau, \omega) : \tau \in \mathbf{R}, \omega \in \Omega\}$ *in* $D(A^\alpha)$, *then we have* $\mathscr{A}(\tau, \omega) \subseteq \widetilde{\mathscr{M}}(\tau, \omega)$ *for all* $\tau \in \mathbf{R}$ *and* $\omega \in \Omega$, *where* $\widetilde{\mathscr{M}}(\tau, \omega)$ *is given by* (11.59).

Proof Given $u \in \mathscr{A}(\tau, \omega)$ and $t_m \to \infty$, by the invariance of $\mathscr{A}$, we find that for every $m \in \mathbb{N}$, there exists $u_m \in \mathscr{A}(\tau - t_m, \theta_{-t_m}\omega)$ such that

$$u = \Phi(t_m, \tau - t_m, \theta_{-t_m}\omega, u_m). \tag{11.65}$$

By (11.64), there exists $u_m^* \in \widetilde{\mathscr{M}}(\tau - t_m, \theta_{-t_m}\omega)$ such that

$$\begin{aligned} &\|\Phi(t_m, \tau - t_m, \theta_{-t_m}\omega, u_m^*) - \Phi(t_m, \tau - t_m, \theta_{-t_m}\omega, u_m)\|_{D(A^\alpha)} \\ &\leq \frac{1}{1-\delta} e^{-\mu t_m} \|Q_n u_m - \widetilde{m}(\tau - t_m, \theta_{-t_m}\omega)(P_n u_m)\|_{D(A^\alpha)}. \end{aligned} \tag{11.66}$$

Since $u_m^* \in \widetilde{\mathscr{M}}(\tau - t_m, \theta_{-t_m}\omega)$, by the invariance of $\widetilde{\mathscr{M}}$, we have $\Phi(t_m, \tau - t_m, \theta_{-t_m}\omega, u_m^*) \in \widetilde{\mathscr{M}}(\tau, \omega)$. By (11.60) we may write $\Phi(t_m, \tau - t_m, \theta_{-t_m}\omega, u_m^*) = x_m + \widetilde{m}(\tau, \omega)(x_m)$ for some $x_m \in P_n H$. This along with (11.65)–(11.66) implies

$$\|(x_m - P_n u) + (\widetilde{m}(\tau, \omega)(x_m) - Q_n u)\|_{D(A^\alpha)} \leq \frac{1}{1-\delta} e^{-\mu t_m} \|Q_n u_m - \widetilde{m}(\tau - t_m, \theta_{-t_m}\omega)(P_n u_m)\|_{D(A^\alpha)}. \tag{11.67}$$

By (11.61), (11.39) and $u_m \in \mathscr{A}(\tau - t_m, \theta_{-t_m}\omega)$ we get

$$\begin{aligned} &\|Q_n u_m - \widetilde{m}(\tau - t_m, \theta_{-t_m}\omega)(P_n u_m)\|_{D(A^\alpha)} \leq \|u_m\|_{D(A^\alpha)} + \|\widetilde{m}(\tau - t_m, \theta_{-t_m}\omega)(P_n u_m)\|_{D(A^\alpha)} \\ &\leq \|u_m\|_{D(A^\alpha)} + \|A^\alpha z(\theta_{-t_m}\omega)\| + \|m(\tau - t_m, \theta_{-t_m}\omega)(P_n(u_m - z(\theta_{-t_m}\omega)))\|_{D(A^\alpha)} \\ &\leq \frac{2-k}{1-k}\|u_m\|_{D(A^\alpha)} + \frac{2-k}{1-k}\|A^\alpha z(\theta_{-t_m}\omega)\| \\ &+ \frac{k}{1-k} \sup_{r \leq 0} e^{\mu r} \|z(\theta_{r - t_m}\omega)\|_{D(A^\alpha)} + \frac{1}{1-k} \int_{-\infty}^0 e^{\lambda_1 s} \|g(s + \tau - t_m)\|_{D(A^\alpha)} ds \\ &\leq \frac{2-k}{1-k}\|\mathscr{A}(\tau - t_m, \theta_{-t_m}\omega)\|_{D(A^\alpha)} + \frac{2-k}{1-k}\|z(\theta_{-t_m}\omega)\|_{D(A^\alpha)} \\ &+ \frac{k}{1-k} e^{\mu t_m} \sup_{s \leq -t_m} e^{\mu s} \|z(\theta_s \omega)\|_{D(A^\alpha)} + \frac{e^{\mu t_m}}{1-k} \int_{-\infty}^{-t_m} e^{\lambda_1 r} \|g(r + \tau)\|_{D(A^\alpha)} ds. \end{aligned} \tag{11.68}$$

Taking the limit of (11.67) as $m \to \infty$, by (11.68), (11.4) and the temperedness of $\mathscr{A}$ and z, we get

$$\lim_{m\to\infty} \|(x_m - P_n u) + (\widetilde{m}(\tau,\omega)(x_m) - Q_n u)\|_{D(A^\alpha)} = 0$$

which implies $x_m \to P_n u$ and $\widetilde{m}(\tau,\omega)(x_m) \to Q_n u$. By the continuity of $\widetilde{m}(\tau,\omega)$, we obtain $Q_n u = \widetilde{m}(\tau,\omega)(P_n u)$ and hence $u \in \widetilde{\mathscr{M}}(\tau,\omega)$ for any $u \in \mathscr{A}(\tau,\omega)$ as desired. □

11.4 Periodicity and Almost Periodicity of Inertial Manifolds

In this section, we assume the external function g in (11.1) is time periodic or almost periodic, and establish the pathwise periodicity or almost periodicity of the random inertial manifolds constructed in the previous section. If $g : \mathbf{R} \to D(A^\alpha)$ is almost periodic, then g must be bounded; that is, $\sup_{t\in\mathbf{R}} \|g(t)\|_{D(A^\alpha)} < \infty$. Therefore, in this case, condition (11.3) is trivially fulfilled.

The main results of this section are given below.

Theorem 11.3 *Suppose* (11.1) *and* (11.17)–(11.18) *hold with* $k \in (0, \frac{1}{2})$. *If* $g : \mathbf{R} \to D(A^\alpha)$ *is almost periodic, then* Φ *has an almost periodic random inertial manifold* $\widetilde{\mathscr{M}}(\tau,\omega)$ *as given by* (11.59).

Proof By the almost periodicity of g, for every $\varepsilon > 0$, there exists a positive number $l = l(\varepsilon)$ such that any interval of length l contains a point τ_0 such that

$$\|g(r+\tau_0) - g(r)\|_{D(A^\alpha)} \le \frac{1}{2}\varepsilon(1-k)\lambda_n, \quad \text{for all } r \in \mathbf{R}. \tag{11.1}$$

Based on (11.1) we will prove

$$\sup_{x\in P_n H} \|\widetilde{m}(\tau+\tau_0,\omega)(x) - \widetilde{m}(\tau,\omega)(x)\|_{D(A^\alpha)} \le \varepsilon, \quad \text{for all } \tau \in \mathbf{R}, \tag{11.2}$$

which will complete the proof. Let $\xi^*(x,\omega,r)$ be the unique fixed point of $\mathscr{I}(\cdot, x, \omega, r)$ given by (11.28) for every fixed $x \in P_n H$, $\omega \in \Omega$ and $r \in \mathbf{R}$. Then by (11.28), (11.5)–(11.7), (11.1) and (11.1) we get for $t \le 0$,

$$e^{\mu t}\|\xi^*(x,\omega,\tau+\tau_0)(t) - \xi^*(x,\omega,\tau)(t)\|_{D(A^\alpha)}$$

$$= e^{\mu t}\|\mathscr{I}(\xi^*(x,\omega,\tau+\tau_0), x,\omega,\tau+\tau_0) - \mathscr{I}(\xi^*(x,\omega,\tau), x,\omega,\tau)\|_{D(A^\alpha)}$$

$$\le \int_t^0 e^{\mu t}\|A^\alpha e^{-A(t-s)} P_n(F(\xi^*(x,\omega,\tau+\tau_0)(s)+z(\theta_s\omega)) - F(\xi^*(x,\omega,\tau)(s)+z(\theta_s\omega)))\|ds$$

$$+\int_{-\infty}^{t} e^{\mu t}\|A^\alpha e^{-A(t-s)}Q_n(F(\xi^*(x,\omega,\tau+\tau_0)(s)+z(\theta_s\omega))-F(\xi^*(x,\omega,\tau)(s)+z(\theta_s\omega)))\|ds$$

$$+e^{\mu t}\int_t^0 \|e^{-A(t-s)}P_nA^\alpha(g(s+\tau+\tau_0)-g(s+\tau))\|ds+e^{\mu t}\int_{-\infty}^t \|e^{-A(t-s)}Q_nA^\alpha(g(s+\tau+\tau_0)-g(s+\tau))\|$$

$$\leq L\|\xi^*(x,\omega,\tau+\tau_0)-\xi^*(x,\omega,\tau)\|_{\mathscr{S}}\left(\lambda_n^\alpha\int_t^0 e^{(\mu-\lambda_n)(t-s)}ds+\int_{-\infty}^t (\frac{\alpha^\alpha}{(t-s)^\alpha}+\lambda_{n+1}^\alpha)e^{(\mu-\lambda_{n+1})(t-s)}ds\right)$$

$$+e^{\mu t}\int_t^0 e^{-\lambda_n(t-s)}\|g(s+\tau+\tau_0)-g(s+\tau)\|_{D(A^\alpha)}ds+\int_{-\infty}^t e^{-\lambda_{n+1}(t-s)}\|g(s+\tau+\tau_0)-g(s+\tau)\|_{D(A^\alpha)}$$

$$\leq L\left(\frac{\lambda_n^\alpha}{\mu-\lambda_n}+\frac{\lambda_{n+1}^\alpha+c_\alpha(\lambda_{n+1}-\mu)^\alpha}{\lambda_{n+1}-\mu}\right)\|\xi^*(x,\omega,\tau+\tau_0)-\xi^*(x,\omega,\tau)\|_{\mathscr{S}}$$

$$+\frac{1}{2}\varepsilon(1-k)\lambda_n e^{\mu t}\int_t^0 e^{-\lambda_n(t-s)}ds+\frac{1}{2}\varepsilon(1-k)\lambda_n\int_{-\infty}^t e^{-\lambda_{n+1}(t-s)}ds. \quad (11.3)$$

Since $t \leq 0$ and $\mu \in (\lambda_n, \lambda_{n+1})$, the last integral in (11.3) is bounded by $\varepsilon(1-k)$, which together with (11.17) and (11.3) implies

$$\|\xi^*(x,\omega,\tau+\tau_0)-\xi^*(x,\omega,\tau)\|_{\mathscr{S}} \leq k\|\xi^*(x,\omega,\tau+\tau_0)-\xi^*(x,\omega,\tau)\|_{\mathscr{S}}+\varepsilon(1-k). \quad (11.4)$$

Thus we get $\|\xi^*(x,\omega,\tau+\tau_0)-\xi^*(x,\omega,\tau)\|_{\mathscr{S}} \leq \varepsilon$ and hence $\|\xi^*(x,\omega,\tau+\tau_0)(0)-\xi^*(x,\omega,\tau)(0)\|_{D(A^\alpha)} \leq \varepsilon$. Since $\xi^*(x,\omega,r)(0) = x+m(r,\omega)x$ for all $r \in \mathbf{R}$, we obtain $\|m(\tau+\tau_0,\omega)(x)-m(\tau,\omega)(x)\|_{D(A^\alpha)} \leq \varepsilon$ for all $x \in P_nH$, which along with (11.61) yields (11.2), and thus completes the proof. □

Finally, we present the pathwise periodicity of random inertial manifolds.

Theorem 11.4 *Suppose* (11.1) *and* (11.17)–(11.18) *hold with* $k \in (0, \frac{1}{2})$. *If* $g : \mathbf{R} \to D(A^\alpha)$ *is periodic with period* $T > 0$, *then* Φ *has a* T*-periodic random inertial manifold* $\widetilde{\mathscr{M}}(\tau,\omega)$ *as given by* (11.59).

Proof Following the arguments of (11.4), we obtain in the present case that

$$\|\xi^*(x,\omega,\tau+T)-\xi^*(x,\omega,\tau)\|_{\mathscr{S}} \leq k\|\xi^*(x,\omega,\tau+T)-\xi^*(x,\omega,\tau)\|_{\mathscr{S}}.$$

Since $k \in (0,1)$ we get $\|\xi^*(x,\omega,\tau+T)-\xi^*(x,\omega,\tau)\|_{\mathscr{S}} = 0$, and hence $\xi^*(x,\omega,\tau+T)(0) = \xi^*(x,\omega,\tau)(0)$. As a consequence, we get $m(\tau+T,\omega) = m(\tau,\omega)$ and thus $\widetilde{m}(\tau+T,\omega) = \widetilde{m}(\tau,\omega)$. □

References

1. Arnold, L.: Random Dynamical Systems. Springer, Berlin (1998)
2. Bensoussan, A., Flandoli, F.: Stochastic inertial manifold. Stoch. Stoch. Rep. **53**, 13–39 (1995)
3. Brune, P., Schmalfuss, B.: Inertial manifolds for stochastic PDE with dynamical boundary conditions. Commun. Pure Appl. Anal. **10**, 831–846 (2011)
4. Caraballo, T., Duan, J., Lu, K., Schmalfuss, B.: Invariant manifolds for random and stochastic partial differential equations. Adv. Nonlinear Stud. **10**, 23–52 (2010)
5. Chueshov, I., Scheutzow, M.: Inertial manifolds and forms for stochastically perturbed retarded semilinear parabolic equations. J. Dyn. Differ. Equ. **13**, 355–380 (2001)
6. Chueshov, I., Scheutzow, M., Schmalfuss, B.: Continuity properties of inertial manifolds for stochastic retarded semilinear parabolic equations. In: Deuschel, J., Greven, A. (eds.) Interacting Stochastic Systems, pp. 353–375. Springer, Berlin (2005)
7. Chueshov, I., Girya, T.: Inertial manifolds for stochastic dissipative dynamical systems. Dokl. Acad. Sci. Ukr. **7**, 42–45 (1994)
8. Chueshov, I., Girya, T.: Inertial manifolds and forms for semilinear parabolic equations subjected to additive white noise. Lett. Math. Phys. **34**, 69–76 (1995)
9. Duan, J., Lu, K., Schmalfuss, B.: Invariant manifolds for stochastic partial differential equations. Ann. Probab. **31**, 2109–2135 (2003)
10. Duan, J., Lu, K., Schmalfuss, B.: Smooth stable and unstable manifolds for stochastic evolutionary equations. J. Dyn. Differ. Equ. **16**, 949–972 (2004)
11. Garrido-Atienza, M.J., Lu, K., Schmalfuss, B.: Unstable invariant manifolds for stochastic PDEs driven by a fractional Brownian motion. J. Differ. Equ. **248**, 1637–1667 (2010)
12. Girya, T., Chueshov, I.: Inertial manifolds and stationary measures for stochastically perturbed dissipative dynamical systems. Sb. Math. **186**, 29–46 (1995)
13. Lian, Z., Lu, K.: Lyapunov exponents and invariant manifolds for infinite-dimensional random dynamical systems in a Banach space. Mem. Am. Math. Soc. **206**(967), 1–106 (2010)
14. Lu, K., Schmalfuss, B.: Invariant manifolds for stochastic wave equations. J. Differ. Equ. **236**, 460–492 (2007)
15. Mohammed, S.-E.A., Scheutzow, M.K.R.: The stable manifold theorem for stochastic differential equations. Ann. Probab. **27**, 615–652 (1999)
16. Wang, B.: Sufficient and necessary criteria for existence of pullback attractors for non-compact random dynamical systems. J. Differ. Equ. **253**, 1544–1583 (2012)
17. Wanner, T.: Linearization of random dynamical systems. Dyn. Rep. **4**, 203–269 (1995)
18. Yoshizawa, T.: Stability Theory and the Existence of Periodic Solutions and Almost Periodic Solutions. Springer, New York (1975)

Chapter 12
Some Properties for Exact Generalized Processes

Jacson Simsen and Érika Capelato

Abstract In this work, we define an exact generalized process and we establish some results such as monotonicity, compactness, and upper semicontinuity for the multivalued process defined by the exact generalized process. The main result is on compactness, invariance, and attraction properties of ω-limit sets.

Keywords Nonautonomous dynamical systems · Pullback attraction · Generalized processes · Multivalued processes

12.1 Introduction

The concept of attraction is fundamental to analyze the asymptotic behavior of solutions of partial differential equations or inclusions. See, for example, [1–11, 19] for autonomous, [12–17] for nonautonomous and [18, 19] for stochastic problems. A concept which has been used a lot in the last years to deal with nonautonomous problems is pullback attraction, which attracts the solutions of the problem from $-\infty$, i.e., the initial time goes to $-\infty$ while the final time remains fixed [16, 17, 20]. The authors in [13] gave results which guarantee the existence of pullback attractors for processes, extending results of the autonomous case. In [21], the authors present abstract results that provide sufficient conditions to the existence of pullback attractors for multivalued processes. Observe that is also possible considering forward attractors for nonautonomous problems [22].

J. Simsen (✉)
Instituto de Matemática e Computaccão, Universidade Federal de Itajubá, Av. BPS n. 1303, Bairro Pinheirinho, Itajubá, MG 37500-903, Brazil
e-mail: jacson@unifei.edu.br

É. Capelato
Departamento de Economia - Faculdade de Ciências e Letras, Universidade Estadual Paulista, Araraquara, SP
148000-901, Brazil

© Springer International Publishing Switzerland 2015

V.A. Sadovnichiy and M.Z. Zgurovsky (eds.), *Continuous and Distributed Systems II*, Studies in Systems, Decision and Control 30,
DOI 10.1007/978-3-319-19075-4_12

For the treatment of autonomous equations without uniqueness of solution, Ball [23] defined the concept of generalized semiflow which was also used by Simsen and Gentile in [10] where the authors extended results from Semigroup Theory on existence and characterization of attractors to the multivalued case. For the treatment of the class of nonautonomous equations, Ball defined in [24] the generalized process and here in this work we add to that definition some extra conditions such as concatenation in order to obtain an exact generalized process.

We organize the work as follows. Section 12.2 contains notations, definitions, and some properties on the multivalued process defined by an exact generalized process. In Sect. 12.3 we obtain results for w-limit sets.

12.2 Notations, Definitions, and Some Properties on the Multivalued Process

Let (X, ρ) be a complete metric space and let $\wp(X)$, $\mathscr{B}(X)$, and $\mathscr{K}(X)$ denote, respectively, nonempty, nonempty and bounded, and nonempty and compact subsets of X. For $x \in X$ and $A, B \in \wp(X)$, and $\varepsilon > 0$ we set

$\rho(x, A) := \inf_{a \in A} \{\rho(x, a)\}$;

$\text{dist}(A, B) := \sup_{a \in A} \{\rho(a, B)\} = \sup_{a \in A} \inf_{b \in B} \{\rho(a, b)\}$;

$\mathscr{O}_\varepsilon(A) := \{z \in X; \rho(z, A) < \varepsilon\}$.

Definition 12.1 A **generalized process** $\mathscr{G} = \{\mathscr{G}(\tau)\}_{\tau \in \mathbb{R}}$ on X is a family of function sets $\mathscr{G}(\tau)$ consisting of maps $\varphi : [\tau, \infty) \to X$, satisfying the conditions:

C1- For each $\tau \in \mathbb{R}$ and $z \in X$, there exists at least one $\varphi \in \mathscr{G}(\tau)$ with $\varphi(\tau) = z$;

C2- If $\varphi \in \mathscr{G}(\tau)$ and $s \geq 0$, then $\varphi^s \in \mathscr{G}(\tau)$ and $\varphi^{+s} \in \mathscr{G}(\tau + s)$, where $\varphi^{+s} := \varphi_{|[\tau+s,\infty)}$ and $\varphi^s(t) := \varphi(t + s)$ for all $t \in [\tau, \infty)$;

C3- If $\{\varphi_j\}_{j \in \mathbb{N}} \subset \mathscr{G}(\tau)$ and $\varphi_j(\tau) \to z$, then there exists a subsequence $\{\varphi_\mu\}_{\mu \in \mathbb{N}}$ of $\{\varphi_j\}_{j \in \mathbb{N}}$ and $\varphi \in \mathscr{G}(\tau)$ with $\varphi(\tau) = z$ such that $\varphi_\mu(t) \to \varphi(t)$ for each $t \geq \tau$.

Remark 12.1 Given $\varphi \in \mathscr{G}(\tau)$, $Domain(\varphi^{+s}) \neq Domain(\varphi^s)$, but $Range(\varphi^{+s}) = Range(\varphi^s)$ since $\varphi^{+s}(\ell) = \varphi(\ell) = \varphi(\ell - s + s) = \varphi^s(\ell - s)\ \forall\ \ell \geq \tau + s$.

Definition 12.2 A generalized process $\mathscr{G} = \{\mathscr{G}(\tau)\}_{\tau \in \mathbb{R}}$ which satisfies the condition

C4- (Concatenation) If $\varphi, \psi \in \mathscr{G}$ with $\varphi \in \mathscr{G}(\tau)$, $\psi \in \mathscr{G}(r)$ and $\varphi(s) = \psi(s)$ for some $s \geq r \geq \tau$, then $\theta \in \mathscr{G}(\tau)$, where $\theta(t) := \begin{cases} \varphi(t), t \in [\tau, s] \\ \psi(t), t \in (s, \infty) \end{cases}$,

is called an **exact (or strict) generalized process**.

Definition 12.3 We say that **an exact generalized process** $\mathscr{G}$ **is continuous** if for each $\tau \in \mathbb{R}$ any $\varphi \in \mathscr{G}(\tau)$ is a continuous map from $[\tau, \infty)$ into X.

Definition 12.4 A **multivalued process** $\{U_{\mathscr{G}}(t, \tau)\}_{t \geq \tau}$ **defined by** $\mathscr{G}$ is a family of multivalued operators $U_{\mathscr{G}}(t, \tau) : \wp(X) \to \wp(X)$ with $-\infty < \tau \leq t < +\infty$, such that for each $\tau \in \mathbb{R}$

$$U_{\mathscr{G}}(t,\tau)E = \{\varphi(t);\, \varphi \in \mathscr{G}(\tau), \text{ with } \varphi(\tau) \in E\}, \ t \geq \tau.$$

Now, we will prove some properties of $\{U_{\mathscr{G}}(t,\tau)\}_{t\geq\tau}$.

Theorem 12.1 *Let $\mathscr{G}$ be an exact generalized process. If $\{U_{\mathscr{G}}(t,\tau)\}_{t\geq\tau}$ is a multivalued process defined by $\mathscr{G}$, then $\{U_{\mathscr{G}}(t,\tau)\}_{t\geq\tau}$ is an exact multivalued process on $\wp(X)$, i.e.,*

(1) $U_{\mathscr{G}}(t,t) = Id_{\wp(X)}$
(2) $U_{\mathscr{G}}(t,\tau) = U_{\mathscr{G}}(t,s)U_{\mathscr{G}}(s,\tau)$ *for all* $-\infty < \tau \leq s \leq t < +\infty$.

Proof Let $E \subset X$ be an arbitrary set.

1. Let $z \in U_{\mathscr{G}}(t,t)E$. By the definition of the multivalued process there exists $\varphi \in \mathscr{G}(t)$ such that $\varphi(t) = z$ and $\varphi(t) \in E$. So, $z = \varphi(t) \in E$. To prove the another inclusion, take $z \in E$. From $C1$ there exists at least one $\varphi \in \mathscr{G}(t)$ such that $\varphi(t) = z$. Therefore, $z \in U_{\mathscr{G}}(t,t)E$.

2. Let $v \in E$ and $z \in U_{\mathscr{G}}(t_1+t_2,\tau)v$. Then there is $\varphi \in \mathscr{G}(\tau)$ such that $\varphi(t_1+t_2) = z$ and $\varphi(\tau) = v$. It is clear that $\varphi(t_2) \in U_{\mathscr{G}}(t_2,\tau)v$. By defining $\psi(t) := \varphi(t), \forall t \geq t_2$, we have $\psi \in \mathscr{G}(t_2)$ and $\psi(t_2) = \varphi(t_2)$ which implies

$$z = \varphi(t_1+t_2) = \psi(t_1+t_2) \in U_{\mathscr{G}}(t_1+t_2,t_2)U_{\mathscr{G}}(t_2,\tau)v, \ \forall\, t_1 \geq 0 \text{ and } t_2 \geq \tau.$$

Therefore,

$$U_{\mathscr{G}}(t_1+t_2,\tau)v \subset U_{\mathscr{G}}(t_1+t_2,t_2)U_{\mathscr{G}}(t_2,\tau)v, \ \forall\, t_1 \geq 0 \text{ and } t_2 \geq \tau.$$

Since $v \in E$ was arbitrary we conclude

$$U_{\mathscr{G}}(t_1+t_2,\tau)E \subset U_{\mathscr{G}}(t_1+t_2,t_2)U_{\mathscr{G}}(t_2,\tau)E, \ \forall\, t_1 \geq 0 \text{ and } t_2 \geq \tau.$$

On the other hand, given $z \in U_{\mathscr{G}}(t_1+t_2,t_2)U_{\mathscr{G}}(t_2,\tau)v$ then there exists $\psi \in \mathscr{G}(t_2)$ with $\psi(t_1+t_2) = z$ and $\psi(t_2) \in U_{\mathscr{G}}(t_2,\tau)v$. Thus, there is $\varphi \subset \mathscr{G}(\tau)$ such that $\varphi(t_2) = \psi(t_2)$ with $\varphi(\tau) = v$. Define $\theta(t) := \begin{cases} \varphi(t), & t \in [\tau,t_2] \\ \psi(t), & t \in (t_2,\infty) \end{cases}$. From $C4$, we have $\theta \in \mathscr{G}(\tau)$ with $\theta(\tau) = \varphi(\tau) = v$. Then, $z = \psi(t_1+t_2) = \theta(t_1+t_2) \in U_{\mathscr{G}}(t_1+t_2,\tau)v$. Therefore,

$$U_{\mathscr{G}}(t_1+t_2,t_2)U_{\mathscr{G}}(t_2,\tau)v \subset U_{\mathscr{G}}(t_1+t_2,\tau)v.$$

□

Remark 12.2 A family $\{U(t,\tau) : X \to \wp(X), -\infty < \tau \leq t < +\infty\}$ of multivalued operators satisfying the properties (1) and (2) in Proposition 12.1 was called by Carvalho and Gentile in [3] a "m—evolution process".

Proposition 12.1 *Let $\mathscr{G}$ be an exact generalized process. If $\{U_{\mathscr{G}}(t,\tau)\}_{t\geq\tau}$ is a multivalued process defined by $\mathscr{G}$, then $\{U_{\mathscr{G}}(t,\tau)\}_{t\geq\tau}$ is monotone with respect to the partial order of set inclusion, i.e., $E\subset F$ implies $U_{\mathscr{G}}(t,\tau)E\subset U_{\mathscr{G}}(t,\tau)F$, for all $-\infty<\tau\leq t<+\infty$.*

Proof Let $z\in U_{\mathscr{G}}(t,\tau)E$. Then there exists $\varphi\in\mathscr{G}(\tau)$ such that $\varphi(t)=z$ and $\varphi(\tau)\in E$. By hypothesis $E\subset F$, then $\varphi(\tau)\in F$. So, by definition, $z\in U_{\mathscr{G}}(t,\tau)F$. □

Proposition 12.2 *Let $\mathscr{G}$ be an exact generalized process. If $\{U_{\mathscr{G}}(t,\tau)\}_{t\geq\tau}$ is a multivalued process defined by $\mathscr{G}$, then $U_{\mathscr{G}}(t,\tau)x$ is compact for each $x\in X$ and $-\infty<\tau\leq t<+\infty$.*

Proof Consider $\{z_n\}_{n\in\mathbb{N}}$ a sequence in $U_{\mathscr{G}}(t,\tau)x$, $x\in X$ fixed, then for each $n\in\mathbb{N}$ there exists $\varphi_n\in\mathscr{G}(\tau)$ such that $\varphi_n(t)=z_n$, $\varphi_n(\tau)=x$. As $\varphi_n(\tau)=x\ \forall\ n$ we have $\varphi_n(\tau)\to x$ as $n\to+\infty$. Then from $C3$ there exist $\{\varphi_\mu\}_{\mu\in\mathbb{N}}$ subsequence of $\{\varphi_n\}_{n\in\mathbb{N}}$ and $\varphi\in\mathscr{G}(\tau)$ with $\varphi(\tau)=x$ such that $\varphi_\mu(t)\to\varphi(t)$ for each $t\geq\tau$. It is clear that $\varphi(t)\in U_{\mathscr{G}}(t,\tau)x$. □

Theorem 12.2 *Let $\mathscr{G}$ be an exact generalized process and $\{U_{\mathscr{G}}(t,\tau)\}_{t\geq\tau}$ a multivalued process defined by $\mathscr{G}$. If K is a compact subset of X and $\{K_n\}_{n\in\mathbb{N}}$ is a sequence of compact subsets of X such that $dist(K_n,K)\to0$ as $n\to\infty$, then*

$$dist(U_{\mathscr{G}}(t,\tau)K_n,U_{\mathscr{G}}(t,\tau)K)\to0\ as\ n\to\infty,$$

for each $-\infty<\tau\leq t<+\infty$.

Proof Suppose that this is not true. Then there would exists an $\varepsilon_0>0$, a subsequence $\{K_\mu\}_{\mu\in\mathbb{N}}$ of $\{K_n\}_{n\in\mathbb{N}}$ and elements $a_\mu\in U_{\mathscr{G}}(t,\tau)K_\mu$ such that

$$\text{dist}(a_\mu,U_{\mathscr{G}}(t,\tau)K)>\varepsilon_0. \tag{12.1}$$

We have $a_\mu=\varphi_\mu(t)$ with $\{\varphi_\mu\}_{\mu\in\mathbb{N}}\subset\mathscr{G}(\tau)$ and $\varphi_\mu(\tau)\in K_\mu$. Since $\text{dist}(K_n,K)\to 0$ as $n\to\infty$ and K is compact there are $z\in K$ and a subsequence of $\{\varphi_\mu\}_{\mu\in\mathbb{N}}$, which we do not relabel, such that $\varphi_\mu(\tau)\to z$. From C3 there are $\psi\in\mathscr{G}(\tau)$ and a subsequence of $\{\varphi_\mu\}_{\mu\in\mathbb{N}}$, which we call the same, such that $\varphi_\mu(t)\to\psi(t)$, $\forall\ t\geq\tau$ and with $\psi(\tau)=z\in K$. Then $\psi(t)\in U_{\mathscr{G}}(t,\tau)K$, which contradicts (12.1). □

Definition 12.5 Let X, Y be metric spaces. A multivalued map $F:X\to\wp(Y)$ is **w-upper semicontinuous** if for all $x\in X$ and any ε-neighborhood of $F(x)$, $\mathscr{O}_\varepsilon(F(x))$, there is $\delta>0$ such that if $\rho(x,z)<\delta$, then $F(z)\subset\mathscr{O}_\varepsilon(F(x))$. If we replace the ε-neighborhood $\mathscr{O}_\varepsilon$ by an arbitrary one $\mathscr{O}$, then F is called **upper semicontinuous**. F is **lower semicontinuous** if for all $x\in X$, $x_n\to x$ and $y\in F(x)$, there is a sequence $\{y_n\}$ such that $y_n\in F(x_n)$ and $y_n\to y$. F is **continuous** if it is upper and lower semicontinuous.

Theorem 12.3 *Let $\mathscr{G}$ be an exact generalized process. If $\{U_{\mathscr{G}}(t,\tau)\}_{t\geq\tau}$ is a multivalued process defined by $\mathscr{G}$, then $U_{\mathscr{G}}(t,\tau):X\to\mathscr{K}(X)$ is an upper semicontinuous map and it has closed graph for all $-\infty<\tau\leq t<+\infty$.*

Proof Suppose that $x \mapsto U_{\mathscr{G}}(t,\tau)x$ is not an upper semicontinuous map. Then, there exist $x_0 \in X$, a neighborhood $\mathscr{O}(U_{\mathscr{G}}(t,\tau)x_0)$ and $\{\xi_n\}_{n\in\mathbb{N}} \subset U_{\mathscr{G}}(t,\tau)x_n$ with $\rho(x_n,x_0) < \frac{1}{n}$ such that $\xi_n \notin \mathscr{O}(U_{\mathscr{G}}(t,\tau)x_0)$. Since $\{\xi_n\}_{n\in\mathbb{N}} \subset U_{\mathscr{G}}(t,\tau)x_n$, there exists $\{\varphi_n\}_{n\in\mathbb{N}} \subset \mathscr{G}(\tau)$ such that $\xi_n = \varphi_n(t)$ and $\varphi_n(\tau) = x_n \to x_0$ as $n \to \infty$. From $C3$ there exist a subsequence of $\{\varphi_n\}$, which we do not relabel, and $\varphi \in \mathscr{G}(\tau)$ such that $x_0 = \varphi(\tau)$ and $\varphi_n(t) \to \varphi(t)$. Then, $\xi = \varphi(t) \in U_{\mathscr{G}}(t,\tau)x_0$, which is a contradiction. Therefore, $x \mapsto U_{\mathscr{G}}(t,\tau)x$ is upper semicontinuous.

To prove the second part, consider $-\infty < \tau \le t < +\infty$ fixed and a sequence (x_n, y_n) such that $y_n \in U(t,\tau)x_n$ with $x_n \to x$ and $y_n \to y$ in X. As $x_n \to x$, given $\delta > 0$ there is $n_0(\delta) > 0$ such that $\rho(x_n,x) < \delta$, $\forall\, n \ge n_0(\delta)$. By upper semicontinuity, given an $\frac{\varepsilon}{2}$–neighborhood of $U_{\mathscr{G}}(t,\tau)x$, $\mathscr{O}_{\varepsilon/2}(U_{\mathscr{G}}(t,\tau)x)$, there exist $\delta_0 > 0$ and $n_0 = n_0(\delta_0)$ such that $U_{\mathscr{G}}(t,\tau)x_n \subset \mathscr{O}_{\varepsilon/2}(U_{\mathscr{G}}(t,\tau)x)$, $\forall\, n \ge n_0$, i.e., $y_n \in \mathscr{O}_{\varepsilon/2}(U_{\mathscr{G}}(t,\tau)x)$, $\forall\, n \ge n_0$. Since $y_n \to y$ then for a given $\varepsilon > 0$, there exists $n_1 = n_1(\varepsilon) > 0$ such that $\rho(y_n,y) < \frac{\varepsilon}{2}$, $\forall\, n \ge n_1$. Considering $N := \max\{n_0,n_1\}$ we have $\rho(y,U_{\mathscr{G}}(t,\tau)x) \le \rho(y,y_N) + \rho(y_N,U_{\mathscr{G}}(t,\tau)x) < \frac{\varepsilon}{2} + \frac{\varepsilon}{2} = \varepsilon$. So, $y \in \mathscr{O}_{\varepsilon}(U_{\mathscr{G}}(t,\tau)x)$. As ε was arbitrary and $U_{\mathscr{G}}(t,\tau)x$ is closed, we obtain that $y \in U_{\mathscr{G}}(t,\tau)x$. □

12.3 Pullback Attraction and Properties on ω-limit Sets

In this section, we shall define ω-limit sets and prove properties on them. $\mathscr{G}$ will always be an exact generalized process.

Definition 12.6 Let $\tau \in \mathbb{R}$ be arbitrary. The **orbit** of $\varphi \in \mathscr{G}(\tau)$ and $E \subset X$, at time t, with $t \ge \tau$, is given by

- $\gamma_\tau^t(\varphi) := \{\varphi(r);\, \tau \le r \le t\}$;
- $\gamma_\mu(t,E) := U_{\mathscr{G}}(t,\mu)E$;
- $\gamma^\xi(t,E) := \bigcup_{s\le\xi} \gamma_s(t,E)$.

Definition 12.7 Let $\tau \in \mathbb{R}$ be arbitrary. If $\varphi \in \mathscr{G}(\tau)$, $E \subset X$ and $t \ge \tau$,

- $\omega(\varphi) := \left\{z \in X;\, \varphi(t_j) \to z, \text{ with } \{t_j\}_{j\in\mathbb{N}} \subset [\tau,\infty),\, t_j \to \infty\right\}$;
- $\omega(t,\varphi) := \left\{z \in X;\, \varphi(t_j) \to z, \text{ with } \{t_j\}_{j\in\mathbb{N}} \subset [\tau,t]\right\}$;
- $\omega(t,E) := \bigcap_{\xi\le t} \overline{\gamma^\xi(t,E)}^X$;
- $\omega(E) := \bigcap_{t\ge 0} \overline{\gamma_0(t,E)}^X = \bigcap_{t\ge 0} \overline{U_{\mathscr{G}}(t,0)E}^X$.

Moreover,

$$\omega_B(t, E) := \Big\{z \in X;\ \exists\, \varphi_j \in \mathscr{G}(\tau_j),\ \{\varphi_j(\tau_j)\}_{j\in\mathbb{N}} \subset E, \tau_j \to -\infty,$$

$$\{\varphi_j(\tau_j)\}_{j\in\mathbb{N}} \in \mathscr{B}(X), \varphi_j(t) \to z, j \to \infty\Big\},$$

$$\omega_B(E) := \Big\{z \in X;\ \exists\, \varphi_j \in \mathscr{G}(0),\ \{\varphi_j(0)\}_{j\in\mathbb{N}} \subset E, \{\varphi_j(0)\}_{j\in\mathbb{N}} \in \mathscr{B}(X)$$

$$\text{and exists } \{t_j\}_{j\in\mathbb{N}} \subset \mathbb{R}^+,\ t_j \to \infty \quad \text{with} \quad \varphi_j(t_j) \to z\Big\}.$$

Proceeding as in [21] it can be shown that

Remark 12.3 The set $\omega(t, E)$ consists of the limits of all converging sequences $\{\xi_n\}_{n\in\mathbb{N}}$ where $\xi_n \in U_{\mathscr{G}}(t, s_n)E, s_n \to -\infty$. So, it is clear that $\omega_B(t, E) \subset \omega(t, E)$ and $\omega_B(E) \subset \omega(E)$. Moreover, $\omega_B(t, E) = \omega(t, E)$ and $\omega_B(E) = \omega(E)$ whenever E is a bounded set.

Definition 12.8 We say that there exists a **complete orbit** through $x \in X$ if there is a map $\psi : \mathbb{R} \to X$ with $\psi(\tau) = x$, for some $\tau \in \mathbb{R}$ and for all $s \in \mathbb{R}$, $\psi^{+s} \in \mathscr{G}(\tau + s)$. In this case, the complete orbit ψ is given by

$$\gamma(\psi) := Im(\psi) = \{\psi(t) : t \in \mathbb{R}\}.$$

We also say that ψ **is a complete orbit through** x **at time** τ.

Definition 12.9 If ψ is a complete orbit, the set α-limit is given by

$$\alpha(\psi) := \{z \in X;\ \psi(t_j) \to z,\ t_j \to -\infty\}.$$

Definition 12.10 We say that a complete orbit $\psi : \mathbb{R} \to X$ is **stationary** if $\psi(t) = z$, for all $t \in \mathbb{R}$, for some $z \in X$. We set

$$Z(\mathscr{G}) := \{z \in X;\ \text{there exists a complete orbit} \quad \psi \quad \text{such that} \quad \psi(t) = z\ \forall\, t \in \mathbb{R}\}.$$

Remark 12.4 We observe that $z \in Z(\mathscr{G})$ can be called a **stationary solution** in $\mathscr{G}$ once we have that if $z \in Z(\mathscr{G})$ then $\exists\, \phi \in \mathscr{G}$ such that $\phi(t) = z$ for all $t \geq \tau_0$, for some $\tau_0 \in \mathbb{R}$.

Indeed, if $z \in Z(\mathscr{G})$ then there is a complete orbit ψ such that $\psi(t) = z\ \forall\, t \in \mathbb{R}$. Taking $\tau, s \in \mathbb{R}$, $\phi := \psi^{+s} \in \mathscr{G}(\tau + s)$, we have $\phi(t) = \psi(t) = z$ for all $t \geq \tau_0 := \tau + s$.

Remark 12.5 When talking about differential equations or inclusions, we also have that if $z \in Z(\mathscr{G})$ then $z \in U_{\mathscr{G}}(t, \tau)z$ for all $t \geq \tau$, for some $\tau \in \mathbb{R}$. So, we can refer to z as an **equilibrium** of $U_{\mathscr{G}}(t, \tau)$.

Indeed, if $z \in Z(\mathscr{G})$ then there exists a complete orbit ψ such that $\psi(t) = z \ \forall\, t \in \mathbb{R}$ and, by definition of complete orbit, $\psi^{+0} \in \mathscr{G}(\tau)$, for some $\tau \in \mathbb{R}$. Considering $\phi := \psi^{+0}$ we have $z = \phi(t)$ with $\phi \in \mathscr{G}(\tau)$ and $\phi(\tau) = z$. So, $z \in U_{\mathscr{G}}(t,\tau)z$ for all $t \geq \tau$.

Definition 12.11 Let $A = \{A(t)\}_{t\in\mathbb{R}}$ be a family of subsets of X.

- A is **positively invariant** if $U_{\mathscr{G}}(t,\tau)A(\tau) \subset A(t)$ for all $-\infty < \tau \leq t < \infty$;
- A is **negatively invariant** if $A(t) \subset U_{\mathscr{G}}(t,\tau)A(\tau)$ for all $-\infty < \tau \leq t < \infty$;
- A is **invariant** if $U_{\mathscr{G}}(t,\tau)A(\tau) = A(t)$ for all $-\infty < \tau \leq t < \infty$;
- A is **quasi-invariant** if for each $z \in A(\tau)$ for some $\tau \in \mathbb{R}$, there exists a complete orbit ψ through z at τ (i.e., $\psi(\tau) = z$) and $\psi(t) \in A(t)$ for all $t \in \mathbb{R}$.

The concept of quasi-invariance has also been called previously in the literature weakly invariance (see, e.g., [25] for the autonomous case or [26] for the nonautonomous case).

With some computations we can show the following

Proposition 12.3 *Invariant $\Rightarrow$ quasi-invariant $\Rightarrow$ negatively invariant.*

Definition 12.12 Let $t \in \mathbb{R}$.

(1) A set $A(t) \subset X$ **pullback attracts** a set $B \in X$ at time t if

$$\mathrm{dist}(U_{\mathscr{G}}(t,s)B, A(t)) \to 0 \quad \text{as} \quad s \to -\infty.$$

(2) A family $A = \{A(t)\}_{t\in\mathbb{R}}$ **pullback attracts bounded sets** of X if $A(\tau) \subset X$ pullback attracts all bounded subsets at τ, for each $\tau \in \mathbb{R}$.

(3) **A set** $A(t) \subset X$ **pullback absorbs bounded subsets** of X at time t if, for each $B \in \mathscr{B}(X)$, there exists $T = T(t,B) \leq t$ such that $U_{\mathscr{G}}(t,\tau)B \subset A(t) \ \forall\, \tau \leq T$.

Definition 12.13

(1) **A family** $\{A(t)\}_{t\in\mathbb{R}}$ **pullback absorbs bounded subsets** of X if, $A(t)$ pullback absorbs bounded sets at time t, for each $t \in \mathbb{R}$.

(2) A family $\{A(t)\}_{t\in\mathbb{R}}$ is called a **pullback attractor** if it is invariant, $A(t)$ is compact for all $t \in \mathbb{R}$, and pullback attracts all bounded subsets of X at time t, for each $t \in \mathbb{R}$.

Remark 12.6 Generally, the interest is on the pullback atractor wich is the maximal family of compact invariant sets and also the minimal closed, i.e., $\{A(t)\}_{t\in\mathbb{R}}$ is minimal in the sense that, if there exists another family of closed bounded sets $\left\{\hat{A}(t)\right\}_{t\in\mathbb{R}}$ which pullback attracts bounded subsets of X, then $A(t) \subseteq \hat{A}(t)$ for all $t \in \mathbb{R}$.

Definition 12.14 $\mathscr{G}$ is called **pullback bounded dissipative** if there exists a family $\{B(t)\}_{t\in\mathbb{R}}$ with $B(t) \in \mathscr{B}(X) \ \forall\, t \in \mathbb{R}$ which pullback absorbs bounded subsets of X. In a completely similar way, we define the notions of **pullback point dissipative** and **pullback compact dissipative**.

Definition 12.15 $\mathscr{G}$ is called **pullback asymptotically compact at time** t if for any sequence $\{\varphi_j\}_{j\in\mathbb{N}}, \varphi_j \in \mathscr{G}(\tau_j)$ with $\{\tau_j\}_{j\in\mathbb{N}} \subset (-\infty, t], \tau_j \to -\infty$ and $\{\varphi_j(\tau_j)\}_{j\in\mathbb{N}} \in \mathscr{B}(X)$ the sequence $\{\varphi_j(t)\}_{j\in\mathbb{N}}$ has a convergent subsequence. Or equivalently, for all $B \in \mathscr{B}(X)$ each sequence $\{\xi_j\}_{j\in\mathbb{N}}$ with $\xi_j \in U_{\mathscr{G}}(t, \tau_j)B\ \forall\ j \in \mathbb{N}$ and $\tau_j \to -\infty$ has a convergent subsequence. We say that $\mathscr{G}$ is **pullback asymptotically compact** if it is at each time $t \in \mathbb{R}$.

Definition 12.16 $\mathscr{G}$ is called **pullback conditionally asymptotically compact at time** t if for all $B \in \mathscr{B}(X)$ such that $\gamma^{\tau_0}(t, B) \in \mathscr{B}(X)$, for some $\tau_0 = \tau_0(B) < t$, each sequence $\{\xi_n\}_{n\in\mathbb{N}}$ with $\xi_n \in U_{\mathscr{G}}(t, t_n)B\ \forall\ n \in \mathbb{N}$ and $t_n \to -\infty$ has a convergent subsequence. We say that $\mathscr{G}$ is **pullback conditionally asymptotically compact** if it is at each time $t \in \mathbb{R}$.

Definition 12.17 $\mathscr{G}$ is **pullback eventually bounded at time** t if for all $B \in \mathscr{B}(X)$ there exists $\tau_0 = \tau_0(B) \leq t$ such that $\gamma^{\tau_0}(t, B) \in \mathscr{B}(X)$.

Proposition 12.4 *Let $\mathscr{G}$ be pullback asymptotically compact at time t. Then $\mathscr{G}$ is pullback eventually bounded at time t.*

Proof Let $a \in X$ and $B \in \mathscr{B}(X)$, and suppose for contradiction that $\gamma^{\tau}(t, B)$ is unbounded for all $\tau \leq t$. Then there exist $\varphi_j \in \mathscr{G}(s_j)$ with $\varphi_j(s_j) \in B$ and $s_j \to -\infty$ as $j \to +\infty$ with $\rho(\varphi_j(t), a) \to +\infty$ as $j \to +\infty$. But $\{\varphi_j(t)\}_{j\in\mathbb{N}}$ has a convergent subsequence by pullback asymptotic compactness at time t. □

Theorem 12.4 *$\mathscr{G}$ is pullback asymptotically compact at time t if and only if $\mathscr{G}$ is pullback eventually bounded at time t and pullback conditionally asymptotically compact at time t.*

Proof If $\mathscr{G}$ is pullback asymptotically compact at time t then Proposition 12.4 ensures that $\mathscr{G}$ is pullback eventually bounded at time t. It is clear that $\mathscr{G}$ pullback asymptotically compact at time t implies that $\mathscr{G}$ is pullback conditionally asymptotically compact at time t.

Reciprocally, suppose that $\mathscr{G}$ is pullback eventually bounded at time t and pullback conditionally asymptotically compact at time t. Let $\varphi_j \in \mathscr{G}(\tau_j)$ with $B := \{\varphi_j(\tau_j)\}_{j\in\mathbb{N}} \in \mathscr{B}(X)$ and $t \geq \tau_j,\ \tau_j \to -\infty$. As $\mathscr{G}$ is pullback eventually bounded at time t there exists $\tau_0 = \tau_0(B) < t$ such that $\gamma^{\tau_0}(t, B) \in \mathscr{B}(X)$. Since $\mathscr{G}$ pullback conditionally asymptotically compact at time t, the sequence $\{\varphi_j(t)\}_{j\in\mathbb{N}}$ has a convergent subsequence. This show that $\mathscr{G}$ is pullback asymptotically compact at time t. □

The next two results follow from Theorem 6 and Lemma 8 in [21].

Proposition 12.5 *Let $t \in \mathbb{R}$ and $B \in \mathscr{B}(X)$. Suppose there exists $D(t, B) \in \mathscr{K}(X)$ such that*

$$lim_{s\to-\infty} dist\,(U_{\mathscr{G}}(t, s)B, D(t, B)) = 0.$$

Then $\omega(t, B) \in \mathscr{K}(X)$ and is the minimal closed set which pullback attracts B at time t.

Proposition 12.6 *$\mathscr{G}$ is pullback asymptotically compact if and only if for each $t \in \mathbb{R}$ and $B \in \mathscr{B}(X)$ there exists a compact set $D(t, B)$ satisfying*

$$lim_{s\to-\infty} dist(U_{\mathscr{G}}(t,s)B, D(t,B)) = 0.$$

As a consequence of Propositions 12.5 and 12.6 we have the following

Corollary 12.1 *If $\mathscr{G}$ is pullback asymptotically compact, then for all $t \in \mathbb{R}$ and $B \in B(X)$, $\omega(t, B) \in \mathscr{K}(X)$ and is the minimal closed set which pullback attracts B at time t.*

The next is our main result

Theorem 12.5 *Suppose $\mathscr{G}$ is pullback asymptotically compact at time t and let $B \in \mathscr{B}(X)$. Then $\omega(t, B) \in \mathscr{K}(X)$ and is the minimal closed set which pullback attracts B at time t. If $\mathscr{G}$ is pullback asymptotically compact, then the family $\{\omega(t, B)\}_{t\in\mathbb{R}}$ is quasi-invariant. If we have besides the hypothesis above that for each $t \in \mathbb{R}$, $U_{\mathscr{G}}(t,r)\omega(r, B) \subset B$ for all $r \leq t$ then $\{\omega(t, B)\}_{t\in\mathbb{R}}$ is invariant.*

Proof Let $t \in \mathbb{R}$ be arbitrary. As $\mathscr{G}$ is pullback asymptotically compact at time t, Corollary 12.1 ensures that $w(t, B) \in \mathscr{K}(X)$ and is the minimal closed set which pullback attracts B at time t.

To prove that $\{\omega(t, B)\}_{t\in\mathbb{R}}$ is quasi-invariant, take $z \in \omega(t, B)$ arbitrarily. By Remark 12.3 there exists $\varphi_j \in \mathscr{G}(\tau_j)$ with $\tau_j < t \ \forall j$ and $\tau_j \to -\infty$ with $\varphi_j(\tau_j) \in B$ and $\varphi_j(t) \to z$. From $C2$, $\varphi_j^{+(t-\tau_j)} \in \mathscr{G}(t) = \mathscr{G}(\tau_j + t - \tau_j)$ for each $j \in \mathbb{N}$. As $\varphi_j^{+(t-\tau_j)}(t) = \varphi_j(t) \to z$, by $C3$ there exist a subsequence of $\left\{\varphi_j^{+(t-\tau_j)}\right\}$, which we do not relabel, and a solution $\psi_0 \in \mathscr{G}(t)$ with $\psi_0(t) = z$ such that $\varphi_j(\ell) = \varphi_j^{+(t-\tau_j)}(\ell) \to \psi_0(\ell)$, $\forall \ \ell \geq t$. Clearly, $\psi_0(\ell) \in \omega(\ell, B) \ \forall \ \ell \geq t$. Now, consider the sequence $\{\varphi_j^{+(t-\tau_j-1)}\}$. From $C2$, $\varphi_j^{+(t-\tau_j-1)} \in \mathscr{G}(t-1) = \mathscr{G}(\tau_j + t - \tau_j - 1)$ for each $j \in \mathbb{N}$. Once $\mathscr{G}$ is pullback asymptotically compact, $\{\varphi_j(t-1)\}$ has a convergent subsequence, which we do not relabel, such that $\varphi_j^{+(t-\tau_j-1)}(t-1) = \varphi_j(t-1) \to z_1$. By $C3$, there exists a subsequence of $\{\varphi_j^{+(t-\tau_j-1)}\}$, which we call the same, and a solution $\psi_1 \in \mathscr{G}(t-1)$ with $\psi_1(t-1) = z_1$ such that $\varphi_j(\ell) = \varphi_j^{+(t-\tau_j-1)}(\ell) \to \psi_1(\ell)$, $\forall \ \ell \geq t-1$. It is easy to see that $\psi_1(\ell) \in \omega(\ell, B) \ \forall \ \ell \geq t-1$ and $\psi_1(\ell) = \psi_0(\ell) \ \forall \ \ell \geq t$. Proceeding inductively, we find for each $r = 1, 2, \ldots$ a solution $\psi_r \in \mathscr{G}(t-r)$ such that $\psi_r(\ell) = \psi_{r-1}(\ell) \ \forall \ r \geq t-(r-1)$ and $\psi_r(\ell) \in \omega(\ell, B) \ \forall \ \ell \geq t-r$. Given $\ell \in \mathbb{R}$, define $\psi(\ell)$ as the common value of $\psi_r(\ell + r)$ for $r \geq t - \ell$. Then ψ is a complete orbit with $\psi(t) = \psi_0(t) = z$ and $\psi(\ell) \in \omega(\ell, B)$ for all $\ell \geq t$.

Finally, supposing that for each $t \in \mathbb{R}$, $U_{\mathscr{G}}(t,r)\omega(r, B) \subset B$ for all $r \leq t$, we will prove that $\{\omega(t, B)\}_{t\in\mathbb{R}}$ is invariant. By Proposition 12.3, $\omega(t, B)$ quasi-invariant implies $\omega(t, B)$ negatively invariant, i.e., $\omega(t, B) \subset U_{\mathscr{G}}(t, t_0)\omega(t_0, B)$ for all $-\infty < t_0 \leq t < \infty$. It remains to prove the opposite inclusion. Let $z \in$

$U_{\mathscr{G}}(t,t_0)\omega(t_0,B)$, then $z=\varphi(t)$ with $\varphi\in\mathscr{G}(t_0)$ and $\varphi(t_0)=:y\in\omega(t_0,B)$. Since $\omega(t_0,B)\subset U_{\mathscr{G}}(t_0,\tau_n)\omega(\tau_n,B)$, $\tau_n\le t_0\le t$, $\forall\ n\in\mathbb{N}$ and $\tau_n\to-\infty$, we have $y=\varphi_n(t_0)$ for some $\varphi_n\in\mathscr{G}(\tau_n)$ and $\varphi_n(\tau_n)\in\omega(\tau_n,B)$. Define the map $\theta_n(s):=\begin{cases}\varphi_n(s), & s\in[\tau_n,t_0]\\ \varphi(s), & s\ge t_0\end{cases}$. As $\varphi(t_0)=y=\varphi_n(t_0)$, by $C4$ $\theta_n\in\mathscr{G}(\tau_n)$. Moreover, $z=\varphi(t)=\theta_n(t)$. So, $z=\lim_{n\to\infty}\theta_n(t)$ with $\theta_n(\tau_n)=\varphi_n(\tau_n)\in\omega(\tau_n,B)\subset U_{\mathscr{G}}(\tau_n,\ell)\omega(\ell,B)\subset B$ for all $\ell\le\tau_n$. Therefore, $\theta_n(t)\in U_{\mathscr{G}}(t,\tau_n)B$ and we conclude that $z\in\omega(t,B)$. □

12.4 Final Remarks

The literature results have already given sufficient conditions on multivalued processes in order to obtain the existence of the pullback attractor, see for example the Theorem 7 in [12], for a weak pullback attractor and Theorems 11 and 18 in [21] for a pullback $\sum_1$ −uniform global attractor.

It would be interesting to investigate more properties for an exact generalized process such as necessary and sufficient conditions for the existence of the pullback attractor and characterizations for it.

References

1. Babin, A.V., Vishik, M.I.: Attractors of evolution equations. Studies in Applied Mathematics, vol. 25. North-Holland Publishing Co., Amsterdam (1992)
2. Caraballo, T., Marín-Rubio, P., Robinson, J.C.: A comparison between two theories for multivalued semiflows and their asymptotic behaviour. Set-Valued Anal. **11**, 297–322 (2003)
3. Carvalho, A.N., Gentile, C.B.: Asymptotic behaviour of non-linear parabolic equations with monotone principal part. J. Math. Anal. Appl. **280**, 252–272 (2003)
4. Cholewa, J.W., Dlotko, T.: Global Attractors in Abstract Parabolic Problems. Cambridge University Press, Cambridge (2000)
5. Chueshov, I.D.: Introduction to the Theory of Infinite-dimensional Dissipative Systems. ACTA Scientific Publishing House, Kharkiv (2002)
6. Hale, J.K.: Asymptotic behavior of dissipative system. mathematical surveys and monographs. in American Mathematical Society, vol. 25. Providence (1988)
7. Kapustyan, A.V., Melnik, V.S., Valero, J., Yasinsky, V.V.: Global Attractors for Multi-Valued Evolution Equations Without Uniqueness. Naukova Dumka, Kiev (2008)
8. Ladyzhenskaya, O.: Attracors for Semigroups an Evolution Equations. Lincei Lectures, vol. 25. Cambridge University Press, Cambridge (1991)
9. Melnik, V.S., Valero, J.: On attractors of multivalued semi-flows and differential inclusions. Set-Valued Anal. **6**, 83–111 (1998)
10. Simsen, J., Gentile, C.B.: On attractors for multivalued semigroups defined by generalized semiflows. Set-Valued Anal. **16**, 105–124 (2008)
11. Temam, R.: Infinite-Dimensional Dynamical Systems in Mechanics and Physics. Springer, New York (1988)
12. Caraballo, T., Kloeden, P.E., Marín-Rubio, P.: Weak pullback attractors of setvalued processes. J. Math. Anal. Appl. **288**, 692–707 (2003)

13. Caraballo, T., Carvalho, A.N., Langa, J.A., Rivero, F.: Existence of pullback attractors for pullback asymptotically compact processes. Nonlinear Anal. **72**, 1967–1976 (2010)
14. Carvalho, A.N., Langa, J.A., Robinson, J.C.: On the continuity of pullback attractors for evolution processes. Nonlinear Anal. **71**, 1812–1824 (2009)
15. Cheban, D.N.: Global Attractors of Non-autonomous Dissipative Dynamical Systems. World Scientific Publishing Ltd., Singapore (2004)
16. Kloeden, P.E., Schmalfuss, B.: Asymptotic behaviour of non-autonomous difference inclusions. Syst. Control Lett. **33**, 275–280 (1998)
17. Kloeden, P.E., Schmalfuβ, B.: Nonautonomous systems, cocycle attractors and variable time-step discretization. Numer. Algorithms **14**, 141–152 (1997)
18. Crauel, H., Debussche, A., Flandoli, F.: Random attractors. J. Dyn. Differ. Equ. **9**(2), 307–341 (1997)
19. Crauel, H., Flandoli, F.: Attractors for random dynamical systems. Probab. Theory Relat. Fields **100**, 365–393 (1994)
20. Kloeden, P.E.: Pullback attractors in nonautonomous difference equations. J. Differ. Equ. Appl. **6**(1), 33–52 (2000)
21. Caraballo, T., Langa, J.A., Melnik, V.S., Valero, J.: Pullback attractors of nonautonomous and stochastic multivalued dynamical systems. Set-Valued Analysis **11**, 153–201 (2003)
22. Melnik, V.S., Valero, J.: On global attractors of multivalued semiprocesses and nonautonomous evolution inclusions. Set-Valued Anal. **8**, 375–403 (2000)
23. Ball, J.M.: Continuity properties and global attractors of generalized semiflows and the Navier-Stokes equations. J. Nonlinear Sci. **7**(5), 475–502 (1997)
24. Ball, J.M.: On the asymptotic behavior of generalized process with applications to nonlinear evolution equations. J. Differ. Equ. **27**, 224–265 (1978)
25. Szegö, G.P., Treccani, G.: Semigruppi di Trasformazioni Multivoche. Springer Lecture Notes in Mathematics, vol. 101 (1969)
26. Kloeden, P.E., Marín-Rubio, P.: Weak pullback attractors of non-autonomous difference inclusions. J. Differ. Equ. Appl. **9**(5), 489–502 (2003)

Chapter 13
Uniform Trajectory Attractors for Nonautonomous Dissipative Dynamical Systems

Mikhail Z. Zgurovsky and Pavlo O. Kasyanov

Abstract For all global weak solutions of the general classes of nonautonomous evolution equations and inclusions that satisfy standard sign and polynomial growth conditions, the multivalued dynamics as time $t \to +\infty$ is studied. The existence of a compact uniform trajectory attractor is justified. The obtained results allow to investigate long-time behavior of distributions of state functions for various mathematical models in geophysics, mechanics, biology, medicine etc.

Keywords Evolution inclusion · Pseudomonotone map · Uniform trajectory attractor · Feedback control

13.1 Introduction and Setting of the Problem

For evolution triple $(V_i; H; V_i^*)$[1] and multivalued map $A_i : \mathbf{R}_+ \times V \rightrightarrows V^*$, $i = 1, 2, \ldots, N$, $N = 1, 2, \ldots$, we consider a problem of long-time behavior of all globally defined weak solutions for nonautonomous evolution inclusion

$$y'(t) + \sum_{i=1}^{N} A_i(t, y(t)) \ni \bar{0}, \tag{13.1}$$

[1] i.e., V_i is a real reflexive separable Banach space continuously and densely embedded into a real Hilbert space H, H is identified with its topologically conjugated space H^*, V_i^* is a dual space to V_i. So, there is a chain of continuous and dense embeddings: $V_i \subset H \equiv H^* \subset V_i^*$ (see, for example, Gajewski et al. [4, Chap. I]).

M.Z. Zgurovsky
National Technical University of Ukraine "Kyiv Politechnic Institute",
Peremogy Ave. 37, Build. 1, Kyiv 03056, Ukraine
e-mail: zgurovsm@hotmail.com

P.O. Kasyanov (✉)
Institute for Applied System Analysis, National Technical University of Ukraine
"Kyiv Politechnic Institute", Peremogy Ave. 37, Build. 35,
Kyiv 03056, Ukraine
e-mail: kasyanov@i.ua

© Springer International Publishing Switzerland 2015
V.A. Sadovnichiy and M.Z. Zgurovsky (eds.), *Continuous and Distributed Systems II*, Studies in Systems, Decision and Control 30,
DOI 10.1007/978-3-319-19075-4_13

as $t \to +\infty$. Let $\langle \cdot, \cdot \rangle_{V_i} : V_i{}^* \times V_i \to \mathbb{R}$ be the pairing in $V_i{}^* \times V_i$ that coincides on $H \times V_i$ with the inner product $(\cdot, \cdot)$ in the Hilbert space H.

Note that Problem (13.1) arises in many important models for distributed parameter control problems and that large class of identification problems enter this formulation. Let us indicate a problem which is one of the motivations for the study of the nonautonomous evolution inclusion (13.1) (see, for example, Migórski and Ochal [22]; Zgurovsky et al. [29] and references therein). In a subset Ω of $\mathbb{R}^3$,, we consider the nonstationary heat conduction equation

$$\frac{\partial y}{\partial t} - \Delta y = f \text{ in } \Omega \times (0, +\infty)$$

with initial conditions and suitable boundary ones. Here $y = y(x, t)$ represents the temperature at the point $x \in \Omega$ and time $t > 0$. It is supposed that $f = f_1 + f_2$, where f_2 is given and f_1 is a known function of the temperature of the form

$$-f_1(x, t) \in \partial j(x, t, y(x, t)) \text{ a.e. } (x, t) \in \Omega \times (0, +\infty).$$

Here $\partial j(x, t, \xi)$ denotes generalized gradient of Clarke (see Clarke [12]) with respect to the last variable of a function $j : \Omega \times \mathbb{R} \to \mathbb{R}$ which is assumed to be locally Lipschitz in ξ (cf. Migórski and Ochal [22] and references therein). The multivalued function $\partial j(x, t, \cdot) : \mathbb{R} \to 2^{\mathbb{R}}$ is generally nonmonotone and it includes the vertical jumps. In a physicist's language, it means that the law is characterized by the generalized gradient of a nonsmooth potential j (cf. Panagiotopoulos [24]). Models of physical interest include also the next (see, for example, Balibrea et al. [2] and references therein): a model of combustion in porous media; a model of conduction of electrical impulses in nerve axons; a climate energy balance model; etc.

To introduce the assumptions on parameters of Problem (13.1), let us introduce additional constructions. A function $\varphi \in L_\gamma^{\text{loc}}(\mathbf{R}_+)$, $\gamma > 1$, is called *translation bounded* in $L_\gamma^{\text{loc}}(\mathbf{R}_+)$, if

$$\sup_{t \geq 0} \int_t^{t+1} |\varphi(s)|^\gamma ds < +\infty;$$

Chepyzhov and Vishik [10, p. 105]. A function $\varphi \in L_1^{\text{loc}}(\mathbf{R}_+)$ is called *translation uniform integrable (t.u.i.)* in $L_1^{\text{loc}}(\mathbf{R}_+)$, if

$$\lim_{K \to +\infty} \sup_{t \geq 0} \int_t^{t+1} |\varphi(s)| \mathrm{i}\{|\varphi(s)| \geq K\} ds = 0.$$

Note that Dunford–Pettis compactness criterion provides that a function $\varphi \in L_1^{\text{loc}}(\mathbf{R}_+)$ is t.u.i. in $L_1^{\text{loc}}(\mathbf{R}_+)$ if and only if for every sequence of elements $\{\tau_n\}_{n \geq 1} \subset \mathbf{R}_+$ the sequence $\{\varphi(\cdot + \tau_n)\}_{n \geq 1}$ contains a subsequence which converges weakly in $L_1^{\mathrm{loc}}(\mathbf{R}_+)$. Note that for any $\gamma > 1$ every translation bounded in $L_\gamma^{\mathrm{loc}}(\mathbf{R}_+)$ function is t.u.i. in $L_1^{\mathrm{loc}}(\mathbf{R}_+)$; Gorban et al. [6].

Throughout this paper, we suppose that the listed below assumptions hold:

Assumption I Let $p_i \geq 2$, $q_i > 1$ are such that $\frac{1}{p_i} + \frac{1}{q_i} = 1$, for each for $i = 1, 2, \ldots, N$, and the embedding $V_i \subset H$ is compact one, for some for $i = 1, 2, \ldots, N$.

Assumption II (Grows Condition) There exist a t.u.i. in $L_1^{\mathrm{loc}}(\mathbf{R}_+)$ function $c_1 : \mathbf{R}_+ \to \mathbf{R}_+$ and a constant $c_2 > 0$ such that

$$\max_{i=1}^{N} \|d_i\|_{V_i^*}^q \leq c_1(t) + c_2 \sum_{i=1}^{N} \|u\|_{V_i}^p$$

for any $u \in V_i$, $d_i \in A_i(t, u)$, $i = 1, 2, \ldots, N$, and a.e. $t > 0$.

Assumption III (Signed Assumption) There exist a constant $\alpha > 0$ and a t.u.i. in $L_1^{\mathrm{loc}}(\mathbf{R}_+)$ function $\beta : \mathbf{R}_+ \to \mathbf{R}_+$ such that

$$\sum_{i=1}^{N} \langle d_i, u \rangle_{V_i} \geq \alpha \sum_{i=1}^{N} \|u\|_{V_i}^p - \beta(t)$$

for any $u \in V_i$, $d_i \in A_i(t, u)$, $i = 1, 2, \ldots, N$, and a.e. $t > 0$.

Assumption IV (Strong Measurability) If $C \subseteq V_i^*$ is a closed set, then the set $\{(t, u) \in (0, +\infty) \times V_i : A_i(t, u) \cap C \neq \emptyset\}$ is a Borel subset in $(0, +\infty) \times V_i$.

Assumption V (Pointwise Pseudomonotonicity) Let for each $i = 1, 2, \ldots, N$ and a.e. $t > 0$ two assumptions hold:

(a) for every $u \in V_i$ the set $A_i(t, u)$ is nonempty, convex, and weakly compact one in V_i^*;

(b) if a sequence $\{u_n\}_{n\geq 1}$ converges weakly in V_i towards $u \in V_i$ as $n \to +\infty$, $d_n \in A_i(t, u_n)$ for any $n \geq 1$, and $\limsup_{n\to+\infty} \langle d_n, u_n - u \rangle_{V_i} \leq 0$, then for any $\omega \in V_i$ there exists $d(\omega) \in A_i(t, u)$ such that

$$\liminf_{n\to+\infty} \langle d_n, u_n - \omega \rangle_{V_i} \geq \langle d(\omega), u - \omega \rangle_{V_i}.$$

Let $0 \leq \tau < T < +\infty$. As a *weak solution* of evolution inclusion (13.1) on the interval $[\tau, T]$, we consider an element $u(\cdot)$ of the space $\cap_{i=1}^N L_{p_i}(\tau, T; V_i)$ such that for some $d_i(\cdot) \in L_{q_i}(\tau, T; V_i^*)$, $i = 1, 2, \ldots, N$, it is fulfilled:

$$-\int_\tau^T (\xi'(t), y(t))dt + \sum_{i=1}^{N} \int_\tau^T \langle d_i(t), \xi(t) \rangle_{V_i} dt = 0 \quad \forall \xi \in C_0^\infty([\tau, T]; V_i), \quad (13.2)$$

and $d_i(t) \in A_i(t, y(t))$ for each $i = 1, 2, \ldots, N$ and a.e. $t \in (\tau, T)$.

13.2 Preliminary Properties of Weak Solutions

For fixed nonnegative τ and T, $\tau < T$, let us consider

$$X_{\tau,T} = \cap_{i=1}^{N} L_{p_i}(\tau, T; V_i), \quad X^*_{\tau,T} = \sum_{i=1}^{N} L_{q_i}(\tau, T; V_i^*), \quad W_{\tau,T} = \{y \in X_{\tau,T} \mid y' \in X^*_{\tau,T}\},$$

$$\mathscr{A}_{\tau,T} : X_{\tau,T} \rightrightarrows X^*_{\tau,T}, \quad \mathscr{A}_{\tau,T}(y) = \{d \in X^*_{\tau,T} \mid d(t) \in A(t, y(t)) \text{ for a.e. } t \in (\tau, T)\},$$

where y' is a derivative of an element $u \in X_{\tau,T}$ in the sense of $\mathscr{D}([\tau, T]; \sum_{i=1}^{N} V_i^*)$ (see, for example, Gajewski et al. [4, Definition IV.1.10]). Note that the space $W_{\tau,T}$ is a reflexive Banach space with the graph norm of a derivative $\|u\|_{W_{\tau,T}} = \|u\|_{X_{\tau,T}} + \|u'\|_{X^*_{\tau,T}}$, $u \in W_{\tau,T}$. Let $\langle \cdot, \cdot \rangle_{X_{\tau,T}} : X^*_{\tau,T} \times X_{\tau,T} \to \mathbb{R}$ be the pairing in $X^*_{\tau,T} \times X_{\tau,T}$ that coincides on $L_2(\tau, T; H) \times X_{\tau,T}$ with the inner product in $L_2(\tau, T; H)$, i.e., $\langle u, v \rangle_{X_{\tau,T}} = \int_\tau^T (u(t), v(t))dt$ for any $u \in L_2(\tau, T; H)$ and $v \in X_{\tau,T}$. Gajewski et al. [4, Theorem IV.1.17] provide that the embedding $W_{\tau,T} \subset C([\tau, T]; H)$ is continuous and dense one. Moreover,

$$(u(T), v(T)) - (u(\tau), v(\tau)) = \int_\tau^T \Big[\langle u'(t), v(t) \rangle_{V_i} + \langle v'(t), u(t) \rangle_{V_i} \Big] dt, \qquad (13.3)$$

for any $u, v \in W_{\tau,T}$.

Migórski [23, Lemma 7, p. 516] (see paper and references therein) and Assumptions I–V provide the existence of multivalued Nemitsky operator $\mathscr{A}_{\tau,T} : X_{\tau,T} \rightrightarrows X^*_{\tau,T}$ for $\sum_{i=1}^{N} A_i$ that satisfies the following properties:

Property I The mapping $\mathscr{A}_{\tau,T}$ transforms an each bounded set in $X_{\tau,T}$ onto bounded subset of $X^*_{\tau,T}$.

Property II There exist positive constants $C_1 = C_1(\tau, T)$ and $C_2 = C_2(\tau, T)$ such that $\langle d, y \rangle_{X_{\tau,T}} \geq C_1 \|y\|^p_{X_{\tau,T}} - C_2$ for any $y \in X_{\tau,T}$ and $d \in \mathscr{A}_{\tau,T}(y)$.

Property III The multivalued mapping $\mathscr{A}_{\tau,T} : X_{\tau,T} \rightrightarrows X^*_{\tau,T}$ is (generalized) pseudomonotone on $W_{\tau,T}$, i.e., (a) for every $y \in X_{\tau,T}$ the set $\mathscr{A}_{\tau,T}(y)$ is a nonempty, convex, and weakly compact one in $X^*_{\tau,T}$; (b) $\mathscr{A}_{\tau,T}$ is upper semicontinuous from every finite dimensional subspace $X_{\tau,T}$ into $X^*_{\tau,T}$ endowed with the weak topology; (c) if a sequence $\{y_n, d_n\}_{n\geq 1} \subset W_{\tau,T} \times X^*_{\tau,T}$ converges weakly in $W_{\tau,T} \times X^*_{\tau,T}$ towards $(y, d) \in W_{\tau,T} \times X^*_{\tau,T}$, $d_n \in \mathscr{A}_{\tau,T}(y_n)$ for any $n \geq 1$, and $\limsup_{n\to+\infty} \langle d_n, y_n - y \rangle_{X_{\tau,T}} \leq 0$, then $d \in \mathscr{A}_{\tau,T}(y)$ and $\lim_{n\to+\infty} \langle d_n, y_n \rangle_{X_{\tau,T}} = \langle d, y \rangle_{X_{\tau,T}}$.

Formula (13.2) and definition of the derivative for an element from $\mathscr{D}([\tau, T]; \sum_{i=1}^{N} V_i^*)$ yield that each weak solution $y \in X_{\tau,T}$ of Problem (13.1) on $[\tau, T]$ belongs to the space $W_{\tau,T}$ and $y' + \mathscr{A}_{\tau,T}(y) \ni \bar{0}$. Vice versa, if $y \in W_{\tau,T}$ satisfies the last inclusion, then y is a weak solution of Problem (13.1) on $[\tau, T]$.

Assumption I, Properties I–III, and Denkowski et al. [13, Theorem 1.3.73] (see also Zgurovsky et al. [27, Chap. 2] and references therein) provide the existence of a weak solution of Cauchy problem (13.1) with initial data $y(\tau) = y^{(\tau)}$ on the interval $[\tau, T]$, for any $y^{(\tau)} \in H$.

For fixed τ and T, such that $0 \le \tau < T < +\infty$, we denote

$$\mathscr{D}_{\tau,T}(y^{(\tau)}) = \{y(\cdot) \mid y \text{ is a weak solution of (13.1) on}[\tau, T],\ y(\tau) = y^{(\tau)}\}, \quad y^{(\tau)} \in H.$$

We remark that $\mathscr{D}_{\tau,T}(y^{(\tau)}) \neq \emptyset$ and $\mathscr{D}_{\tau,T}(y^{(\tau)}) \subset W_{\tau,T}$, if $0 \le \tau < T < +\infty$ and $y^{(\tau)} \in H$. Moreover, the concatenation of Problem (13.1) weak solutions is a weak solutions too, i.e., if $0 \le \tau < t < T$, $y^{(\tau)} \in H$, $y(\cdot) \in \mathscr{D}_{\tau,t}(y^{(\tau)})$, and $v(\cdot) \in \mathscr{D}_{t,T}(y(t))$, then

$$z(s) = \begin{cases} y(s), & s \in [\tau, t], \\ v(s), & s \in [t, T], \end{cases}$$

belongs to $\mathscr{D}_{\tau,T}(y^{(\tau)})$; cf. Zgurovsky et al. [29, pp. 55–56].

Gronwall lemma provides that for any finite time interval $[\tau, T] \subset \mathbf{R}_+$ each weak solution y of Problem (13.1) on $[\tau, T]$ satisfies estimates

$$\|y(t)\|_H^2 - 2\int_0^t \beta(\xi)d\xi + 2\alpha \sum_{i=1}^{N} \int_s^t \|y(\xi)\|_{V_i}^p d\xi \le \|y(s)\|_H^2 - 2\int_0^s \beta(\xi)d\xi, \tag{13.4}$$

$$\|y(t)\|_H^2 \le \|y(s)\|_H^2 e^{-2\alpha\gamma(t-s)} + 2\int_s^t (\beta(\xi) + \alpha\gamma)e^{-2\alpha\gamma(t-\xi)}d\xi, \tag{13.5}$$

where $t, s \in [\tau, T]$, $t \ge s$; γ is a constant that does not depend on y, s, and t; cf. Zgurovsky et al. [29, p. 56]. In the proof of (13.5), we used the inequality $\|u\|_H^2 - 1 \le \|u\|_H^p$ for any $u \in H$.

Therefore, any weak solution y of Problem (13.1) on a finite time interval $[\tau, T] \subset \mathbf{R}_+$ can be extended to a global one, defined on $[\tau, +\infty)$. For arbitrary $\tau \ge 0$ and $y^{(\tau)} \in H$ let $\mathscr{D}_\tau(y^{(\tau)})$ be the set of all weak solutions (defined on $[\tau, +\infty)$) of Problem (13.1) with initial data $y(\tau) = y^{(\tau)}$. Let us consider the family $\mathscr{K}_\tau^+ = \cup_{y^{(\tau)} \in H} \mathscr{D}_\tau(y^{(\tau)})$ of all weak solutions of Problem (13.1) defined on the semi-infinite time interval $[\tau, +\infty)$.

13.3 Uniform Trajectory Attractor and Main Result

Consider the Fréchet space $C^{\mathrm{loc}}(\mathbb{R}_+; H)$. We remark that the sequence $\{f_n\}_{n\geq 1}$ converges in $C^{\mathrm{loc}}(\mathbb{R}_+; H)$ towards $f \in C^{\mathrm{loc}}(\mathbb{R}_+; H)$ as $n \to +\infty$ iff the sequence $\{\Pi_{t_1,t_2} f_n\}_{n\geq 1}$ converges in $C([t_1, t_2]; H)$ towards $\Pi_{t_1,t_2} f$ as $n \to +\infty$ for any finite interval $[t_1, t_2] \subset \mathbf{R}_+$, where Π_{t_1,t_2} is the restriction operator to the interval $[t_1, t_2]$; Chepyzhov and Vishik [8, p. 918]. We denote $T(h)y(\cdot) = y_h(\cdot)$, where $y_h(t) = y(t + h)$ for any $y \in C^{\mathrm{loc}}(\mathbb{R}_+; H)$ and $t, h \geq 0$.

In the autonomous case, when $A(t, y)$ does not depend on t, the long-time behavior of all globally defined weak solutions for Problem (13.1) is described by using trajectory and global attractors theory; Kasyanov [15, 16], Zgurovsky et al. [29, Chap. 2] and references therein; see also Balibrea et al. [2]. In this situation, the set $\mathscr{K}^+ := \mathscr{K}_0^+$ is *translation invariant*, i.e., $T(h)\mathscr{K}^+ \subseteq \mathscr{K}^+$ for any $h \geq 0$. As trajectory attractor it is considered a classical global attractor for translation semigroup $\{T(h)\}_{h\geq 0}$, that acts on $\mathscr{K}^+$.

In the nonautonomous case, we notice that $T(h)\mathscr{K}_0^+ \nsubseteq \mathscr{K}_0^+$. Therefore, we need to consider *united trajectory space* that includes all globally defined on any $[\tau, +\infty) \subseteq \mathbf{R}_+$ weak solutions of Problem (13.1) shifted to $\tau = 0$:

$$\mathscr{K}^+ = \mathrm{cl}_{C^{\mathrm{loc}}(\mathbb{R}_+;H)} \left[\bigcup_{\tau\geq 0} \{y(\cdot + \tau) : y \in \mathscr{K}_\tau^+\} \right],$$

where $\mathrm{cl}_{C^{\mathrm{loc}}(\mathbb{R}_+;H)}[\,\cdot\,]$ is the closure in $C^{\mathrm{loc}}(\mathbb{R}_+; H)$. Note that $T(h)\{y(\cdot + \tau) : y \in \mathscr{K}_\tau^+\} \subseteq \{y(\cdot + \tau + h) : y \in \mathscr{K}_{\tau+h}^+\}$ for any $\tau, h \geq 0$. Moreover,

$$T(h)\mathscr{K}^+ \subseteq \mathscr{K}^+ \text{ for any } h \geq 0,$$

because

$$\rho_{C^{\mathrm{loc}}(\mathbb{R}_+;H)}(T(h)u, T(h)v) \leq \rho_{C^{\mathrm{loc}}(\mathbb{R}_+;H)}(u, v) \text{ for any } u, v \in C^{\mathrm{loc}}(\mathbb{R}_+; H),$$

where $\rho_{C^{\mathrm{loc}}(\mathbb{R}_+;H)}$ is a standard metric on Fréchet space $C^{\mathrm{loc}}(\mathbb{R}_+; H)$; Chepyzhov and Vishik [8].

A set $\mathscr{P} \subset C^{\mathrm{loc}}(\mathbb{R}_+; H) \cap L_\infty(\mathbb{R}_+; H)$ is said to be a *uniformly attracting set* (cf. Chepyzhov and Vishik [8, p. 921]) for the united trajectory space $\mathscr{K}^+$ of Problem (13.1) in the topology of $C^{\mathrm{loc}}(\mathbb{R}_+; H)$, if for any bounded in $L_\infty(\mathbb{R}_+; H)$ set $\mathscr{B} \subseteq \mathscr{K}^+$ and any segment $[t_1, t_2] \subset \mathbf{R}_+$ the following relation holds:

$$\mathrm{dist}_{C([t_1,t_2];H)}(\Pi_{t_1,t_2} T(t)\mathscr{B}, \Pi_{t_1,t_2}\mathscr{P}) \to 0, \quad t \to +\infty, \tag{13.6}$$

where $\mathrm{dist}_{C([t_1,t_2];H)}$ is the Hausdorff semimetric.

A set $\mathscr{U} \subset \mathscr{K}^+$ is said to be a *uniform trajectory attractor* (cf. Chepyzhov and Vishik [8, p. 921]) of the translation semigroup $\{T(t)\}_{t\geq 0}$ on $\mathscr{K}^+$ in the induced topology from $C^{\text{loc}}(\mathbb{R}_+; H)$, if

(i) $\mathscr{U}$ is a compact set in $C^{\text{loc}}(\mathbb{R}_+; H)$ and bounded in $L_\infty(\mathbb{R}_+; H)$;
(ii) $\mathscr{U}$ is strictly invariant with respect to $\{T(h)\}_{h\geq 0}$, i.e., $T(h)\mathscr{U} = \mathscr{U}$ $\forall h \geq 0$;
(iii) $\mathscr{U}$ is a minimal uniformly attracting set for $\mathscr{K}^+$ in the topology of $C^{\text{loc}}(\mathbb{R}_+; H)$, i.e., $\mathscr{U}$ belongs to any compact uniformly attracting set $\mathscr{P}$ of $\mathscr{K}^+$: $\mathscr{U} \subseteq \mathscr{P}$.

Note that uniform trajectory attractor of the translation semigroup $\{T(t)\}_{t\geq 0}$ on $\mathscr{K}^+$ in the induced topology from $C^{\text{loc}}(\mathbb{R}_+; H)$ coincides with the classical trajectory attractor for the continuous semigroup $\{T(t)\}_{t\geq 0}$ defined on $\mathscr{K}^+$ (see, for example, Chepyzhov and Vishik [9, Definition 1.1]).

Presented construction is coordinated with the theory of uniform trajectory attractors for nonautonomous problems of the form

$$\partial_t u(t) = A_{\sigma(t)}(u(t)), \tag{13.7}$$

where $\sigma(s)$, $s \geq 0$, is a functional parameter called the time symbol of equation (13.7) (t is replaced by s). In applications to mathematical physics equations, a function $\sigma(s)$ consists of all time-dependent terms of the equation under consideration: external forces, parameters of mediums, interaction functions, control functions, etc.; Chepyzhov and Vishik [7, 8, 11]; Sell [25]; Zgurovsky et al. [29] and references therein; see also Hale [14]; Ladyzhenskaya [19]; Mel'nik and Valero [21]; Kapustyan et al. [18]. In the above-mentioned papers and books, it is assumed that the symbol σ of equation (13.7) belongs to a Hausdorff topological space Ξ_+ of functions defined on $\mathbf{R}_+$ with values in some complete metric space. Usually, in applications, the topology in the space Ξ_+ is a local convergence topology on any segment $[t_1, t_2] \subset \mathbf{R}_+$. Further, they consider the family of equation (13.7) with various symbols $\sigma(s)$ belonging to a set $\Sigma \subseteq \Xi_+$. The set Σ is called the symbol space of the family of equation (13.7). It is assumed that the set Σ, together with any symbol $\sigma(s) \in \Sigma$, contains all positive translations of $\sigma(s)$: $\sigma(t \mid s) = T(t)\sigma(s) \in \Sigma$ for any $t, s \geq 0$. The symbol space Σ is invariant with respect to the translation semigroup $\{T(t)\}_{t\geq 0}$: $T(t)\Sigma \subseteq \Sigma$ for any $t \geq 0$. To prove the existence of uniform trajectory attractor, they suppose that the symbol space Σ with the topology induced from Ξ_+ is a compact metric space. Mostly in applications, as a symbol space Σ it is natural to consider the hull of translation-compact function $\sigma_0(s)$ in an appropriate Hausdorff topological space Ξ_+. The direct realization of this approach for Problem (13.1) is problematic without any additional assumptions for parameters of Problem (13.1) and requires the translation compactness of the symbol $\sigma(s) = A(s, \cdot)$ in some compact Hausdorff topological space of measurable multivalued mappings acts from $\mathbf{R}_+$ to some metric space of pseudomonotone operators from ($V_i \to 2^{V_i^*}$) satisfying grows and signed assumptions. To avoid these technical difficulties, we present the alternative approach for the existence and construction of the uniform trajectory attractor for all weak solutions for Problem (13.1). Note that Assumptions (I)–(V) are natural and guaranty, in the general case, only existence of weak solution for

Cauchy problem on any finite time interval $[\tau, T] \subset \mathbf{R}_+$ and for any initial data form H; see, for example, Denkowski et al. [13]; Gasinski and Papageorgiou [5] etc.

The main result of this paper has the following form.

Theorem 13.1 *Let Assumptions (I)–(V) hold. Then there exists a uniform trajectory attractor* $\mathscr{U} \subset \mathscr{K}^+$ *of the translation semigroup* $\{T(t)\}_{t\geq 0}$ *on* $\mathscr{K}^+$ *in the induced topology from* $C^{\text{loc}}(\mathbb{R}_+; H)$*. Moreover, there exists a compact in* $C^{\text{loc}}(\mathbb{R}_+; H)$ *uniformly attracting set* $\mathscr{P} \subset C^{\text{loc}}(\mathbb{R}_+; H) \cap L_\infty(\mathbb{R}_+; H)$ *for the united trajectory space* $\mathscr{K}^+$ *of Problem (13.1) in the topology of* $C^{\text{loc}}(\mathbb{R}_+; H)$ *such that* $\mathscr{U}$ *coincides with* ω*-limit set of* $\mathscr{P}$*:*

$$\mathscr{U} = \bigcap_{t\geq 0} \text{cl}_{C^{\text{loc}}(\mathbb{R}_+;H)} \left[\bigcup_{h\geq t} T(h)\mathscr{P} \right]. \tag{13.8}$$

13.4 Proof of Theorem 13.1

Before the proof of Theorem 13.1, we provide some auxiliary constructions.

Assumptions (II) and (III) yield that there exist a positive constant $\alpha' > 0$ and a t.u.i. function c' in $L_1^{\text{loc}}(\mathbf{R}_+)$ such that $A(t, u) \subseteq \mathbb{A}_{c'(t)}(u)$ for each $u \in \cap_{i=1}^N V_i$ and a.e. $t > 0$, where

$$\mathbb{A}_{c'(t)}(u) := \left\{ \sum_{i=1}^N p_i \; : \; p_i \in V_i^*, \; \sum_{i=1}^N \langle p_i, u \rangle_{V_i} \geq \alpha' \max_{i=1}^N \left\{ \|u\|_{V_i}^p; \|p\|_{V_i^*}^q \right\} - c'(t) \right\}.$$

Let $\mathscr{H}(c')$ be the hull of t.u.i. function c' in $L_{1,w}^{\text{loc}}(\mathbf{R}_+)$, i.e., $\mathscr{H}(c') = \text{cl}_{L_1^{\text{loc}}(\mathbf{R}_+)}\{c'(\cdot + h) \; : \; h \geq 0\}$. This is a weakly compact set in $L_1^{\text{loc}}(\mathbf{R}_+)$; Gorban et al. [6].

Let us consider the family of problems

$$y' = \mathbb{A}_\sigma(y), \quad \sigma \in \Sigma := \mathscr{H}(c'). \tag{13.9}$$

To each $\sigma \in \Sigma$ there corresponds a space of all globally defined on $[0, +\infty)$ weak solutions $\mathscr{K}_\sigma^+ \subset C^{\text{loc}}(\mathbb{R}_+; H)$ of Problem (13.9). We set $\mathscr{K}_\Sigma^+ = \cup_{\sigma\in\Sigma} \mathscr{K}_\sigma^+$.

We remark that any element from $\mathscr{K}_\Sigma^+$ satisfies prior estimates.

Lemma 13.1 *There exist positive constants* c_3 *and* c_4 *such that for any* $\sigma \in \Sigma$ *and* $y \in \mathscr{K}_\sigma^+$ *the inequalities hold:*

$$\|y(t)\|_H^2 - 2\int_0^t \sigma(\xi)d\xi + 2\alpha' \sum_{i=1}^N \int_s^t \|y(\xi)\|_{V_i}^p d\xi \leq \|y(s)\|_H^2 - 2\int_0^s \sigma(\xi)d\xi, \tag{13.10}$$

$$\|y(t)\|_H^2 \le \|y(s)\|_H^2 e^{-c_3(t-s)} + c_4 \int_s^t \sigma(\xi) e^{-c_3(t-\xi)} d\xi, \tag{13.11}$$

for any $t \ge s \ge 0$.

Proof The proof naturally follows from conditions for the parameters of Problem (13.9) and Gronwall lemma. □

Let us provide the result characterizing the compactness properties of solutions for the family of Problems (13.9).

Theorem 13.2 *Let* $\{y_n\}_{n\ge1} \subset \mathscr{K}_\Sigma^+$ *be an arbitrary sequence, that is bounded in* $L_\infty(\mathbb{R}_+; H)$. *Then there exist a subsequence* $\{y_{n_k}\}_{k\ge1} \subset \{y_n\}_{n\ge1}$ *and an element* $y \in \mathscr{K}_\Sigma^+$ *such that*

$$\max_{t\in[\tau,T]} \|y_{n_k}(t) - y(t)\|_H \to 0, \quad k \to +\infty, \tag{13.12}$$

for any finite time interval $[\tau, T] \subset (0, +\infty)$.

Proof For any $n \ge 1$ there exists $\sigma_n \in \Sigma$ such that $y_n \in \mathscr{K}_\sigma^+$. Furthermore, the definition of weak solution of evolution inclusion yields that for any $n \ge 1$ and $i = 1, 2, \ldots, N$, there exists $d_{n,i} \in L_{q_i}^{\text{loc}}(\mathbf{R}_+; V_i^*)$ such that $y_n'(t) + \sum_{i=1}^N d_{n,i}(t) = \bar{0}$ for a.e. $t > 0$. The definition of $\mathbb{A}_\sigma$ and estimates (13.10) and (13.11) provide that the sequence $\{y_n, y_n', d_{n,i}\}_{n\ge1}$ is bounded in $\cap_{i=1}^N L_{p_i}^{\text{loc}}(\mathbf{R}_+; V_i) \times \sum_{i=1}^N L_{q_i}^{\text{loc}}(\mathbf{R}_+; V_i^*) \times L_{q_i}^{\text{loc}}(\mathbf{R}_+; V_i^*)$, $i = 1, 2, \ldots, N$. Since Σ is a weakly compact set in $L_1^{\text{loc}}(\mathbf{R}_+)$, Banach–Alaoglu theorem (cf. Zgurovsky et al. [27, Chap. 1]; Kasyanov [15]) yields that there exist a subsequence $\{y_{n_k}, d_{n_k,i}\}_{k\ge1} \subset \{y_n, d_n\}_{n\ge1}$ and elements $d_i \in L_{q_i}^{\text{loc}}(\mathbf{R}_+; V_i^*)$, $y \in \cap_{i=1}^N L_{p_i}^{\text{loc}}(\mathbf{R}_+; V_i)$, and $\sigma \in \Sigma$, such that $y' \in \sum_{i=1}^N L_q^{\text{loc}}(\mathbf{R}_+; V_i^*)$ and for each $i = 1, 2, \ldots, N$ the following convergence hold:

$$\begin{array}{lll} y_{n_k} \to y & \text{weakly in } \cap_{i=1}^N L_{p_i}^{\text{loc}}(\mathbf{R}_+; V_i), \\ y_{n_k}' \to y' & \text{weakly in } \sum_{i=1}^N L_{q_i}^{\text{loc}}(\mathbf{R}_+; V_i^*), \\ d_{n_k,i} \to d_i & \text{weakly in } L_{q_i}^{\text{loc}}(\mathbf{R}_+; V_i^*), \\ y_{n_k} \to y & \text{weakly in } C^{\text{loc}}(\mathbf{R}_+; H), \\ y_{n_k} \to y & \text{in } L_2^{\text{loc}}(\mathbf{R}_+; H), \\ y_{n_k}(t) \to y(t) & \text{in } H \text{ for a.e. } t > 0, \\ \sigma_{n_k} \to \sigma & \text{weakly in } L_1^{\text{loc}}(\mathbf{R}_+), \quad k \to +\infty. \end{array} \tag{13.13}$$

Formula (13.12) follows from Zgurovsky et al. [29, Steps 1 and 5, p. 58]. We remark that in the proof we need to consider continuous and nonincreasing (by Lemma 13.1) functions on $\mathbf{R}_+$:

$$J_k(t) = \|y_{n_k}(t)\|_H^2 - 2\int\limits_0^t \sigma_{n_k}(\xi)d\xi, \;\; J(t) = \|y(t)\|_H^2 - 2\int\limits_0^t \sigma(\xi)d\xi, \;\; k \geq 1. \tag{13.14}$$

The two last statements in (13.13) imply $J_k(t) \to J(t)$, as $k \to +\infty$, for a.e. $t > 0$.

The definition of a weak solution of evolution inclusion (cf. Zgurovsky et al. [29, p. 58]) and (13.13) yield $y'(t) = -\sum_{i=1}^N d_i(t)$ for a.e. $t > 0$. To finish the proof it is necessary to provide that

$$\sum_{i=1}^N d_i(t) \in \mathbb{A}_{\sigma(t)}(y(t)) \text{ for a.e. } t > 0. \tag{13.15}$$

Let $\varphi \in C_0^\infty((0, +\infty))$, $\varphi \geq 0$. Then

$$\int_{\mathbf{R}_+} \varphi(t)\big(\alpha' \max_{i=1}^N \left\{\|y(t)\|_{V_i}^p; \|d_i(t)\|_{V_i^*}^q\right\} - \sigma(t)\big)dt \leq$$

$$\liminf_{k\to+\infty} \int_{\mathbf{R}_+} \varphi(t)\big(\alpha' \max_{i=1}^N \left\{\|y_{n_k}(t)\|_{V_i}^p; \|d_{n_k,i}(t)\|_{V_i^*}^q\right\} - \sigma_{n_k}(t)\big)dt \leq$$

$$\lim_{k\to+\infty} \int_{\mathbf{R}_+} \varphi(t) \sum_{i=1}^N \langle d_{n_k,i}(t), y_{n_k}(t)\rangle_V dt = \lim_{k\to+\infty} \frac{1}{2}\int_{\mathbf{R}_+} \|y_{n_k}(t)\|_H^2 \frac{d}{dt}\varphi(t)dt =$$

$$\frac{1}{2}\int_{\mathbf{R}_+} \|y(t)\|_H^2 \frac{d}{dt}\varphi(t)dt = \sum_{i=1}^N \int_{\mathbf{R}_+} \varphi(t)\langle d_i(t), y(t)\rangle_{V_i} dt,$$

where the first inequality holds, because the convex functional

$$(y, d) \to \int_{\mathbf{R}_+} \varphi(t)\big(\alpha' \max_{i=1}^N \left\{\|y(t)\|_{V_i}^p; \|d_i(t)\|_{V_i^*}^q\right\}\big)dt$$

is weakly lower semicontinuous on $\cap_{i=1}^N L_{p_i}^{\text{loc}}(\mathbf{R}_+; V_i) \times L_{q_1}^{\text{loc}}(\mathbf{R}_+; V_1{}^*) \times L_{q_2}^{\text{loc}}(\mathbf{R}_+; V_2{}^*) \times \ldots \times L_{q_N}^{\text{loc}}(\mathbf{R}_+; V_N{}^*)$; the second inequality follows from the definition of $\mathbb{A}_\sigma$; the first and the third equalities follow from formula (13.3), because $y_{n_k}'(t) + \sum_{i=1}^N d_{n_k,i}(t) = y'(t) + \sum_{i=1}^N d_i(t) = \bar{0}$ for any $k \geq 1$ and a.e. $t > 0$; the second equality holds, because $y_{n_k} \to y$ in $L_2^{\text{loc}}(\mathbf{R}_+; H)$, as $k \to +\infty$. As a nonnegative function $\varphi \in C_0^\infty((0, +\infty))$ is an arbitrary, then, by definition of $\mathbb{A}_\sigma$, formula (13.15) holds.

Proof of Theorem 13.1 First, let us show that there exists a uniform trajectory attractor $\mathscr{U} \subset \mathscr{K}^+$ of the translation semigroup $\{T(t)\}_{t\geq 0}$ on $\mathscr{K}^+$ in the induced topology from $C^{\text{loc}}(\mathbb{R}_+; H)$. Lemma 13.1 and Theorem 13.2 yields that the translation semigroup $\{T(t)\}_{t\geq 0}$ has a compact absorbing (and, therefore, an uniformly attracting) set in the space of trajectories $\mathscr{K}_\Sigma^+$; Kasyanov [15, p. 215]. This set can be constructed as follows: (1) consider $\mathscr{P}$, the intersection of $\mathscr{K}_\Sigma^+$ with a ball in the space of bounded continuous functions on $\mathbf{R}_+$ with values in H, $C_b(\mathbf{R}_+; H)$, of sufficiently large radius; (2) shift the resulting set by any fixed distance $h > 0$. Thus, we obtain $T(h)\mathscr{P}$, a set with the required properties. Recall that the semigroup $\{T(t)\}_{t\geq 0}$ is continuous. Therefore, the set $\mathscr{P}_1 := \mathscr{P} \cap \mathscr{K}^+$ is a compact absorbing (and, therefore, an uniformly attracting) in the space $\mathscr{K}^+$ with the induced topology of $C^{\mathrm{loc}}(\mathbb{R}_+; H)$. Then we can apply, for example, Theorem 1.1 from Melnik and Valero [20, p. 197]. In this case, the spaces E and E_0 coincide with H. In fact, here one can apply the classical theorem on the global attractor of a (unique) continuous semigroup in a complete metric space, the semigroup in question having a compact attracting (and, in particular, absorbing) set (see, for example, Babin and Vishik [1]; Temam [26]). In particular, formula (13.8) holds; cf. Babin and Vishik [1]; Melnik and Valero [20], Temam [26] etc.

13.5 Conclusions

For the class of nonautonomous differential-operator inclusions with pointwise pseudomonotone dependence between the defining parameters of the problem, the dynamics as $t \to +\infty$ of all global weak solutions defined on $[0, +\infty)$ is studied. The existence of a compact uniform trajectory attractor is proved. The results obtained allow one to study the dynamics of solutions for new classes of evolution inclusions related to nonlinear mathematical models of geophysical and socioeconomic processes and for fields with interaction functions of pseudomonotone type satisfying the power growth and sign conditions. For applications, one can consider new classes of problems with degeneracy, feedback control problems, problems on manifolds, problems with delay, stochastic partial differential equations, etc. (see Balibrea et al. [2]; Hu and Papageorgiou [3]; Gasinski and Papageorgiou [5]; Kasyanov [15]; Kasyanov, Toscano, and Zadoianchuk [17]; Mel'nik and Valero [21]; Denkowski, Migórski, and Papageorgiou [13]; Gasinski and Papageorgiou [5]; Zgurovsky et al. [29, 30]; etc.) involving differential operators of pseudomonotone type and the corresponding choice of the phase spaces. This paper is a continuation of Zgurovsky and Kasyanov [28].

References

1. Babin, A.V., Vishik, M.I.: Attractors of Evolution Equations. Nauka, Moscow (1989) [in Russian]
2. Balibrea, F., Caraballo, T., Kloeden, P.E., Valero, J.: Recent developments in dynamical systems: three perspectives. Int. J. Bifurc. Chaos (2010). doi:10.1142/S0218127410027246
3. Hu, S., Papageorgiou, N.S.: Handbook of Multivalued Analysis. Volume II: Applications. Kluwer, Dordrecht (2000)
4. Gajewski, H., Gröger, K., Zacharias, K.: Nichtlineare operatorgleichungen und operatordifferentialgleichungen. Akademie-Verlag, Berlin (1978)

5. Gasinski, L., Papageorgiou, N.S.: Nonlinear Analysis. Series in Mathematical Analysis and Applications, vol. 9. Chapman & Hall/CRC, Boca Raton (2005)
6. Gorban, N.V., Kapustyan, O.V., Kasyanov, P.O.: Uniform trajectory attractor for non-autonomous reaction-diffusion equations with Caratheodory's nonlinearity. Nonlinear Anal. Theory Methods Appl. **98**, 13–26 (2014). doi:10.1016/j.na.2013.12.004
7. Chepyzhov, V.V., Vishik, M.I.: Trajectory attractors for evolution equations. C. R. Acad. Sci. Paris. Ser. I **321**, 1309–1314 (1995)
8. Chepyzhov, V.V., Vishik, M.I.: Evolution equations and their trajectory attractors. J. Math. Pures Appl. **76**, 913–964 (1997)
9. Chepyzhov, V.V., Vishik, M.I.: Trajectory and global attractors for 3D Navier-Stokes system. Mat. Zametki. (2002). doi:10.1023/A:1014190629738
10. Chepyzhov, V.V., Vishik, M.I.: Attractors for Equations of Mathematical Physics. American Mathematical Society, Providence (2002)
11. Chepyzhov, V.V., Vishik, M.I.: Trajectory attractor for reaction-diffusion system with diffusion coefficient vanishing in time. Discret. Contin. Dyn. Syst. **27**(4), 1498–1509 (2010)
12. Clarke, F.H.: Optimization and Nonsmooth Analysis. Wiley, New York (1983)
13. Denkowski, Z., Migórski, S., Papageorgiou, N.S.: An Introduction to Nonlinear Analysis: Applications. Kluwer Academic/Plenum Publishers, Boston (2003)
14. Hale, J.K.: Asymptotic Behavior of Dissipative Systems. AMS, Providence (1988)
15. Kasyanov, P.O.: Multivalued dynamics of solutions of autonomous operator differential equations with pseudomonotone nonlinearity. Math. Notes **92**, 205–218 (2012)
16. Kasyanov, P.O.: Multivalued dynamics of solutions of an autonomous differential-operator inclusion with pseudomonotone nonlinearity. Cybern. Syst. Anal. **47**, 800–811 (2011)
17. Kasyanov, P.O., Toscano, L., Zadoianchuk, N.V.: Regularity of weak solutions and their attractors for a parabolic feedback control problem. Set-Valued Var. Anal. (2013). doi:10.1007/s11228-013-0233-8
18. Kapustyan, O.V., Kasyanov, P.O., Valero, J.: Pullback attractors for a class of extremal solutions of the 3D Navier-Stokes equations. J. Math. Anal. Appl. (2011). doi:10.1016/j.jmaa.2010.07.040
19. Ladyzhenskaya, O.A.: Attractors for Semigroups and Evolution Equations. Cambridge University Press, Cambridge (1991)
20. Melnik, V.S., Valero, J.: On attractors of multivalued semi-flows and generalized differential equations. Set-Valued Anal. **6**(1), 83–111 (1998)
21. Mel'nik, V.S., Valero, J.: On global attractors of multivalued semiprocesses and nonautonomous evolution inclusions. Set-Valued Anal. (2000). doi:10.1023/A:1026514727329
22. Migórski, S., Ochal, A.: Optimal control of parabolic hemivariational inequalities. J. Glob. Optim. **17**, 285–300 (2000)
23. Migórski, S.: Boundary hemivariational inequalities of hyperbolic type and applications. J. Glob. Optim. **31**(3), 505–533 (2005)
24. Panagiotopoulos, P.D.: Inequality Problems in Mechanics and Applications Convex and Nonconvex Energy Functions. Birkhauser, Basel (1985)
25. Sell, G.R.: Global attractors for the three-dimensional Navier-Stokes equations. J. Dyn. Differ. Equ. **8**(12), 1–33 (1996)
26. Temam, R.: Infinite-Dimensional Dynamical Systems in Mechanics and Physics. Applied Mathematical Sciences, vol. 68. Springer, New York (1988)
27. Zgurovsky, M.Z., Mel'nik, V.S., Kasyanov, P.O.: Evolution Inclusions and Variation Inequalities for Earth Data Processing II. Springer, Berlin (2011)
28. Zgurovsky, M.Z., Kasyanov, P.O.: Evolution Inclusions in Nonsmooth Systems with Applications for Earth Data Processing, Advances in Global Optimization. In: Proceedings in Mathematics & Statistics, vol. 95. Springer (2014). doi:10.1007/978-3-319-08377-3_29
29. Zgurovsky, M.Z., Kasyanov, P.O., Kapustyan, O.V., Valero, J., Zadoianchuk, N.V.: Evolution Inclusions and Variation Inequalities for Earth Data Processing III. Springer, Berlin (2012)
30. Zgurovsky, M.Z., Kasyanov, P.O.: Long-time behavior of solutions for quasilinear hyperbolic hemivariational inequalities with application to piezoelectricity problem. Appl. Math. Lett. **25**(10), 1569–1574 (2012)

Chapter 14
Lyapunov Functions for Differential Inclusions and Applications in Physics, Biology, and Climatology

Mark O. Gluzman, Nataliia V. Gorban and Pavlo O. Kasyanov

Abstract We investigate additional regularity properties of all globally defined weak solutions, their global and trajectory attractors for classes of semi-linear parabolic differential inclusions with initial data from the natural phase space. The main contributions in this note are: (i) sufficient conditions for the existence of a Lyapunov function for a class of parabolic feedback control problems; (ii) convergence results for all weak solutions in the strongest topologies; and (iii) new structure and regularity properties for global and trajectory attractors. Results applied to the long-time behavior of state functions for the following problems: (a) a model of combustion in porous media; (b) a model of conduction of electrical impulses in nerve axons; and (c) a climate energy balance model.

14.1 Introduction and Regularity of All Weak Solutions

Let $(V; H; V')$ be evolution triple, where V be a real Hilbert space, such that $V \subset H$ with compact imbedding. Let $A : V \to V'$ be a linear symmetric operator such that $\exists c > 0 : \langle Av, v\rangle_V \geq c\|v\|_V^2$, for each $v \in V$ and let $D(A) = \{u \in V : Au \in H\}$. We note that the mapping $v \to \|Av\|_H$ defines the equivalent norm on $D(A)$; Temam [36, Chap. III].

Let $J_i : H \to \mathbb{R}$ be a convex, lower semi-continuous function such that the following assumptions hold: (i) (growth condition) there exists $c_1 > 0$ such that $\|y\|_H \leq c_1(1 + \|u\|_H)$, for each $u \in H$ and $y \in \partial J_i(u)$ and $i = 1, 2$; (ii) (sign condition) there exist $c_2 > 0$, $\lambda \in (0, c)$ such that $(y_1 - y_2, u)_H \geq -\lambda\|u\|_H^2 - c_2$,

M.O. Gluzman · N.V. Gorban (✉) · P.O. Kasyanov
Institute for Applied System Analysis, National Technical University of Ukraine "Kyiv Polytechnic Institute", Peremogy Ave., 37, Build, 35, Kyiv 03056, Ukraine
e-mail: nata_gorban@i.ua

M.O. Gluzman
e-mail: gluzmanmark@gmail.com

P.O. Kasyanov
e-mail: kasyanov@i.ua

© Springer International Publishing Switzerland 2015

V.A. Sadovnichiy and M.Z. Zgurovsky (eds.), *Continuous and Distributed Systems II*, Studies in Systems, Decision and Control 30,
DOI 10.1007/978-3-319-19075-4_14

for each $y_i \in \partial J_i(u)$, $u \in H$, where $\partial J_i(u)$ is the subdifferential of $J_i(\cdot)$ at a point u. Note that $u^* \in \partial J_i(u)$ if and only if $u^*(v-u) \le J_i(v) - J_i(u)\ \forall v \in H; i = 1, 2$.

We consider the differential-operator inclusion:

$$\frac{du}{dt} + Au(t) + \partial J_1(u(t)) - \partial J_2(u(t)) \ni 0 \text{ on } (-\infty < \tau < T < +\infty). \tag{14.1}$$

The function $u(\cdot) \in L^2(\tau, T; V)$ is called a *weak solution* of Problem (14.1) on $[\tau, T]$, if there exist Bochner measurable functions $d_i : (\tau, T) \to H; i = 1, 2$, such that

$$d_i(t) \in \partial J_i(u(t)) \text{ for a.e. } t \in (\tau, T),\ \ i = 1, 2; \text{ and} \tag{14.2}$$

$$\int_\tau^T \big[-\langle u, v\rangle \xi'(t) + \langle Au, v\rangle \xi(t) + \langle d_1, v\rangle \xi(t) - \langle d_2, v\rangle \xi(t)\big]\, dt = 0, \tag{14.3}$$

for all $\xi \in C_0^\infty(\tau, T)$ and for all $v \in V$, where $\langle\cdot,\cdot\rangle$ denotes the pairing in the space V.

Theorem 14.1 *Let $-\infty < \tau < T < +\infty$ and $u_\tau \in H$. Problem (14.1) has at least one weak solution $u(\cdot) \in L^2(\tau, T; V)$ on $[\tau, T]$ such that $u(\tau) = u_\tau$. Moreover, if $u(\cdot)$ is a weak solution of Problem (14.1) on $[\tau, T]$, then $u(\cdot) \in C([\tau+\varepsilon, T]; V) \cap L^2(\tau+\varepsilon, T; D(A))$ and $\frac{du}{dt}(\cdot) \in L^2(\tau+\varepsilon, T; H)$ for any $\varepsilon \in (0, T-\tau)$.*

Proof We note that for any $u_\tau \in H$ there exists at least one weak solution of Problem (14.1) on $[\tau, T]$ with initial condition $u(\tau) = u_\tau$; see Kasyanov [27] and references therein. In the general case, Problem (14.1) on $[\tau, T]$ with such initial condition has no unique weak solution with $u_\tau \in H$; Balibrea et al. [2, p. 2600] and references therein.

Let us prove the second part of Theorem (14.1). Let $u(\cdot)$ be an arbitrary weak solution of Problem (14.1) on $[\tau, T]$. According to the definition of a weak solution of Problem (14.1) on $[\tau, T]$, there exist $d_i \in L^2(\tau, T; H)$, $i = 1, 2$, such that $u(\cdot) \in L^2(\tau, T; V)$, $d_1(\cdot)$ and $d_2(\cdot)$ satisfy (14.2)–(14.3). Note that the set

$$\mathscr{D} := \{s \in (\tau, T) \mid u(s) \in V\}$$

is dense in $[\tau, T]$. For an arbitrary fixed $s \in \mathscr{D}$, we remark that $u(\cdot)$ is the unique weak solution on $[\tau, T]$ of the problem

$$\begin{cases} \frac{dz}{dt} + Az(t) = g(t) \text{ on } (s, T), \\ z(s) = u(s). \end{cases} \tag{14.4}$$

where $g(\cdot) = -d_1(\cdot) + d_2(\cdot) \in L^2(\tau, T; H)$.

Therefore, $u(\cdot) \in L^2(s, T; D(A)) \cap C([s, T]; V)$ and $\frac{du}{dt}(\cdot) \in L^2(s, T; H)$, $s \in \mathscr{D}$ (cf. [35, Chap. 4.I], [36, Chap. III] and references therein). Thus $u(\cdot) \in C([\tau+\varepsilon, T]; V) \cap L^2(\tau+\varepsilon, T; D(A))$ and $\frac{du}{dt}(\cdot) \in L^2(\tau+\varepsilon, T; H)$ for any $\varepsilon \in (0, T-\tau)$.

We remark that Theorem 14.1 implies that each weak solution of Problem (14.1) on $[\tau, T]$ is *regular*, that is $u(\cdot) \in L^2(\varepsilon, T; D(A)) \cap C([\varepsilon, T]; V)$ and $\frac{du}{dt}(\cdot) \in L^2(\varepsilon, T; H)$, for each $\varepsilon \in (0, T - \tau)$. We also note that each weak solution of Problem (14.1) on a finite interval can be extended to a global one, defined on $[\tau, \infty)$.

Sufficient conditions for the existence of a Lyapunov function for autonomous evolution inclusions of hyperbolic type were considered in [29, 42, 44, 45]. We remark that the existence of a Lyapunov function for a class of parabolic feedback control problems and its applications were already announced in [15]. The global and trajectory attractors for such kind of systems in the natural phase and exstanded phase spaces were provided in [1, 7, 8, 16–25, 28, 30, 39–41]. Topological properties of strong and weak solutions were considered in [14, 31–34]. Regularity properties of global and trajectory attractors were provided in [18, 24–26].

14.2 A Lyapunov Type Function and Strongest Convergence Results for All Weak Solutions

Denote by $\mathscr{K}_+$ the family of all, globally defined on $[0, +\infty)$, weak solutions of Problem (14.1). The set $\mathscr{K}_+$ is *translation invariant* one, that is $u(\cdot + h) \in \mathscr{K}_+$ for each $u(\cdot) \in \mathscr{K}_+$ and $h \geq 0$.

Let us consider Problem (14.1) on the entire time axis. A function $u \in L^\infty(\mathbb{R}; H)$ is called a *complete trajectory* of Problem (14.1), if $\Pi_+ u(\cdot + h) \in \mathscr{K}_+$ for each $h \geq 0$, where Π_+ is the restriction operator to the interval $[0, +\infty)$. The family of all complete trajectories of Problem (14.1), we denote as $\mathscr{K}$. A complete trajectory $u(\cdot) \in \mathscr{K}$ is *stationary* if there is $z \in D(A)$ such that $u(t) = z$ for all $t \in \mathbb{R}$. Each such z is called a *rest point*. We denote the set of all rest points by Z.

Definition 14.1 The function $E : V \to \mathbb{R}$ is called a *Lyapunov type one* for $\mathscr{K}_+$, if the following conditions hold:

(a) E is continuous on V;
(b) $E(u(t)) \leq E(u(s))$ whenever $u \in \mathscr{K}_+$ and $t \geq s > 0$;
(c) if $E(u(\cdot)) \equiv \text{const}$, for some $u \in \mathscr{K}$, then u is stationary complete trajectory.

Let us set

$$E(u) = \frac{1}{2}\langle Au, u\rangle + J_1(u) - J_2(u), \quad u \in V. \tag{14.5}$$

Theorem 14.2 *The function $E : V \to \mathbb{R}$, defined in (14.5), is a Lyapunov type function for $\mathscr{K}_+$. Moreover, for each $u \in \mathscr{K}_+$ and all τ and T, $0 < \tau < T < \infty$, the energy equality holds*

$$E(u(T)) - E(u(\tau)) = -\int_\tau^T \|\frac{du}{ds}(s)\|_H^2 ds. \tag{14.6}$$

Proof Statement (a) of Definition 14.1 holds, because the function E is continuous on V.

Let us prove statement (b) of Definition 14.1. Suppose $u(\cdot) \in \mathscr{K}_+$ be arbitrary and fixed and let $0 < \tau < T < +\infty$. To simplify conclusions, let the symbol $u(\cdot)$ denotes the restriction of $u(\cdot)$ on $[\tau, T]$. Theorem 14.1 implies that $u(\cdot) \in C([\tau, T]; V) \cap L^2(\tau, T; D(A))$ and $\frac{du}{dt}(\cdot) \in L^2(\tau, T; H)$, because $\tau > 0$. Barbu [5, Lemma 2.1, p. 189] yields that the functions $J_i(u(\cdot))$, $i = 1, 2$, are absolutely continuous on $[\tau, T]$ and the equality holds:

$$\frac{d}{dt} J_i(u(t)) = \langle h_i(t), \frac{du}{dt}(t)\rangle_H, \text{ for a.e. } t \in (\tau, T), \tag{14.7}$$

for all $h_i(\cdot) \in L^2(\tau, T; H)$ such that $h_i(t) \in \partial J_i(s)|_{s=u(t)}$ for a.e. $t \in (\tau, T)$, $i = 1, 2$.

We remark that the mapping $t \to \langle Au(t), u(t)\rangle_V$ is absolutely continuous on $[\tau, T]$ and the equality holds:

$$\frac{d}{dt}\langle Au(t), u(t)\rangle = 2\langle Au(t), \frac{du}{dt}(t)\rangle_H, \text{ for a.e. } t \in (\tau, T) \tag{14.8}$$

Thus, the function $E(u(\cdot))$ is absolutely continuous on $[\tau, T]$ as the linear combination of absolutely continuous on $[\tau, T]$ functions. According to formulae (14.7) and (14.8), $\frac{d}{dt} E(u(t)) = -\|\frac{du}{dt}(t)\|_H^2$ for a.e. $t \in (\tau, T)$. The last statement implies (14.6). In particular, $E(u(t)) \le E(u(s))$ whenever $T \ge t \ge s \ge \tau > 0$. Since $u(\cdot) \in \mathscr{K}_+$ and $0 < \tau < T < \infty$ are arbitrary, statement (b) of Definition 14.1 and the energy equality (14.6) hold.

To finish the proof we note that if $E(u(\cdot)) \equiv \text{const}$, for some $u \in \mathscr{K}$, then, according to energy equality (14.6), u is stationary.

Kasyanov et al. [26, p. 274] implies, that there exists $C > 0$ such that for any $\tau < T$ and for each weak solution $u(\cdot)$ of Problem (14.1) on $[\tau, T]$ the inequality holds

$$(t-\tau)\|u(t)\|_V^2 + \int_\tau^t (s-\tau)\|u(s)\|_{D(A)}^2 ds \le C(1+\|u(\tau)\|_H^2 + (t-\tau)^2) \quad \forall t \in (\tau, T]. \tag{14.9}$$

For any $u_\tau \in H$ we set

$$\mathscr{D}_{\tau,T}(u_\tau) = \{u(\cdot) \in L^2(\tau, T; V) \big| \; u(\cdot) \text{ is a weak solution of Problem (14.1) and } u(\tau) = u_\tau\}.$$

Let us provide the main convergence result for all weak solutions in strongest topologies.

Theorem 14.3 *Let $\tau < T$, $u_{\tau,n} \to u_\tau$ weakly in H, $u_n(\cdot) \in D_{\tau,T}(u_{\tau,n})$ for any $n \geq 1$. Then there exists a subsequence $\{u_{n_k}(\cdot)\}_{k\geq 1} \subseteq \{u_n(\cdot)\}_{n\geq 1}$ and $u(\cdot) \in D_{\tau,T}(u_\tau)$ such that*

$$\sup_{t\in[\tau+\varepsilon,T]} \|u_{n_k}(t) - u(t)\|_V \to 0, \tag{14.10}$$

$$\int_{\tau+\varepsilon}^{T} \|\frac{du_{n_k}}{dt}(t) - \frac{du}{dt}(t)\|_H^2 dt \to 0, \tag{14.11}$$

as $k \to +\infty$, for all $\varepsilon \in (0, T - \tau)$.

Proof The inequality (14.9), Kasyanov et al. [26, Theorem 3], Banach–Alaoglu theorem, and Cantor diagonal arguments yield that there exist a subsequence $\{u_{n_k}(\cdot)\}_{k\geq 1} \subseteq \{u_n(\cdot)\}_{n\geq 1}$ and $u(\cdot) \in D_{\tau,T}(u_\tau)$ such that the following statements hold:

(a) the restrictions of $u_{n_k}(\cdot)$ and $u(\cdot)$ on $[\tau+\varepsilon, T]$ belong to $C([\tau+\varepsilon, T]; V) \cap L^2(\tau+\varepsilon, T; D(A))$ and $\frac{du_{n_k}}{dt}(\cdot), \frac{du}{dt}(\cdot) \in L^2(\tau+\varepsilon, T; H)$;

(b) the following convergence hold:

$$\begin{aligned} &u_{n_k}(\cdot) \to u(\cdot) \text{ weakly in } L^2(\tau+\varepsilon, T; D(A)),\\ &u_{n_k}(\cdot) \to u(\cdot) \text{ strongly in } C([\tau+\varepsilon, T]; V),\\ &\tfrac{du_{n_k}}{dt}(\cdot) \to \tfrac{du}{dt}(\cdot) \text{ weakly in } L^2(\tau+\varepsilon, T; H), \end{aligned} \tag{14.12}$$

as $k \to \infty$, for each $\varepsilon \in (0, T - \tau)$, that imply statement (14.10). Let us prove (14.11). Theorem 14.2 yields the following energy equalities

$$\int_{\tau+\varepsilon}^{T} \|\frac{du}{dt}(t)\|_H^2 dt = E(u(\tau+\varepsilon)) - E(u(T)), \tag{14.13}$$

$$\int_{\tau+\varepsilon}^{T} \|\frac{du_{n_k}}{dt}(t)\|_H^2 dt = E(u_{n_k}(\tau+\varepsilon)) - E(u_{n_k}(T)), \tag{14.14}$$

$k \geq 1$, $\varepsilon \in (0, T - \tau)$. The continuity of E on V and (14.10) imply

$$E(u_{n_k}(\tau+\varepsilon)) - E(u_{n_k}(T)) \to E(u(\tau+\varepsilon)) - E(u(T)),\ m \to \infty. \tag{14.15}$$

Therefore, formulae (14.13)–(14.15) yield

$$\int_{\tau+\varepsilon}^{T} \|\frac{du_{n_k}}{dt}(t)\|_H^2 dt \to \int_{\tau+\varepsilon}^{T} \|\frac{du}{dt}(t)\|_H^2 dt, \tag{14.16}$$

as $k \to \infty$, for each $\varepsilon \in (0, T - \tau)$. Since, $L^2(\tau+\varepsilon; T)$ is a Hilbert space, (14.12) and (14.16) imply (14.11).

14.3 Structure Properties and Regularity of Global and Trajectory Attractors

Define real Banach space $W(M_1, M_2) = \{u(\cdot) \in C([M_1, M_2]; V) : \frac{du}{dt}(\cdot) \in L^2(M_1, M_2; H)\}$ with the norm $\|u\|_{W(M_1,M_2)} = \|u\|_{C([M_1,M_2];V)} + \|\frac{du}{dt}\|_{L^2(M_1,M_2;H)}$, $u \in W(M_1, M_2)$, $-\infty < M_1 < M_2 < +\infty$.

We denote the set of all nonempty (nonempty bounded) subsets of H by $P(H)$ ($\mathscr{B}(H)$). Let us define the strict m-semiflow $G : \mathbb{R}_+ \times H \to P(H)$ in the following way: $G(t, u_0) = \{u(t) : u(\cdot) \in \mathscr{K}_+,\ u(0) = u_0\}$. We recall that the multivalued map $G : \mathbb{R}_+ \times H \to P(H)$ is said to be a *strict multivalued semiflow (strict m-semiflow)* if:

(a) $G(0, \cdot) = \mathrm{Id}$ (the identity map);
(b) $G(t + s, x) = G(t, G(s, x))\ \forall x \in H, t, s \in \mathbb{R}_+$.

We recall that the set $\mathscr{A} \subseteq H$ is said to be *an invariant global attractor* of G if:

(1) $\mathscr{A}$ is invariant (that is $\mathscr{A} = G(t, \mathscr{A})\ \forall t \geq 0$);
(2) $\mathscr{A}$ is attracting set, that is,

$$\mathrm{dist}_H(G(t, B), \mathscr{A}) \to 0, \quad t \to +\infty \quad \forall B \in \mathscr{B}(H), \tag{14.17}$$

where $\mathrm{dist}_H(C, D) = \sup_{c\in C} \inf_{d\in D} \|c - d\|_H$ is the Hausdorff semidistance;
(3) for any closed set $Y \subseteq H$ satisfying (14.17), we have $\mathscr{A} \subseteq Y$ (minimality).

Let $\{T(h)\}_{h\geq 0}$ be the translation semigroup acting on $\mathscr{K}_+$, that is $T(h)u(\cdot) = u(\cdot + h)$, $h \geq 0$, $u(\cdot) \in \mathscr{K}_+$. On $\mathscr{K}_+$ we consider the topology induced from the Fréchet space $C^{loc}(\mathbb{R}_+; H)$. Note that $f_n(\cdot) \to f(\cdot)$ in $C^{loc}(\mathbb{R}_+; H)$ if and only if $\forall M > 0\ \Pi_M f_n(\cdot) \to \Pi_M f(\cdot)$ in $C([0, M]; H)$, where Π_M is the restriction operator to the interval $[0, M]$; Chepyzhov, Vishik [6, p. 18].

A set $\mathscr{U} \subset \mathscr{K}_+$ is said to be *trajectory attractor* in the trajectory space $\mathscr{K}_+$ with respect to the topology of $C^{loc}(\mathbb{R}_+; H)$, if $\mathscr{U} \subset \mathscr{K}_+$ is a global attractor for the translation semigroup $\{T(h)\}_{h\geq 0}$ acting on $\mathscr{K}_+$; Kasyanov et al. [26, Sect. 3].

The following theorem completely describes the long-time behavior of all weak solutions, as time $t \to +\infty$, for Problem (14.1). The structure properties of global and trajectory attractors and the strongest convergence results of solutions are provided.

Theorem 14.4 *The following statements hold:*

(i) *the strict m-semiflow $G : \mathbb{R}_+ \times H \to P(H)$ has the invariant global attractor $\mathscr{A}$;*
(ii) *there exists the trajectory attractor $\mathscr{U} \subset \mathscr{K}_+$ in the space $\mathscr{K}_+$;*
(iii) *the following equalities hold: $\mathscr{U} = \Pi_+\mathscr{K} = \{y \in \mathscr{K}_+ \mid y(t) \in \mathscr{A}\ \forall t \in \mathbb{R}_+\}$;*
(iv) *$\mathscr{A}$ is a compact subset of V;*
(v) *for each $B \in \mathscr{B}(H)$ $\mathrm{dist}_V(G(t, B), \mathscr{A}) \to 0$, as $t \to \infty$;*
(vi) *$\mathscr{U}$ is a bounded subset of $L^\infty(\mathbb{R}_+; V)$ and compact subset of $W^{\mathrm{loc}}(\mathbb{R}_+)$, that is $\Pi_M\mathscr{U}$ is compact in $W(0, M)$ for each $M > 0$;*

(vii) *for any bounded in* $L^\infty(\mathbb{R}_+; H)$ *set* $\mathbf{B} \subset \mathscr{K}_+$ *and any* $M \geq 0$ *the following relation holds:* $\text{dist}_{W(0,M)}(\Pi_M T(t)\mathbf{B}, \Pi_M \mathscr{U}) \to 0, \quad t \to +\infty$;

(viii) $\mathscr{K}$ *is a bounded subset of* $L^\infty(\mathbb{R}; V)$ *and compact subset of* $W^{\text{loc}}(\mathbb{R})$, *that is* $\Pi_{M_1,M_2}\mathscr{U}$ *is compact in* $W(M_1, M_2)$ *for each* M_1, M_2, $-\infty < M_1 < M_2 < +\infty$;

(ix) *for each* $u \in \mathscr{K}$ *the limit sets*

$$\alpha(u) = \{z \in V \mid u(t_j) \to z \text{ in } V \text{ for some sequence } t_j \to -\infty\},$$

$$\omega(u) = \{z \in V \mid u(t_j) \to z \text{ in } V \text{ for some sequence } t_j \to +\infty\}$$

are connected subsets of Z *on which* E *is constant. If* Z *is totally disconnected (in particular, if* Z *is countable) the limits in* V

$$z_- = \lim_{t\to-\infty} u(t), \quad z_+ = \lim_{t\to+\infty} u(t) \tag{14.18}$$

exist and z_-, z_+ *are rest points; furthermore,* $u(t)$ *tends in* V *to a rest point as* $t \to +\infty$ *for every* $u \in \mathscr{K}_+$.

Proof Statements (i)–(v) of Theorem 14.4 follow from Kasyanov et al. [26, Theorems 4–6]. Statements (vi)–(viii) of Theorem 14.4 follow from Theorem 14.3 and Kasyanov et al. [26, Theorem 6]. Statement (ix) of Theorem 14.4 follows from Theorem 14.2 and Ball [4, Theorem 2.7].

14.4 Faedo–Galerkin Approximation for the Global and Trajectory Attractors

Let $\{h_i\}_{i\geq 1} \subset V$ be a special basis, that is $\langle Ah_i, h_j\rangle_V = \lambda_j(h_i, h_j) = \lambda_j\delta_{ij}$, $i, j = 1, 2, \ldots$, where $\delta_{ij} = \begin{cases} 0, \text{ if } i \neq j \\ 1, \text{ if } i = j \end{cases}$ is the Kronecker delta. Let H_n be the linear span of the set $\{h_i\}_{i=1}^n$, $H_n = span\{h_i\}_{i=1}^n$, equipped with the scalar product induced from H; $n = 1, 2, \ldots$.

For an arbitrary $n \geq 1$, let $I_n \in \pounds(L^2(\tau, T; H_n); L^2(\tau, T; V'))$ be the canonical imbedding of $L^2(\tau, T; H_n)$ into $L^2(\tau, T; V')$, that is $I_n(z) = z$, for all $z \in L^2(\tau, T; V')$. Let I_n' be the adjoint operator to I_n.

Let $P_n : H \to H_n \subset H$ be the orthogonal projection operator. Note that P_n satisfied the following assumptions:

$$\|P_n\|_{\pounds(H,H)} \leq 1. \tag{14.19}$$

Consider the following "approximate" Problems of the Problem (14.1) :

$$y_n' + A_n y_n + \partial(J_1 \circ I_n)(y_n) - \partial(J_2 \circ I_n)(y_n) \ni 0, n = 1, 2, \ldots, \tag{14.20}$$

where $A_n = I_n' A I_n : L^2(\tau, T; H_n) \to L^2(\tau, T; H_n)$.

We remark that Problem (14.20) satisfies Theorems 14.1 and 14.4 assumptions.

A weak solution $y(\cdot)$ of Problem (14.1) on $[\tau, T]$ is called a *Faedo–Galerkin weak solution* of this problem on $[\tau, T]$, if y is a weak limit in $L^2(\tau, T; V)$ of some subsequence $\{y_{n_k}\}_{k=1,2,\ldots}$ of a sequence $\{y_n\}_{n=1,2,\ldots}$, such that the following statements hold: (i) $y_n \in L^2(\tau, T; H_n)$ satisfies (14.20), $n = 1, 2, \ldots$; ii) $y_{n_k} \to y$ in $L^2(\tau+\varepsilon, T; D(A)) \cap C([\tau+\varepsilon, T]; V)$ and $\frac{dy_{n_k}}{dt} \to \frac{dy}{dt}$ in $L^2(\tau, T; H)$, as $k \to \infty$.

The family of all Faedo–Galerkin weak solutions of Problem (14.1) on $[0, \infty)$ were denoted by $\mathscr{K}_+^{(FG)}$. Global solutions of Problem (14.20), defined on $[0, \infty)$, were denoted by $\mathscr{K}_+^{(n)}$, $n = 1, 2, \ldots$. According to [43, Chap. 7], $\mathscr{K}_+^{(FG)} \neq \emptyset$. Moreover, $\overline{\mathrm{Lim}}_{n\to\infty} \mathscr{A}_n \subseteq \mathscr{A}$ and $\overline{\mathrm{Lim}}_{n\to\infty} \mathscr{U}_n \subseteq \mathscr{U}$, where $\mathscr{A}_n$ and $\mathscr{U}_n$ are the global and trajectory attractors, respectively, for Problem (14.20), $n = 1, 2, \ldots$.

14.5 Applications

Let us concentrate on the following four types of applications: (i) a feedback control problem; (ii) a model of combustion in porous media; (iii) a model of conduction of electrical impulses in nerve axons; and (iv) a climate energy balance model.

Example 14.1 (A feedback control problem). We consider the nonstationary heat conduction system of equations:

$$\frac{d\overline{y}}{dt} + A\overline{y} = \overline{g} \text{ in } \Omega \times (0, +\infty)$$

Suppose that $\overline{g} = g_1 + g_2$, where $g_2 \in H$ is given and g_1 is a known function of the temperature of the form

$$-g_1(x, t) \in \partial J_1(x, y(t)) - \partial J_2(x, y(t)) \text{ a.e. } (x, t) \in \Omega \times (0, +\infty).$$

In a physicist's language, it means that the law is characterized by the generalized gradient of a nonsmooth potential $j = J_1 - J_2$.

If growth and sign conditions hold, then all statements of Theorems 14.2 and 14.4 hold. Statement (ix) of Theorem 14.4 provides sufficient conditions for stabilization of the considered feedback control problem.

Example 14.2 (A model of combustion in porous media). We consider the following problem:

$$\begin{cases} \frac{\partial u}{\partial t} - \frac{\partial^2 u}{\partial x^2} - f(u) \in \lambda H(u-1), & (t, x) \in \mathbb{R}_+ \times (0, \pi), \\ u(t, 0) = u(t, \pi) = 0, & t \in \mathbb{R}_+, \end{cases} \tag{14.21}$$

where $f : \mathbb{R} \to \mathbb{R}$ is a continuous and nondecreasing function satisfying growth and sign assumptions, $\lambda > 0$, and $H(0) = [0, 1]$, $H(s) = \mathbf{I}\{s > 0\}$, $s \neq 0$; Feireisl and Norbury [13]. Then all statements of Theorems 14.2 and 14.4 hold.

Example 14.3 (A model of conduction of electrical impulses in nerve axons). Consider the problem:

$$\begin{cases} \frac{\partial u}{\partial t} - \frac{\partial^2 u}{\partial x^2} + u \in \lambda H(u-a), & (t,x) \in (0,T) \times (0,\pi), \\ u(t,0) = u(t,\pi) = 0, & t \in \mathbb{R}_+, \end{cases} \tag{14.22}$$

where $a \in \left(0, \frac{1}{2}\right)$; Terman [37, 38]. Since Problem (14.22) is a particular case of Problem (14.1), all statements of Theorems 14.2 and 14.4 hold.

Example 14.4 (A climate energy balance model). Formulate the problem:

$$\begin{cases} \frac{\partial u}{\partial t} - \frac{\partial^2 u}{\partial x^2} + Bu \in QS(x)\beta(u) + h(x), & (t,x) \in \mathbb{R}_+ \times (-1,1), \\ u_x(t,-1) = u_x(t,1) = 0, & t \in \mathbb{R}_+, \end{cases} \tag{14.23}$$

where B, $Q > 0$ are constants, S, $h \in L^\infty(-1,1)$, $u_0 \in L^2(-1,1)$ and β is a maximal monotone graph in $\mathbb{R}^2$. Assume that: (a) there exist m, $M \in \mathbb{R}$ such that $\forall s \in \mathbb{R}$, $\forall z \in \beta(s)$ $m \le z \le M$; (b) for a.e. $x \in (-1,1)$ $0 < S_0 \le S(x) \le S_1$. This energy balance climate model was proposed in Budyko [3] and researched also in Díaz et al. [9–11]. The unknown $u(t,x)$ represents the average temperature of the Earth's surface, Q is a solar constant, $S(x)$ is an insolation function, given the distribution of solar radiation falling on upper atmosphere, and β represents the ratio between absorbed and incident solar energy at the point x of the Earth's surface (so-called co-albedo function). All statements of Theorems 14.2 and 14.4 hold, because Problem (14.23) is a particular case of Problem (14.1).

Acknowledgments This research was partially supported by the Ukrainian State Fund for Fundamental Research under grant GP/F49/070, and by grant 2273/14 from the National Academy of Sciences of Ukraine.

References

1. Arrieta, J.M., Rodrígues-Bernal, A., Valero, J.: Dynamics of a reaction-diffusion equation with discontinuous nonlinearity. Int. J. Bifurc. Chaos **16**, 2695–2984 (2006)
2. Balibrea, F., Caraballo, T., Kloeden, P.E., Valero, J.: Recent developments in dynamical systems: three perspectives. Int. J. Bifurc. Chaos (2010). doi:10.1142/S0218127410027246
3. Budyko, M.I.: The effects of solar radiation variations on the climate of the earth. Tellus **21**, 611–619 (1969)
4. Ball, J.M.: Global attractors for damped semilinear wave equations. DCDS **10**, 31–52 (2004)
5. Barbu, V.: Nonlinear Semigroups and Differential Equations in Banach Spaces. Editura Academiei, Bucuresti (1976)

6. Chepyzhov, V.V., Vishik, M.I.: Trajectory and global attractors of three-dimensional Navier-Stokes systems. Math. Notes **71**, 177–193 (2002). doi:10.1023/A:1014190629738
7. Chepyzhov, V.V., Vishik, M.I.: Trajectory attractor for reaction-diffusion system with diffusion coefficient vanishing in time. Discret. Contin. Dyn. Syst.—Ser. A **27**, 1493–1509 (2013)
8. Chepyzhov, V.V., Conti, M., Pata, V.: A minimal approach to the theory of global attractors. Discret. Contin. Dyn. Syst. **32**, 2079–2088 (2012)
9. Díaz, H., Díaz, J.: On a stochastic parabolic PDE arising in climatology. Rev. R. Acad. Cien. Ser. A Mat. **96**, 123–128 (2002)
10. Díaz, J., Herníndez, J., Tello, L.: On the multiplicity of equilibrium solutions to a nonlinear diffusion equation on a manifold arising in climatology. J. Math. Anal. Appl. **216**, 593–613 (1997)
11. Díaz, J., Herníndez, J., Tello, L.: Some results about multiplicity and bifurcation of stationary solutions of a reaction diffusion climatological model. Rev. R. Acad. Cien. Ser. A. Mat. **96**(3), 357–366 (2002)
12. Díaz, J., Tello, L.: Infinitely many stationary solutions for a simple climate model via a shooting method. Math. Methods Appl. Sci. **25**, 327–334 (2002)
13. Feireisl, E., Norbury, J.: Some existence and nonuniqueness theorems for solutions of parabolic equations with discontinuous nonlinearities. Proc. R. Soc. Edinb. A. **119**(1–2), 1–17 (1991)
14. Gajewski, H., Gröger, K., Zacharias, K.: Nichtlineare operatorgleichungen und operatordifferentialgleichungen. Akademie-Verlag, Berlin (1974)
15. Gluzman, M.O., Gorban, N.V., Kasyanov, P.O.: Lyapunov type functions for classes of autonomous parabolic feedback control problems and applications.Appl. Math. Lett. (2014). http://dx.doi.ord/10.1016/j.aml.2014.08.006
16. Goldstein, G.R., Miranville, A.: A Cahn-Hilliard-Gurtin model with dynamic boundary conditions. Discret. Contin. Dyn. Syst.—Ser. S **6** (2013)
17. Gorban, N.V., Kapustyan, O.V., Kasyanov, P.O.: Uniform trajectory attractor for non-autonomous reaction-diffusion equations with Caratheodory's nonlinearity. Nonlinear Anal. Theory, Methods Appl. **98**, 13–26 (2014). doi:10.1016/j.na.2013.12.004
18. Gorban, N.V., Kasyanov, P.O.: On regularity of all weak solutions and their attractors for reaction-diffusion inclusion in unbounded domain. Solid Mech. Appl. **211**, 205–220 (2014)
19. Gorban, N.V., Kapustyan, O.V., Kasyanov, P.O., Paliichuk, L.S.: On global attractors for autonomous damped wave equation with discontinuous nonlinearity. Solid Mech. Appl. **211**, 221–237 (2014)
20. Efendiev, M., Miranville, A., Zelik, S.: Exponential attractors for a nonlinear reaction-diffusion system in R^3. C. R. de l'Academie des Sci.-Ser. I—Math. **330**, 713–718 (2000)
21. Kalita, P., Lukaszewicz, G.: Global attractors for multivalued semiflows with weak continuity properties. Nonlinear Anal.: Theory, Methods Appl. **101**, 124–143 (2014)
22. Kalita, P., Lukaszewicz, G.: Attractors for Navier-Stokes flows with multivalued and non-monotone subdifferential boundary conditions. Nonlinear Anal.: Real World Appl. **19**, 75–88 (2014)
23. Kapustyan, O.V., Kasyanov, P.O., Valero, J., Zgurovsky, M.Z.: Structure of uniform global attractor for general non-autonomous reaction-diffusion system. Solid Mech. Appl. **211**, 163–180 (2014)
24. Kapustyan, O.V., Kasyanov, P.O., Valero, J.: Regular solutions and global attractors for reaction-diffusion systems without uniqueness. Commun. Pure Appl. Anal. **13**, 1891–1906 (2014). doi:10.3934/cpaa.2014.13.1891
25. Kapustyan, O.V., Kasyanov, P.O., Valero, J.: Structure and regularity of the global attractor of a reaction-diffusion equation with non-smooth nonlinear term. Commun. Pure Appl. Anal. **34**, 4155–4182 (2014). doi:10.3934/dcds.2014.34.4155
26. Kasyanov, P.O., Toscano, L., Zadoianchuk, N.V.: Regularity of weak solutions and their attractors for a parabolic feedback control problem. Set-Valued Var. Anal. **21**, 271–282 (2013). doi:10.1007/s11228-013-0233-8
27. Kasyanov, P.O.: Multivalued dynamics of solutions of an autonomous differential-operator inclusion with pseudomonotone nonlinearity. Cybern. Syst. Anal. **47**, 800–811 (2011)

28. Kasyanov, P.O.: Multivalued dynamics of solutions of autonomous operator differential equations with pseudomonotone nonlinearity. Math. Notes **92**, 205–218 (2012)
29. Kasyanov, P.O., Toscano, L., Zadoianchuk, N.V.: Long-time behaviour of solutions for autonomous evolution hemivariational inequality with multidimensional reaction-displacement. Law, Abstr. Appl. Anal. 2012 (2012). doi:10.1155/2012/450984
30. Melnik, V.S., Valero, J.: On attractors of multivalued semiflows and differential inclusions. Set Valued Anal. **6**, 83–111 (1998). doi:10.1023/A:1008608431399
31. Migórski, S.: On the existence of solutions for parabolic hemivariational inequalities. J. Comput. Appl. Math. **129**, 77–87 (2001)
32. Migórski, S., Ochal, A.: Optimal control of parabolic hemivariational inequalities. J. Glob. Optim. **17**, 285–300 (2000)
33. Otani, M., Fujita, H.: On existence of strong solutions for $\frac{du}{dt}(t)+\partial\varphi^1(u(t))-\partial\varphi^2(u(t)) \ni f(t)$, J. Fac. Sci. Univ. Tokyo. Sect. 1 A, Math. **24**d(3) (1977) 575–605
34. Panagiotopoulos, P.D.: Inequality Problems in Mechanics and Applications. Convex and Nonconvex Energy Functions. Birkhauser, Basel (1985)
35. Sell, G.R., You, Yu.: Dynamics of Evolutionary Equations. Springer, New York (2002)
36. Temam, R.: Infinite-Dimensional Dynamical Systems in Mechanics and Physics. Springer, New York (1988)
37. Terman, D.: A free boundary problem arising from a bistable reaction-diffusion equation. SIAM J. Math. Anal. **14**, 1107–1129 (1983)
38. Terman, D.: A free boundary arising from a model for nerve conduction. J. Differ. Equ. **58**(3), 345–363 (1985)
39. Valero, J.: Attractors of parabolic equations without uniqueness. J. Dyn. Differ. Equ. **13**, 711–744 (2001). doi:10.1023/A:1016642525800
40. Valero, J., Kapustyan, A.V.: On the connectedness and asymptotic behaviour of solutions of reaction-diffusion systems. J. Math. Anal. Appl. (2006). doi:10.1016/j.jmaa.2005.10.042
41. Vishik, M.I., Zelik, S.V., Chepyzhov, V.V.: Strong trajectory attractor for dissipative reaction-diffusion system. Doclady Math. (2010). doi:10.1134/S1064562410060086
42. Zadoianchuk, N.V., Kasyanov, P.O.: Dynamics of solutions of a class of second-order autonomous evolution inclusions. Cybern. Syst. Anal. **48**, 414–428 (2012)
43. Zgurovsky, M.Z., Kasyanov, P.O., Kapustyan, O.V., Valero, J., Zadoianchuk, N.V.: Evolution Inclusions and Variation Inequalities for Earth Data Processing III. Springer. Berlin (2012). doi:10.1007/978-3-642-28512-7
44. Zgurovsky, M.Z., Kasyanov, P.O.: Multivalued dynamics of solutions for autonomous operator differential equations in strongest topologies. Solid Mech. Appl. **211**, 149–162 (2014)
45. Zgurovsky, M.Z., Kasyanov, P.O., Zadoianchuk, N.V.: Long-time behavior of solutions for quasilinear hyperbolic hemivariational inequalities with application to piezoelectricity problem. Appl. Math. Lett. **25**, 1569–1574 (2012). doi:10.1016/j.aml.2012.01.016

Part III
Optimization, Control and Decision Sciences for Continuum Mechanics Problems

Chapter 15
Robust Stability, Minimax Stabilization and Maximin Testing in Problems of Semi-Automatic Control

Victor A. Sadovnichiy, Vladimir V. Alexandrov, Stephan S. Lemak, Dmitry I. Bugrov, Katerina V. Tikhonova and Raul Temoltzi Avila

Abstract The conditions about the problem of robust stability for systems of differential equations with bounded external perturbations are obtained from the solution of the problem of maximum deviation using an extension of the definition of Duboshin and Malkin on stability under permanent perturbations. Further, a linear controlled system with close-loop control is under consideration. Solutions of minimax stabilization problem and a problem of maximin testing in a stability region with given stability factor are described both for finite and infinite stabilization time.

V.A. Sadovnichiy (✉)
Lomonosov Moscow State University, GSP-1, Leninskie Gory, Moscow 119991, Russian Federation
e-mail: rector@msu.ru

V.V. Alexandrov
Mechanics and Mathematics Faculty, Department of Applied Mechanics and Control, Lomonosov Moscow State University, GSP-1, Leninskie Gory, Moscow 119991, Russian Federation
e-mail: vladimiralexandrov366@hotmail.com

S.S. Lemak · D.I. Bugrov · K.V. Tikhonova
Mechanics and Mathematics Faculty, Lomonosov Moscow State University, GSP-1, Leninskie Gory, Moscow 119991, Russian Federation
e-mail: lemaks2004@mail.ru

D.I. Bugrov
e-mail: d.i.bugrov@gmail.com

K.V. Tikhonova
e-mail: katerina.tikhonova@innopraktika.ru

R. Temoltzi Avila
Mathematics Faculty, Universidad Autonoma Del Estado de Hidalgo, Pachuca, Mexico
e-mail: raul_math@hotmail.com

© Springer International Publishing Switzerland 2015

V.A. Sadovnichiy and M.Z. Zgurovsky (eds.), *Continuous and Distributed Systems II*, Studies in Systems, Decision and Control 30,
DOI 10.1007/978-3-319-19075-4_15

15.1 Introduction

An object of investigation in this paper is a dynamical system in the form

$$\begin{cases} \dot{\mathbf{x}} = \mathbf{f}(\mathbf{x}, u, v) \\ u = K\mathbf{x} + w \\ v(\cdot) \in V \\ w(\cdot) \in W \\ \mathbf{x}(0) \in X_0 \end{cases} \tag{15.1}$$

where $\mathbf{x} = (x_1, \ldots, x_n)^T$ is n-dimensional generalized coordinates vector, $\mathbf{f}$ is a continuous vector function, u is a perturbed control in the closed-loop form, w, v are perturbations, V and W are convex closed bounded sets, the beginning conditions $\mathbf{x}(0)$ are known to reside in the convex closed bounded set X_0. We suppose $\mathbf{f}(0, 0, 0) \equiv \mathbf{0}$. One can get a linear model corresponding to (15.1) for small values $\mathbf{x}$

$$\begin{cases} \dot{\mathbf{x}} = (A + BK)\mathbf{x} + Bw + Cv \\ w(\cdot) \in W \\ v(\cdot) \in V \\ |\mathbf{x}(0)| \le R > 0 \end{cases} \tag{15.2}$$

Now we can consider some problems for the system (15.2) stabilization neglecting different terms in equation. These problems provide robust stability of linear systems (Eq. (15.2) at $w \equiv 0$, $K = 0$, i.e., at $u \equiv 0$); minimax stabilization and antagonistic game (Eq. (15.2) at $w \equiv 0$, $v \equiv 0$).

15.2 Robust Stability of Linear Systems

1.1. The maximum deviation problem for linear systems is discussed in [1], which determines maximum deviations of system solutions of differential equations with bounded additional perturbation at each time $T > 0$. From the results of Bulgakov problem for estimating the behavior of the solution of system:

$$\begin{aligned} &\dot{\mathbf{x}} = A\mathbf{x} + \mathbf{b}v, \quad \mathbf{x}(0) = \mathbf{0}, \\ &v(\cdot) \in V = \{v(t) \in L_\infty : |v(t)| \le \delta_1\}, \end{aligned} \quad 0 \le t \le T, \tag{15.3}$$

where L_∞ is the set of measurable functions bounded almost anywhere, A is a $n \times n$ Hurwitz matrix, $\mathbf{b}$ is a $n \times 1$ vector and rang$[\mathbf{b}, A\mathbf{b}, \ldots, A^{n-1}\mathbf{b}] = n$, under the vector norm defined by $\|\mathbf{y}\| = \max\limits_{1 \le i \le n} |y_i|$, the following equality occurs [2, 3]

$$\|\mathbf{x}(v(\cdot), T)\| = \max_{1\le j\le n} \left|x_j(\mathbf{v}(\cdot), T)\right| \le \delta_1 \cdot \max_{1\le j\le n} \int_0^T \left|\mathbf{e}_j^\top e^{At}\mathbf{b}\right| dt,$$

here $\mathbf{e}_j$ is the jth canonical vector corresponding to the x_j coordinate system. If for $\varepsilon > 0$ the estimated desired is $\|\mathbf{x}(\mathbf{v}(t), T)\| \le \varepsilon$, then it is sufficient to choose $\delta_1 = \varepsilon/\chi$, where

$$\chi = \max_{1\le j\le n} \int_0^T |\mathbf{e}_j^\top e^{At}\mathbf{b}|\, dt.$$

In this case, the criteria for robust stability for systems with initial conditions is introduced in [3, 4] which is defined as

$$\chi = \sup_{0<\varepsilon<\infty} \frac{\varepsilon}{\delta_1(\varepsilon)}.$$

For this value, the estimate defines the n-dimensional cube that contains the attainability set of system. Sets of this type are the best possible for at least two faces of the cube have common points with the attainability set.

1.2. The dynamics of a system governed by the differential equation with an additional perturbation is considered

$$\dot{\mathbf{x}} = A\mathbf{x} + \mathbf{b}u,$$
$$u(\cdot) \in \mathscr{U} = \{u(t) \in PC : |u(t)| \le \delta_1\},$$

where $\mathbf{x}(t) = (x_1(t), \ldots, x_n(t))^\top$ is the n-dimensional vector of phase coordinates of the system, A is a Hurwitz matrix constant of size $n \times n$, $\mathbf{b}$ is a constant vector n-dimensional, and $u(\cdot)$ is a bounded piecewise constant function, where PC denotes the set of piecewise continuous functions (additional disturbance). It is assumed that the matrix A has complex eigenvalues with negative real part and the pair $\{A, \mathbf{b}\}$ is completely controllable. The complete controllability of the system allows to express the value given in the linear differential equation with constant coefficients system according to the results in [2]:

$$y_1^{(n)} + a_1 y_1^{(n-1)} + \cdots + a_{n-1}\dot{y}_1 + a_n y_1 = u, \quad u(\cdot) \in \mathscr{U},$$

which is expressed as a system of linear differential equations by considering the change of variables $y_1 = y, y_2 = \dot{y}, \ldots, y_n = y^{(n-1)}$:

$$\dot{\mathbf{y}} = A_0\mathbf{y} + \mathbf{e}_n u, \quad u(\cdot) \in \mathscr{U}, \tag{15.4}$$

where $\mathbf{y}(t) = (y_1(t), \ldots, y_n(t))^\top$ are the coordinates phase and

$$A_0 = \begin{bmatrix} 0 & 1 & \cdots & 0 \\ \vdots & \vdots & \ddots & \vdots \\ 0 & 0 & \cdots & 1 \\ -a_n & -a_{n-1} & \cdots & -a_1 \end{bmatrix}, \quad \mathbf{e}_n = \begin{bmatrix} 0 \\ \vdots \\ 0 \\ 1 \end{bmatrix}.$$

The pair $\{A_0, \mathbf{e}_n\}$ is completely controllable because its controllability matrix $U = [\mathbf{e}_n, A_0\mathbf{e}_n, \ldots, A_0^{n-1}\mathbf{e}_n]$ has maximum rank since $\det U = (-1)^n$.

Since the eigenvalues of the matrix A_0 have negative real part, the trivial solution $\mathbf{y} \equiv \mathbf{0}$ of system (15.4) (i.e., the trivial solution to (15.4) for $u(\cdot) \equiv 0$), is asymptotically stable when $\delta_1 \equiv 0$; in the case where $\delta_1 > 0$, the concept of stability of the trivial solution under permanent perturbations, the first definition was introduced by I. G. Malkin and G. N. Duboshin, see [5]. In this context, we consider the definition of robust stability considered in [3, 4]:

Definition 15.1 The trivial solution $\mathbf{y} \equiv \mathbf{0}$ of system (15.4) is stable under time-varying perturbations $u(\cdot) \in \mathcal{U}$ if for $\varepsilon > 0$ there exists $\delta_1 = \delta_1(\varepsilon) > 0$ and $\delta_2 = \delta_2(\varepsilon) > 0$ such that $\|\mathbf{y}(t)\| \leq \varepsilon$ for any $t \geq t_0$ with $\|\mathbf{y}(t_0)\| \leq \delta_1$ and $|u(t)| \leq \delta_2$ for almost all $t \geq t_0$ and for all $u(\cdot) \in \mathcal{U}$.

The study of the robust stability of system (15.4) is in the sense of Definition 15.1 to consider solutions of the problem of determining the worst additional perturbations $u(\cdot) \in \mathcal{U}$ such that the functionals $\varphi_i[\mathbf{y}(t)] = |y_i(t)|$, $i = 1, \ldots, n$, give rise to the problem of optimization

$$\varphi_i[\mathbf{y}(t)] \to \sup_{u(\cdot)\in\mathcal{U}}, \quad i = 1, 2, \ldots, n.$$

The value obtained defines the maximum deviations of the coordinates $y_i(t)$ of system (15.4). Each of these values determine the magnitude of the additional perturbation $u(\cdot)$ to ensure the robust stability of (15.4). Different approaches of the maximum deviation problems are shown in [6, 7].

The following result shows the conditions under which the system (15.4) can be decoupled at m subsystems controllable of order two when $n = 2m$ and A has m different complex conjugate eigenvalues with negative real part.

Lemma 15.1 *If the matrix A_0 of (15.4) has $2m$ different complex conjugate eigenvalues with negative real part denoted by $-\varepsilon_1 \pm i\omega_1, \ldots, -\varepsilon_m \pm i\omega_m$, where $n = 2m$, then the system (15.4) is equivalent to the system*

$$\dot{\mathbf{x}} = B\mathbf{x} + \mathbf{b}u, \quad u(\cdot) \in \mathcal{U}, \tag{15.5}$$

where $\mathbf{b} = (0, 1, \ldots, 0, 1)^\top$ *and* $B = \mathrm{diag}[B_1, \ldots, B_m]$,

$$B_j = \begin{bmatrix} 0 & 1 \\ -\varepsilon_j^2 - \omega_j^2 & -2\varepsilon_j \end{bmatrix}.$$

Proof Let $\mathbf{u}_1 \pm i\mathbf{v}_1, \ldots, \mathbf{u}_m \pm i\mathbf{v}_m$ the corresponding eigenvectors associated. As the multiplicity of each eigenvalue is one, $Q = [\mathbf{v}_1, \mathbf{u}_1, \ldots, \mathbf{v}_m, \mathbf{u}_m]$ is invertible, and is known from [8] that

$$Q^{-1}A_0Q = \mathrm{diag}[A_1, \ldots, A_m],$$

with

$$A_j = \begin{bmatrix} -\varepsilon_j & \omega_j \\ -\omega_j & -\varepsilon_j \end{bmatrix}, \quad j = 1, \ldots, m.$$

If $\mathbf{q} = Q^{-1}\mathbf{e}_n = (q_1, \ldots, q_n)^\top$, then the system (15.4), under the transformation $\mathbf{y} = Q\bar{\mathbf{y}}$, is equivalent to the system

$$\dot{\bar{\mathbf{y}}} = Q^{-1}A_0Q\bar{\mathbf{y}} + \mathbf{q}u, \quad u(\cdot) \in \mathscr{U},$$

which is completely controllable with controllability matrix $Q^{-1}U$. As this system is decoupled, which follows from the structure of the matrix $Q^{-1}A_0Q$, each of its subsystems is completely controllable, and thus each subsystem can be transformed into canonical form following the development of [2] by the transformation given by $\bar{\mathbf{y}} = \bar{U}P\mathbf{x}$, where

$$\bar{U} = \mathrm{diag}\left\{\bar{U}_1, \ldots, \bar{U}_m\right\},$$
$$P = \mathrm{diag}\left\{P_1, \ldots, P_m\right\},$$

with

$$U_j = \begin{bmatrix} q_{2j-1} & -\varepsilon_j q_{2j-1} + \omega_j q_{2j} \\ q_{2j} & -\omega_j q_{2j-1} - \varepsilon_j q_{2j} \end{bmatrix}, \quad P_j = \begin{bmatrix} 2\varepsilon_j & 1 \\ 1 & 0 \end{bmatrix}.$$

It follows that $\mathbf{b} = (Q\tilde{U}P)^{-1}\mathbf{e_n}$. Therefore, the transformation $\mathbf{y} = S\mathbf{x}$ where $S = Q\bar{U}P$, is not degenerated. This shows that (15.4) and (15.5) are equivalent.

Therefore, the system (15.4) is equivalent to a system of m differential equations of order two with an additional perturbation:

$$\begin{cases} \ddot{x}_{2j-1} + 2\varepsilon_j\dot{x}_{2j-1} + (\varepsilon_j^2 + \omega_j^2)x_{2j-1} = u, \\ u(\cdot) \in \mathscr{U}, \end{cases} \quad j = 1, \ldots, m. \tag{15.6}$$

The maximum deviations of the coordinates of system that satisfy the optimization problems are given by [4, 9, 10]:

$$\sup_{u(\cdot)\in\mathscr{U}} |x_i(t)| \to \alpha_i^* := \delta_1\alpha_i, \quad i = 1, \ldots, n,$$

where

$$\alpha_{2j-1} = \frac{1}{\varepsilon_j^2+\omega_j^2} \coth \frac{\pi\varepsilon_j}{2\omega_j},$$

$$\alpha_{2j} = \frac{1}{\sqrt{\varepsilon_j^2+\omega_j^2}} \left(1 + \coth \frac{\pi\varepsilon_j}{2\omega_j}\right) e^{-\frac{\varepsilon_j}{\omega_j}\left(\frac{\pi}{2} - \arctan \frac{\varepsilon_j}{\omega_j}\right)},$$

by means of which a square R_{2j-1} in $\mathbf{R}^2$ is formed (see Fig. 15.1).

If the initial condition $\mathbf{x}_0 = (x_1^0, \ldots, x_n^0)^\top$ satisfies the constraint $\left|x_i^0\right| \leq \alpha_i^*$, $i = 1, \ldots, n$, the coordinates of the solution of system (15.5) satisfy the relations $|x_i(t)| \leq \alpha_i^*$ for all $u(\cdot) \in \mathscr{U}$ and $t \geq 0$. Therefore, there exists an n-dimensional cube $P_{\mathbf{x}}$ in $\mathbf{R}^n$ whose 2^n vertices $\{\delta_1\mathbf{x}_1, \delta_1\mathbf{x}_2, \ldots, \delta_1\mathbf{x}_{2^n}\}$ depend on the maximum deviation of (15.5), where

$$\mathbf{x}_1 = (\alpha_1, \alpha_2, \ldots, \alpha_n)^\top, \mathbf{x}_2 = (\alpha_1, -\alpha_2, \ldots, \alpha_n)^\top,$$
$$\ldots, \mathbf{x}_{2^n} = (\alpha_1, -\alpha_2, \ldots, -\alpha_n)^\top,$$

and such that $\mathbf{x}(t) \in P_{\mathbf{x}}$ for all $t \geq 0$. It follows that

$$\|\mathbf{x}(t)\| \leq \delta_1 \|\mathbf{x}_i\|, \quad i \in \left\{1, \ldots, 2^n\right\}.$$

It is known of [4, 10] that maximization problem for the deviation yields the worst perturbation $u^o(t) = \delta_1 \operatorname{sign} \dot{x}_{2j-1}(t)$ to the system (15.5), where $j \in \{1, \ldots, m\}$, and that the coordinates $(x_{2j-1}(t), \dot{x}_{2j-1}(t))$ of the solution $\mathbf{x}(t)$, tend to a unique stable limit cycle maximum C_{2j-1}, that have common points with two faces of the n-dimensional cube $P_{\mathbf{x}}$. The parametric equations of C_{2j-1} are

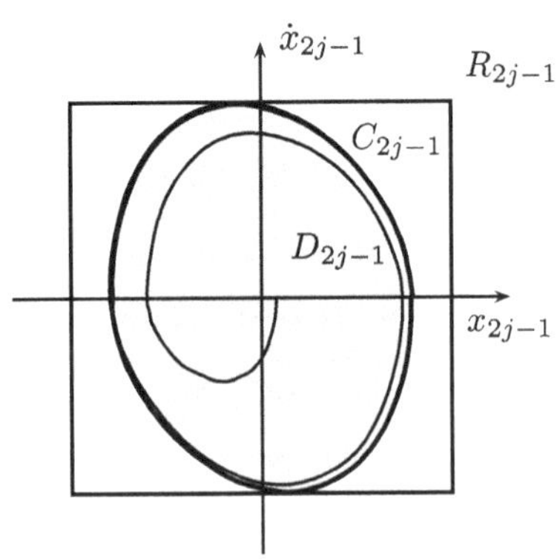

Fig. 15.1 Graphical representation of the maximum deviations of the coordinates x_{2j-1} and $\dot{x}_{2j-1}$ and the sets that are generated from them

$$x_{2j-1}(t) = \mp\left(\tfrac{\delta_1}{\varepsilon_j^2+\omega_j^2} + \alpha_{2j-1}^*\right)\cdot$$
$$\cdot\, e^{-\varepsilon_j t}\left(\cos\omega_j t + \tfrac{\varepsilon_j}{\omega_j}\sin\omega_j t\right) \pm \tfrac{\delta_1}{\varepsilon_j^2+\omega_j^2}, \qquad 0 \le t \le \tfrac{\pi}{\omega_j},$$
$$\dot{x}_{2j-1}(t) = \pm\tfrac{\varepsilon_j^2+\omega_j^2}{\omega_j}\left(\tfrac{\delta_1}{\varepsilon_j^2+\omega_j^2} + \alpha_{2j_0-1}^*\right)e^{-\varepsilon_j t}\sin\omega_j t,$$

The maximum limit cycle C_{2j-1} coincides with the boundary of the attainability set D_{2j-1} of the differential equation with an additional perturbation:

$$\ddot{x}_{2j-1} + 2\varepsilon_j\dot{x}_{2j-1} + (\varepsilon_j^2+\omega_j^2)x_{2j-1} = u,\, u(\cdot)\in\mathcal{U},$$
$$\left(x_{2j-1}(0), \dot{x}_{2j-1}(0)\right) \in D_{2j-1},\quad 1 \le j \le m.$$

Moreover, since the remaining coordinates of system (15.6) are periodic with period $T = \frac{\pi}{\omega_j}$, it follows that the system (15.5) has a limit cycle that depends on the worst perturbation $u_i^o(t)$, $i = 1, \ldots, n$.

Thus $D_{\mathbf{x}} = D_1 \times D_3 \times \cdots \times D_{2m-1}$ approximates the attainability set of the system (15.5). It follows from the construction that $D_{\mathbf{x}} \subset P_{\mathbf{x}}$.

Since the transformation $\mathbf{y} = S\mathbf{x}$ is linear and nondegenerate, it is concluded that the solutions $\mathbf{y}(t)$ of the system (15.4) for each $u(\cdot)\in\mathcal{U}$ are contained in the parallelepiped

$$P_{\mathbf{y}} = \left\{\mathbf{y}\in\mathbf{R}^n : \mathbf{y} = S\mathbf{x},\ \ \mathbf{x}\in P_{\mathbf{x}}\right\}, \tag{15.7}$$

and the attainability set is estimated by

$$D_{\mathbf{y}} = \left\{\mathbf{y}\in\mathbf{R}^n : \mathbf{y} = S\mathbf{x},\ \ \mathbf{x}\in D_{\mathbf{x}}\right\}. \tag{15.8}$$

The robust stability of the trivial solution $\mathbf{y} \equiv \mathbf{0}$ of the system (15.4) under constant perturbations gives the estimated robust quality; i.e., for $\varepsilon > 0$ the estimated $\|\mathbf{y}(t)\| \le \varepsilon$ for all $u(\cdot) \subset \mathcal{U}$, is obtained by taking $\delta_1 = \varepsilon/\chi$, where

$$\chi := \frac{\varepsilon}{\delta_1(\varepsilon)} = \max\left\{\|S\mathbf{x}_1\|, \ldots, \|S\mathbf{x}_{2^n}\|\right\},$$

provided that the initial condition of the system satisfies the inequality

$$\|\mathbf{y}(0)\| \le \delta_2 := \delta_1(\varepsilon)\cdot\max\left\{\|S\mathbf{x}_1\|, \ldots, \|S\mathbf{x}_{2^n}\|\right\}.$$

The estimated of the robust quality that characterizes the maximum deviations of the system (15.4) is given by

$$\chi = \sup_{0<\varepsilon<\infty}\frac{\varepsilon}{\delta_1(\varepsilon)}. \tag{15.9}$$

Theorem 15.1 *If A_0 satisfies the conditions of Lemma 15.1, then the system (15.4) is robustly stable under the estimated disturbance (15.9). Moreover, $\mathbf{y}(t) \in D_{\mathbf{y}} \subset P_{\mathbf{y}}$ for each $u(\cdot) \in \mathscr{U}$, where $P_{\mathbf{y}}$ and $D_{\mathbf{y}}$ is defined by (15.7) and (15.8), respectively.*

The following example shows the results obtained

Example 15.1 Consider the system

$$\dot{\mathbf{y}} = A_0\mathbf{y} + \mathbf{e}_n\varphi(t, \mathbf{y}). \tag{15.10}$$

The matrix A_0 and the vector $\mathbf{e}_n$ are defined as in (15.4), and assume that $\phi(t, \mathbf{y})$ is a function of bounded module: $|\varphi(t, \mathbf{y})| \leq \delta_1$. It is assumed that the uncertainty in the choice of the function $\varphi(t, \mathbf{y})$, and that the analysis of the dynamics of the system (15.10) can be determined if it is known that the system (15.4) is robustly stable. If the coefficients of the matrix A_0 of the system (15.10) are expressed by $a_1 = 2(\varepsilon_1 + \varepsilon_2)$, $a_2 = 4\varepsilon_1\varepsilon_2 + \varepsilon_1^2 + \omega_1^2 + \varepsilon_2^2 + \omega_2^2$, $a_3 = 2\varepsilon_1(\varepsilon_2^2 + \omega_2^2) + 2\varepsilon_2(\varepsilon_1^2 + \omega_1^2)$ and $a_4 = (\varepsilon_1^2 + \omega_1^2)(\varepsilon_2^2 + \omega_2^2)$, then the eigenvalues of the matrix A_0 are $\lambda_{1,2} = \varepsilon_1 \pm i\omega_1$ and $\lambda_{1,2} = \varepsilon_1 \pm i\omega_1$. The transformation $\mathbf{y} = S\mathbf{x}$, where

$$S^{-1} = \begin{bmatrix} \varepsilon_2^2 + \omega_2^2 & 2\varepsilon_2 & 1 & 0 \\ 0 & \varepsilon_2^2 + \omega_2^2 & 2\varepsilon_2 & 1 \\ \varepsilon_1^2 + \omega_1^2 & 2\varepsilon_1 & 1 & 0 \\ 0 & \varepsilon_1^2 + \omega_1^2 & 2\varepsilon_1 & 1 \end{bmatrix}$$

giving the system (15.10) in the equivalent form

$$\begin{aligned} &\dot{\mathbf{x}} = B\mathbf{x} + \mathbf{b}\varphi(t, S\mathbf{x}), \\ &B = \begin{bmatrix} 0 & 1 & 0 & 0 \\ -\varepsilon_1^2 - \omega_1^2 & -2\varepsilon_1 & 0 & 0 \\ 0 & 0 & 0 & 1 \\ 0 & 0 & -\varepsilon_2^2 - \omega_2^2 & -2\varepsilon_2 \end{bmatrix}, \\ &\mathbf{b} = \begin{bmatrix} 0 \ 1 \ 0 \ 1 \end{bmatrix}^\top. \end{aligned} \tag{15.11}$$

As a particular case for the matrix coefficients A_0 are chosen $a_1 = 2.6000$, $a_2 = 2.8600$, $a_3 = 1.0260$ and $a_4 = 0.2405$. If solutions are sought in the system (15.10) such that $\|\mathbf{y}(t)\| \leq 1$, then it is sufficient to choose $\delta_1 = 0.1180$. In the associated system is obtained the estimated $\|\mathbf{x}(t)\| \leq 1.1630$. Figure 15.2 shows the projection of the solutions of systems (15.10) and (15.11) in the coordinates $y_1y_2y_3$ and $x_1x_2x_3$, respectively. In the coordinates $y_1y_2y_3$, the estimation of the attainability set $D_{\mathbf{y}}$ is formed by an oblique cylinder (which is obtained from the fact that C_1 and C_3 are limit cycles), while in the coordinates $x_1x_2x_3$, the estimate $D_{\mathbf{x}}$ is a cylinder. Similar graphs were obtained in the remaining projections.

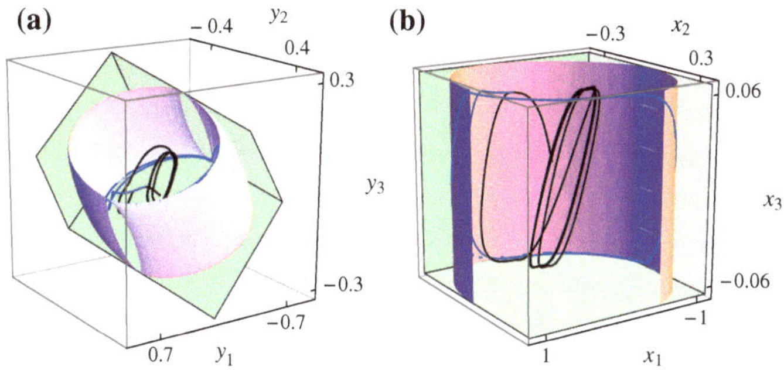

Fig. 15.2 **a** Representation of the projection of the parallelepiped $P_{\mathbf{y}}$ and the estimate of the attainability set $D_{\mathbf{y}}$ for the system (15.10) in the coordinates $y_1 y_2 y_3$. **b** Representation of the projection of the parallelepiped $P_{\mathbf{x}}$ and the estimate the attainability set $D_{\mathbf{x}}$ for the transformed system (15.11) in the coordinates $y_1 y_2 y_3$

15.3 Minimax Stabilization and Antagonistic Game

2.1. Let us consider a controlled system

$$\dot{\mathbf{x}} = A\mathbf{x} + \mathbf{b}u_1, \tag{15.12}$$

where $\mathbf{x} = (x_1, \ldots, x_n)^T$ is n-dimensional generalized coordinates vector, $\det(\mathbf{b}, A\mathbf{b}, A^2\mathbf{b}, \ldots, A^{n-1}\mathbf{b}) \neq 0$, and a criterion in the form

$$\varphi(u_1, \mathbf{x}(t_0)) = \int_{t_0}^{t_1} (\mathbf{x}^T G\mathbf{x} + s_0 u_1^2)\, d\tau + \mathbf{x}^T(t_1) S_1 \mathbf{x}(t_1), \tag{15.13}$$

where $t_1 \leq \infty$, $G^T = G \geq 0$, $s_0 > 0$, $S_1^T = S_1 \geq 0$, $|\mathbf{x}(t_0)| \leq 1$.

We suppose the control $u_1(\cdot)$ has a closed-loop form $u_1 = \mathbf{k}^T\mathbf{x}, \mathbf{k} = (k_1, \ldots, k_n)^T$, $\mathbf{k} \in Q_0$, Q_0 is a convex closed bounded subset of $\{\mathbf{k} \in R_n : Re\lambda_j \leq -\alpha_0, \alpha_0 > 0, \det(\lambda_j E_n - A - bk^T) = 0, j = 1, \ldots, n\}$, R_n is a Euclidean space, E_n is identity matrix. So criterion (15.13) depends on $\mathbf{k}$ and $\mathbf{x}(0)$,

$$\varphi(u_1, \mathbf{x}(t_0)) = \varphi(\mathbf{k}, \mathbf{x}(t_0)) = \int_{t_0}^{t_1} \mathbf{x}^T S_2(k)\mathbf{x}\, d\tau + \mathbf{x}^T(t_1) S_1 \mathbf{x}(t_1), \tag{15.14}$$

$S_2^T = S_2 = G + \mathbf{k}^T s_0 \mathbf{k} > 0$.

Now one can consider two problems:
a problem of minimax stabilization

$$\max_{|\mathbf{x}(t_0)| \leq 1} \varphi(\mathbf{k}, \mathbf{x}(t_0)) \rightarrow \min_{\mathbf{k} \in Q_0} \tag{15.15}$$

and a problem of maximin testing [11]

$$\min_{\mathbf{k}\in Q_0} \varphi(\mathbf{k}, \mathbf{x}(t_0)) \to \max_{|\mathbf{x}(t_0)|\leq 1}. \tag{15.16}$$

2.2. Case $t_1 = \infty$.
In this case criterion (15.14) transforms to

$$\varphi(\mathbf{f}, \mathbf{x}(t_0)) = \int_{t_0}^{\infty} \mathbf{x}^T S_2(k)\mathbf{x}\, d\tau = \mathbf{x}^T(t_0)H\mathbf{x}(t_0), \tag{15.17}$$

where matrix $H^T = H(\mathbf{k})$ is the solution of the Lyapunov linear matrix equation [12]

$$(A^T + \mathbf{k}\mathbf{b}^T)H + H(A + \mathbf{b}\mathbf{k}^T) = -S_2(\mathbf{k}). \tag{15.18}$$

So

$$\max_{|\mathbf{x}(t_0)|\leq 1} \varphi(\mathbf{k}, \mathbf{x}(t_0)) = \max_{|\mathbf{x}(t_0)|\leq 1} \mathbf{x}^T(t_0)H(\mathbf{k})\mathbf{x}(t_0) = \max_{|\mathbf{x}(t_0)|\leq 1} \mu_{max}(\mathbf{k})|\mathbf{x}(t_0)|^2 = \mu_{max}(\mathbf{k}).$$

To solve the problem of minimax stabilization (15.15) one is to find the maximal eigenvalue $\mu_{max}(\mathbf{k})$ of the matrix $H(\mathbf{k})$ and to minimize it at $\mathbf{k} \in Q_0$. One can use tent-method [13], for example, to minimize $\mu_{max}(\mathbf{k})$. $\mathbf{k}^0 : \mu_{max}(\mathbf{k}^0) = \min_{\mathbf{k}\in Q_0} \mu_{max}(\mathbf{k})$ is the solution of the minimax stabilization problem, and eigenvectors $\mathbf{x}^0$ of the matrix $H(\mathbf{k}^0)$ corresponding to the maximal eigenvalue $\mu_{max}(\mathbf{k})$ are the points of maximum at $\mathbf{x}(t_0)$

$$\min_{\mathbf{k}\in Q_0} \max_{|\mathbf{x}(t_0)|\leq 1} \varphi(\mathbf{k}, \mathbf{x}(t_0)) = \mu_{max}(\mathbf{k}^0) = \varphi(\mathbf{k}^0, \mathbf{x}^0).$$

One can see that if $\mathbf{x}^0$ is $\varphi(\mathbf{k}^0, \mathbf{x}(t_0))$ maximum point then $-\mathbf{x}^0$ is $\varphi(\mathbf{k}^0, \mathbf{x}(t_0))$ maximum point too.

To solve the problem of maximin testing (15.16) one is to find $\min_{\mathbf{k}\in Q_0} \varphi(\mathbf{k}, \mathbf{x}(t_0))$ it is necessary to solve Kalman–Letov problem

$$\min_{u_1(\cdot)} \varphi(u_1, \mathbf{x}(t_0)) = \min_{u_1(\cdot)} \int_{t_0}^{\infty} (\mathbf{x}^T G\mathbf{x} + s_0 u_1^2)\, d\tau = \mathbf{x}^T(t_0)L_0\mathbf{x}(t_0), \tag{15.19}$$

where $L_0^T = L_0 > 0$ is the unique solution of the algebraic Riccati equation

$$A^T L + LA + G - L\mathbf{b}s_0^{-1}\mathbf{b}^T L = 0, \tag{15.20}$$

and the matrix L_0 does not depend on $\mathbf{x}(t_0)$.

Then $\min\limits_{u_1(\cdot)} \varphi(u_1, \mathbf{x}(t_0)) = \varphi(u_1^0, \mathbf{x}(t_0))$, where $u_1^0 = \mathbf{k}_0\mathbf{x}$, $\mathbf{k}_0 = -s_0^{-1}\mathbf{b}^T L_0$. So if $\mathbf{k}_0 \in Q_0$ then

$$\max_{|\mathbf{x}(t_0)|\le 1}\min_{\mathbf{k}\in Q_0}\varphi(\mathbf{k}, \mathbf{x}(t_0)) = \max_{|\mathbf{x}(t_0)|\le 1}\min_{u_1(\cdot)}\varphi(u_1, \mathbf{x}(t_0)) = \max_{|\mathbf{x}(t_0)|\le 1}\mathbf{x}^T(t_0)L_0\mathbf{x}(t_0) =$$
$$= \max_{|\mathbf{x}(t_0)|\le 1}\nu_{max}|\mathbf{x}(t_0)|^2 = \nu_{max},$$

where ν_{max} is the maximal eigenvalue of the matrix L_0. Eigenvectors $\mathbf{x}_0$ of the matrix L_0 corresponding to the maximal eigenvalue ν_{max} are the points of maximum at $\mathbf{x}(t_0)$. Similar to the problem of minimax stabilization if $\mathbf{x}_0$ is $\mathbf{x}^T(t_0)L_0\mathbf{x}(t_0)$ maximum point then $-\mathbf{x}_0$ is $\mathbf{x}^T(t_0)L_0\mathbf{x}(t_0)$ maximum point too. In such a way we get the solution of maximin testing problem in the case $\mathbf{k}_0 \in Q_0$. In this case L_0 is the solution of the Lyapunov linear matrix equation (15.18), $L_0 = H(\mathbf{k}^0)$. We can write

$$\max_{|\mathbf{x}(t_0)|\le 1}\mathbf{x}^T(t_0)H(\mathbf{k}^0)\mathbf{x}(t_0) \ge \min_{\mathbf{k}\in Q_0}\max_{|\mathbf{x}(t_0)|\le 1}\mathbf{x}^T(t_0)H(\mathbf{k})\mathbf{x}(t_0). \tag{15.21}$$

In general case we have an inequality

$$\varphi(\mathbf{k}_0, \mathbf{x}_0) = \max_{|\mathbf{x}(t_0)|\le 1}\min_{\mathbf{k}\in Q_0}\mathbf{x}^T(t_0)H(\mathbf{k})\mathbf{x}(t_0) \le$$
$$\le \min_{\mathbf{k}\in Q_0}\max_{|\mathbf{x}(t_0)|\le 1}\mathbf{x}^T(t_0)H(\mathbf{k})\mathbf{x}(t_0) = \varphi(\mathbf{k}^0, \mathbf{x}^0). \tag{15.22}$$

Guaranteeing testing, i.e., maximin testing is one of estimation methods for robust controlled systems. The presence of the saddle point in antagonist game is the base of this method and in this case objective quality estimation takes place. If the saddle point takes place, i.e., inequality (15.21) transforms to equality then in accordance with maximin testing method, one may check the quality of stabilization algorithm under worst beginning conditions $\mathbf{x}^0 = \mathbf{x}_0$. Combining (15.21) and (15.22) one can conclude that in the case $\mathbf{k}_0 \in Q_0$ the saddle point takes place and maximin testing procedure gives objective quality estimation for any close-loop control law in the system (15.12).

Example 15.2 Let us consider a controlled system (15.12), where

$$A = \begin{pmatrix} 0 & 1 \\ 0 & 0 \end{pmatrix}, \quad \mathbf{b} = \begin{pmatrix} 0 \\ 1 \end{pmatrix},$$

and the functional (15.13) at $G = E$—identity matrix, $s_0 = 1$, $u_1 = \mathbf{k}^T x$, $t_1 = \infty$. Let a set Q_0 defines a stability region for (15.12) with stability factor $\alpha_0 > 0$. Then the boundaries of Q_0 are two rays $\{k_2 = -2\alpha_0, k_1 \le -\alpha_0^2\}$ and $\{k_1 - \alpha_0 k_2 = \alpha_0^2, k_1 \le -\alpha_0^2\}$. Let's take $\alpha_0 = 0.1$.

Our object is to find the maximin strategy. In accordance with (15.19) and (15.20) the inner minimum is reached at the control $u_1 = -\mathbf{b}^T L_0 \mathbf{x} = \mathbf{k}_0^T \mathbf{x}$, where $\mathbf{k}_0 = (-1, -\sqrt{3})^T$,

$$L_0 = \begin{pmatrix} \sqrt{3} & 1 \\ 1 & \sqrt{3} \end{pmatrix}.$$

To find the external maximum point $(\mathbf{x}_0)$ in the initial conditions one can use $H(\mathbf{k}_0) = L_0$. Further we get eigenvectors $\mathbf{x}_0 = (\pm\sqrt{2}/2, \pm\sqrt{2}/2)^T$ related to maximal eigenvalue. So maximin strategy is defined. We can see that the point $\mathbf{k}_0 = (-1, -\sqrt{3})^T$ found as a solution of algebraic Riccati equation always lies within Q_0 at $\alpha_0 < \sqrt{3}/2$. So we conclude this case the saddle point takes place, the solution of the minimax stabilization problem is the same, $\mathbf{k}^0 = \mathbf{k}_0$, $\mathbf{x}^0 = \mathbf{x}_0$ and maximin testing procedure gives objective quality estimation for any close-loop control law in the system (15.12).

2.3. Case $t_1 < \infty$.
Criterion (15.14) can be written as

$$\varphi(\mathbf{k}, \mathbf{x}(t_0)) = \int_{t_0}^{t_1} \mathbf{x}^T S_2(\mathbf{k})\mathbf{x}\, d\tau + \mathbf{x}^T(t_1) S_1 \mathbf{x}(t_1) = \mathbf{x}^T(t_0) H(\mathbf{k}) \mathbf{x}(t_0), \quad (15.23)$$

where $H(\mathbf{k}) = H_1(\mathbf{k}) + H_2(\mathbf{k})$,

$$H_1(\mathbf{k}) = \int_{t_0}^{t_1} e^{A_c^T(\mathbf{k})(\tau - t_0)} S_2 e^{A_c(\mathbf{k})(\tau - t_0)}\, d\tau,$$

$$H_2(\mathbf{k}) = e^{A_c^T(\mathbf{k})(t_1 - t_0)} S_1 e^{A_c(\mathbf{k})(t_1 - t_0)},$$

$$A_c = A + \mathbf{b}\mathbf{k}^T.$$

It is known [12] matrix H_1 is the solution of the Lyapunov linear matrix equation

$$A_c^T H_1 + H_1 A_c = -S(\mathbf{k}),$$

$$S(\mathbf{k}) = S_2 - \left(e^{A_c^T(\mathbf{k})(t_1 - t_0)} S_2 e^{A_c(\mathbf{k})(t_1 - t_0)}\right).$$

We assume Q_0 and t_0, t_1: $S(\mathbf{k}) \geq 0 \quad \forall \mathbf{k} \in Q_0$.

One can find another expression for matrix H in the form

$$H = P + e^{A_c^T(\mathbf{k})(t_1 - t_0)} (S_1 - P) e^{A_c(\mathbf{k})(t_1 - t_0)},$$

where P is the solution of the Lyapunov linear matrix equation

$$A_c^T(\mathbf{k}) P + P A_c(\mathbf{k}) = -S_2.$$

Similar to 2.1 to solve the problem of minimax stabilization one is to find the maximal eigenvalue $\mu_{max}(\mathbf{k})$ of the matrix $H(\mathbf{k})$ and to minimize it at $\mathbf{k} \in Q_0$. The result $\mathbf{k}^0$ of minimization is the solution of the minimax stabilization problem, and eigenvectors $\mathbf{x}^0$ of the matrix $H(\mathbf{k}^0)$ corresponding to the maximal eigenvalue $\mu_{max}(\mathbf{k})$ are the points of maximum at $\mathbf{x}(t_0)$.

To solve the problem of maximin testing, i.e., to find $\min\limits_{\mathbf{k}\in Q_0} \varphi(\mathbf{k}, \mathbf{x}(t_0))$ we follow 2.1 again.

$$\min_{u_1(\cdot)} \varphi(u_1, \mathbf{x}(t_0)) = \min_{u_1(\cdot)} \int_{t_0}^{t_1} (\mathbf{x}^T G\mathbf{x} + s_0 u_1^2)\, d\tau + \mathbf{x}^T(t_1) S_1 \mathbf{x}(t_1) = \mathbf{x}^T(t_0) L\mathbf{x}(t_0),$$

where $L^T = L > 0$ is the unique solution of the differential Riccati equation

$$\dot{L} = -A^T L - LA - G + L\mathbf{b}s_0^{-1}\mathbf{b}^T L, \tag{15.24}$$

with the boundary condition

$$L(t_1) = S_1. \tag{15.25}$$

We assume $S_1 = L_0$ satisfies the algebraic Riccati equation

$$A^T S_1 + S_1 A + G - S_1 \mathbf{b}s_0^{-1}\mathbf{b}^T S_1 = 0,$$

so $L = S_1$ is a stationary solution of the Eq. (15.24) with the boundary condition (15.25).

Then $\min\limits_{u_1(\cdot)} \varphi(u_1, \mathbf{x}(t_0)) = \varphi(u_1^0, \mathbf{x}(t_0))$, where $u_1^0 = \mathbf{k}^0\mathbf{x}$, $\mathbf{k}^0 = -s_0^{-1}\mathbf{b}^T S_1$. So if $\mathbf{k}^0 \in Q_0$ then

$$\max_{|\mathbf{x}(t_0)|\le 1} \min_{\mathbf{k}\in Q_0} \varphi(\mathbf{k}, \mathbf{x}(t_0)) = \nu_{max},$$

where ν_{max} is the maximal eigenvalue of the matrix S_1. Eigenvectors $\mathbf{x}^0$ of the matrix S_1 corresponding to the maximal eigenvalue ν_{max} are the points of maximum at $\mathbf{x}(t_0)$.

The important question is the stability of the solution $L(t) \equiv S_1$ for (15.24) and (15.25). Let L_0 being the stationary solution of the Riccati equation

$$A^T L_0 + L_0 A + G - L_0 \mathbf{b}s_0^{-1}\mathbf{b}^T L_0 = 0,$$

We assume $L = L_0 + \Delta L$, $\Delta L = \Delta L^T > 0$. Then nonlinear differential equation governing ΔL is

$$\Delta\dot{L} = -(A^T - L_0\mathbf{b}s_0^{-1}\mathbf{b}^T)\Delta L - \Delta L(A - \mathbf{b}s_0^{-1}\mathbf{b}^T L_0) + \Delta L\mathbf{b}s_0^{-1}\mathbf{b}^T \Delta L,$$

and linear variational equation has the form

$$\Delta\dot{L} = -(A^T - L_0\mathbf{b}s_0^{-1}\mathbf{b}^T)\Delta L - \Delta L(A - \mathbf{b}s_0^{-1}\mathbf{b}^T L_0).$$

The solution of this differential equation is [14]

$$\Delta L = e^{-(A^T - L_0 \mathbf{b} s_0^{-1} \mathbf{b}^T)t} C e^{-(A - \mathbf{b} s_0^{-1} \mathbf{b}^T L_0)t}, \qquad \Delta L(0) = C,$$

and it is unstable in forward time (asymptotically stable in backward time) under the condition of asymptotical stability of linear systems with matrix $A - \mathbf{b} s_0^{-1} \mathbf{b}^T L_0$.

So we can conclude that asymptotical stability of linear system with matrix $A - \mathbf{b} s_0^{-1} \mathbf{b}^T S_1$ is the sufficient condition for asymptotical stability of the Riccati equation solution $L = S_1$ in backward time.

As a result we have both in the case $t_1 < \infty$ and $t_1 = \infty$ if $\mathbf{k}_0 \in Q_0$ then the solutions of the minimax stabilization problem and maximin testing coincide, $\mathbf{k}^0 = \mathbf{k}_0$, $\mathbf{x}^0 = \mathbf{x}_0$ and maximin testing procedure gives objective quality estimation for any close-loop control law in the system (15.12). To get the solution of the minimax stabilization problem, one can use the solution of maximin testing problem based on the solution of the Riccati equation.

15.4 Maximin Testing of Quality of Control Algorithm

Let us consider the problem of testing the quality of robust stabilization for a desire motion of controlled system (15.2).

Control algorithms will be tested with linear equations in deviations from the desired motion written as

$$\begin{cases} \dfrac{dx}{dt} = A(t)x + B(t)u + C(t)v, \\ u(\cdot) \in U = \{u(\cdot) \in L_2^s \big| \, u(t) \in P \subset R^s\}; \\ q(\cdot) \in V = \{v(\cdot) \in PC \big| \, v(t) \in Q \subset R^m\}, \; x(t_0) \in R_0 \subset R^n. \end{cases} \tag{15.26}$$

Here $x(t)$ is the n-dimensional vector of trajectory deviations, $u(\cdot) \in U$ is the s-dimensional vector function of stabilizing controls, $v(\cdot) \in V$ is the time-varying disturbance, $x(t_0) \in R_0$ is the initial deviation.

We shall estimate the accuracy of stabilization by the performance index

$$J(u, v) = x^{\top}(t_k) S x(t_k), \tag{15.27}$$

where the matrix S is symmetric and positive semi-definite, and the moment of test termination t_k is fixed. The best accuracy J_0 (or the best result in terms of accuracy of robust stabilization) is defined as a solution the problem

$$J_0(u^0, v^0) = \sup_{w \in W} \inf_{u(\cdot) \in U} J. \tag{15.28}$$

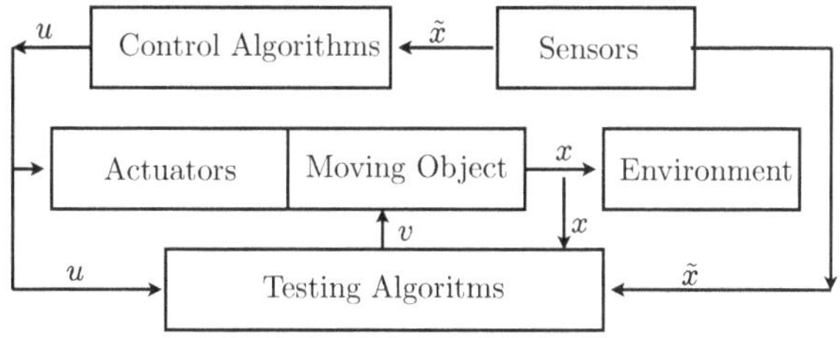

Fig. 15.3 Functional diagram of the test bench

Here $w^\top = \{x_0, v(\cdot)\}$ is the set of initial and time-varying disturbances $v(t)$, and $W = R_0 \times V$. The testing process is objective if upper and lower bounds are attainable and a saddle point (u^0, v^0) of differential game is in existence:

$$J_0 = J(u^0, w^0)) = \max_{w \in W} \min_{u(\cdot) \in U} J(u, w)) = \min_{u(\cdot) \in U} \max_{w \in W} J(y(u, w)) = J^0. \quad (15.29)$$

We estimate the control algorithm on a special test bench [2], which is represented by the functional diagram shown in Fig. 15.3.

In this diagram, the units of the operation mechanisms, the moving object, sensors, and the environment can be implemented in the form of computer models. Testing algorithm forms an estimation for the operation of the control algorithm and worst perturbations.

The following three stages can be singled out in the maximin testing problem [2].

The first, preliminary stage. At this stage, one finds the least (best) estimate of the control performance index J_0 and the optimal behavior strategy $v^0(\cdot)$ of external perturbations with the use of a computer solution of the maximin problem.

The second, main stage. At this stage, one carries out the testing process in the form of computer simulation implementing the testing strategy with the use of the optimal perturbation strategy found at the first stage. A real estimate of the control performance $J^r(u^r, w^0)$ is found as a result of simulation, where $u^r \in U$ is an arbitrary control strategy.

The third, concluding stage. At this stage, one compares the best J_0 and real J^r estimates and issues recommendations for further training, diagnostics, calibration, and correction.

15.4.1 Program Strategy of Testing

Let us consider time-depended strategies $v(t)$ for first player (perturbations) and for second player $u(t)$ (control algorithm).

The proposed algorithm searching for a saddle point employs equivalence of the dynamic game and a geometrical game on the sets of attainability of (15.26). Let us represent write x as $x = y - z$. The system (15.26) can then be rewritten as two subsystems:

$$\dot{y} = A(t)y + C(t)v \qquad y(t_0) = x(t_0) \in R_0, \tag{15.30}$$

$$\dot{z} = A(t)z - B(t)u, \qquad z(t_0) = 0. \tag{15.31}$$

For a fixed $x(t_0) \in R_0$, the set of all possible functions of time $u(t) \in U$ and $w(t) \in W$ on the interval $t \in [t_0, t_k]$ generates the sets of attainability $D_v(t_0, t_k)$ and $D_u(t_0, t_k)$.

The necessary and sufficient conditions for existence of a saddle point in the geometrical game are obtained in [2, 15]. It is shown that in a saddle point $x^0 = y(t_k, v^0) - z(t_k, u^0)$, the tangent hyperplanes for Ω_v and Ω_u (or supporting hyperplanes if $y(t_k, v^0)$ or $z(t_k, u^0)$ is a corner point) are both orthogonal to the vector x^0, i.e., they are parallel.

15.4.2 Closed-Loop Strategy of Testing

We suppose that initial deviation $x(t_0) \in R_0$ is fixed. Consider a situation when a saddle point (15.29) of geometrical game for time-depended strategies does not exist

$$J_0(u^0, v^0) = \max_{v(\cdot)\in V} \min_{u(\cdot)\in U} J(u, v) < \min_{u(\cdot)\in U} \max_{v(\cdot)\in V} J(u, v)) = J^0. \tag{15.32}$$

In this case it is impossible to obtain an objective state the value of quality of control algorithm. Thus closed-loop strategy of testing $v(x, t)$ is used to solve this problem (N. Krasovskii [16]). Existence of a saddle point of game for closed-loop strategies and convex functional (15.27) is shown by Krasovskii [16]. He proposed the extreme shift method for optimal closed-loop strategies search.

Now we suppose that a differential game for system (15.26) is regular. Therefore, a unique solution $(u^0(t), v^0(t))$ of maximin problem exist. Let the sets of allowable values of control $u(t) \in P$ and perturbations $v(t) \in Q$ are convex and closed. It follows [17] that the sets of attainability $D_v(t_0, t_k)$ and $D_u(t_0, t_k)$ are convex and closed too.

Denote $J^* = J(u^*(x, t), v^*(x, t))$ as game value and corresponding saddle point for closed-loop strategies as (u^*, v^*).

We have inequalities

$$J_0 \leq J(u^*(x, t), v^*(x, t)) \leq J^0,$$

where $J_0(u^0, v^0)$ and J^0 are lower and upper values of game for time-depended strategies (15.32).

Following statement holds true.

Theorem 15.2 *Game value J^* in the class of closed-loop strategies is equal lower game value J_0 in the class of time-depended strategies, i.e., $J(u^*(x,t), v^*(x,t)) = J_0(u^0(t), v^0(t))$.*

Proof We use splitting initial system (15.26) into two subsystems (15.30) and (15.31). It follows that cost functional $J = x^\top(t_k)Sx(t_k) = (y(t_k)-z(t_k))^\top S(y(t_k)-z(t_k)) = \rho^2(y(t_k), z(t_k))$ may be represented as the distance squared (in metric S) between points of attainable sets $y \in D_v(t_0, t_k)$ and $z \in D_u(t_0, t_k)$.

Let us consider a maximin point $(y^0(t_k), z^0(t_k))$ where the lower value of game $J^0(u^0, v^0) = \rho^2(y^0, z^0) = \rho_0^2$ is reached. Program strategies $u^0(t), v^0(t), t \in [t_0, t_k]$ correspond to points y^0, z^0 (see Fig. 15.4).

Let $t_1 \in (t_0, t_k)$ be an intermediate timepoint. Evolutionary property [17] of attainable sets means that

$$D_u(t_0, t_k) = D_u(t_1, t_k, D_u(t_0, t_1)), \quad D_v(t_0, t_k) = D_v(t_1, t_k, D_v(t_0, t_1)),$$

therefore $D_u(t_1, t_k, z(t_1)) \subset D_u(t_0, t_k)$ and $D_v(t_1, t_k, y(t_1)) \subset D_v(t_0, t_k)$ (see Fig. 15.4).

One can show the optimal strategy u^* of second player (control) to submit a following condition: on interval $\tau \in [t_0, t_1]$ fulfilled $u^*(\tau) = u^0(\tau)$. In this case the point $z^0(t_k)$ belongs to attainable set $z^0 \in D_u(t_1, t_k, z(t_1))$.

Otherwise, the first player (perturbations) can save the boundary point $y^0(t_k) \in D_v(t_1, t_k, y(t_1))$ if he choose the strategy $v^*(\tau) = v^0(\tau)$, $\tau \in [t_0, t_1]$.

Then the value of game $J = \rho^2(y^0, z^1)$ increases

$$\rho(y^0, z^1) = \min_{z \in D_u(t_1, t_k, z(t_1))} \rho(y^0, z) > \rho(y^0, z^0)$$

Fig. 15.4 Extremal shift method

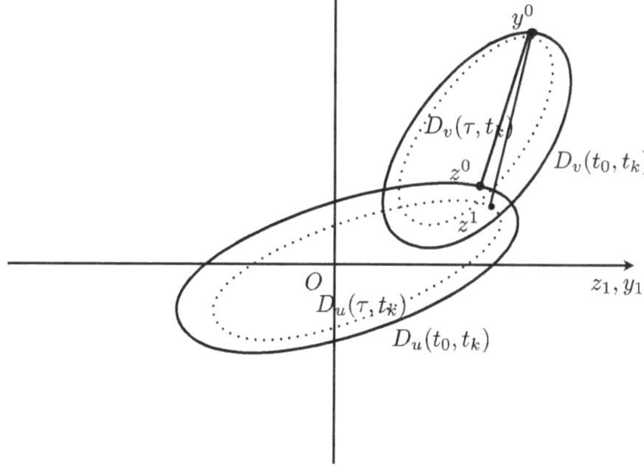

because the attainable set $D_u(t_1, t_k, z(t_1))$ is convex. It does not satisfy the second player's interests. Therefore, the optimal strategy u^* must save the point z^0.

Similarly, it is possible to show that the strategy $v^*(\tau) = v^0(\tau)$ at $\tau \in [t_0, t_1]$ is optimal for the first player.

From here it follows an "extremal shift" method. For any intermediate moment, the optimal strategy tries to shift the final position of subsystems (15.30) and (15.31) to minimax point (y^0, z^0). According to Krasovskii's theorem, the saddle point in a class of closed-loop (positional) strategies is reached. Therefore, we have $J(u^*, v^*) = J_0$. This completes the proof of Theorem 15.2.

Optimal position strategy can be formed according to the following scheme. We choose the partition of interval $[t_0, t_k] = \bigcup_{i=0}^{N} [t_i, t_{i+1}]$ so a diameter of partition $d = \max |t_{i+1} - t_i|$ is small enough that with sufficient accuracy conditions $u^*(\tau) = \text{const}$ and $v^*(\tau) = \text{const}$ at $\tau \in [t_i, t_{i+1}]$ are satisfied.

We solve the maximin problem

$$J_i(u_i^0, v_i^0) = \max_{v(\cdot)\in V} \min_{u(\cdot)\in U} J(u, v)$$

at $t \in [t_i, t_k]$ for each current state $x(t_i) = y(t_i) - z(t_i)$, $i = 0, 1 \ldots, N-1$.

For each player, the position strategy is formed by the rule: $u^*(x(t_i)) = u_i^0(t_i)$, $v^*(x(t_i) = v_i^0(t_i)$, $i = 0, 1 \ldots, N-1$.

15.5 Conclusions

In this report, we propose the application of three methods to estimate the influence of the beginning conditions and additive permanent perturbations on linear dynamical systems. In the first section, the estimation of the quality of m-DOF oscillating system robust stabilization is obtained on the base of the limit cycle synthesis for oscillating 1-DOF system. In the second section, the problem of minimax stabilization is solved both for asymptotic (infinite time) and finite time versions using the result of R. Kalman. In the third section, it is shown that in the framework of N. Krasovskii's extremal shift method for the solution of the antagonistic differential game in positional strategies, lower value of the game is reached.

Acknowledgments This work was partially supported by the Russian Foundation for Basic Researches (project No. 13-01-00515) and Russian Scientific Foundation (project No. 14-50-00029).

References

1. Bulgakov, B.V.: On accumulation of perturbations in the linear oscillatory controls with constant parameters. Dokl. Akad. Nauk SSSR **51**, 339–342 (1946)
2. Aleksandrov, V.V., Boltianski, V.G., Lemak, S.S., Parusnikov, N.A., Tikhomirov, B.M.: Optimal Control of Movement (in Russian). FIZMATLIT, Moscow (2005)
3. Aleksandrov, V.V., Zueva, I.O., Sidorenko, GYu.: Robust stability of third-order control systems. Mosc. Univ. Mech. Bull. **69**(1), 10–15 (2014)
4. Aleksandrov, V.V., Sidorenko, GYu., Temoltzi-Avila, R.: Robust stability of control systems (in Russian). X Conferencia Internacional Chetaev, Mecanica Analtica, Estabilidad y Contro **1**, 42–52 (2012)
5. Elsgoltz, L.: Ecuaciones diferenciales y calculo variacional (in Russian). Mir, Moscow (1969)
6. Aleksandrov, V.V.: On the accumulation of disturbances in linear systems on two coordinates (in Russian). Mosc. Univ. Mech. Bull. **3**, 67–76 (1968)
7. Zhermolenko, V.N.: On maximal deviation of linear system. Autom. Remote Control **73**, 1117–1125 (2012)
8. Hirsch, M.W., Smale, S.: Differential Equations, Dynamical Systems and Linear Algebra. Academic Press, New York (1974)
9. Zhermolenko, V.N.: Maximum deviation of oscillating system of the second order with external and parametric disturbances. J. Comput. Syst. Sci. Int. **46**, 407–411 (2007)
10. Aleksandrov, V.V., Alexandrova, O., Prikhod′ko, I.P., Telmotzi-Auila, R.: Mosc. Univ. Mech. Bull. **62**, 65–68 (2007)
11. Sadovnichii, V.A., Alexandrov, V.V., Lebedev, A.V., Lemak, S.S.: Mixed strategies in maximin testing of robust stabilization performance. Differ. Equ. **45**, 1823–1829 (2010)
12. Polyak, B.T., Sherbakov, P.S.: Robust Stability and Control. Editorial Nauka, Moscow (2002). (in Russian)
13. Boltiansky, V.G., Poznyak, A.S.: The Robust Maximum Principle. Theory and Applications. Birkhauser, New York (2012)
14. Bellman, R.: Introduction to Matrix Analysis. McGraw Hill, New York (1960)
15. Alexandrov, V.V., Blagennova-Mikulich, LJu., Gutieres-Arias, I.M., Lemak, S.S.: Maximin testing of stabilization accuracy and saddle points in geometric games. Mosc. Univ. Mech. Bull. **2005**(1), 43–49 (2005)
16. Krasovskii, N.N., Subbotin, A.I.: Game-Theoretical Control Problems. Springer, New York (1988)
17. Chernousko, F.L.: Estimation of phase vector of dynamical system. Ellipsoid Method (in Russian). Science, Moscow (1988)

Chapter 16
Dynamics of Solutions for Controlled Piezoelectric Fields with Multivalued "Reaction-Displacement" Law

Mikhail Z. Zgurovsky, Pavlo O. Kasyanov, Liliia S. Paliichuk and Alla M. Tkachuk

Abstract We consider the second-order evolution inclusion that describes a class of Continuum Mechanics controlled processes, in particular, the controlled piezoelectric fields with multivalued "reaction-displacement" law. Discontinuous interaction function is represented as the difference of subdifferentials of convex functionals for more flexible control. This case is actual for automatic feedback control problems. We study the dynamics of weak solutions of the investigated problem in terms of the theory of trajectory and global attractors for multivalued semiflows generated by weak solutions of given problem. A priory estimates for weak solutions are obtained. The existence of global and trajectory attractors is proved. The relationship between the global attractor, trajectory attractor, and the space of complete trajectories are provided. Results of this study allow us to direct the states of investigated system to the desired asymptotic levels and may be used for the development of technical equipment based on piezoelectric effect, in particular for design of piezoelectric positioning controller etc.

M.Z. Zgurovsky
National Technical University of Ukraine "Kyiv Polytechnic Institute",
Peremogy Ave., 37, Build, 1, Kyiv 03056, Ukraine
e-mail: zgurovsm@hotmail.com

P.O. Kasyanov · L.S. Paliichuk (✉)
Institute for Applied System Analysis, National Technical University
"Kyiv Polytechnic Institute", Peremogy Ave., 7, Kyiv 03056, Ukraine
e-mail: kasyanov@i.ua

L.S. Paliichuk
e-mail: lili262808@gmail.com

A.M. Tkachuk
National University of Food Technologies, Volodymyrska St., 68, Kyiv 01601, Ukraine
e-mail: tkachukam@ukr.net

© Springer International Publishing Switzerland 2015

V.A. Sadovnichiy and M.Z. Zgurovsky (eds.), *Continuous and Distributed Systems II*, Studies in Systems, Decision and Control 30,
DOI 10.1007/978-3-319-19075-4_16

16.1 Introduction and the Main Problem

Investigation of dynamics of the controlled piezoelectric processes and fields is an actual problem because of wide use of the technical equipment with piezoelements, in particular for image stabilization, positioning accuracy. In many cases, impulse or smooth control is possible (or convenient) for control of such systems. To achieve by system the desired states, the choice of method for constructing interaction function that as accurately as possible approximate the necessary states is very important point. As is known from the analysis, any continuous function on a closed interval can be approximated by a polynomial up to a parameter ε. But to improve the accuracy of interpolation, the interpolation nodes must be added that increase the order of the polynomial. In turn, this leads to an increase in error. More effective method is splines, which are polynomials of low order on each interval between the interpolation nodes. But for the cubic splines, the additional information about the restored function is necessary to maintain the accuracy of approximation. Such interpolation can be performed only in small intervals for several nodes.

In this investigation, we introduce the gaps and consider the piecewise constant functions. The interaction function is presented as the difference of subdifferentials for convex functionals:

$$\partial J_1(\cdot) - \partial J_2(\cdot), \tag{16.1}$$

where $J_i : H \to \mathbf{R}$, $i = 1, 2$ are locally Lipschitz functionals; $\partial J_i(\cdot)$ are the Clarke subdifferentials of $J_i(\cdot)$, $i = 1, 2$; H is a Hilbert space. Suppose that $J_i(\cdot)$, $i = 1, 2$ satisfy the next conditions:

- functionals $J_i : H \to \mathbf{R}$, $i = 1, 2$ are locally Lipschitz and regular [1], i.e., $\forall x, v \in H$ there exist the usual one-sided directional derivatives $J_i'(x; v) = \lim\limits_{t\to 0} \frac{J_i(x+tv)-J_i(x)}{t}$, and $\forall x, v \in H$ $J_i'(x; v) = J_i^\circ(x; v)$, where $J_i^\circ(x; v) = \overline{\lim\limits_{y\to x,\, t\to 0}} \frac{J_i(y+tv)-J_i(y)}{t}$, $i = 1, 2$;
- for $i = 1, 2$ there exist $c_i > 0$ such that $\|l\|_H \le c_i(1 + \|v\|_H)$ $\forall l \in \partial J_i(v)$ and $\forall v \in H$;
- there exists $c_2^* > 0$ such that $(l, v)_H \le \lambda \|v\|_H^2 + c_2^*$ $\forall l \in \partial J_2(v)$ and $\forall v \in H$, where $\partial J_i(u) = \{p \in H \,|\, (p, w)_H \le J_i^\circ(u; w)\ \forall w \in H$ denote the Clarke subdifferentials for $J_i(\cdot)$, $i = 1, 2$ in $u \in H$ [1]; $\lambda \in (0, \lambda_1)$, $\lambda_1 > 0$: $c_A \|v\|_V^2 \ge \lambda_1 \|v\|_H^2$ $\forall v \in V$, where V is a Hilbert space.

Denote $J(u) = J_1(u) - J_2(u), u \in H$. It follows from Lebourg mean value theorem [1, Chap. 2] that there exist constants $c_3, c_4 > 0$ and $\mu \in (0, \lambda_1)$ such that $\forall u \in H$ the next conditions hold:

$$|J(u)| \le c_3(1 + \|u\|_H^2), \quad J(u) \ge -\frac{\mu}{2}\|u\|_H^2 - c_4.$$

To justify the accuracy of this approximation, it is necessary to show that the field under investigation with the control law in the form (16.1) have the finitedimensional dynamics of solutions within a small parameter.

For this goal let us consider the autonomous second-order differential inclusion of hyperbolic type with discontinuous nonlinearity that describes not only the piezoelectric processes, fields, and different wave effects, but a wide class of Continuum Mechanics processes. Feedback control for driving of piezoceramic element is convenient and effective for such systems [2, 3]. Feedback control allows to obtain a quick change of position as a result of rapid change of the control voltage. Feedback control problems with a long memory are investigated in [4–6]. Some types of subdifferential form for such control are studied in [7–10]. Because of skin effects such as friction, adhesion, in this paper we consider the nonmonotone, possibly multivalued control law between forces, and respective displacements in the form (16.1) which facilitates more flexible control of the investigated system.

16.2 Setting of the Problem and the Main Results

Let $A : H \to H$ and $B : V \to V^*$ be the linear symmetric operators. Assume that there exists $\beta > 0$ such that $(Av, v)_H = \beta \|v\|_H^2$ for each $v \in H$. Suppose that there exists $c_B > 0$ such that $\langle Bv, v\rangle_V \geq c_B \|v\|_V^2$ for each $v \in V$. $(V; H; V^*)$ is an evolution triple.

Consider the problem with multivalued "reaction-displacement" law:

$$u_{tt}(t) + Au_t(t) + Bu(t) + \partial J_1(u(t)) - \partial J_2(u(t)) \ni \bar{0} \text{ for a.a. } t. \tag{16.2}$$

$X = V \times H$ is the phase space of Problem (16.2). The functionals $J_i(\cdot)$ satisfy the described above conditions. By $(\cdot, \cdot)_X$ and $\|\cdot\|_X$ denote the inner product and corresponding norm for Hilbert space X, respectively. Suppose for simplicity that $(u, v)_V = \langle Bu, v\rangle_V$, $\|v\|_V^2 = \langle Bu, v\rangle_V$, $\beta(u, v)_H = (Au, v)_H$, $\beta \|v\|_H^2 = (Av, v)_H$ for each $u, v \in V$.

To investigate the long-term behavior of solutions for evolution inclusions with discontinuous right-hand side, the methods and principles of the theory of multivalued semiflow for global and trajectory attractors are developed. In particular, the asymptotic behavior of solutions for such type problems are studied in [11–15]. In the nonautonomous case, the existence of weak solutions for investigated model is obtained in [16]. Asymptotic behavior of solutions for Problem (16.2) with $J_2 \equiv 0$ is studied in [15, 17]. Corresponding scalar case is considered in [18].

By [19, Lemmas 2.4 and 2.6], it follows that the asymptotic compactness of weak solutions of Problem (16.2) is equivalent to flattening property, which implies that the dynamics of weak solutions of Problem (16.2) is finite dimensional within a small parameter.

First, let us define the notion of weak solutions for Problem (16.2). Let $-\infty < \tau < T < +\infty$. The function $(u(\cdot), u_t(\cdot))^T \in L_\infty[\tau, T; X]$ is called *a weak solution* for Problem (16.2) on $[\tau, T]$ if there exist $l_i(\cdot) \in L_2(\tau, T; H), i = 1, 2, l_i(t) \in \partial J_i(u(t))$ for a.e. $t \in (\tau, T)$ such that

$$- \int_\tau^T (u_t(t), \psi)_H \eta_t(t) dt +$$
$$\int_\tau^T [(u_t(t), \psi)_H + (u(t), \psi)_H + (l_1(t), \psi)_H - (l_2(t), \psi)_H] \eta(t) dt = 0,$$

for each $\psi \in V$ and $\eta \in C_0^\infty(\tau, T)$.

References [17, Lemma 2] and [16, Theorem 4.1] imply the existence of a weak solution $(u(\cdot), u_t(\cdot))^{\mathrm{T}} \in \mathrm{L}_\infty(\tau, T; X)$ on $[\tau, T]$ for Problem (16.2) with initial data $u(\tau) = a$, $u_t(\tau) = b$, where $a \in V$ and $b \in H$ are arbitrary fixed. Moreover, $(u(\cdot), u_t(\cdot))^{\mathrm{T}} \in C([\tau, T]; X), u_{tt}(\cdot) \in L_2(\tau, T)$. Reference [18, Lemma 16.4] yields that the translates and concatenation of weak solutions are also weak solutions [17]. For each $\varphi_\tau = (a, b)^T \in X$ let us set $\mathscr{D}_{\tau,T}(\varphi_\tau) = \{(u(\cdot), u_t(\cdot))^T | (u, u_t)^T$ is a weak solution of Problem (16.2) on $[\tau, T], u(\tau) = a, u_t(\tau) = b\}$. Note that $\mathscr{D}_{\tau,T}(\varphi_\tau) \subset C([\tau, T]; X)$.

Remark also that each weak solution of Problem (16.2) on $[\tau, T]$ can be extended to a global one, defined on $[0, +\infty)$ [20, Lemma 4]. For each $\varphi_0 \in X$ let $\mathscr{D}(\varphi_0)$ be the set of all weak solutions of Problem (16.2) defined on $[0, +\infty)$ with initial data $\varphi(0) = \varphi_0$. For each $\varphi_0 \in X$ and $\varphi(\cdot) \in \mathscr{D}(\varphi_0)$, the next inequality holds $\forall t > 0$: $\|\varphi(t)\|_X^2 \le \frac{\lambda_1 + 2c_3}{\lambda_1 - \mu} \|\varphi(0)\|_X^2 + \frac{2(c_3 + c_4)\lambda_1}{\lambda_1 - \mu}$.

For each $\varphi = (a, b)^T \in X$ let us define the function $\mathscr{V}(\varphi) = \frac{1}{2} \|\varphi\|_X^2 + J_1(a) - J_2(a)$. By Lemma 3 in [20], it follows that for each $\varphi_\tau \in X$ and $\varphi(\cdot) = (u(\cdot), u_t(\cdot))^T \in \mathscr{D}_{\tau,T}(\varphi_\tau)$, the function $\mathscr{V} \circ \varphi : [\tau, T] \to \mathbf{R}$ is absolutely continuous, and $\frac{d}{dt} \mathscr{V}(\varphi(t)) = -\beta \|u_t(t)\|_H^2$ for a.e. $t \in (\tau, T)$.

Moreover, from [20, Theorems 1 and 2] the following convergence properties for weak solutions of Problem (16.2) hold: let $\{\varphi_n(\cdot)\}_{n\ge1} \subset C([\tau, T]; X)$ be an arbitrary sequence of weak solutions of Problem (16.2) on $[\tau, T]$ such that:

- $\varphi_n(\tau) \to \varphi_\tau$ weakly in X as $n \to +\infty$, and let $\{t_n\}_{n\ge1} \subset [\tau, T]$ be a sequence such that $t_n \to t_0$ as $n \to +\infty$, then there exists $\varphi(\cdot) \in \mathscr{D}_{\tau,T}(\varphi_\tau)$ such that within a some subsequence the sequence $\varphi_n(t_n) \to \varphi(t_0)$ weakly in X as $n \to +\infty$;
- $\varphi_n(\tau) \to \varphi_\tau$ strongly in X as $n \to +\infty$, then there exists $\varphi(\cdot) \in \mathscr{D}_{\tau,T}(\varphi_\tau)$ such that within a some subsequence the sequence $\varphi_n(\cdot) \to \varphi(\cdot)$ strongly in $C([\tau, T]; X)$ as $n \to +\infty$.

Let us define the m-semiflow $\mathscr{G}$ as

$$\mathscr{G}(t, \xi_0) = \{\xi(t) \ | \xi(\cdot) \in \mathscr{D}(\xi_0)\}, \quad t \ge 0.$$

Let $P(X)$ be the set of all nonempty subsets of X and $\beta(X)$ be the set of all nonempty bounded subsets of X. Note that the multivalued map $\mathscr{G} : \mathbf{R}_+ \times X \to P(X)$ is a strict m-semiflow (see for detail [15, p. 5]). Lemma 3 in [20] implies the existence of a Lyapunov-type function (see [21, p. 486]) for m-semiflow $\mathscr{G}$. Moreover, $\mathscr{V}$ is a

Lyapunov function for $\mathscr{G}$. The proof of Theorem 3 in [20] provides the asymptotic compactness of m-semiflow $\mathscr{G}$ generated by all weak solutions of Problem (16.2). Let $B_r(x)$ be a closed ball centered in $x \in X$ with the radius $r > 0$. Hence, from [19, Lemmas 2.4 and 2.6], the asymptotic compactness of $\mathscr{G}$ and the separability of considered Hilbert spaces, it follows that the next theorem holds.

Theorem 16.1 *[22, Theorem 1] Let the assumptions on the parameters of Problem (16.2) hold. Then the multivalued semiflow $\mathscr{G}$ satisfies the following condition: for every bounded set $B \subset X$ and $\varepsilon > 0$, there exist a moment of time $t_0(B, \varepsilon)$ and finite dimensional subspace E in X such that for some bounded projector $P : X \to E$, the set $P\left(\bigcup_{t \geq t_0} \mathscr{G}(t, B)\right)$ is bounded in X, and $(I - P)\left(\bigcup_{t \geq t_0} \mathscr{G}(t, B)\right) \subset B_\varepsilon(\bar{0})$, where I is an identically map in X.*

Theorem 16.1 justifies that dynamics of all weak solutions of studied problem is finitedimensional within a small parameter.

Moreover, the above results allow to obtain the existence of the global and trajectory attractors for multivalued semiflow $\mathscr{G}$, which describe the whole dynamics of investigated system.

Let $\mathscr{K}_+ = \cup_{u_0 \in X} \mathscr{D}(u_0)$ be the family of all weak solutions of inclusion (16.2) defined on $[0, +\infty)$. Remark that $\mathscr{K}_+$ is translation invariant one, i.e., for each $u(\cdot) \in \mathscr{K}_+$ and $h \geq 0$, $u_h(\cdot) \in \mathscr{K}_+$, where $u_h(s) = u(h + s)$, $s \geq 0$. Consider the translation semigroup $\{T(h)\}_{h \geq 0}$ on $\mathscr{K}_+$, $T(h)u(\cdot) = u_h(\cdot)$, where $u \in \mathscr{K}_+$. Note that $T(h)\mathscr{K}_+ \subset \mathscr{K}_+$ as $h \geq 0$. On $\mathscr{K}_+$ let us consider a topology induced from the Fre′chet space $C^{loc}(\mathbf{R}_+; X)$. By Π_M (Π_+) denote the restriction operator to the interval $[0, M]$ ($[0, +\infty)$). Note that $f_n(\cdot) \to f(\cdot)$ in $C^{loc}(\mathbf{R}_+; X)$ if and only if for each $M > 0$ $\Pi_M f_n(\cdot) \to \Pi_M f(\cdot)$ in $C([0, M]; X)$ [23, p. 179]. Note that the space $C^{loc}(\mathbf{R}; X)$ is endowed with the topology of local uniform convergence on each interval $[-M, M] \subset \mathbf{R}$ (cf. [23, p. 180]). By $\mathscr{K}$ denote a family of all complete trajectories of Problem (16.2) (see for detail [23, p. 180]). Note that for each $h \in \mathbf{R}$ and $u(\cdot) \in \mathscr{K}$, $u_h(\cdot) \in \mathscr{K}$. Define the set of the rest points of $\mathscr{G}$ as [21, p. 486]

$$Z(\mathscr{G}) = \{(\bar{0}, u) \mid u \in V, \;\; B(u) + \partial J_1(u) - \partial J_2(u) \ni \bar{0}\}.$$

The main assumptions on the parameters of Problem (16.2) provide that the set $Z(\mathscr{G})$ is bounded in X.

From [24, Theorem 2.7], the existence and properties of the weak solutions of Problem (16.2), the asymptotic compactness of multivalued semiflow $\mathscr{G}$ and the existence of a Lyapunov-type function for $\mathscr{G}$, it follows that multivalued semiflow $\mathscr{G}$ has an invariant compact in the phase space X global attractor $\mathscr{A}$ (see for detail [21, 25]). Note, that the structural properties of a global attractor $\mathscr{A}$ are the similar ones as in [20, Theorem 16.4].

Therefore, by [22], it follows that there exists a trajectory attractor (see [15, Definition 1.10]) in the extended phase space for investigated problem, i.e., the following theorem holds.

Theorem 16.2 *If there exists a global attractor $\mathscr{A}$ for the multivalued semiflow generated by all weak solutions of Problem (16.2), then there exists the trajectory attractor $\mathscr{U} \subset \mathscr{K}_+$ in $\mathscr{K}_+$. Herewith, the next relation holds:*

$$\mathscr{U} = \Pi_+\mathscr{K} = \Pi_+\{u(\cdot) \in \mathscr{K} \mid u(t) \in \mathscr{A} \ \ \forall t \in \mathbf{R}\}. \tag{16.3}$$

Proof For any $t \geq 0$ consider the multivalued semiflow $\mathscr{G}$ generated by weak solutions of Problem (16.2). As mentioned above, there exists the global attractor $\mathscr{A}$ for Problem (16.2). Let us show that condition of Theorem 1.12 in [15] holds, i.e.: for each sequence $\{\varphi_n(\cdot)\}_{n\geq 1} \subset \mathscr{K}_+$, that satisfies $\varphi_n(0) \to \varphi_0 \in \mathscr{A}$ in X, there exists $\varphi(\cdot) \in \mathscr{K}_+$, $\varphi(0) = \varphi_0$ such that within a some subsequence the sequence $\varphi_n(t) \to \varphi(t)$ in X for all $t \geq 0$. Reference [20, Theorem 2] implies that this condition holds for arbitrary sequence of weak solutions of Problem (16.2) defined on $[\tau, T]$ such that $\varphi_n(\tau) \to \varphi(\tau)$ in X. Therefore, we can choose the subsequence $\{\varphi_{n,1}(\cdot)\}_{n\geq 1} \subset \{\varphi_n(\cdot)\}_{n\geq 1}$ of weak solutions of Problem (16.2) defined on $[0, 1]$, $\varphi_{n,1}(0) \to \varphi_0$, such that there exists $\widehat{\varphi}_1(\cdot) \in \mathscr{K}_+$: $\widehat{\varphi}_1(0) = \varphi_0$ and $\varphi_{n,1}(t) \to \widehat{\varphi}_1(t)$ in X. Then let us choose the subsequence $\{\varphi_{n,2}(\cdot)\}_{n\geq 1} \subset \{\varphi_{n,1}(\cdot)\}_{n\geq 1}$ of weak solutions of Problem (16.2) defined on $[0, 2]$, $\varphi_{n,2}(0) \to \varphi_0$. Reference [20, Theorem 2] implies that there exists $\widehat{\varphi}_2(\cdot) \in \mathscr{K}_+$: $\widehat{\varphi}_2(0) = \varphi_0$ such that $\varphi_{n,2}(t) \to \widehat{\varphi}_2(t)$ in X for all $t \in [0, 2]$. On the other hand, $\varphi_{n,2}(t) \to \widehat{\varphi}_1(t)$ in X for all $t \in [0, 1]$. Therefore, $\widehat{\varphi}_1(t) = \widehat{\varphi}_2(t)$ for all $t \in [0, 1]$. Similarly let us choose the subsequences $\{\varphi_{n,k}(\cdot)\}_{n\geq 1} \subset \{\varphi_{n,k-1}(\cdot)\}_{n\geq 1} \subset \ldots \subset \{\varphi_{n,1}(\cdot)\}_{n\geq 1} \subset \{\varphi_n(\cdot)\}_{n\geq 1}$, $k = 1, 2, \ldots$, of weak solutions of Problem (16.2) defined on $[0, +\infty)$, $\varphi_{n,k}(0) \to \varphi_0$ such that $\varphi_{n,k}(t) \to \varphi(t)$ in X, where $\varphi(\cdot) \in \mathscr{K}_+$ and $\varphi(0) = \varphi_0$. Cantor diagonal method provides a choice of the necessary subsequence $\{\varphi_{n_k}\}_{k\geq 1} : \varphi_{n_k}(t) \to \varphi(t)$ for $t \geq 0$.

Hence, there exists a trajectory attractor $\mathscr{U} \in \mathscr{K}_+$. Moreover, relation (16.3), which provides the relationship between a trajectory attractor, a global attractor, and the space of complete trajectories for investigated problem, holds.

Therefore, by Theorem 16.1, it follows that solution dynamics of fields, described by autonomous second-order differential inclusion of hyperbolic type with control law in the form (16.1), is finite dimension within a small parameter. Moreover, the existence of a Lyapunov-type function, the global and trajectory attractors for the investigated problem allow us to construct the functionals $J_i(\cdot)$, $i = 1, 2$, which satisfy the main assumptions on the parameters of Problem (16.2), to achieve the desired behavior in some cases of controlled system (16.2). Therefore, we can direct the solutions of the studied problem to the set of stationary states.

Example 16.1 Consider a mathematical model which describes the contact between a piezoelectric body and a foundation [20]. Let $\mathbf{R}^d$ be a d-dimensional real linear space and $\mathbf{S}^d$ be the linear space of second-order symmetric tensors on $\mathbf{R}^d$ with the inner product $\sigma{:}\tau{=}\sum_{ij} \sigma_{ij}\tau_{ij}$ and the corresponding norm $\| \ \tau \|^2_{\mathbf{S}^d}{=}\tau{:}\tau$, σ_{ij}, $\tau_{ij} \in \mathbf{S}^d$.

Consider a plane electro-elastic material which in its undeformed state occupies an open a bounded domain $\Omega \subset \mathbf{R}^d$, $d{=}\ 2$. This domain as a result of volume forces and boundary friction can contact with rigid or elastic support. Let the boundary of piezoelectric body Ω be Lipschitz continuous. Assume that the boundary Γ, on the one hand, consists of two disjoint measurable parts Γ_D and Γ_N, $m(\Gamma_D) > 0$, and, on the other hand, consists of two disjoint measurable parts Γ_a and Γ_b, $m(\Gamma_a) > 0$. Suppose that the body is clamped on Γ_D, so the displacement field $u:Q \to \mathbf{R}^d$, $u=u(x,t)$, where $Q=\Omega \times (0,+\infty)$, vanishes there. Moreover, a surface traction of density g act on Γ_N, and the electric potential $\varphi:\Omega \to \mathbf{R}$ vanishes on Γ_a. The body Ω is lying on "support" medium, which introduces frictional effects. The interaction between the body and the support is described, due to the adhesion or skin friction, by a nonmonotone possibly multivalued law between the bonding forces and the corresponding displacements.

The body forces of density f consist of force f_e which is prescribed external loading and force f_s which is the reaction of constrains introducing the skin effects, i.e., $f=f_e+f_s$. Here f_s is a possibly multivalued function of the displacement u.

To describe the contact between a piezoelectric body Ω and a foundation let us consider the basic piezoelectric equations: equation of motion, equilibrium equation, strain-displacement equation, equation of electric field-potential and other constitutive relations [16].

We suppose that the process is dynamic. Let us set the constant mass density $\rho = 1$. Then we have the equation of motion for the stress field and the equilibrium equation for the electric displacement field, respectively:

$$u_{tt} - Div\sigma = f - \gamma u_t \quad in \quad Q, \tag{16.4}$$

$$divD = 0 \quad in \quad Q,$$

where $\gamma \in L^\infty(\Omega)$ is the nonnegative function of viscosity; $\sigma:Q \to \mathbf{S}^d$, $\sigma=(\sigma_{ij})$ is the stress tensor; $D:\Omega \to \mathbf{R}^d$, $D=(D_i)$, $i,j=1,2$ is the electric displacement field; $Div\sigma=(\sigma_{ij,j})$ is the divergence operator for tensor-valued functions; $divD=(D_{i,i})$ is the divergence operator for vector valued. Equation (16.4) regulates the change in time of the mechanical state of the piezoelectric body.

The stress-charge form of piezoelectric constitutive relations describes the behavior of the material and are following:

$$\sigma = \mathscr{A}\varepsilon(u) - \mathscr{P}^{\mathrm{T}}E(\varphi) \quad in \quad Q,$$

$$D = \mathscr{P}\varepsilon(u) + \mathscr{B}E(\varphi) \quad in \quad Q,$$

where $\mathscr{A}:\Omega \times \mathbf{S}^d \to \mathbf{S}^d$ is a linear elasticity operator with the elasticity tensor $a=(a_{ijkl})$, $i,j,k,l=1,2$; $\mathscr{P}:\Omega \times \mathbf{S}^d \to \mathbf{R}^d$ is a linear piezoelectric operator represented by the piezoelectric coefficients $p=(p_{ijk})$, $i,j,k=1,2$; $\mathscr{P}^{\mathrm{T}}:\Omega \times \mathbf{R}^d \to \mathbf{S}^d$ is transpose to $\mathscr{P}$ operator represented by $p^{\mathrm{T}}=(p^{\mathrm{T}}_{ijk})=(p_{kij})$, $i,j,k=1,2$; $\mathscr{B}:\Omega \times \mathbf{R}^d \to \mathbf{R}^d$ is a linear electric permittivity operator with the dielectric constants $\beta= (\beta_{ij}), i, j= 1, 2$; $\varepsilon(u) = (\varepsilon_{ij}(u)), i, j= 1, 2$ is the linear strain tensor; $E(\varphi) = (E_i(\varphi)), i= 1, 2$ the electric vector field.

The elastic strain–displacement and electric field-potential relations are given by

$$\varepsilon(u) = 1/2(\nabla u+(\nabla u)^{\mathrm{T}}) \quad in \quad Q,$$

$$E(\varphi) = -\nabla\varphi \quad in \quad Q.$$

We consider the reaction-displacement law of the form:

$$-f_s(x,t) \in \partial G_1(x, u(x,t)) - \partial G_2(x, u(x,t)) \quad in \quad Q,$$

where $G_i{:}\Omega\times\mathbf{R}^d \to \mathbf{R}, i= 1, 2$ are measurable in (x, u), convex in u for a.e. $x \in \Omega$ functionals; $\partial J_i(x,\cdot), i= 1, 2$ are their subdifferentials [1].

Let u_0 be the initial displacement and u_1 be the initial velocity. The classical formulation of the mechanical model can be stated as follows: find a displacement field u on $Q\times\mathbf{R}^d$ and an electric potential φ on $\Omega\times\mathbf{R}$ such that:

$$\begin{aligned} u_{tt} - Div\sigma &= f_e + f_s - \gamma u_t \text{ in } Q,\\ divD &= 0 \text{ in } Q,\\ \sigma &= \mathscr{A}\varepsilon(u) - \mathscr{P}^T E(\varphi) \text{ in } Q,\\ D &= \mathscr{P}\varepsilon(u) + \mathscr{B}E(\varphi) \text{ in } Q,\\ -\, & f_s(x,t) \in \partial G_1(x, u(x,t)) - \partial G_2(x, u(x,t)) \text{ in } Q,\\ u &= 0 \text{ on } \Gamma_D\times(0,T), \quad n = g \text{ on } \Gamma_N\times(0,T),\\ \varphi &= 0 \text{ on } \Gamma_a\times(0,T), \quad Dn = 0 \text{ on } \Gamma_b\times(0,T),\\ u(0) &= u_0, \quad u_t(0) = u_1, \end{aligned} \tag{16.5}$$

where n denotes the outward unit normal to Γ.

We now turn to the variational formulation of Problem (16.5). Let us consider the space $V = \{v \in H^1(\Omega;\mathbf{R}^d) : v = 0 \text{ on } \Gamma_D\} \subset H^1(\Omega;\mathbf{R}^d)$. Let $H = L^2(\Omega;\mathbf{R}^d)$, $\mathscr{H} = (\Omega;\mathbf{R}^d)$ be Hilbert spaces equipped with the inner products $\langle u,v\rangle_H = \int_\Omega uvdx$ and $\langle\sigma,\tau\rangle_{\mathscr{H}} = \int_\Omega \sigma : \tau dx$, respectively. Then (V, H, V^*) be an evolution triple of spaces. Then $\langle u,v\rangle_V = \langle\varepsilon(u),\varepsilon(v)\rangle_{\mathscr{H}}$, $\|v\|_V = \|\varepsilon(v)\|_{\mathscr{H}}$, $u, v \in V$ is the inner product and the corresponding norm on V. Therefore $(V, \|\cdot\|_V)$ is Hilbert space.

Assume that $G_i : \Omega\times\mathbf{R}^d \to \mathbf{R}$, $i = 1, 2$, satisfies standard Carathéodory's conditions and there exist $c^{(i)} \in L_1(\Omega)$ and $\alpha^{(i)} > 0$ and that $\|d^{(i)}\|_{\mathbf{R}^d} \leq c^{(i)}(x) + \alpha^{(i)}\|u\|^d_{\mathbf{R}}$ for a.e. $x \in \Omega$ and any $u \in \mathbf{R}^d$, $d^{(i)} \in \partial G_i(x,u)$. Moreover, α^2 is sufficiently small.

Let us set the following hypotheses for the constitutive tensors:

(i) $a = (a_{ijkl})$, $a_{ijkl} \in L^\infty(\Omega)$, $a_{ijkl} = a_{klij}$, $a_{ijkl} = a_{jikl}$, $a_{ijkl} = a_{ijlk}$, $a_{ijkl}(x)\tau_{ij}\tau_{kl} \geq \alpha\tau_{ij}\tau_{ij}$ for a.e. $x \in \Omega$, $\forall\tau = (\tau_{ij}) \in \mathbf{S}^d_+$, $\alpha > 0$,

(ii) $p = (p_{ijk}),\ p_{ijk} \in L^\infty(\Omega)$,
(iii) $\beta = (\beta_{ij}),\ \beta_{ij} = \beta_{ji} \in L^\infty(\Omega),\ \beta_{ij}(x)\zeta_i\zeta_j \geq m_\beta \|\zeta\|^2_{\mathbf{R}^d}$ for a.e. $x \in \Omega$, $\forall \zeta = (\zeta_i) \in \mathbf{R}^d,\ m_\beta > 0$.

Without loss of generality, let us consider $g \equiv 0$ and $f_e \equiv 0$. Hence, Problem (16.5) can be presented in the form (16.2) (see [16]).

Therefore, under the above listed assumptions on parameters of Problem (16.5), all statements of Theorems 16.1 and 16.2 hold for Problem (16.5). The results obtained in this paper for Problem (16.5) is more general than in [18].

References

1. Clarke, F.H.: Optimization and Nonsmooth Analysis. Wiley, New York (1983)
2. Moheimani, S.O.R., Fleming, A.J.: Piezoelectric Transducers for Vibration Control and Damping. Springer, Berlin (2006)
3. Jalili, N.: Piezoelectric-Based Vibration Control. From Macro to Micro/Nano Scale Systems. Springer, New York (2010)
4. Burns, J., King, B.: Representation of feedback operators for hyperbolic systems. Comput. Control IV. Prog. Syst. Control Theory **20**, 57–73 (1995)
5. Khalil, H.: Nonlinear Systems. Prentice Hall, New Jersey (2002)
6. Rowley, C., Batten, B.: Dynamic and closed-loop control. In: Joslin, R.D., Miller, D.N. (eds.) Fundamentals and Applications of Modern Flow Control, pp. 115–148. American Institute of Aeronautics and Astronautics, Reston (2009)
7. Naniewicz, Z., Panagiotopoulos, P.: Mathematical theory of hemivariational inequalities and applications. Nonconvex optimization and its applications. Pure and Applied Mathematics. A Series of Monographs and Textbooks. Marcel Dekker Inc., New York (1995)
8. Dem'yanov, V., Stavroulakis, G., Polyakova, L., Panagiotopoulos, P.: Quasidifferentiability and nonsmooth modeling in Mechanics, Engineering and Economics. Nonconvex Optimization and Its Applications. Kluwer Academic Publishers, Dordrecht (1996)
9. Panagiotopoulos, P.D., Koltsakis, E.K.: The nonmonotone skin effects in plane elasticity problems obeying to linear elastic and subdifferential laws. Zeitschrift fur Angewandte Mathematik und Mechanik. **70**(1), 13–21 (1990)
10. Panagiotopoulos, P.D.: Inequality Problems in Mechanics and Applications: Convex and Nonconvex Energy Functions. Birkhäuser, Boston (1985)
11. Kasyanov, P.O., Mel'nik, V.S., Toscano, S.: Solutions of Cauchy and periodic problems for evolution inclusions with multi-valued w_{λ_0}-pseudomonotone maps. J. Differ. Equ. **249**(6), 1258–1287 (2010)
12. Kasyanov, P.O.: Multivalued dynamic of solutions of autonomous differential-operator inclusion with pseudomonotone nonlinearity. Cybern. Syst. Anal. **47**(5), 800–811 (2011)
13. Zadoyanchuk, N.V., Kas'yanov, P.O.: Faedo Galerkin method forsecond-order evolution inclusions with W_λ-pseudomonotone mappings. Ukr. Math. J. **61**(2), 236–258 (2009)
14. Zadoyanchuk, N.V., Kasyanov, P.O.: Analysis and control of second-order differential-operator inclusions with +-coercive damping. Cybern. Syst. Anal. **46**(2), 305–313 (2010)
15. Zgurovsky, M.Z., Kasyanov, P.O., Kapustyan, O.V., Valero, J., Zadoianchuk, N.V.: Evolution Inclusions and Variation Inequalities for Earth Data Processing III. Long-Time Behavior of Evolution Inclusions Solutions in Earth Data Analysis. Advances in Mechanics and Mathematics. Springer, Berlin (2012)
16. Liu, Z., Migorski, S.: Noncoercive damping in dynamic hemivariational inequality with application to problem of piezoelectricity. Discret. Contin. Dyn. Syst. **9**(1), 129–143 (2008)

17. Zgurovsky, M.Z., Kasyanov, P.O., Zadoianchuk, N.V.: Long-time behavior of solutions for quasilinear hyperbolic hemivariational inequalities with application to piezoelectricity problem. Appl. Math. Lett. **25**, 1569–1574 (2012)
18. Gorban, N.V., Kapustyan, V.O., Kasyanov, P.O., Paliichuk, L.S.: On global attractors for autonomous damped wave equation with discontinuous nonlinearity. In: Zgurovsky, M.Z., Sadovnichiy, V.A. (eds.) Continuous and Distributed Systems: Theory and Applications. Solid Mechanics and Its Applications, vol. 211, pp. 221–237 (2014)
19. Kalita, P., Łukaszewicz, G.: Global attractors for multivalued semiflows with weak continuity properties. Nonlinear Anal. **101**, 124–143 (2014)
20. Zgurovsky, M.Z., Kasyanov, P.O., Paliichuk, L.S.: Automatic feedback control for one class of contact piezoelectric problems. Syst. Res. Inf. Technol. **1**, 56–68 (2014)
21. Ball, J.M.: Continuity properties and global attractors of generalized semiflows and the Navier-Stokes equations. J. Nonlinear Sci. **7**(5), 475–502 (1997)
22. Kasyanov, P.O., Paliichuk, L.S., Tkachuk, A.N.: Method of multivalued operator semigroup to investigate the long-termforecasts for controlled piezoelectric fields. Chebyshevskii Sb. **15**(2), 21–32 (2014)
23. Vishik, M., Chepyzhov, V.: Trajectory and global attractors of three-dimensional Navier-Stokes systems. Math. Notes **71**(1–2), 177–193 (2002)
24. Ball, J.M.: Global attractors for damped semilinear wave equations. DCDS **10**, 31–52 (2004)
25. Melnik, V.S., Valero, J.: On attractors of multivalued semi-flows and differential inclusions. Set-Valued Anal. **6**(1), 83–111 (1998)

Chapter 17
Minimax Estimates for Solutions of Parabolic-Hyperbolic Equations with Nonlocal Boundary Conditions

V.O. Kapustyan and I.O. Pyshnograiev

Abstract In the paper, we consider the minimax estimation problems for solutions of parabolic–hyperbolic equations with nonlocal boundary conditions for solutions. The feature of the boundary value problem is that under fixed perturbations its solutions depend on value of perturbation at fixed time point and these solutions must be classic. This leads to restrictions on unknown perturbations and they must belong to the class of absolutely continuous functions. We construct and substantiate boundary value problems for minimax estimates of solutions of parabolic-hyperbolic equations with nonlocal boundary conditions.

17.1 Introduction

The study of equations of mixed type is one of the most important trends in the theory of partial differential equations. Necessity of research boundary value problems for equations of mixed type is dictated by numerous practical applications in gas dynamics, the theory of infinitesimal bendings of surfaces, the theory of shells, magnetohydrodynamics, the theory of electron scattering, mathematical biology, and other fields [1, 2]. Some optimal control problems for parabolic–hyperbolic equations with nonlocal boundary conditions have been considered by the authors [3, 4].

Nowadays, problems for equations of this type are becoming more complicated. And this article discusses the problem of minimax estimates for parabolic–hyperbolic equations with distributed and divided observations.

V.O. Kapustyan · I.O. Pyshnograiev (✉)
NTUU "KPI", Prosp. Peremohy, 37, Kyiv 03056, Ukraine
e-mail: pyshnograiev@gmail.com

V.O. Kapustyan
e-mail: kapustyanv@ukr.net

© Springer International Publishing Switzerland 2015

V.A. Sadovnichiy and M.Z. Zgurovsky (eds.), *Continuous and Distributed Systems II*, Studies in Systems, Decision and Control 30,
DOI 10.1007/978-3-319-19075-4_17

17.2 The Problem with Distributed Observation

17.2.1 Problem Statement

Let the system is described by problem

$$\frac{dy_0(t)}{dt} = f_0(0) + \int_0^t \xi_0(\tau)d\tau, t > 0,$$
$$\frac{d^2y_0(t)}{dt^2} = f_{0,-}(t), t < 0,$$
$$y_0(-\alpha) = \varphi_0; \tag{17.1}$$

$$\frac{dy_{2k-1}(t)}{dt} + \lambda_k^2 y_{2k-1}(t) = f_{2k-1}(0) + \int_0^t \xi_{2k-1}(\tau)d\tau, t > 0,$$
$$\frac{d^2y_{2k-1}(t)}{dt^2} + \lambda_k^2\, y_{2k-1}(t) = f_{2k-1,-}(t), t < 0,$$
$$y_{2k-1}(-\alpha) = \varphi_{2k-1}, \lambda_k = 2k\pi, \tag{17.2}$$

$$\frac{dy_{2k}(t)}{dt} + \lambda_k^2 y_{2k}(t) = -2\lambda_k y_{2k-1}(t) + f_{2k}(0) + \int_0^t \xi_{2k}(\tau)d\tau, t > 0,$$
$$\frac{d^2y_{2k}(t)}{dt^2} + \lambda_k^2 y_{2k}(t) = -2\lambda_k y_{2k-1}(t) + f_{2k,-}(t), t < 0,$$
$$f_i(t) = f_i(0) + \int_0^t \xi_i(\tau)\, d\tau, i \geq 0,$$
$$y_{2k}(-\alpha) = \varphi_{2k}, k = 1, 2, ..., y_i(t) \in C^1(-\alpha, T), i \geq 0, \tag{17.3}$$

where

$$y_i(t) = (y(.,t), Y_i(.))_{L_2(0,1)},\ \varphi_i = (\varphi(.), Y_i(.))_{L_2(0,1)},$$
$$f_{i,-}(t) = (f_-(.,t), Y_i(.))_{L_2(0,1)}, f_i(t) = (f(.,t), Y_i(.))_{L_2(0,1)}, i \geq 0.$$

The system of differential equations of the such type considered in [4]. Observation is as follows:

$$\theta_i(t) = y_i(t) + F_i(t), t \in [-\alpha, T], i \geq 0, \tag{17.4}$$

where

$$F_i(t) = (F(.,t), Y_i(.))_{L_2(0,1)}.$$

Perturbations, which influence on the system (17.1)–(17.4), are constrained by conditions

$$\sum_{i=0}^{\infty}[\hat{\alpha}\,\varphi_i^2 + \hat{\beta} f_i^2(0) + \hat{\gamma}\int_{-\alpha}^{0} f_{i,-}^2(t)dt +$$

$$+\hat{\mu}\int_{0}^{T}\xi_i^2(t)dt + \hat{\nu}\int_{-\alpha}^{T}F_i^2(t)dt] \le 1, \qquad (17.5)$$

where $\hat{\alpha}, \hat{\beta}, \hat{\gamma}, \hat{\mu}, \ \hat{\nu} > 0$.

On the problem's solution (17.1)–(17.3), we define the linear functional

$$l(y) = \sum_{i=0}^{\infty}\int_{-\alpha}^{T} q_i(t)y_i(t)dt, \qquad (17.6)$$

where $q_i(t) = (q(.,t), X_i(.))_{L_2(0,1)}$.
Estimate of the functional (17.6) we will find in the class of linear estimates of the form

$$\hat{\hat{l}}(y) = -\sum_{i=0}^{\infty}\int_{-\alpha}^{T}\hat{u}_i(t)\theta_i(t)dt, \qquad (17.7)$$

where $\hat{u}_i(t) = (\hat{u}(.,t), X_i(.))_{L_2(0,1)}$.
The estimation of $\hat{l}(y)$ from class (17.7) will be called apriory minimax estimation [5], if it satisfies the equality

$$\sigma^2 = |l(y) - \hat{l}(y)|^2 = \inf_{\hat{u}}\sup_{\varphi,f,F}|l(y) - \hat{\hat{l}}(y)|^2. \qquad (17.8)$$

17.2.2 Formal Solution of the Problem

Let us reduce the problem of minimax estimation (17.1)–(17.8) to some problem of optimal control. For this purpose, we find the representation of the linear form $l(y) - \hat{\hat{l}}(y)$ in the form with unknown disturbances.

$$\Sigma(y) = l(y) - \hat{\hat{l}}(y) = \sum_{i=0}^{\infty} \int_{-\alpha}^{T} (q_i(t)\; y_i(t) + \hat{u}_i(t)(y_i(t) + F_i(t)))dt \; +$$

$$+ \int_{-\alpha}^{0} \psi_0(t)\Big(\frac{d^2 y_0(t)}{dt^2} - f_{0,-}(t)\Big)dt + \int_0^T \psi_0(t)\left(\frac{dy_0(t)}{dt} - f_0(0) - \int_0^t \xi_0(\tau)d\tau\right)dt +$$

$$+\sum_{i=1}^{\infty}\Bigg[\int_{-\alpha}^{0} \psi_{2i-1}(t)\Big(\frac{d^2 y_{2i-1}(t)}{dt^2} + \lambda_i^2 y_{2i-1}(t) - f_{2i-1,-}(t)\Big)dt +$$

$$+ \int_0^T \psi_{2i-1}(t)\left(\frac{dy_{2i-1}(t)}{dt} + \lambda_i^2 y_{2i-1}(t) - f_{2i-1}(0) - \int_0^t \xi_{2i-1}(\tau)d\tau\right) dt +$$

$$+ \int_{-\alpha}^{0} \psi_{2i}(t)\Big(\frac{d^2 y_{2i}(t)}{dt^2} + \lambda_i^2 y_{2i}(t) + 2\,\lambda_i\, y_{2i-1}(t) - f_{2i,-}(t)\Big)dt +$$

$$+ \int_0^T \psi_{2i}(t)\Big(\frac{dy_{2i}(t)}{dt} + \lambda_i^2 y_{2i}(t) + 2\,\lambda_i\, y_{2i-1}(t) - f_{2i}(0) - \int_0^t \xi_{2i}(\tau)d\tau\Big)dt\Bigg], \tag{17.9}$$

where $\psi_k(t) = (\psi(.,t), X_k(.))_{L_2(0,1)}$.

We integrate the representation (17.9) by parts.

$$\sum_{i=0}^{\infty}\Big(\int_{-\alpha}^{0} \psi_i(t)\frac{d^2 y_i(t)}{dt^2}dt + \int_0^T \psi_i(t)\frac{dy_i(t)}{dt}dt\Big) =$$

$$= -\Big(\frac{d\psi_0(0-)}{dt} + \psi_0(0+)\Big)\; y_0(0+) - \sum_{i=1}^{\infty}\Bigg(\Big(\frac{d\psi_{2i-1}(0-)}{dt} + (1+\lambda_i^2)\;\psi_{2i-1}(0+) +$$

$$+2\lambda_i\,\psi_{2i}(0+)\Big)y_{2i-1}(0+) + \Big(\frac{d\psi_{2i}(0-)}{dt} + (1+\lambda_i^2)\;\psi_{2i}(0+)\Big)y_{2i}(0+)\Bigg) +$$

$$+\sum_{i=0}^{\infty}\Bigg(\int_{-\alpha}^{0} \frac{d^2\psi_i(t)}{dt^2}y_i(t)dt - \int_0^T \frac{d\psi_i(t)}{dt}y_i(t)dt -$$

$$-\psi_i(-\alpha)\frac{dy_i(-\alpha)}{dt} + \frac{d\psi_i(-\alpha)}{dt}\varphi_i + \psi_i(T)\; y_i(T) + \psi_i(0-)\, f_i(0)\Bigg).$$

We define the functions $\psi_i(t)$ as solutions of boundary value problems

$$\frac{d\psi_0(t)}{dt} = q_0(t) + u_0(t),\ t > 0,$$
$$\frac{d^2\psi_0(t)}{dt^2} = -q_0(t) - v_0(t),\ t < 0,$$
$$\psi_0(-\alpha) = 0,\ \frac{d\psi_0(0-)}{dt} + \psi_0(0+) = 0,\ \psi_0(T) = 0; \tag{17.10}$$

$$\frac{d\psi_{2i-1}(t)}{dt} - \lambda_i^2\ \psi_{2i-1}(t) - 2\lambda_i\psi_{2i}(t) = q_{2i-1}(t) + u_{2i-1}(t),\ t > 0,$$
$$\frac{d^2\psi_{2i-1}(t)}{dt^2} + \lambda_i^2\ \psi_{2i-1}(t) + 2\lambda_i\psi_{2i}(t) = -q_{2i-1}(t) - v_{2i-1}(t),\ t < 0,$$
$$\psi_{2i-1}(-\alpha) = 0,$$
$$\frac{d\psi_{2i-1}(0-)}{dt} + (1 + \lambda_i^2)\psi_{2i-1}(0+) + 2\lambda_i\ \psi_{2i}(0+) = 0,\ \psi_{2i-1}(T) = 0; \tag{17.11}$$

$$\frac{d\psi_{2i}(t)}{dt} - \lambda_i^2\ \psi_{2i}(t) = q_{2i}(t) + u_{2i}(t),\ t > 0,$$
$$\frac{d^2\psi_{2i}(t)}{dt^2} + \lambda_i^2\ \psi_{2i}(t) = -q_{2i}(t) - v_{2i}(t),\ t < 0,$$
$$\psi_{2i}(-\alpha) = 0,$$
$$\frac{d\psi_{2i}(0-)}{dt} + (1 + \lambda_i^2)\psi_{2i}(0+) = 0,\ \psi_{2i}(T) = 0. \tag{17.12}$$

Then linear form (17.9) takes the following view

$$\Sigma(y) = \sum_{i=0}^{\infty}\Bigg[\int_{-\alpha}^{T}\hat{u}_i(t)F_i(t)dt - \int_{-\alpha}^{0}\psi_i(t)f_{i,-}(t)dt + \Bigg(\psi_i(0-) - \int_0^T\psi_i(t)dt\Bigg)f_i(0) - \int_0^T\Bigg(\int_t^T\psi_i(\tau)d\tau\Bigg)\xi_i(t)dt + \frac{d\psi_i(-\alpha)}{dt}\varphi_i\Bigg], \tag{17.13}$$

and the problem of minimax estimation reduces to the optimal control problem [5]: to minimize the quality criterion

$$J(u) = \sum_{i=0}^{\infty}\Bigg[\hat{\nu}^{-1}\int_{-\alpha}^{T}\hat{u}_i^2(t)dt + \hat{\gamma}^{-1}\int_{-\alpha}^{0}\psi_i^2(t)\,dt + \hat{\beta}^{-1}\Bigg(\psi_i(0-) - \int_0^T\psi_i(t)dt\Bigg)^2 + \hat{\mu}^{-1}\int_0^T\Bigg(\int_t^T\psi_i(\tau)d\tau\Bigg)^2dt + \hat{\alpha}^{-1}\Bigg(\frac{d\psi_i(-\alpha)}{dt}\Bigg)^2\Bigg], \tag{17.14}$$

which is defined on the solution of the boundary value problem (17.10)–(17.12).

The above stated infinite dimensional optimal control problem is equivalent to a sequence of finite dimensional problems:

(1) to find the minimum of the functional

$$J_0(\hat{u}_0) = \hat{\nu}^{-1} \int_{-\alpha}^{T} \hat{u}_0^2(t)dt + \hat{\gamma}^{-1} \int_{-\alpha}^{0} \psi_0^2(t)dt + \hat{\beta}^{-1}(\psi_0(0-) - \\ - \int_0^T \psi_0(t)dt)^2 + \hat{\mu}^{-1} \int_0^T \int_0^T \mathscr{K}(t,\tau)\psi_0(\tau)\psi_0(t)d\tau dt + \hat{\alpha}^{-1}\Big(\frac{d\psi_0(-\alpha)}{dt}\Big)^2, \tag{17.15}$$

on solutions of problem (17.10);
(2) to find the minimum of the functional

$$J_k(\hat{u}_{2k-1}, \hat{u}_{2k}) = \sum_{i=2k-1}^{2k} \Bigg[\hat{\nu}^{-1} \int_{-\alpha}^{T} \hat{u}_i^2(t)dt + \hat{\gamma}^{-1} \int_{-\alpha}^{0} \psi_i^2(t)dt + \hat{\beta}^{-1}\Bigg(\psi_i(0-) - \\ - \int_0^T \psi_i(t)dt\Bigg)^2 + \hat{\mu}^{-1} \int_0^T \int_0^T \mathscr{K}(t,\tau)\psi_i(\tau)\psi_i(t)d\tau dt + \hat{\alpha}^{-1}\Big(\frac{d\psi_i(-\alpha)}{dt}\Big)^2\Bigg], \tag{17.16}$$

on solutions of problem (17.11)–(17.12), where

$$\mathscr{K}(t,\tau) = \begin{cases} \tau, \ \mathrm{t} \geq \tau, \\ t, \ \mathrm{t} < \tau. \end{cases} \tag{17.17}$$

From necessary and sufficient conditions of optimality for the problem (17.15), (17.10), we find

$$\begin{aligned} v_0(t) &= -\hat{\nu} p_0(t), t < 0, \\ u_0(t) &= -\hat{\nu} p_0(t), t > 0, \end{aligned} \tag{17.18}$$

where function $p_0(t), t \in [-\alpha, T]$ is a solution of the boundary value problem

$$\begin{aligned} \frac{d\psi_0(t)}{dt} &= q_0(t) - \hat{\nu}\ p_0(t), t > 0, \\ \frac{d^2\psi_0(t)}{dt^2} &= -q_0(t) + \hat{\nu} p_0(t), t < 0, \\ \psi_0(-\alpha) = 0, \frac{d\psi_0(0-)}{dt} &+ \psi_0(0+) = 0, \psi_0(T) = 0; \end{aligned} \tag{17.19}$$

$$\frac{dp_0(t)}{dt} = \hat{\beta}^{-1}\left(\psi_0(0-) - \int_0^T \psi_0(t)dt\right) - \hat{\mu}^{-1}\int_0^T \mathscr{K}(t,\tau)\psi_0(\tau)d\tau,\ t > 0,$$
$$\frac{d^2 p_0(t)}{dt^2} = -\hat{\gamma}^{-1}\,\psi_0(t), t < 0,$$
$$p_0(0+) = p_0(0-),\ \frac{dp_0(0-)}{dt} = \hat{\beta}^{-1}\left(\psi_0(0-) - \int_0^T \psi_0(t)dt\right),$$
$$p_0(-\alpha) = \hat{\alpha}^{-1}\frac{d\psi_0(-\alpha)}{dt}. \tag{17.20}$$

From necessary and sufficient conditions of optimality for the problem (17.11), (17.12) and (17.16), we find

$$v_i(t) = -\hat{v}\,p_i(t), t < 0,$$
$$u_i(t) = -\hat{v}\,p_i(t), t > 0, \tag{17.21}$$

where functions $p_i(t), t \in [-\alpha, T], i = \overline{2k-1, 2k}$ are a solution of the boundary value problem

$$\frac{d\psi_{2k-1}(t)}{dt} - \lambda_k^2\,\psi_{2k-1}(t) - 2\lambda_k\psi_{2k}(t) = q_{2k-1}(t) - \hat{v}\,p_{2k-1}(t), t > 0,$$
$$\frac{d^2\psi_{2k-1}(t)}{dt^2} + \lambda_k^2\,\psi_{2k-1}(t) + 2\lambda_k\psi_{2k}(t) = -\,q_{2k-1}(t) + \hat{v}\,p_{2k-1}(t), t < 0,$$
$$\psi_{2k-1}(-\alpha) = 0,$$
$$\frac{d\psi_{2k-1}(0-)}{dt} + (1+\lambda_k^2)\psi_{2k-1}(0+) + 2\lambda_k\,\psi_{2k}(0+) = 0,\ \psi_{2k-1}(T) = 0; \tag{17.22}$$

$$\frac{d\psi_{2k}(t)}{dt} - \lambda_k^2\,\psi_{2k}(t) = q_{2k}(t) - \hat{v}\,p_{2k}(t), t > 0,$$
$$\frac{d^2\psi_{2k}(t)}{dt^2} + \lambda_k^2\,\psi_{2k}(t) = -q_{2k}(t) + \hat{v}\,p_{2k}(t), t < 0,$$
$$\psi_{2k}(-\alpha) = 0,$$
$$\frac{d\psi_{2k}(0-)}{dt} + (1+\lambda_k^2)\psi_{2k}(0+) = 0,\ \psi_{2k}(T) = 0; \tag{17.23}$$

$$\frac{dp_{2k-1}(t)}{dt} + \lambda_k^2 p_{2k-1}(t) = \hat{\beta}^{-1}\left(\psi_{2k-1}(0-) - \int_0^T \psi_{2k-1}(t)dt\right) -$$
$$-\hat{\mu}^{-1}\int_0^T \mathcal{K}(t,\tau)\,\psi_{2k-1}(\tau)d\tau, t > 0,$$
$$\frac{d^2 p_{2k-1}(t)}{dt^2} + \lambda_k^2\, p_{2k-1}(t) = -\hat{\gamma}^{-1}\psi_{2k-1}(t), t < 0,$$
$$p_{2k-1}(0+) = (1+\lambda_k^2)p_{2k-1}(0-),\ \frac{dp_{2k-1}(0-)}{dt} = \hat{\beta}^{-1}\left(\psi_{2k-1}(0-) -\right.$$
$$\left.-\int_0^T \psi_{2k-1}(t)dt\right),\ p_{2k-1}(-\alpha) = \hat{\alpha}^{-1}\frac{d\psi_{2k-1}(-\alpha)}{dt}; \tag{17.24}$$

$$\frac{dp_{2k}(t)}{dt} + \lambda_k^2 p_{2k}(t) + 2\lambda_k p_{2k-1}(t) = \hat{\beta}^{-1}\left(\psi_{2k}(0-) - \int_0^T \psi_{2k}(t)dt\right) -$$
$$-\hat{\mu}^{-1}\int_0^T \mathcal{K}(t,\tau)\,\psi_{2k}(\tau)d\tau, t > 0,$$
$$\frac{d^2 p_{2k}(t)}{dt^2} + \lambda_k^2 p_{2k}(t) + 2\lambda_k p_{2k-1}(t) = -\hat{\gamma}^{-1}\,\psi_{2k}(t), t < 0,$$
$$p_{2k}(0+) = (1+\lambda_k^2)p_{2k}(0-) + 2\,\lambda_k p_{2k-1}(0-),\ \frac{dp_{2k}(0-)}{dt} =$$
$$= \hat{\beta}^{-1}\left(\psi_{2k}(0-) - \int_0^T \psi_{2k}(t)dt\right),\ p_{2k}(-\alpha) = \hat{\alpha}^{-1}\frac{d\psi_{2k}(-\alpha)}{dt}. \tag{17.25}$$

Functionals (17.6)–(17.8) can be formally represented as

$$l(y) = l_0(y_0) + \sum_{k=1}^{\infty} l_k(y_{2k-1}, y_{2k}),$$
$$\hat{\hat{l}}(y) = \hat{\hat{l}}_0(y_0) + \sum_{k=1}^{\infty} \hat{\hat{l}}_k(y_{2k-1}, y_{2k}),$$
$$\sigma^2 = J_0(\hat{u}_0) + \sum_{k=1}^{\infty} J_k(\hat{u}_{2k-1}, \hat{u}_{2k}), \tag{17.26}$$

where

$$l_0(y_0) = \int_{-\alpha}^{T} q_0(t) y_0(t) dt, \hat{\hat{l}}_0(y_0) = -\int_{-\alpha}^{T} \hat{u}_0(t) \theta_0(t) dt,$$

$$l_k(y_{2k-1}, y_{2k}) = \int_{-\alpha}^{T} (q_{2k-1}(t) y_{2k-1}(t) + q_{2k}(t) y_{2k}(t)) dt,$$

$$\hat{\hat{l}}_k(y_{2k-1}, y_{2k}) = -\int_{-\alpha}^{T} (\hat{u}_{2k-1}(t) \theta_{2k-1}(t) + \hat{u}_{2k}(t) \theta_{2k}(t)) dt,$$

functionals $J_0(.)$, $J_k(.,.)$ are given by (17.15)–(17.16).

We will show that we have the representation

$$J_0(\hat{u}_0) = l_0(p_0). \tag{17.27}$$

Indeed,

$$\hat{\gamma}^{-1} \int_{-\alpha}^{0} \psi_0^2(t)\, dt + \hat{\mu}^{-1} \int_0^T \int_0^T \mathscr{K}(t,\tau) \psi_0(\tau) \psi_0(t)\, d\tau dt = -\int_{-\alpha}^{0} \frac{d^2 p_0(t)}{dt^2} \psi_0(t) dt +$$

$$+ \int_0^T \Big(-\frac{dp_0(t)}{dt} + \frac{dp_0(0-)}{dt} \Big) \psi_0(t) dt = -\frac{dp_0(t)}{dt} \psi_0(t)|_{-\alpha}^{0} + \frac{d\psi_0(t)}{dt} p_0(t)|_{-\alpha}^{0} -$$

$$-\psi_0(t) p_0(t)|_0^T + \int_{-\alpha}^{0} q_0(t) p_0(t) dt - \hat{\mathrm{v}}^{-1} \int_{-\alpha}^{0} v_0^2(t) dt + \int_0^T q_0(t) p_0(t) dt -$$

$$-\hat{\mathrm{v}}^{-1} \int_0^T u_0^2(t) dt + \frac{dp_0(0-)}{dt} \int_0^T \psi_0(t) dt = -\hat{\alpha}^{-1} \Big(\frac{d\psi_0(-\alpha)}{dt} \Big)^2 - \hat{\beta}^{-1} (\psi_0(0-) -$$

$$- \int_0^T \psi_0(t) dt)^2 - \hat{\mathrm{v}}^{-1} \int_{-\alpha}^{T} \hat{u}_0^2(t) dt + l(p_0).$$

In addition, we have the representation

$$\hat{l}_0(y_0) = l_0(\hat{\vartheta}_0), \tag{17.28}$$

if function $\hat{\vartheta}_0(t), t \in [-\alpha, T]$ is a solution of the boundary value problem

$$\frac{d\hat{p}_0(t)}{dt} = \hat{\nu}(\theta_0(t) - \hat{\vartheta}_0(t)), t > 0,$$
$$\frac{d^2\hat{p}_0(t)}{dt^2} = -\hat{\nu}\ (\theta_0(t) - \hat{\vartheta}_0(t)), t < 0,$$
$$\hat{p}_0(-\alpha) = 0, \frac{d\hat{p}_0(0-)}{dt} + \hat{p}_0(0+) = 0, \hat{p}_0(T) = 0; \tag{17.29}$$

$$\frac{d\hat{\vartheta}_0(t)}{dt} = \hat{\beta}^{-1}(\hat{p}_0(0-) - \int_0^T \hat{p}_0(t)dt) - \hat{\mu}^{-1}\int_0^T \mathscr{K}(t,\tau)\hat{p}_0(\tau)d\tau,\ t > 0,$$
$$\frac{d^2\hat{\vartheta}_0(t)}{dt^2} = -\hat{\gamma}^{-1}\ \hat{p}_0(t), t < 0,$$
$$\hat{\vartheta}_0(0+) = \hat{\vartheta}_0(0-), \frac{d\hat{\vartheta}_0(0-)}{dt} = \hat{\beta}^{-1}(\hat{p}_0(0-) - \int_0^T \hat{p}_0(t)dt),$$
$$\hat{\vartheta}_0(-\alpha) = \hat{\alpha}^{-1}\frac{d\hat{p}_0(-\alpha)}{dt}. \tag{17.30}$$

Indeed,

$$\hat{l}_0(y_0) = \hat{\nu}\int_{-\alpha}^T p_0(t)\theta_0(t)dt = \int_0^T p_0(t)\Big(\frac{d\hat{p}_0(t)}{dt} + \hat{\nu}\hat{\vartheta}(t)\Big)dt - \int_{-\alpha}^0 p_0(t)\times$$
$$\times\Big(\frac{d^2\hat{p}_0(t)}{dt^2} - \hat{\nu}\hat{\vartheta}(t)\Big)dt = p_0(t)\hat{p}_0(t)|_0^T - \hat{\beta}^{-1}\Big(\psi_0(0-) - \int_0^T \psi_0(t)dt\Big)\int_0^T \hat{p}_0(t)dt +$$
$$+\hat{\mu}^{-1}\int_0^T\int_0^T \mathscr{K}(t,\tau)\psi_0(\tau)\hat{p}_0(t)d\tau dt - p_0(t)\frac{d\hat{p}_0(t)}{dt}|_{-\alpha}^0 + \frac{dp_0(t)}{dt}\hat{p}_0(t)|_{-\alpha}^0 +$$
$$+\hat{\gamma}^{-1}\int_{-\alpha}^0 \psi_0(t)\hat{p}_0(t)dt + \hat{\nu}\int_{-\alpha}^T p_0(t)\hat{\vartheta}(t)dt = \hat{\alpha}^{-1}\frac{d\psi_0(-\alpha)}{dt}\ \frac{d\hat{p}_0(-\alpha)}{dt} +$$
$$+\hat{\beta}^{-1}\Big(\psi_0(0-) - \int_0^T \psi_0(t)dt)(\hat{p}_0(0-) - \int_0^T \hat{p}_0(t)dt\Big) +$$

$$+\hat{\mu}^{-1}\int_0^T\int_0^T \mathcal{K}(t,\tau)\psi_0(\tau)\,\hat{p}_0(t)d\tau dt+\hat{\gamma}^{-1}\int_{-\alpha}^{0}\psi_0(t)\,\hat{p}_0(t)dt+\hat{\nu}\int_{-\alpha}^{T}p_0(t)\hat{\vartheta}(t)dt=$$

$$=\hat{\alpha}^{-1}\frac{d\psi_0(-\alpha)}{dt}\frac{d\hat{p}_0(-\alpha)}{dt}+\hat{\beta}^{-1}\Big(\psi_0(0-)\;-\int_0^T\psi_0(t)dt\Big)\Big(\hat{p}_0(0-)-\int_0^T\hat{p}_0(t)dt\Big)+$$

$$+\hat{\beta}^{-1}\Big(\hat{p}_0(0-)-\int_0^T\hat{p}_0(t)dt\Big)\int_0^T\psi_0(t)dt-\hat{\vartheta}_0(t)\psi_0(t)|_0^T+\int_0^T q_0(t)\hat{\vartheta}_0(t)dt-$$

$$-\hat{\nu}\int_0^T p_0(t)\,\hat{\vartheta}_0(t)dt-\psi_0(t)\frac{d\hat{\vartheta}_0(t)}{dt}|_{-\alpha}^{0-}+\frac{d\psi_0(t)}{dt}\hat{\vartheta}_0(t)|_{-\alpha}^{0-}+\int_{-\alpha}^{0}q_0(t)\hat{\vartheta}_0(t)dt\;-$$

$$-\hat{\nu}\int_{-\alpha}^{0}p_0(t)\,\hat{\vartheta}_0(t)dt+\hat{\nu}\int_{-\alpha}^{T}p_0(t)\hat{\vartheta}_0(t)dt=l_0(\hat{\vartheta}_0).$$

Similarly, we can show that the following representations are correct

$$J_k(\hat{u}_{2k-1},\hat{u}_{2k})=\;l_k(p_{2k-1},p_{2k}),\hat{l}_k(y_{2k-1},y_{2k})=\;l_k(\hat{\vartheta}_{2k-1},\hat{\vartheta}_{2k}),k>0,\tag{17.31}$$

where functions $\hat{\vartheta}_{2k-1}(t),\hat{\vartheta}_{2k}(t),t\in[-\alpha,T]$ are solutions of the boundary value problem

$$\begin{aligned}
&\frac{d\hat{p}_{2k-1}(t)}{dt}-\lambda_k^2\,\hat{p}_{2k-1}(t)-2\lambda_k\hat{p}_{2k}(t)=\;\hat{\nu}(\theta_{2k-1}(t)-\hat{\vartheta}_{2k-1}(t)),t\;>0,\\
&\frac{d^2\hat{p}_{2k\;\,1}(t)}{dt^2}+\lambda_k^2\,\hat{p}_{2k-1}(t)+2\lambda_k\hat{p}_{2k}(t)=-\;\hat{\nu}(\theta_{2k-1}(t)-\hat{\vartheta}_{2k-1}(t)),t<0,\\
&\hat{p}_{2k-1}(-\alpha)=0,\\
&\frac{d\hat{p}_{2k-1}(0-)}{dt}+(1+\;\lambda_k^2)\hat{p}_{2k-1}(0+)+2\lambda_k\;\hat{p}_{2k}(0+)=0,\,\hat{p}_{2k-1}(T)=\;0;
\end{aligned}\tag{17.32}$$

$$\begin{aligned}
&\frac{d\hat{p}_{2k}(t)}{dt}-\lambda_k^2\,\hat{p}_{2k}(t)=\hat{\nu}(\theta_{2k}(t)-\;\hat{\vartheta}_{2k}(t)),t>0,\\
&\frac{d^2\hat{p}_{2k}(t)}{dt^2}+\lambda_k^2\,\hat{p}_{2k}(t)=-\hat{\nu}(\theta_{2k}(t)-\;\hat{\vartheta}_{2k}(t)),t<0,\\
&\hat{p}_{2k}(-\alpha)=0,\\
&\frac{d\hat{p}_{2k}(0-)}{dt}+(1+\;\lambda_k^2)\hat{p}_{2k}(0+)=0,\,\hat{p}_{2k}(T)=\;0;
\end{aligned}\tag{17.33}$$

$$\begin{aligned}
&\frac{d\hat{\vartheta}_{2k-1}(t)}{dt} + \lambda_k^2\, \hat{\vartheta}_{2k-1}(t) = \hat{\beta}^{-1}\left(\hat{p}_{2k-1}(0-) - \int_0^T \hat{p}_{2k-1}(t)\, dt\right) - \\
&\qquad -\hat{\mu}^{-1}\int_0^T \mathcal{K}(t,\tau)\, \hat{p}_{2k-1}(\tau) d\tau,\, t > 0, \\
&\frac{d^2\hat{\vartheta}_{2k-1}(t)}{dt^2} + \lambda_k^2 \hat{\vartheta}_{2k-1}(t) = -\,\hat{\gamma}^{-1}\hat{p}_{2k-1}(t),\, t < 0, \\
&\hat{\vartheta}_{2k-1}(0+) = (1+\lambda_k^2)\, \hat{\vartheta}_{2k-1}(0-),\ \frac{d\hat{\vartheta}_{2k-1}(0-)}{dt} = \hat{\beta}^{-1}\Bigg(\hat{p}_{2k-1}(0-) - \\
&\qquad -\int_0^T \hat{p}_{2k-1}(t) dt\Bigg),\ \hat{\vartheta}_{2k-1}(-\alpha) = \hat{\alpha}^{-1}\frac{d\hat{p}_{2k-1}(-\alpha)}{dt};
\end{aligned} \tag{17.34}$$

$$\begin{aligned}
&\frac{d\hat{\vartheta}_{2k}(t)}{dt} + \lambda_k^2 \hat{\vartheta}_{2k}(t) + 2\lambda_k \hat{p}_{2k-1}(t) = \hat{\beta}^{-1}\left(\hat{p}_{2k}(0-) - \int_0^T \hat{p}_{2k}(t) dt\right) - \\
&\qquad -\hat{\mu}^{-1}\int_0^T \mathcal{K}(t,\tau)\, \hat{p}_{2k}(\tau) d\tau,\, t > 0, \\
&\frac{d^2\hat{\vartheta}_{2k}(t)}{dt^2} + \lambda_k^2 \hat{\vartheta}_{2k}(t) + 2\lambda_k\, \hat{\vartheta}_{2k-1}(t) = -\hat{\gamma}^{-1}\, \hat{p}_{2k}(t),\, t < 0, \\
&\hat{\vartheta}_{2k}(0+) = (1+\lambda_k^2)\, \hat{\vartheta}_{2k}(0-) + 2\lambda_k\, \hat{\vartheta}_{2k-1}(0-),\ \frac{d\hat{\vartheta}_{2k}(0-)}{dt} = \\
&= \hat{\beta}^{-1}\left(\hat{p}_{2k}(0-) - \int_0^T \hat{p}_{2k}(t) dt\right),\ \hat{\vartheta}_{2k}(-\alpha) = \hat{\alpha}^{-1}\frac{d\hat{p}_{2k}(-\alpha)}{dt}.
\end{aligned} \tag{17.35}$$

17.3 The Problem with Divided Observation

17.3.1 Problem Statement

Let the state of the system is described by the problem (17.1)–(17.3). Observation is

$$\theta(t) = \sum_{i=0}^{\infty} g_i y_i(t) + F(t), t \in [-\alpha, T], \tag{17.36}$$

where $g_i = (g(.), X_i(.))_{L_2(0,1)}$, $F(t) \in L_2(-\alpha, T)$.
Perturbations, which influence on the system (17.1)–(17.4), are constrained by conditions

$$\sum_{i=0}^{\infty} \Bigg[\hat{\alpha}\, \varphi_i^2 + \hat{\beta} f_i^2(0) + \hat{\gamma} \int_{-\alpha}^{0} f_{i,-}^2(t)dt + \\ +\hat{\mu} \int_0^T \xi_i^2(t)dt \Bigg] + \hat{\nu} \int_{-\alpha}^{T} F^2(t)dt \le 1, \tag{17.37}$$

where $\hat{\alpha}, \hat{\beta}, \hat{\gamma}, \hat{\mu}, \hat{\nu} > 0$.

On the problem's solution (17.1)–(17.3), we define the linear functional (17.6). The estimate of this functional will be found in the class of linear estimates in the form

$$\hat{\hat{l}}(y) = -\int_{-\alpha}^{T} \hat{u}(t)\theta(t)dt, \tag{17.38}$$

where $\hat{u}_i(t) = (\hat{u}(., t), X_i(.))_{L_2(0,1)}$.

17.3.2 Formal Solution of the Problem

Let us find an a priori minimax estimate of the functional (17.6) in class (17.38), which is define by condition (17.8).

From Sect. 17.2, we know that the problem of minimax estimation is formally reduced to the following optimal control problem: find the control $v(t) \in C(-\alpha, 0)$, $u(t) \in C[0, T]$, on which functional

$$J(u) = \hat{\nu}^{-1} \int_{-\alpha}^{T} \hat{u}^2(t)dt + \sum_{i=0}^{\infty} \Bigg[\hat{\gamma}^{-1} \int_{-\alpha}^{0} \psi_i^2(t)\, dt + \hat{\beta}^{-1} \Bigg(\psi_i(0-) - \\ - \int_0^T \psi_i(t)dt \Bigg)^2 + \hat{\mu}^{-1} \int_0^T \Bigg(\int_t^T \psi_i(\tau)d\tau \Bigg)^2 dt + \hat{\alpha}^{-1} \Big(\frac{d\psi_i(-\alpha)}{dt} \Big)^2 \Bigg] \tag{17.39}$$

takes the smallest possible value, and the functions $\psi_i(t), t \in [-\alpha, T], i \geq 0$, are solutions of problem

$$\begin{gathered} \frac{d\psi_0(t)}{dt} = q_0(t) + g_0 u(t),\ t > 0, \\ \frac{d^2\psi_0(t)}{dt^2} = -q_0(t) - g_0 v(t), t < 0, \\ \psi_0(-\alpha) = 0, \frac{d\psi_0(0-)}{dt} + \psi_0(0+) = 0, \psi_0(T) = 0; \end{gathered} \tag{17.40}$$

$$\begin{gathered} \frac{d\psi_{2i-1}(t)}{dt} - \lambda_i^2\, \psi_{2i-1}(t) - 2\lambda_i \psi_{2i}(t) = q_{2i-1}(t) + g_{2i-1}u(t), t > 0, \\ \frac{d^2\psi_{2i-1}(t)}{dt^2} + \lambda_i^2\, \psi_{2i-1}(t) + 2\lambda_i \psi_{2i}(t) = -\, q_{2i-1}(t) - g_{2i-1}v(t), t < 0, \\ \psi_{2i-1}(-\alpha) = 0, \\ \frac{d\psi_{2i-1}(0-)}{dt} + (1 + \lambda_i^2)\psi_{2i-1}(0+) + 2\lambda_i\, \psi_{2i}(0+) = 0, \psi_{2i-1}(T) = 0; \end{gathered} \tag{17.41}$$

$$\begin{gathered} \frac{d\psi_{2i}(t)}{dt} - \lambda_i^2\, \psi_{2i}(t) = q_{2i}(t) + g_{2i}u(t), t > 0, \\ \frac{d^2\psi_{2i}(t)}{dt^2} + \lambda_i^2\, \psi_{2i}(t) = -q_{2i}(t) - g_{2i}v(t), t < 0, \\ \psi_{2i}(-\alpha) = 0, \\ \frac{d\psi_{2i}(0-)}{dt} + (1 + \lambda_i^2)\psi_{2i}(0+) = 0, \psi_{2i}(T) = 0. \end{gathered} \tag{17.42}$$

From necessary and sufficient conditions of optimality for the problem (17.39)–(17.42) we find

$$\begin{gathered} v(t) = -\hat{v} \sum_{j=0}^{\infty} g_j\, p_j(t), t < 0, \\ u(t) = -\hat{v} \sum_{j=0}^{\infty} g_j p_j(t), t > 0, \end{gathered} \tag{17.43}$$

where functions $p_i(t), t \in [-\alpha, T]$ are solutions of the problem

$$\frac{d\psi_0(t)}{dt} = q_0(t) - \hat{v}\, g_0 \sum_{j=0}^{\infty} g_j p_j(t), t > 0,$$

$$\frac{d^2\psi_0(t)}{dt^2} = -q_0(t) + \hat{v} g_0 \sum_{j=0}^{\infty} g_j p_j(t),\ t < 0,$$

$$\psi_0(-\alpha) = 0,\ \frac{d\psi_0(0-)}{dt} + \psi_0(0+) = 0,\ \psi_0(T) = 0; \qquad (17.44)$$

$$\frac{dp_0(t)}{dt} = \hat{\beta}^{-1}\left(\psi_0(0-) - \int_0^T \psi_0(t)dt\right) - \hat{\mu}^{-1} \int_0^T \mathscr{K}(t,\tau)\psi_0(\tau)d\tau,\ t > 0,$$

$$\frac{d^2 p_0(t)}{dt^2} = -\hat{\gamma}^{-1}\ \psi_0(t), t < 0,$$

$$p_0(0+) = p_0(0-),\ \frac{dp_0(0-)}{dt} = \hat{\beta}^{-1}\left(\psi_0(0-) - \int_0^T \psi_0(t)dt\right),$$

$$p_0(-\alpha) = \hat{\alpha}^{-1}\frac{d\psi_0(-\alpha)}{dt}, \qquad (17.45)$$

$$\frac{d\psi_{2k-1}(t)}{dt} - \lambda_k^2\ \psi_{2k-1}(t) - 2\lambda_k\psi_{2k}(t) = q_{2k-1}(t) -$$

$$-\hat{v} g_{2k-1} \sum_{j=0}^{\infty} g_j p_j(t), t > 0,$$

$$\frac{d^2\psi_{2k-1}(t)}{dt^2} + \lambda_k^2\ \psi_{2k-1}(t) + 2\lambda_k\psi_{2k}(t) = -\ q_{2k-1}(t) +$$

$$+\hat{v} g_{2k-1} \sum_{j=0}^{\infty} g_j p_j(t), t < 0,\ \psi_{2k-1}(-\alpha) = 0,$$

$$\frac{d\psi_{2k-1}(0-)}{dt} + (1 + \lambda_k^2)\psi_{2k-1}(0+) + 2\lambda_k\ \psi_{2k}(0+) = 0,$$

$$\psi_{2k-1}(T) = 0; \qquad (17.46)$$

$$\frac{d\psi_{2k}(t)}{dt} - \lambda_k^2\ \psi_{2k}(t) = q_{2k}(t) - \hat{v} g_{2k} \sum_{j=0}^{\infty} g_j p_j(t), t > 0,$$

$$\frac{d^2\psi_{2k}(t)}{dt^2} + \lambda_k^2\ \psi_{2k}(t) = -q_{2k}(t) + \hat{v} g_{2k} \sum_{j=0}^{\infty} g_j p_j(t), t < 0,$$

$$\psi_{2k}(-\alpha) = 0,$$

$$\frac{d\psi_{2k}(0-)}{dt} + (1 + \lambda_k^2)\psi_{2k}(0+) = 0,\ \psi_{2k}(T) = 0; \qquad (17.47)$$

$$\frac{dp_{2k-1}(t)}{dt} + \lambda_k^2 p_{2k-1}(t) = \hat{\beta}^{-1}\left(\psi_{2k-1}(0-) - \int_0^T \psi_{2k-1}(t)dt\right) -$$
$$-\hat{\mu}^{-1}\int_0^T \mathcal{K}(t,\tau)\,\psi_{2k-1}(\tau)d\tau, t > 0,$$
$$\frac{d^2 p_{2k-1}(t)}{dt^2} + \lambda_k^2\, p_{2k-1}(t) = -\hat{\gamma}^{-1}\psi_{2k-1}(t), t < 0,$$
$$p_{2k-1}(0+) = (1+\lambda_k^2)p_{2k-1}(0-), \ \frac{dp_{2k-1}(0-)}{dt} = \hat{\beta}^{-1}\left(\psi_{2k-1}(0-) -\right.$$
$$\left. - \int_0^T \psi_{2k-1}(t)dt\right),\ p_{2k-1}(-\alpha) = \hat{\alpha}^{-1}\frac{d\psi_{2k-1}(-\alpha)}{dt}; \tag{17.48}$$

$$\frac{dp_{2k}(t)}{dt} + \lambda_k^2 p_{2k}(t) + 2\lambda_k p_{2k-1}(t) = \hat{\beta}^{-1}\left(\psi_{2k}(0-) - \int_0^T \psi_{2k}(t)dt\right) -$$
$$-\hat{\mu}^{-1}\int_0^T \mathcal{K}(t,\tau)\,\psi_{2k}(\tau)d\tau, t > 0,$$
$$\frac{d^2 p_{2k}(t)}{dt^2} + \lambda_k^2 p_{2k}(t) + 2\lambda_k p_{2k-1}(t) = -\hat{\gamma}^{-1}\,\psi_{2k}(t), t < 0,$$
$$p_{2k}(0+) = (1+\lambda_k^2)p_{2k}(0-) + 2\,\lambda_k p_{2k-1}(0-),\ \frac{dp_{2k}(0-)}{dt} =$$
$$= \hat{\beta}^{-1}\left(\psi_{2k}(0-) - \int_0^T \psi_{2k}(t)dt\right),\ p_{2k}(-\alpha) = \hat{\alpha}^{-1}\frac{d\psi_{2k}(-\alpha)}{dt}. \tag{17.49}$$

We show that the representations are true:

$$\sigma^2 = J(\hat{u}) = l(p),$$
$$\hat{l}(y) = l(\hat{\vartheta}), \tag{17.50}$$

where $p(t)$-infinite dimensional vector with components $p_i(t)$, which are determined by solving the boundary value problem (17.44)–(17.49), and the components of an infinite dimensional vector $\hat{\vartheta}(t)$ are solutions of the boundary value problem

$$\frac{d\hat{p}_0(t)}{dt} = \hat{\nu} g_0\left(\theta(t) - \sum_{i=0}^{\infty} g_i \hat{\vartheta}_i(t)\right), t > 0,$$

$$\frac{d^2\hat{p}_0(t)}{dt^2} = -\hat{\nu} g_0\left(\theta(t) - \sum_{i=0}^{\infty} g_i\, \hat{\vartheta}_i(t)\right), t < 0,$$

$$\hat{p}_0(-\alpha) = 0, \frac{d\hat{p}_0(0-)}{dt} + \hat{p}_0(0+) = 0, \ \hat{p}_0(T) = 0; \tag{17.51}$$

$$\frac{d\hat{\vartheta}_0(t)}{dt} = \hat{\beta}^{-1}\left(\hat{p}_0(0-) - \int_0^T \hat{p}_0(t)dt\right) - \hat{\mu}^{-1}\int_0^T \mathscr{K}(t,\tau)\hat{p}_0(\tau)d\tau, t > 0,$$

$$\frac{d^2\hat{\vartheta}_0(t)}{dt^2} = -\,\hat{\gamma}^{-1}\hat{p}_0(t), t < 0,$$

$$\hat{\vartheta}_0(0+) = \hat{\vartheta}_0(0-), \frac{d\hat{\vartheta}_0(0-)}{dt} = \hat{\beta}^{-1}\left(\hat{p}_0(0-) - \int_0^T \hat{p}_0(t)dt\right),$$

$$\hat{\vartheta}_0(-\alpha) = \hat{\alpha}^{-1}\frac{d\hat{p}_0(-\alpha)}{dt}. \tag{17.52}$$

$$\frac{d\hat{p}_{2k-1}(t)}{dt} - \lambda_k^2\, \hat{p}_{2k-1}(t) - 2\lambda_k \hat{p}_{2k}(t) = \hat{\nu} g_{2k-1}\left(\theta(t) - \sum_{i=0}^{\infty} g_i \hat{\vartheta}_i(t)\right), t > 0,$$

$$\frac{d^2\hat{p}_{2k-1}(t)}{dt^2} + \lambda_k^2\, \hat{p}_{2k-1}(t) + 2\lambda_k \hat{p}_{2k}(t) = -\,\hat{\nu} g_{2k-1}\left(\theta(t) - \sum_{i=0}^{\infty} g_i \hat{\vartheta}_i(t)\right), t < 0,$$

$$\hat{p}_{2k-1}(-\alpha) = 0,$$

$$\frac{d\hat{p}_{2k-1}(0-)}{dt} + (1 + \lambda_k^2)\hat{p}_{2k-1}(0+) + 2\lambda_k\, \hat{p}_{2k}(0+) = 0, \hat{p}_{2k-1}(T) = 0; \tag{17.53}$$

$$\frac{d\hat{p}_{2k}(t)}{dt} - \lambda_k^2\, \hat{p}_{2k}(t) = \hat{\nu} g_{2k}\left(\theta(t) - \sum_{i=0}^{\infty} g_i \hat{\vartheta}_i(t)\right), t > 0,$$

$$\frac{d^2\hat{p}_{2k}(t)}{dt^2} + \lambda_k^2\, \hat{p}_{2k}(t) = -\hat{\nu} g_{2k}\left(\theta(t) - \sum_{i=0}^{\infty} g_i \hat{\vartheta}_i(t)\right), t < 0,$$

$$\hat{p}_{2k}(-\alpha) = 0,$$

$$\frac{d\hat{p}_{2k}(0-)}{dt} + (1 + \lambda_k^2)\hat{p}_{2k}(0+) = 0, \hat{p}_{2k}(T) = 0; \tag{17.54}$$

$$\frac{d\hat{\vartheta}_{2k-1}(t)}{dt} + \lambda_k^2\, \hat{\vartheta}_{2k-1}(t) = \hat{\beta}^{-1} \left(\hat{p}_{2k-1}(0-) - \int_0^T \hat{p}_{2k-1}(t)\, dt\right) - $$
$$-\hat{\mu}^{-1} \int_0^T \mathcal{K}(t,\tau)\, \hat{p}_{2k-1}(\tau) d\tau,\, t > 0,$$
$$\frac{d^2\hat{\vartheta}_{2k-1}(t)}{dt^2} + \lambda_k^2 \hat{\vartheta}_{2k-1}(t) = -\,\hat{\gamma}^{-1} \hat{p}_{2k-1}(t),\, t < 0,$$
$$\hat{\vartheta}_{2k-1}(0+) = (1+\lambda_k^2)\, \hat{\vartheta}_{2k-1}(0-),\ \frac{d\hat{\vartheta}_{2k-1}(0-)}{dt} = \hat{\beta}^{-1}\Bigg(\hat{p}_{2k-1}(0-) -$$
$$- \int_0^T \hat{p}_{2k-1}(t) dt\Bigg),\, \hat{\vartheta}_{2k-1}(-\alpha) = \hat{\alpha}^{-1} \frac{d\hat{p}_{2k-1}(-\alpha)}{dt}; \tag{17.55}$$

$$\frac{d\hat{\vartheta}_{2k}(t)}{dt} + \lambda_k^2 \hat{\vartheta}_{2k}(t) + 2\lambda_k \hat{p}_{2k-1}(t) = \hat{\beta}^{-1} \left(\hat{p}_{2k}(0-) - \int_0^T \hat{p}_{2k}(t) dt\right) -$$
$$-\hat{\mu}^{-1} \int_0^T \mathcal{K}(t,\tau)\, \hat{p}_{2k}(\tau) d\tau,\, t > 0,$$
$$\frac{d^2\hat{\vartheta}_{2k}(t)}{dt^2} + \lambda_k^2 \hat{\vartheta}_{2k}(t) + 2\lambda_k\, \hat{\vartheta}_{2k-1}(t) = -\hat{\gamma}^{-1}\, \hat{p}_{2k}(t),\, t < 0,$$
$$\hat{\vartheta}_{2k}(0+) = (1+\lambda_k^2)\, \hat{\vartheta}_{2k}(0-) + 2\lambda_k\, \hat{\vartheta}_{2k-1}(0-),\ \frac{d\hat{\vartheta}_{2k}(0-)}{dt} =$$
$$= \hat{\beta}^{-1}\Bigg(\hat{p}_{2k}(0-) - \int_0^T \hat{p}_{2k}(t) dt\Bigg),\ \hat{\vartheta}_{2k}(-\alpha) = \hat{\alpha}^{-1} \frac{d\hat{p}_{2k}(-\alpha)}{dt}. \tag{17.56}$$

Indeed,

$$\sum_{i=0}^{\infty} \left[\hat{\gamma}^{-1} \int_{-\alpha}^{0} \psi_i^2(t) dt + \hat{\mu}^{-1} \int_0^T \int_0^T \mathcal{K}(t,\tau)\, \psi_i(t)\psi_i(\tau) d\tau dt\right] =$$
$$= -\int_{-\alpha}^{0} \frac{d^2 p_0(t)}{dt^2} \psi_0(t) dt - \sum_{i=1}^{\infty} \int_{-\alpha}^{0} \left[\left(\frac{d^2 p_{2i-1}(t)}{dt^2} + \lambda_i^2 p_{2i-1}(t)\right)\psi_{2i-1}(t) + \right.$$

$$+\Big(\frac{d^2 p_{2i}(t)}{dt^2} + \lambda_i^2 \; p_{2i}(t) + 2\lambda_i p_{2i-1}(t)\Big)\psi_{2i}(t)\Big]dt + \int_0^T \Big[-\frac{dp_0(t)}{dt} +$$

$$+\hat{\beta}^{-1}\Big(\psi_0(0-) - \int_0^T \psi_0(\tau)d\tau\Big)\Big]\psi_0(t)dt + \sum_{i=1}^{\infty}\int_0^T \Big[\Big(-\frac{dp_{2i-1}(t)}{dt} - \lambda_i^2 p_{2i-1}(t) +$$

$$+\hat{\beta}^{-1}(\psi_{2i-1}(0-) - \int_0^T \psi_{2i-1}(\tau)d\tau\Big)\Big)\psi_{2i-1}(t) + \Big(-\frac{dp_{2i}(t)}{dt} - \lambda_i^2 p_{2i}(t) -$$

$$-2\lambda_i p_{2i-1}(t) + \hat{\beta}^{-1}\;(\psi_{2i}(0-) - \int_0^T \psi_{2i}(\tau)d\tau)\Big)\psi_{2i}(t)\Big]dt =$$

$$= -\frac{dp_0(t)}{dt}\psi_0(t)|_{-\alpha}^0 + \frac{d\psi_0(t)}{dt}p_0(t)|_{-\alpha}^0 + \int_{-\alpha}^0 (q_0(t) + g_0 u(t))p_0(t)dt -$$

$$-\sum_{i=1}^{\infty}\Big[\frac{dp_{2i-1}(t)}{dt}\psi_{2i-1}(t)|_{-\alpha}^0 - \frac{d\psi_{2i-1}(t)}{dt}p_{2i-1}(t)|_{-\alpha}^0 + \frac{dp_{2i}(t)}{dt}\psi_{2i}(t)|_{-\alpha}^0 -$$

$$-\frac{d\psi_{2i}(t)}{dt}p_{2i}(t)|_{-\alpha}^0 + \int_{-\alpha}^0 (-(\lambda_i^2\psi_{2i-1}(t) + 2\lambda_i\psi_{2i}(t) + q_{2i-1}(t) +$$

$$+g_{2i-1}u(t))p_{2i-1}(t) + \lambda_i^2 p_{2i-1}(t)\psi_{2i-1}(t) + \lambda_i^2 p_{2i}(t)\psi_{2i}(t) +$$

$$+2\lambda_i p_{2i-1}(t)\psi_{2i}(t) - (\lambda_i^2\psi_{2i}(t) + q_{2i}(t) \; + g_{2i}u(t))p_{2i}(t))dt\Big] - p_0(t)\psi_0(t)|_0^T +$$

$$+\int_0^T (q_0(t) + g_0 u(t))p_0(t)dt + \hat{\beta}^{-1}\Big(\psi_0(0-) - \int_0^T \psi_0(t)dt\Big)\int_0^T \psi_0(t)dt +$$

$$+\sum_{i=1}^{\infty}\Big[- p_{2i-1}(t)\psi_{2i-1}(t)|_0^T - p_{2i}(t)\psi_{2i}(t)|_0^T + \int_0^T ((\lambda_i^2\psi_{2i-1}(t) + 2\,\lambda_i\psi_{2i}(t) +$$

$$+q_{2i-1}(t) + g_{2i-1}u(t))p_{2i-1}(t) - \lambda_i^2 p_{2i-1}(t)\psi_{2i-1}(t) + \; \hat{\beta}^{-1}\Big(\psi_{2i-1}(0-) -$$

$$-\int_0^T \psi_{2i-1}(\tau)d\tau\Big)\psi_{2i-1}(t) + (\lambda_i^2\psi_{2i}(t) + \; q_{2i}(t) + g_{2i}u(t))p_{2i}(t) -$$

$$-\lambda_i^2 p_{2i}(t)\psi_{2i}(t) - 2\lambda_i p_{2i-1}(t)\psi_{2i}(t) + \hat{\beta}^{-1}(\psi_{2i}(0-) - \int_0^T \psi_{2i}(\tau)d\tau)\psi_{2i}(t))dt\Big] =$$

$$= -\sum_{i=0}^{\infty}\Big[\hat{\beta}^{-1}\Big(\psi_i(0-) - \int_0^T \psi_i(\tau)d\tau\Big)^2 + \hat{\alpha}^{-1}\Big(\frac{d\psi_i(-\alpha)}{dt}\Big)^2\Big] - \hat{\nu}^{-1}\int_{-\alpha}^T \hat{u}^2(t)dt + l(p).$$

Validity of the representation (17.50) set similarly by using the boundary value problems (17.51)–(17.56).

References

1. Egorov, A.I.: Optimal control of thermal and diffusion processes [Russian]. Nauka, Kyiv (1978)
2. Gelfand, I.M.: Some questions of analysis and differential equations [Russian]. UMJ **14**(3), 3–19 (1959)
3. Kapustyan, V.O., Pyshnograiev, I.O.: Approximate optimal control problem for a parabolic-hyperbolic equation with nonlocal boundary conditions and semidefinite quality criterion [Russian]. J. Comput. Appl. Math. **4**, 24–36 (2013)
4. Kapustyan V.O., Pyshnograiev I.O.: The conditions of existence and uniqueness of the solution of a parabolic-hyperbolic equation with nonlocal boundary conditions [Ukrainian]. Sci. News NTU "KPI". **4**, 72–86 (2012)
5. Nakonechny, O.G.: Optimal control and estimates in equations in partial derivatives [Ukrainian]. KNU, Kyiv (2004)

Chapter 18
The Optimal Control Problem for Parabolic Equation with Nonlocal Boundary Conditions in Circular Sector

V.O. Kapustyan, O.A. Kapustian, O.V. Kapustyan and O.K. Mazur

Abstract We consider the linear-quadratic optimal control problem for parabolic equation with nonlocal boundary conditions in a circular sector with quadratic cost functional. Using the biorthonormal basis systems of functions and Fourier–Bessel series, we prove the classical solvability of such problem in special classes of distributed controls and initial functions.

18.1 Introduction

Among a variety of classical and modern methods for analysis of infinite-dimensional optimal control problems [1–4], Fourier method remains a powerful tool to solve linear-quadratic problems for distributed systems. In many cases, this method allows to decompose initial problem and reduce it to countable number of one-dimensional optimal control problems.

In this paper, we consider a minimization problem for quadratic cost functional on the solutions of linear parabolic equation with nonlocal boundary conditions in a circular sector. A classical solvability of such boundary value problem for Laplace equation was proved in the paper [5], using biorthonormal basis systems of functions.

V.O. Kapustyan
NTUU "KPI", Prosp. Peremohy, 37, Kyiv 03056, Ukraine
e-mail: kapustyanv@ukr.net

O.A. Kapustian
Taras Shevchenko National University of Kyiv, Volodymyrska Str., 64, Kyiv 01601, Ukraine
e-mail: olena_kap@gmail.com

O.V. Kapustyan (✉)
Taras Shevchenko National University of Kyiv, National Technical University of Ukraine "KPI", Kyiv, Ukraine
e-mail: kapustyanav@gmail.com

O.K. Mazur
National University of Food Technologies, Ukraine Volodymyrska Str., 68, Kyiv 01601, Ukraine
e-mail: okmazur@ukr.net

© Springer International Publishing Switzerland 2015

V.A. Sadovnichiy and M.Z. Zgurovsky (eds.), *Continuous and Distributed Systems II*, Studies in Systems, Decision and Control 30,
DOI 10.1007/978-3-319-19075-4_18

Corresponding optimal control problem for elliptic equation was solved in [6]. Here, in contrast to the elliptic case, the problem is reduced to a countable sequence of infinite-dimensional problems which, moreover, are not independent.

Using Fourier–Bessel series [7–9] for a special class of initial data, we proved the solvability of the corresponding initial-boundary value problem for a fixed control. We also established the solvability of the original optimal control problem under additional condition on control function.

18.2 Setting of the Problem

In domain $Q = (0, T) \times \Omega$, $\Omega = \{(r, \theta)| r \in (0, 1),\ \theta \in (0, \pi)\}$ we consider the following problem: to find a state function $y = y(t, r, \theta)$ and control $u = u(t, \theta)$ such that

$$\begin{cases} \frac{\partial y}{\partial t} = \Delta y + g(r)u(t, \theta),\ (t, r, \theta) \in Q, \\ y(t, 1, \theta) = 0,\ t \in (0, T),\ \theta \in (0, \pi), \\ y(t, r, 0) = 0,\ t \in (0, T),\ r \in (0, 1), \\ \frac{\partial y}{\partial \theta}(t, r, 0) = \frac{\partial y}{\partial \theta}(t, r, \pi),\ t \in (0, T),\ r \in (0, 1), \\ y(0, r, \theta) = h(r)p(\theta), \end{cases} \tag{18.1}$$

$$J(y, u) = \int_0^1 r\|y(T, r)\|_D^2 dr + \int_0^1 \|u(r)\|_D^2 dr \to \inf, \tag{18.2}$$

where $\Delta y := \frac{1}{r}\frac{\partial}{\partial r}(r\frac{\partial y}{\partial r}) + \frac{1}{r^2}\frac{\partial^2 y}{\partial \theta^2}$ is Laplace operator in polar coordinates, g, h, p are given functions, $\|\cdot\|_D$ is a norm in $L^2(0, \pi)$, which is equivalent to standard one and is given by the equality

$$\forall v \in L^2(0, \pi)\ \ \|v\|_D = \left(\sum_{n=0}^{\infty} v_n^2\right)^{1/2},$$

where for each $n \geq 1$ $v_n = \int_0^\pi v(\theta)\psi_n(\theta)d\theta$, the systems of functions

$$\Psi = \{\psi_0 = \frac{2}{\pi^2},\ \psi_{2n} = \frac{4}{\pi^2}(\pi - \theta)\sin 2n\theta,\ \psi_{2n-1}(\theta) = \frac{4}{\pi^2}\cos 2n\theta\},$$

$$\Phi = \{\varphi_0 = \theta,\ \varphi_{2n} = \sin 2n\theta,\ \varphi_{2n-1} = \theta\cos 2n\theta\}$$

are biorthonormal and complete in $L^2(0, \pi)$ [5].

Suppose $g, h \in C([0,1])$,

$$p \in C^1([0,\pi]),\ p(0)=0,\ \frac{\partial p}{\partial \theta}(0)=\frac{\partial p}{\partial \theta}(\pi). \tag{18.3}$$

We will find the solution of the problem (18.1) for fixed function $u \in C([0,T] \times [0,\pi])$ in the form

$$y(t,r,\theta)=y_0(t,r)\varphi_0(\theta)+\sum_{n=1}^{\infty}\left(y_{2n-1}(t,r)\varphi_{2n-1}(\theta)+y_{2n}(t,r)\varphi_{2n}(\theta)\right), \tag{18.4}$$

where the functions $\{y_n(t,r)\}_{n=0}^{\infty}$ are defined from the following initial-boundary value problems in domain $\Pi=(0,T)\times(0,1)$:

$$\begin{cases} \frac{\partial y_0}{\partial t}=\frac{1}{r}\frac{\partial}{\partial r}(r\frac{\partial y_0}{\partial r})+u_0(t)g(r),\ (t,r)\in\Pi, \\ y_0(t,1)=0,\ t\in(0,T), \\ y_0(0,r)=p_0h(r),\ r\in(0,1), \end{cases} \tag{18.5}$$

$$\begin{cases} \frac{\partial y_{2n-1}}{\partial t}=\frac{1}{r}\frac{\partial}{\partial r}(r\frac{\partial y_{2n-1}}{\partial r})-(\frac{2n}{r})^2 y_{2n-1}+u_{2n-1}(t)g(r),\ (t,r)\in\Pi, \\ y_{2n-1}(t,1)=0,\ t\in(0,T), \\ y_{2n-1}(0,r)=p_{2n-1}h(r),\ r\in(0,1), \end{cases} \tag{18.6}$$

$$\begin{cases} \frac{\partial y_{2n}}{\partial t}=\frac{1}{r}\frac{\partial}{\partial r}(r\frac{\partial y_{2n}}{\partial r})-(\frac{2n}{r})^2 y_{2n}-\frac{4n}{r^2}y_{2n-1}+u_{2n}(t)g(r),\ (t,r)\in\Pi, \\ y_{2n}(t,1)=0,\ t\in(0,T), \\ y_{2n}(0,r)=p_{2n}h(r),\ r\in(0,1), \end{cases} \tag{18.7}$$

where $\forall n\geq 0\ \ u_n(t)=\int\limits_0^{\pi} u(t,\theta)\psi_n(\theta)d\theta,\ \ p_n=\int_0^{\pi}p(\theta)\cdot\psi_n(\theta)d\theta.$

Therefore, the original optimal control problem (18.1) and (18.2) is reduced to the following one: among admissible pairs $\{u_n(t), y_n(t,r)\}_{n=0}^{\infty}$ of the problem (18.5)–(18.7) one should minimize the cost functional

$$J(y,u)=\sum_{n=0}^{\infty}\left(\int_0^1 ry_n^2(T,r)dr+\int_0^T u_n^2(t)dt\right)=J_0+\sum_{n=1}^{\infty}J_n, \tag{18.8}$$

where

$$J_0=\int_0^1 ry_0^2(T,r)dr+\int_0^T u_0^2(t)dt,$$

$$J_n=\int\limits_0^1 r\left(y_{2n-1}^2(T,r)+y_{2n}^2(T,r)\right)dr+\int_0^1\left(u_{2n-1}^2(t)+u_{2n}^2(t)\right)dt.$$

Herewith, the optimal process $\{\tilde{u}_n(t), \tilde{y}_n(t)\}_{n=0}^{\infty}$ should be such that the formula

$$\tilde{u}(t,\theta) = \sum_{n=0}^{\infty} \tilde{u}_n(t)\varphi_n(\theta) \tag{18.9}$$

defines a function from $C([0,T] \times [0,\pi])$, and the formula (18.4) defines a function from $C(\bar{Q})$ for which

$$\frac{\partial \tilde{y}}{\partial \theta} \in C(\bar{Q}), \ \frac{\partial \tilde{y}}{\partial t}, \frac{\partial^2 \tilde{y}}{\partial r^2}, \frac{\partial^2 \tilde{y}}{\partial r \partial \theta}, \frac{\partial^2 \tilde{y}}{\partial \theta^2} \in C(Q).$$

To solve such a problem, we make additional assumption on parameters: let function p be an element of finite-dimensional subspace, generated by $\{\varphi_i\}_{i=0}^{2N}$, i.e., there exists $N \geq 0$ such that

$$p \in L_N = \text{span}\{\varphi_0, \varphi_1, ..., \varphi_{2N}\}. \tag{18.10}$$

Under realization of assumption (18.10) for $n > N$ we have $p_{2n-1} = 0$, $p_{2n} = 0$, so the minimum value of the functional J_n equals zero and is achieved at values $u_{2n-1}(t) \equiv 0$, $y_{2n-1}(t,r) \equiv 0$, $u_{2n}(t) \equiv 0$, $y_{2n}(t,r) \equiv 0$.

In this manner, under assumption (18.10) we have that on the optimal process the series (18.4), (18.8), (18.9) contain a finite number of nonzero members, so they are convergent. Continuity and smoothness properties of functions $\tilde{u}$ and $\tilde{y}$ follow from the corresponding properties of functions $\tilde{u}_n(t)\varphi_n(\theta)$ and $\tilde{y}_n(t,r)\varphi_n(\theta)$. Therefore, we need to solve the problem (18.5)–(18.8) only for $n \in \overline{0,N}$.

18.3 Some Facts from the Theory of Fourier-Bessel Series

When solving the problem (18.5)–(18.7) for fixed $\{u_i(t)\}_{i=0}^{N}$, we come to Fourier–Bessel series, as will be seen below. Here are the necessary facts from the theory of Bessel functions, based on the papers [7–9].

For $n \geq 0$ n-order Bessel function

$$J_n(x) = \sum_{k=0}^{\infty} (-1)^k \frac{1}{\Gamma(k+1)\Gamma(k+n+1)} \cdot (\frac{x}{2})^{2k+n}$$

is a solution of differential equation

$$y'' + \frac{1}{x}y' + (1 - \frac{n^2}{x^2})y = 0,$$

satisfies integral formula

$$J_n(x) = \frac{1}{\pi} \int_0^{\pi} cos(n\varphi - x sin\varphi) d\varphi,$$

recurrent formulas

$$2n \frac{J_n(x)}{x} = J_{n-1}(x) + J_{n+1}(x),$$

$$2J_n'(x) = J_{n-1}(x) - J_{n+1}(x),$$

$$\frac{d}{dx}(x^n \cdot J_n(x)) = x^n J_{n-1}(x), \ \frac{d}{dx}\left(\frac{J_n(x)}{x^n}\right) = -\frac{J_{n+1}(x)}{x^n},$$

and asymptotic formula

$$J_n(x) = \sqrt{\frac{2}{\pi x}} \left(cos(x - \frac{\pi}{2} n - \frac{\pi}{4}) + \frac{K \cdot \theta(x)}{x} \right),$$

where $K = K(n) > 0$ is a constant that depends on number n only, $0 \leq \theta(x) \leq 1$, and orthogonality property

$$\int_0^1 x \cdot J_n(\lambda_k^{(n)} x) J_n(\lambda_m^{(n)} x) dx = \begin{cases} 0, \ k \neq m, \\ \frac{1}{2}(J_n'(\lambda_m^{(n)})^2, \ k = m, \end{cases}$$

where $\{\lambda_k^{(n)}\}_{k=1}^{\infty}$ is positive, increasing sequence of solutions of equation $J_n(\lambda) = 0$. For such solutions, the following asymptotic formula holds:

$$\lambda_k^{(n)} = k \cdot \pi + q + \frac{L \cdot \theta(k)}{k}, \ k \geq 1,$$

where $q = q(n) \in \mathbf{Z}, \ L = L(n) > 0$ are the constants that depend on number n only, $0 \leq \theta(k) \leq 1$.

Definition 18.1 The series

$$\sum_{m=1}^{\infty} A_m^{(n)}(f) \cdot J_n(\lambda_m^{(n)} r)$$

is called Fourier–Bessel series, where Fourier–Bessel coefficients $A_m^{(n)}(f)$ are given by the formula

$$A_m^{(n)}(f) = \frac{\int_0^1 x \cdot f(x) J_n(\lambda_m^{(n)} x) dx}{\int_0^1 x \cdot J_n^2(\lambda_m^{(n)} x) dx} = \frac{2}{(J_n'(\lambda_m^{(n)}))^2} \int_0^1 x \cdot f(x) J_n(\lambda_m^{(n)} x) dx.$$

Theorem 18.1 *Let $f \in C^2(0,1)$, $|f(0)| < \infty$, $f(1) = 0$. Then the Fourier–Bessel series converges absolutely and uniformly on every $[a,b] \subset (0,1)$. Moreover, the Bessel inequality holds*

$$\sum_{m=1}^{\infty} (A_m^{(n)}(f))^2 \cdot \int_0^1 r J_n^2(\lambda_m^{(n)} r) dr \le \int_0^1 r f^2(r) dr.$$

Theorem 18.2 *Let $f \in C^1([0,1])$. Then for $p \ge 0$ there exists the constant $C > 0$ that depends on Bessel function J_p such that*

$$\forall \lambda > 0 \left| \int_0^1 f(x) x J_p(\lambda x) dx \right| \le \frac{C}{\lambda^{\frac{3}{2}}} \max_{x \in [0,1]} \left(|f(x)| + |f'(x)| \right).$$

Proof It is known [10] that $f(x) = v(x) - \varphi(x)$, where $v(x) = V_0^x[f]$ is a total variation of function f on $[0,x]$, and functions v and φ are absolutely continuous and increasing. Then from the second mean value theorem for integration $\exists \xi \in (0,1)$:

$$\int_0^1 v(x) \sqrt{x} \sqrt{x} J_p(\lambda x) dx = v(1) \int_\xi^1 \sqrt{x} J_p(\lambda x) dx = V_0^1[f] \cdot \int_\xi^1 \sqrt{x} J_p(\lambda x) dx.$$

From the asymptotic formula for $J_p(x)$ we get that there exists a constant $C_1 = C_1(p) > 0$ that depends on J_p such that

$$\forall t > 0 \left| \int_0^t \sqrt{x} J_p(x) dx \right| \le C_1.$$

Then from inequality $V_0^1[f] \le \max_{x \in [0,1]} |f'(x)|$ we obtain

$$\left|\int_0^1 v(x)xJ_p(\lambda x)dx\right| \le \max_{x\in[0,1]}|f'(x)|\left|\int_\xi^1 \sqrt{x}J_p(\lambda x)dx\right| \le$$

$$\le 2\max_{x\in[0,1]}|f'(x)|\frac{1}{\lambda^{\frac{3}{2}}}\left|\int_0^\lambda \sqrt{x}J_p(x)dx\right| \le \frac{2C_1}{\lambda^{\frac{3}{2}}}\max_{x\in[0,1]}|f'(x)|.$$

Similarly, for $\varphi(x) = v(x) - f(x)$ we get

$$\left|\int_0^1 \varphi(x)\sqrt{x}\sqrt{x}J_p(\lambda x)dx\right| \le |v(1) - f(1)|\cdot\frac{2C_1}{\lambda^{\frac{3}{2}}} \le$$

$$\le \frac{2C_1}{\lambda^{\frac{3}{2}}}\max_{x\in[0,1]}(|f'(x)| + |f(x)|).$$

The next theorem follows from Theorem 18.2 and from integration by parts in integral $\int_0^1 xf(x)J_n(\lambda_m^{(n)})dx$ with using recurrent formulas.

Theorem 18.3 *Let $f \in C^{p+1}([0,1])$, $f^{(k)}(0) = f^{(k)}(1) = 0, k = \overline{0, p-1}, p \ge 1$. Then the Fourier–Bessel series converges absolutely and uniformly on $[0,1]$. In this case, there exists a constant $C = C(n) > 0$ that depends on J_n such that*

$$\forall m \ge 0\; |A_m^{(n)}(f)| \le \frac{C}{(\lambda_m^{(n)})^{p+\frac{3}{2}}}\cdot\max_{x\in[0,1]}\sum_{i=0}^{p+1}|f^{(i)}(x)|.$$

18.4 Existence of Classical Solution of the Problem (18.1) with Fixed Control

For $n \ge 0$ let us consider a problem

$$\begin{cases} \frac{\partial z}{\partial t} = \frac{1}{r}\frac{\partial}{\partial r}(r\frac{\partial z}{\partial r}) - (\frac{2n}{r})^2 z, \; (t,r) \in \Pi, \\ z(t,1) = 0, \; z(0,r) = z_0(r), \end{cases} \tag{18.11}$$

where $z_0 \in C^2([0,1])$, $z_0(0) = z_0(1) = 0$.

Using Fourier method, we obtain that the solution of the problem (18.11) is defined by formula

$$z(t,r) = \sum_{m=1}^{\infty} A_m^{(2n)}(z_0)\cdot J_{2n}(\lambda_m^{(2n)}r)e^{-(\lambda_m^{(2n)})^2 t}. \tag{18.12}$$

Since from Theorem 18.3

$$\exists K(n) > 0\ \forall m \geq 1\ |A_m^{(2n)}(z_0)| \leq \frac{K(n)}{(\lambda_m^{(2n)})^{\frac{3}{2}}}, \tag{18.13}$$

then from the asymptotic formula for $\lambda_m^{(2n)}$ we get that the series (18.12) converges uniformly on compact $\overline{\Pi}$, so $z \in C(\overline{\Pi})$. Moreover, because of the exponent the series $\frac{\partial z}{\partial t}$, $\frac{\partial z}{\partial r}$, $\frac{\partial^2 z}{\partial r^2}$ converge uniformly on every compact from Π. Thus, the formula (18.12) defines classical solution of (18.11).

Now for $n \geq 0$, we consider a problem

$$\begin{cases} \frac{\partial z}{\partial t} = \frac{1}{r}\frac{\partial}{\partial r}(r\frac{\partial z}{\partial r}) - (\frac{2n}{r})^2 z + \sum\limits_{m=1}^{\infty} c_m^{(2n)}(t) J_{2n}(\lambda_m^{(2n)} r), \\ z(t,1) = 0,\ z(0,r) = 0, \end{cases} \tag{18.14}$$

where $\forall m \geq 1\ c_m^{(2n)} \in C([0,T])$. In the right-hand part of the equation, the series converges uniformly on any compact from Π and the following condition holds

$$\exists C = C(n) > 0\ \forall m \geq 1\ \forall t \in [0,T]\ |c_m^{(2n)}(t)| \leq \frac{C}{\lambda_m^{(2n)}}. \tag{18.15}$$

Using Fourier method, we obtain that the solution of the problem (18.14) is defined by formula

$$z(t,r) = \sum_{m=1}^{\infty} d_m^{(2n)}(t) \cdot J_{2n}(\lambda_m^{(2n)} r), \tag{18.16}$$

where

$$d_m^{(2n)}(t) = \int_0^t c_m^{(2n)}(s) e^{-(\lambda_m^{(2n)})^2 (t-s)} ds.$$

Formal differentiation of (18.16) and using recurrent formulas give us the following:

$$\frac{\partial z}{\partial t} = \sum_{m=1}^{\infty} \left(-(\lambda_m^{(2n)})^2 d_m^{(2n)}(t) + c_m^{(2n)}(t)\right) J_{2n}(\lambda_m^{(2n)} r),$$

$$\frac{\partial z}{\partial r} = \sum_{m=1}^{\infty} \frac{\lambda_m^{(2n)}}{2} \cdot d_m^{(2n)}(t) \left(J_{2n-1}(\lambda_m^{(2n)} r) - J_{2n+1}(\lambda_m^{(2n)} r)\right),$$

$$\frac{\partial^2 z}{\partial r^2} = \sum_{m=1}^{\infty} \frac{(\lambda_m^{(2n)})^2}{4} \cdot d_m^{(2n)}(t) \left(J_{2n-2}(\lambda_m^{(2n)} r) - 2J_{2n}(\lambda_m^{(2n)} r) + J_{2n+3}(\lambda_m^{(2n)} r)\right).$$

Under the condition (18.15), the following estimate holds

$$\forall m \geq 1 \ \forall t \in [0, T] \ |d_m^{(2n)}(t)| \leq \frac{C}{(\lambda_m^{(2n)})^3}.$$

Since $\sup\limits_{p\geq 0, x\geq 0} |J_p(x)| \leq 1$ then from the asymptotic formula for $\lambda_m^{(2n)}$, we obtain that the series (18.16) converge uniformly on $\overline{\Pi}$, so $z \in C(\overline{\Pi})$. Moreover, from the asymptotic formula for $J_{2n}(x)$ we get:

$$\forall \delta > 0 \ \exists C_\delta > 0 \ \forall m \geq 1 \ \forall r \in [\delta, 1] \ |J_{2n}(\lambda_m^{(2n)} r)| \leq \frac{C_\delta}{\sqrt{\lambda_m^{(2n)}}}.$$

This means that on any compact from Π the series $\frac{\partial z}{\partial t}$, $\frac{\partial z}{\partial r}$, $\frac{\partial^2 z}{\partial r^2}$ converge uniformly. Thus, under the condition (18.15) the formula (18.16) defines classical solution of (18.14).

We introduce some notation:

$$g_m^{(2n)} = A_m^{(2n)}(g), \ h_m^{(2n)} = A_m^{(2n)}(h).$$

Suppose the functions g, h satisfy the condition

$$g, \ h \in C^2([0, 1]), \ g(0) = h(0) = g(1) = h(1) = 0. \tag{18.17}$$

Then under Theorem 18.3 the following estimate holds

$$\exists K = K(n) > 0, \ \forall m \geq 1 \ |g_m^{(2n)}| + |h_m^{(2n)}| \leq \frac{K}{(\lambda_m^{(2n)})^{\frac{3}{2}}}. \tag{18.18}$$

The condition (18.17), estimate (18.18) and previous considerations allow us to state that the formulas

$$\begin{aligned}
y_0(t, r) &= p_0 \sum_{m=1}^{\infty} h_m^{(0)} J_0(\lambda_m^{(0)} r) e^{-(\lambda_m^{(0)})^2 t} + \\
&+ \sum_{m=1}^{\infty} g_m^{(0)} J_0(\lambda_m^{(0)} r) \int_0^t u_0(s) e^{-(\lambda_m^{(0)})^2 (t-s)} ds, \\
y_{2n-1}(t, r) &= p_{2n-1} \sum_{m=1}^{\infty} h_m^{(2n)} J_{2n}(\lambda_m^{(2n)} r) e^{-(\lambda_m^{(2n)})^2 t} + \\
&+ \sum_{m=1}^{\infty} g_m^{(2n)} J_{2n}(\lambda_m^{(0)} r) \int_0^t u_{2n-1}(s) e^{-(\lambda_m^{(2n)})^2 (t-s)} ds
\end{aligned} \tag{18.19}$$

define the classical solutions of the problems (18.5) and (18.6) for fixed $u_0, u_{2n-1} \in C([0, T])$.

Under considerations above, the solution of the problem (18.7) has a form

$$y_{2n}(t,r) = p_{2n} \sum_{m=1}^{\infty} h_m^{(2n)} J_{2n}(\lambda_m^{(2n)} r) e^{-(\lambda_m^{(2n)})^2 t} + \\ + \sum_{m=1}^{\infty} g_m^{(2n)} J_{2n}(\lambda_m^{(0)} r) \int_0^t u_{2n}(s) e^{-(\lambda_m^{(2n)})^2 (t-s)} ds + z(t,r), \tag{18.20}$$

where $z(t,r)$ is the solution of problem

$$\begin{cases} \frac{\partial z}{\partial t} = \frac{1}{r}\frac{\partial}{\partial r}(r\frac{\partial z}{\partial r}) - (\frac{2n}{r})^2 z - \frac{4n}{r^2} y_{2n-1}(t,r), \\ z(t,1) = 0, \ z(0,r) = 0. \end{cases} \tag{18.21}$$

Let us show that the problem (18.21) is a particular case of the problem (18.14). Consider functions

$$\alpha_k^{(2n)}(t) = -4np_{2n-1} \cdot h_k^{(2n)} e^{-(\lambda_k^{(2n)})^2 t} - 4n \cdot g_k^{(2n)} \cdot \int_0^t u_{2n-1}(s) e^{-(\lambda_k^{(2n)})^2 (t-s)} ds,$$

$$f_k^{(2n)}(r) = \frac{1}{r^2} J_{2n}(\lambda_k^{(2n)} r).$$

Then the right-hand side of Eq. (18.21) has a form

$$f(t,r) = \sum_{k=1}^{\infty} \alpha_k^{(2n)} f_k^{(2n)}(r).$$

Using the recurrent formulas, we obtain

$$f_k^{(2n)}(r) = (\lambda_k^{(2n)})^2 (a_n \cdot J_{2n+2}(\lambda_k^{(2n)} r) + \\ + b_n \cdot J_{2n}(\lambda_k^{(2n)} r) + c_n J_{2n-2}(\lambda_k^{(2n)} r)), \tag{18.22}$$

where

$$a_n = \frac{1}{4n(4n+2)}, \quad b_n = \frac{2}{(4n+2)(4n-2)}, \quad c_n = \frac{1}{4n(4n-2)}.$$

The following condition is stronger than (18.17):

$$h \in C^4([0,1]), \ h^{(p)}(0) = h^{(p)}(1) = 0, \ p = \overline{0,2}. \tag{18.23}$$

Then from Theorem 18.3 it follows that $\exists C = C(n) > 0 \ \forall k \geq 1$:

$$|h_k^{(2n)}| \leq \frac{C}{(\lambda_k^{(2n)})^{\frac{7}{2}}}. \tag{18.24}$$

From the estimate (18.24), we obtain existence of constant $C = C(n) > 0$ that depends on n and u_{2n-1} such that

$$\forall k \geq 1 \ \forall t \in [0, T] \ |h_k^{(2n)}(t)| \leq \frac{C}{(\lambda_k^{(2n)})^{\frac{7}{2}}}. \tag{18.25}$$

Then from (18.22) and inequality $\sup_{p\geq 0, x\geq 0} |J_p(x)| \leq 1$ we get that the series for $f(t, r)$ is uniformly convergent on $\overline{\Pi}$, so $f \in C(\overline{\Pi})$. In particular, $\forall t \in [0, T]$ $f(t, \cdot) \in C([0, 1])$, $f(t, 1) = 0$. Moreover, since for each $t \in [0, T]$ the function $y_{2n-1}(t, \cdot)$ belongs to class $C^2(0, 1)$ then $f(t, \cdot) \in C^2(0, 1)$. By Theorem 18.1, the function $f(t, \cdot)$ is decomposed into absolutely and uniformly convergent Fourier–Bessel series on any compact from $(0, 1)$. Let us find such decomposition and show that it satisfies the conditions of the problem (18.14).

Under (18.22) for fixed $k \geq 1$ the function $f_k^{(2n)}(r)$ is decomposed into absolutely and uniformly convergent Fourier–Bessel series

$$f_k^{(2n)}(r) = \sum_{m=1}^{\infty} A_m^{(2n)}(f_k^{(2n)}) \cdot J_{2n}(\lambda_m^{(2n)} r).$$

Then for $f(t, r)$ we have a formal decomposition

$$f(t, r) = \sum_{m=1}^{\infty} \left(\sum_{k=1}^{\infty} \alpha_k^{(2n)}(t) \cdot A_m^{(2n)}(f_k^{(2n)}) \right) \cdot J_{2n}(\lambda_m^{(2n)} r). \tag{18.26}$$

Permutation of series by the index m and k will be valid as soon as we will prove the absolute convergence of the series (18.26). First, consider a case $n > 1$. Then $f_k^{(2n)}(0) = f_k^{(2n)}(1) = 0$ and using the recurrent formulas, we obtain from (18.22):

$$\begin{aligned}
(f_k^{(2n)}(r))' &= (\lambda_k^{(2n)})^3 \cdot \sum_{i=-3}^{3} \gamma_i J_{2n+i}(\lambda_k^{(2n)} r), \\
(f_k^{(2n)}(r))'' &= (\lambda_k^{(2n)})^4 \cdot \sum_{i=-4}^{4} \tilde{\gamma}_i J_{2n+i}(\lambda_k^{(2n)} r), \\
\frac{f_k^{(2n)}(r)}{r} &= (\lambda_k^{(2n)})^3 \cdot \sum_{i=-3}^{3} \beta_i J_{2n+i}(\lambda_k^{(2n)} r), \\
\left(\frac{f_k^{(2n)}(r)}{r} \right)' &= (\lambda_k^{(2n)})^4 \cdot \sum_{i=-4}^{4} \tilde{\beta}_i J_{2n+i}(\lambda_k^{(2n)} r),
\end{aligned} \tag{18.27}$$

where constants $\{\gamma_i\}$, $\{\tilde{\gamma}_i\}$, $\{\beta_i\}$, $\{\tilde{\beta}_i\}$ depend on number n only.

Hence, integrating by parts, we get

$$\int_0^1 r f_k^{(2n)}(r) J_{2n}(\lambda_m^{(2n)} r) dr = -\frac{1}{\lambda_m^{(2n)}} \int_0^1 r (f_k^{(2n)}(r))' J_{2n+1}(\lambda_m^{(2n)} r) dr + \\ + \frac{2n}{\lambda_m^{(2n)}} \int_0^1 r \cdot \frac{f_k^{(2n)}(r)}{r} \cdot J_{2n+1}(\lambda_m^{(2n)} r) dr. \tag{18.28}$$

Applying Theorem 18.2 and formulas (18.27) to equality (18.28), we obtain that there exists a constant $C = C(n) > 0$ such that $\forall m \geq 1 \;\; \forall k \geq 1$

$$\left| \int_0^1 r f_k^{(2n)}(r) J_{2n}(\lambda_m^{(2n)} r) dr \right| \leq \frac{C \cdot (\lambda_k^{(2n)})^4}{(\lambda_m^{(2n)})^{\frac{5}{2}}}. \tag{18.29}$$

Using the asymptotic formula [7, 8]

$$\frac{2}{J_{2n}'} = \lambda_m^{(2n)} \cdot \pi + K(n) \cdot \theta(n, \lambda_m^{(2n)}),$$

where $K = K(n) > 0$ depends on n and $0 \leq \theta(n, \lambda_m^{(2n)}) \leq 1$, we get that there exists a constant $C = C(n) > 0$ such that

$$\forall m \geq 1 \; \forall k \geq 1 \; |A_m^{(2n)}(f_k^{(2n)})| \leq \frac{C \cdot (\lambda_k^{(2n)})^4}{(\lambda_m^{(2n)})^{\frac{3}{2}}}, \tag{18.30}$$

and C depends on J_{2n} only.

Now let us strengthen the conditions (18.17) and (18.23) to the following ones:

$$\begin{gathered} g \in C^4([0,1]), \;\; h \in C^6([0,1]), \\ g^{(p)}(0) = g^{(p)}(1) = 0, \;\; p = \overline{0,2}, \\ h^{(p)}(0) = h^{(p)}(1) = 0, \;\; h = \overline{0,4}. \end{gathered} \tag{18.31}$$

Then, by Theorem 18.3 we obtain that there exists a constant $C = C(n) > 0$ that depends on n and u_{2n-1} such that

$$\forall k \geq 1 \; \forall t \in [0,T] \; |\alpha_k^{(2n)}(t)| \leq \frac{C}{(\lambda_m^{(2n)})^{5+\frac{1}{2}}}. \tag{18.32}$$

Therefore, if we denote

$$c_m^{(2n)}(t) := \sum_{k=1}^{\infty} \alpha_k^{(2n)}(t) \cdot A_m^{(2n)}(f_k^{(2n)}), \tag{18.33}$$

then from (18.30) and (18.32) we get that the series (18.33) is absolutely and uniformly convergent on $[0, T]$. In particular, $c_m^{(2n)} \in C([0, T])$, and

$$\forall m \geq 1 \; \forall t \in [0, T] \; |c_m^{(2n)}(t)| \leq \frac{C}{(\lambda_m^{(2n)})^{\frac{3}{2}}}.$$

Hence, from (18.16) it follows that the solution of the problem (18.21) has a form

$$z(t, r) = \sum_{m=1}^{\infty} J_{2n}(\lambda_m^{(2n)} r) \cdot \int_0^t c_m^{(2n)}(s) e^{-(\lambda_m^{(2n)})^2 (t-s)} ds, \qquad (18.34)$$

where $c_m^{(2n)}$ is defined from the equality (18.33).

Let us prove the equality (18.34) for the case $n = 1$. Here

$$\begin{aligned} f_k^{(2)}(r) = \tfrac{1}{r^2} J_2(\lambda_k^{(2)} r) = (\lambda_k^{(2)})^2 \cdot (a_2 \cdot J_4(\lambda_k^{(2)} r) + \\ + b_2 \cdot J_2(\lambda_k^{(2)} r) + c_2 J_0(\lambda_k^{(2)} r)), \\ f_k^{(2)}(1) = 0, \; f_k^{(2)}(0) = c_2 (\lambda_k^{(2)})^2. \end{aligned} \qquad (18.35)$$

Integrating by parts, we get

$$\begin{aligned} \int_0^1 r f_k^{(2)}(r) J_2(\lambda_m^{(2)} r) dr = \frac{2 f_k^{(2)}(0)}{\lambda_m^{(2)}} \int_0^1 J_3(\lambda_m^{(2)} r) dr - \\ - \frac{1}{\lambda_m^{(2)}} \int_0^1 r (f_k^{(2)}(r))' J_3(\lambda_m^{(2)} r) dr + \frac{2}{\lambda_m^{(2)}} \int_0^1 r \cdot \frac{f_k^{(2)}(r) - f_k^{(2)}(0)}{r} \cdot r J_3(\lambda_m^{(2)} r) dr. \end{aligned} \qquad (18.36)$$

Substituting (18.35) for $f_k^{(2)}(r)$ in (18.36), we can estimate summands with functions J_2 and J_4 analogously to (18.29). Let us estimate the summands with function J_0. Using asymptotic formula

$$\int_0^1 J_3(\lambda_m^{(2)} r) dr = \frac{1}{\lambda_m^{(2)}} + \frac{K \cdot \theta(\lambda_m^{(2)})}{(\lambda_m^{(2)})^{\frac{3}{2}}},$$

where the constant $K > 0, \; 0 \leq \theta(\lambda_m^{(2)}) \leq 1$, we get

$$\left| \frac{2 J_0(0) c_2 \cdot (\lambda_k^{(2)})^2}{(\lambda_k^{(2)})^2} \int_0^1 J_3(\lambda_m^{(2)} r) dr \right| \leq \frac{2 c_2 \cdot K (\lambda_k^{(2)})^2}{(\lambda_m^{(2)})^2}. \qquad (18.37)$$

Estimating another summand, we obtain that there exists a constant $C > 0$ such that $\forall m \geq 1, \; \forall k \geq 1$

$$\left| \frac{1}{\lambda_m^{(2)}} \int_0^1 r \cdot c_2 (\lambda_k^{(2)})^2 \cdot \lambda_k^{(2)} J_1(\lambda_k^{(2)} r) J_3(\lambda_m^{(2)} r) dr \right| = \\ = \frac{c_2 \cdot (\lambda_k^{(2)})^3}{\lambda_m^{(2)}} \left| \int_0^1 r \cdot J_1(\lambda_k^{(2)} r) J_3(\lambda_m^{(2)} r) dr \right| \leq \frac{C \cdot (\lambda_k^{(2)})^4}{(\lambda_m^{(2)})^{\frac{5}{2}}}. \tag{18.38}$$

Finally, since

$$J_0(\lambda x) - 1 = \sum_{k=1}^{\infty} \frac{(-1)^k}{(k!)^2} \left(\frac{\lambda x}{2} \right)^{2k},$$

then the function

$$f(x) = \frac{J_0(\lambda x) - 1}{x} \in C^1([0, 1]),$$

and using the mean value theorem, we obtain that $\exists \theta \in (0, 1)$:

$$f'(x) = \frac{-\lambda x J_1(\lambda x) - (J_0(\lambda x) - 1)}{x^2} = -\frac{\lambda^2}{2}(J_0(\lambda x) + J_2(\lambda x)) + \\ + \frac{\theta \lambda^3}{2}(J_0(\theta \lambda x) + J_2(\theta \lambda x)).$$

Thus, from Theorem 18.3 it follows that

$$\left| \frac{2}{\lambda_m^{(2)}} \int_0^1 \frac{J_0(\lambda_k^{(2)} r) - 1}{r} \cdot r J_3(\lambda_m^{(2)} r) dr \right| \leq \frac{C \cdot (\lambda_k^{(2)})^3}{(\lambda_m^{(2)})^{\frac{5}{2}}} \tag{18.39}$$

with some constant $C > 0$.

So, there exists a constant $C > 0$ that depends only on J_2 such that

$$\forall m \geq 1 \; \forall k \geq 1 \; |A_m^{(2)}(f_k^{(2)})| \leq \frac{C (\lambda_k^{(2)})^4}{\lambda_m^{(2)}}. \tag{18.40}$$

From (18.40), we get that for $n = 1$

$$\forall m \geq 1 \; \forall t \in [0, T] \; |c_m^{(2)}(t)| \leq \frac{C}{\lambda_m^{(2)}}.$$

Therefore, for $n = 1$ the solution of (18.21) has the form (18.34).

Thus, the following theorem was proved.

Theorem 18.4 *Let p be an initial condition and $u(t, \theta)$ be continuous on $[0, T] \times [0, \pi]$ function of control for the problem (18.1). Suppose p and $u(t, \theta)$ satisfy the* *condition (18.10) for each $t \in [0, T]$. Moreover, assume that the functions g and h satisfy the condition (18.31).*

Then the classical solution of (18.1) has a form

$$y(t, r, \theta) = y_0(t, r)\varphi_0(\theta) + \sum_{n=1}^{N}(y_{2n-1}(t, r)\varphi_{2n-1}(\theta) + y_{2n}(t, r)\varphi_{2n}(\theta)),$$

where

$$\begin{aligned} y_0(t, r) &= p_0 \sum_{m=1}^{\infty} h_m^{(0)} J_0(\lambda_m^{(0)} r) e^{-(\lambda_m^{(0)})^2 t} + \\ &+ \sum_{m=1}^{\infty} g_m^{(0)} J_0(\lambda_m^{(0)} r) \int_0^t u_0(s) e^{-(\lambda_m^{(0)})^2 (t-s)} ds, \end{aligned} \tag{18.41}$$

$$\begin{aligned} y_{2n-1}(t, r) &= p_{2n-1} \sum_{m=1}^{\infty} h_m^{(2n)} J_{2n}(\lambda_m^{(2n)} r) e^{-(\lambda_m^{(2n)})^2 t} + \\ &+ \sum_{m=1}^{\infty} g_m^{(2n)} J_{2n}(\lambda_m^{(2n)} r) \int_0^t u_{2n-1}(s) e^{-(\lambda_m^{(2n)})^2 (t-s)} ds, \end{aligned} \tag{18.42}$$

$$\begin{aligned} y_{2n}(t, r) &= p_{2n} \sum_{m=1}^{\infty} h_m^{(2n)} J_{2n}(\lambda_m^{(2n)} r) e^{-(\lambda_m^{(2n)})^2 t} + \\ &+ \sum_{m=1}^{\infty} g_m^{(2n)} J_{2n}(\lambda_m^{(2n)} r) \int_0^t u_{2n}(s) e^{-(\lambda_m^{(2n)})^2 (t-s)} ds - \\ -4np_{2n-1} \cdot &\sum_{m=1}^{\infty} J_{2n}(\lambda_m^{(2n)} r) \sum_{k=1}^{\infty} h_k^{(2n)} A_m^{(2n)}(f_k^{(2n)}) \int_0^t e^{-(\lambda_k^{(2n)})^2 s} \cdot e^{-(\lambda_m^{(2n)})^2 (t-s)} ds - \\ &-4n \sum_{m=1}^{\infty} J_{2n}(\lambda_m^{(2n)} r) \sum_{k=1}^{\infty} g_k^{(2n)} A_m^{(2n)}(f_k^{(2n)}) \times \\ &\times \int_0^t \int_0^s u_{2n-1}(\tau) e^{-(\lambda_k^{(2n)})^2 (s-\tau)} \cdot e^{-(\lambda_m^{(2n)})^2 (t-s)} d\tau ds. \end{aligned} \tag{18.43}$$

18.5 The Optimal Control Problem (18.1) and (18.2)

For $n \geq 0$ let us consider the following functions

$$\Phi_n(t, r) = \sum_{m=1}^{\infty} h_m^{(2n)} J_{2n}(\lambda_m^{(2n)} r) e^{-(\lambda_m^{(2n)})^2 t},$$

$$\Psi_n(t,s,r) = \sum_{m=1}^{\infty} g_m^{(2n)} J_{2n}(\lambda_m^{(2n)} r) e^{-(\lambda_m^{(2n)})^2 (t-s)},$$

$$\gamma_n(t,r) = -4np_{2n-1} \cdot \sum_{m=1}^{\infty} J_{2n}(\lambda_m^{(2n)} r) \sum_{k=1}^{\infty} h_k^{(2n)} A_m^{(2n)}(f_k^{(2n)}) \cdot \int_0^t$$

$$e^{-(\lambda_k^{(2n)})^2 s - (\lambda_m^{(2n)})^2 (t-s)} ds,$$

$$z_n(t,s,r) = -4n \sum_{m=1}^{\infty} J_{2n}(\lambda_m^{(2n)} r) \sum_{k=1}^{\infty} g_k^{(2n)} A_m^{(2n)}(f_k^{(2n)}) \times$$
$$\times \int_s^t e^{-(\lambda_k^{(2n)})^2 (\tau - s) - (\lambda_m^{(2n)})^2 (t-\tau)} d\tau$$

that are continuous under the set of values.

From (18.41), equating the Freshet derivative of the functional J_0 to zero, we deduce that the optimal control $\tilde{u}_0$ is defined as a solution of Fredholm integral equation

$$u_0(t) = -\int_0^T \left(\int_0^1 r\Psi_0(T,t,r)\Psi_0(T,s,r)dr \right) u_0(s)ds -$$

$$- p_0 \int_0^1 r\Phi_0(T,r)\Psi_0(T,t,r)dr. \tag{18.44}$$

For $n \geq 1$, we consider the optimal control problem (18.5)–(18.7) with cost functional J_n with respect to unknown vector-function $\begin{pmatrix} u_{2n-1} \\ u_{2n} \end{pmatrix}$.

Then, changing the order of integration in the last summand of (18.43), we obtain

$$y_{2n-1}(t,r) = p_{2n-1} \cdot \Phi_n(t,r) + \int_0^t \Psi_n(t,s,r) u_{2n-1}(s) ds,$$

$$y_{2n}(t,r) = p_{2n} \cdot \Phi_n(t,r) + \int_0^t \Psi_n(t,s,r) u_{2n}(s) ds +$$
$$+ \gamma_n(t,r) + \int_0^t z_n(t,s,r) u_{2n-1}(s) ds.$$

Therefore, equating the Freshet derivative of the functional J_n to zero, we get that optimal control $\begin{pmatrix} \tilde{u}_{2n-1} \\ \tilde{u}_{2n} \end{pmatrix}$ is defined as a solution of Fredholm integral equation

$$\begin{pmatrix} u_{2n-1}(t) \\ u_{2n}(t) \end{pmatrix} = -\int_0^T K_n(t,s) \begin{pmatrix} u_{2n-1}(s) \\ u_{2n}(s) \end{pmatrix} ds - b_n(t), \tag{18.45}$$

where

$$b_n(t) = \begin{pmatrix} p_{2n-1} \cdot \int_0^1 r\Phi_n(T,r)\Psi_n(T,t,r)dr \\ \int_0^1 r(p_{2n} \cdot \Phi_n(T,r) + \gamma_n(T,r))z_n(T,t,r)dr \end{pmatrix},$$

and matrix $K_n(t,s)$ equals to

$$\begin{pmatrix} \int_0^1 r\Big(\Psi_n(T,t,r)\Psi_n(T,s,r) + z_n(T,t,r)z_n(T,s,r)\Big)dr & \int_0^1 rz_n(T,t,r)\Psi_n(T,s,r)dr \\ \int_0^1 r\Psi_n(T,t,r)z_n(T,s,r)dr & \int_0^1 r\Psi_n(T,t,r)\Psi_n(T,s,r)dr \end{pmatrix}.$$

Equations (18.45) and (18.44) has a unique solution in the class of continuous functions if the initial data are sufficiently small. More precisely, we obtain the following result.

Theorem 18.5 *Let the function p satisfies the condition (18.10) in the problems (18.1) and (18.2),*

$$g, h \in C^6([0,1]),\ g^{(p)}(0) = h^{(p)}(0) = g^{(p)}(1) = h^{(p)}(1) = 0,\ p = \overline{0,4},$$

and the following inequality holds: $\forall n = \overline{0, N}$

$$\begin{aligned} &2\int_0^1 rg^2(r)dr + 16n^2T^2 \cdot C \max_{x\in[0,1]} (\sum_{i=0}^{6} |g^{(i)}(x)|^2) + \\ &+ 8nT\sqrt{\int_0^1 rg^2(r)dr} \cdot \sqrt{C \cdot \max_{x\in[0,1]} (\sum_{i=0}^{6} |g^{(i)}(x)|^2)} < \frac{1}{T}, \end{aligned} \tag{18.46}$$

where the constant $C = C(n)$ depends only on J_{2n}.

Then the optimal control problems (18.1) and (18.2) has the unique solution $\widetilde{u}(t,\theta)$, and

$$\widetilde{u}(t,\theta) = \sum_{n=0}^{N} \widetilde{u}_n(t)\varphi_n(\theta),$$

where functions $\{\widetilde{u}_n\}_{n=0}^N$ are defined from the Eqs. (18.44) and (18.45).

18.6 Conclusions

In this paper, we considered the optimal control problem for linear parabolic equation with nonlocal boundary conditions in circular sector and quadratic cost functional. We proved a solvability of such problem for a special class of the initial data and distributed controls.

References

1. Lions, J.-L.: Optimal Control in PDE Systems. Mir, Moscow (1972)
2. Egorov, A.I.: Optimal Control in Heat and Diffusion Processes. Nauka, Moscow (1978)
3. Belozerov, V.E., Kapustyan, V.E.: Geometrical Methods of Modal Control. Naukova Dumka, Kyiv (1999)
4. Zgurovsky, M.Z., Mel'nik, V.S.: Nonlinear Analysis and Control of Physical Processes and Fields. Springer, Berlin (2004)
5. Moiseev, E.I., Ambarzumyan, V.E.: About resolvability of non-local boundary-value problem with equality of fluxes. Differ. Equ. (Russian) **46**(5), 718–725 (2010)
6. Kapustyan, V.O., Kapustian, O.A., Mazur, O.K.: Distributed optimal control in one non-self-adjoint boundary value problem. Contin. Distrib. Syst. **31**(12), 45–52 (2014)
7. Watson, G.N.: Theory of Bessel Functions. Cambridge University Press, Cambridge (1945)
8. Scherberg, M.G.: The degree of convergence of a series of Bessel functions. Trans. Am. Math. Soc. **35**, 172–183 (1933)
9. Tihonov, A.N., Samarsky, A.A.: Equations of Mathematical Physics. Nauka, Moscow (1966)
10. Kolmogorov, A.N., Fomin, S.V.: Elements of Function Theory and Functional Analysis. Nauka, Moscow (1976)

Chapter 19
On the Existence of Weak Optimal Controls in the Coefficients for a Degenerate Anisotropic p-Laplacian

Olha P. Kupenko and Günter Leugering

Abstract We consider an optimal control problem for nonlinear degenerate elliptic problems involving an anisotropic p-Laplacian and Dirichlet boundary conditions. We take the matrix-valued coefficients $A(x)$ of such system as a control in $L^{p/2}(\Omega; \mathbb{R}^{\frac{N(N+1)}{2}})$. One of the important features of the admissible controls is the fact that eigenvalues of the coefficient matrices may vanish in Ω. Equations of this type may exhibit the Lavrentiev phenomenon and nonuniqueness of weak solutions. Using the concept of convergence in variable spaces and following the direct method in the calculus of variations, we establish the solvability of this optimal control problem in the class of weak solutions.

19.1 Introduction

The aim of this article is to study the existence of so-called weak optimal controls in the coefficients for a nonlinear anisotropic elliptic equation with homogeneous Dirichlet boundary conditions. The controls are taken as the matrix of the coefficients A in the main part of the elliptic operator. The most important feature of such controls is the fact that eigenvalues of the matrix A may either vanish on subsets with zero Lebesgue measure or be unbounded. In this case the precise answer to the question

O.P. Kupenko (✉)
Dnipropetrovsk Mining University, Karl Marks Av., 19,
Dnipropetrovsk 49005, Ukraine
e-mail: kogut_olga@bk.ru

O.P. Kupenko
Institute for Applied System Analysis, National Technical University
of Ukraine "Kiev Polytechnic Institute", Peremogy Av., 37, Building 35,
Kiev 03056, Ukraine

G. Leugering
Department Mathematik, Friedrich-Alexander-Universität
Erlangen-Nürnberg Lehrstuhl AMII, Cauerstr. 11, 91058
Erlangen, Germany
e-mail: guenter.leugering@fau.de

© Springer International Publishing Switzerland 2015
V.A. Sadovnichiy and M.Z. Zgurovsky (eds.), *Continuous and Distributed Systems II*, Studies in Systems, Decision and Control 30,
DOI 10.1007/978-3-319-19075-4_19

of existence or nonexistence of optimal solutions heavily depends on the class of admissible controls chosen. Using the direct method in the calculus of variations, we discuss the solvability of this optimal control problem (OCP) in a class of weak solutions with degenerate $L^{p/2}$-controls in coefficients. In contrast to [7], we do not make use of any relaxations for the original optimal control problem.

We deal with an OCP for the following state system:

$$\begin{cases} -\operatorname{div}\big(|(A(x)\nabla y, \nabla y)_{\mathbb{R}^N}|^{(p-2)/2} A\nabla y\big) = f & \text{in } \Omega, \\ y = 0 & \text{on } \partial\Omega, \end{cases} \tag{19.1}$$

where $f \in L^q(\Omega)$ is a given function and a measurable nonnegative symmetric matrix $A \in L^{p/2}(\Omega; \mathbb{R}^{\frac{N(N+1)}{2}})$ $(p \geq 2)$ is adopted as a control. Following Alessandrini & Sigalotti, we denote the elliptic partial differential equation as anisotropic p-Laplace equation (see [1] and references therein). We define a class of admissible controls $\mathfrak{A}_{ad}$ as a nonempty compact subset of $L^{p/2}(\Omega; \mathbb{R}^{\frac{N(N+1)}{2}})$ such that for every $A \in \mathfrak{A}_{ad}$ we have

$$\zeta_{ad}(x)I \leq A(x) \leq \beta(x)I \quad \text{a. e. in } \Omega, \qquad \int_\Omega A(x)\,dx = M,$$

where $M \in \mathbb{S}^N_{sym}$ is a given nonzero matrix, $\beta \in L^{p/2}(\Omega)$, $\beta \geq \zeta_{ad}$, and $\zeta_{ad}^{-1} \in L^1(\Omega)$. In what follows, we assume that the functions ζ_{ad} and β are smooth and positive around the boundary $\partial\Omega$.

Equation (19.1) can be viewed as the Euler equation for the variational integral

$$J(y) = \int_\Omega |(A\nabla y, \nabla y)_{\mathbb{R}^N}|^{p/2}\,dx$$

and its interest is motivated by various recent applications related to composite materials such as nonlinear dielectric composites, the nonlinearity of which is modeled by a power law.

In the case $p = 2$, the boundary value problem (19.1) becomes linear

$$-\operatorname{div}(A(x)\nabla y) = f.$$

Optimal control problems for linear degenerated objects with L^1-matrix controls were studied in [14, 17]. The case $A = \rho(x)I$ for linear and nonlinear boundary value problems was studied in [15, 16, 22, 28].

There are numerous articles (see, for instance, [6, 8, 21, 22, 28] and references therein) which are devoted to variational and non-variational approaches to problems related to (19.1). As for the optimal control problems in coefficients for degenerate elliptic equations, we can refer to [3–5, 14–17]. Because the boundary value problem (19.1) for the locally integrable matrix-valued function A can exhibit the Lavrentiev phenomenon and, e.g., nonuniqueness of weak solutions, we cannot expect the existence of weak solutions to (19.1) for all admissible initial data $f \in L^q(\Omega)$ and $A \in L^{p/2}(\Omega; \mathbb{R}^{N\times N})$. Indeed, it is clear that the anisotropic p-Laplacian, which can be defined via the pairing

$$\langle \mathscr{A} y, v\rangle_{W(\Omega, A\,dx)} = \left\langle -\operatorname{div}\left(|(A\nabla y, \nabla y)_{\mathbb{R}^N}|^{(p-2)/2} A\nabla y\right), v\right\rangle_{W(\Omega, A\,dx)}$$
$$= \int_\Omega \left(|A^{1/2}\nabla y|^{p-2}_{\mathbb{R}^N} A\nabla y, \nabla v\right)_{\mathbb{R}^N} dx,$$

turns out to be noncoercive. Following the idea of Kogut & Leugering (see [14]) we consider the OCP in the framework of variable Lebesgue spaces $L^p(\Omega, A\,dx)$, where the degenerate anisotropic elliptic operator enjoys monotonicity, semi-continuity, and coercivity. We remark that, even if the original elliptic equation is nondegenerate, i.e., the admissible controls $A(x)$ are such that

$$\alpha\|\xi\|^2_{\mathbb{R}^N} \le (\xi, A(x)\xi)_{\mathbb{R}^N} \le \beta\|\xi\|^2_{\mathbb{R}^N} \quad \forall \xi \in \mathbb{R}^N,$$

with $\alpha > 0$, the optimal control problem may not have any solution in general (see, for instance, [20]). Hence, the problem of existence of solutions to the anisotropic OCP is nontrivial and important for further investigations such as derivation and substantiation of first-order optimality conditions, attainability of optimal solutions, etc.

The paper is organized as follows. Section 19.2 concerns some notation and preliminaries. In Sect. 19.3 we deal with the concepts of weak and strong convergence in variable Lebesgue spaces $L^p(\Omega, A\,dx)$, where we devote some attention to the limiting procedure in such spaces. In Sects. 19.4 and 19.5 we discuss some specific properties of the class of admissible controls and give the precise setting of the optimal control problem in coefficients. In Sect. 19.6, in spite of the fact that the original boundary value problem is ill-possed in general, we show that the corresponding extremal problem is well posed. In particular, we prove that the set of admissible pairs Ξ_w to the considered problem is sequentially closed and the optimization problem itself is solvable in the class of weak solutions.

19.2 Notation and Preliminaries

Let Ω be a bounded open subset of $\mathbb{R}^N$ ($N \ge 2$) with Lipschitz boundary. Let p be a real number such that $2 \le p < \infty$ and q be its dual, namely, $1/p + 1/q = 1$. We define the Banach space $W_0^{1,1}(\Omega)$ as the closure of $C_0^\infty(\Omega)$ in the classical Sobolev space $W^{1,1}(\Omega)$. For any subset $E \subset \Omega$ we denote by $|E|$ its N-dimensional Lebesgue measure $\mathscr{L}^N(E)$.

Symmetric matrices with degenerate eigenvalues. We denote by $\mathbb{S}^N := \mathbb{R}^{\frac{N(N+1)}{2}}$ the set of all symmetric matrices $\boldsymbol{\xi} = [\xi_{ij}]_{i,j=1}^N$, $(\xi_{ij} = \xi_{ji})$. We suppose that $\mathbb{S}^N$ is endowed with the euclidian scalar product $\boldsymbol{\xi} \cdot \boldsymbol{\eta} = \mathrm{tr}(\boldsymbol{\xi}\,\boldsymbol{\eta}) = \xi_{ij}\eta_{ij}$ and with the corresponding euclidian norm $\|\boldsymbol{\xi}\|_{\mathbb{S}^N} = (\boldsymbol{\xi} \cdot \boldsymbol{\xi})^{1/2}$. Let

$$L^{p/2}(\Omega)^{\frac{N(N+1)}{2}} = L^{p/2}\big(\Omega; \mathbb{S}^N\big)$$

be the space of $p/2$-integrable matrix-valued functions with symmetric matrices as values.

Let $\alpha \in \mathbb{R}$ be a fixed positive value. Let $\zeta_{ad} \in L^{p/2}(\Omega)$ be a given function satisfying the properties

$$\zeta_{ad}^{-1} \in L^1(\Omega), \quad \zeta_{ad}^{-1} \notin L^\infty(\Omega).$$

Let $\zeta_{ad}: \Omega \to \mathbb{R}^1_+$ be smooth along the boundary $\partial\Omega$ and

$$\beta(x) \geq \zeta_{ad}(x) \geq \alpha > 0 \quad \text{on} \quad \partial\Omega. \tag{19.2}$$

By $\mathfrak{M}_\alpha^\beta(\Omega)$ we denote the set of all matrices $A(x) = [a_{i\,j}(x)\,] \in \mathbb{S}^N$ such that

$$\zeta_{ad}(x)I \leq A(x) \leq \beta(x)I \quad \text{a. e. in } \Omega, \tag{19.3}$$

Here $\beta \in L^{p/2}(\Omega)$ is a given function such that $\beta(x) > 0$ a.e. in Ω, I is the identity matrix in $\mathbb{R}^{N\times N}$ and (19.3) should be considered in the sense of quadratic forms. Therefore, (19.3) implies the following inequalities:

$$\text{if } A \in L^{p/2}(\Omega; \mathbb{S}^N), \text{ then } \|A(x)\|_{L^{p/2}(\Omega;\mathbb{S}^N)} \leq \|\beta\|_{L^{p/2}(\Omega)} < +\infty, \tag{19.4}$$

$$\zeta_{ad}(x)\|\xi\|^2_{\mathbb{R}^N} \leq (A(x)\xi, \xi)_{\mathbb{R}^N} \quad \text{a. e. in } \Omega, \ \forall \xi \in \mathbb{R}^N. \tag{19.5}$$

Remark 19.1 Since every measurable matrix-valued function $A: \Omega \to \mathbb{S}^N$ can be associated with the collection of its eigenvalues $\{\lambda_1^A, \dots, \lambda_N^A\}$, where each $\lambda_k^A = \lambda_k^A(x)$ is counted with its multiplicity, $(19.3)_1$ means that eigenvalues of matrices $A \in \mathfrak{M}_\alpha^\beta(\Omega)$ may vanish on subdomains of Ω with zero Lebesgue measure. Because of this, these matrices are sometime referred to as matrices with degenerate spectrum.

Weighted Sobolev Spaces. To each matrix $A \in \mathfrak{M}_\alpha^\beta(\Omega)$ we will associate two weighted Sobolev spaces:

$$W_A(\Omega) = W(\Omega, A\,dx) \text{ and } H_A(\Omega) = H(\Omega, A\,dx),$$

where $W_A(\Omega)$ is the set of functions $y \in W_0^{1,1}(\Omega)$ for which the norm

$$\|y\|_A = \left(\int_\Omega \left(y^p + (\nabla y, A(x)\nabla y)_{\mathbb{R}^N}^{p/2}\right) dx\right)^{1/p} \tag{19.6}$$

is finite, and $H_A(\Omega)$ is the closure of $C_0^\infty(\Omega)$ in $W_A(\Omega)$. Note that due to the inequality (19.5) and estimates

$$\int_\Omega |y|\,dx \le \Big(\int_\Omega |y|^p\,dx\Big)^{1/p} |\Omega|^{1/q} \le C\|y\|_A, \tag{19.7}$$

$$\int_\Omega \|\nabla y\|_{\mathbb{R}^N}\,dx \le \Big(\int_\Omega \|\nabla y\|^p_{\mathbb{R}^N}\zeta_{ad}^{p/2}\,dx\Big)^{1/p}\Big(\int_\Omega \zeta_{ad}^{-q/2}\,dx\Big)^{1/q}$$
$$\le \Big(\int_\Omega (\nabla y, A(x)\nabla y)^{p/2}_{\mathbb{R}^N}\,dx\Big)^{1/p}\bigg(\Big(\int_\Omega \zeta_{ad}^{-1}\,dx\Big)^{q/2}|\Omega|^{(2-q)/2}\bigg)^{1/q}$$
$$\le C\Big(\int_\Omega (\nabla y, A(x)\nabla y)^{p/2}_{\mathbb{R}^N}\,dx\Big)^{1/p} \le C\|y\|_A, \tag{19.8}$$

the space $W_A(\Omega)$ is complete with respect to the norm $\|\cdot\|_A$. It is clear that $H_A(\Omega) \subset W_A(\Omega)$, and $W_A(\Omega)$, $H_A(\Omega)$ are Lebesgue spaces. If the eigenvalues $\{\lambda_1^A, \dots, \lambda_N^A\}$ of $A\colon \Omega \to \mathbb{S}^N$ are bounded between two positive constants, then it is easy to verify that $W_A(\Omega) = H_A(\Omega)$ and the norm (19.6) is equivalent to the classical norm of the space $W_0^{1,p}(\Omega)$. However, for a "typical" weight matrix $A \in \mathfrak{M}_\alpha^\beta(\Omega)$ the space of smooth functions $C_0^\infty(\Omega)$ is not dense in $W_A(\Omega)$. Hence the identity $W_A(\Omega) = H_A(\Omega)$ is not always valid (for the corresponding examples in the case when $A(x) = \rho(x)I$, we refer to [9, 25]).

Weak Compactness Criterion in $L^1(\Omega; \mathbb{S}^N)$. Throughout the paper we will often use the concept of weak and strong convergence in $L^1(\Omega; \mathbb{S}^N)$. Let $\{A_\varepsilon\}_{\varepsilon>0}$ be a bounded sequence of matrices in $L^{p/2}(\Omega; \mathbb{S}^N)$, then it is obviously bounded in $L^1(\Omega; \mathbb{S}^N)$. We recall that $\{A_\varepsilon\}_{\varepsilon>0}$ is called equi-integrable on Ω, if for any $\delta > 0$ there is a $\tau = \tau(\delta)$ such that $\int_S \|A_\varepsilon\|_{\mathbb{S}^N}\,dx < \delta$ for every measurable subset $S \subset \Omega$ of Lebesgue measure $|S| < \tau$. Then the following assertions are equivalent for $L^1(\Omega; \mathbb{S}^N)$-bounded sequences:

(i) a sequence $\{A_k\}_{k\in\mathbb{N}}$ is weakly compact in $L^1(\Omega; \mathbb{S}^N)$;
(ii) the sequence $\{A_k\}_{k\in\mathbb{N}}$ is equi-integrable.

Theorem 19.1 (Lebesgue's Theorem) *If a sequence $\{A_k\}_{k\in\mathbb{N}} \subset L^1(\Omega; \mathbb{S}^N)$ is equi-integrable and $A_k \to A$ almost everywhere in Ω then $A_k \to A$ in $L^1(\Omega; \mathbb{S}^N)$.*

Functions with Bounded Variation. Let $f\colon \Omega \to \mathbb{R}$ be a function of $L^1(\Omega)$. Define

$$\int_\Omega |Df| = \sup\bigg\{\int_\Omega f\,\mathrm{div}\varphi\,dx \;:$$
$$\varphi = (\varphi_1, \dots, \varphi_N) \in C_0^1(\Omega; \mathbb{R}^N),\ |\varphi(x)| \le 1 \text{ for } x \in \Omega\bigg\},$$

where $\mathrm{div}\varphi = \sum_{i=1}^N \frac{\partial \varphi_i}{\partial x_i}$.

Definition 19.1 A function $f \in L^1(\Omega)$ is said to have a bounded variation in Ω if $\int_\Omega |Df| < +\infty$. By $BV(\Omega)$ we denote the space of all functions in $L^1(\Omega)$ with bounded variation.

Under the norm $\|f\|_{BV(\Omega)} = \|f\|_{L^1(\Omega)} + \int_\Omega |Df|$, $BV(\Omega)$ is a Banach space. The following compactness result for BV-functions is well known:

Proposition 19.1 *Uniformly bounded sets in BV-norm are relatively compact in* $L^1(\Omega)$.

Definition 19.2 A sequence $\{f_k\}_{k=1}^\infty \subset BV(\Omega)$ weakly converges to some $f \in BV(\Omega)$, and we write $f_k \rightharpoonup f$ iff the two following conditions hold: $f_k \to f$ strongly in $L^1(\Omega)$, and $Df_k \rightharpoonup Df$ weakly-$*$ in the space of Radon measures $M(\Omega; \mathbb{R}^N)$, i.e.,

$$\lim_{k\to\infty} \int_\Omega (\varphi, Df_k)_{\mathbb{R}^N} = \int_\Omega (\varphi, Df)_{\mathbb{R}^N} \qquad \forall \varphi \in C_0(\mathbb{R}^N)^N, \tag{19.9}$$

In the proposition below we give a compactness result related to this convergence, together with lower semi-continuity (see [11]):

Proposition 19.2 *Let* $\{f_k\}_{k=1}^\infty$ *be a sequence in* $BV(\Omega)$ *strongly converging to some* f *in* $L^1(\Omega)$ *and satisfying* $\sup_{k\in\mathbb{N}} \int_\Omega |Df_k| < +\infty$. *Then*

(i) $f \in BV(\Omega)$ *and* $\displaystyle\int_\Omega |Df| \le \liminf_{k\to\infty} \int_\Omega |Df_k|$;
(ii) $f_k \rightharpoonup f$ *in* $BV(\Omega)$.

19.3 $\mathbb{S}^N$-Valued Radon Measures and Weak Convergence in Variable L^p-Spaces

By a nonnegative Radon measure on Ω we mean a nonnegative Borel measure which is finite on every compact subset of Ω. The space of all nonnegative Radon measures on Ω will be denoted by $M_+(\Omega)$. According to the Riesz theory, each Radon measure $\mu \in M_+(\Omega)$ can be interpreted as an element of the dual of the space $C_0(\Omega)$ of all continuous functions with compact support. Let $M(\Omega; \mathbb{S}^N)$ denote the space of all $\mathbb{S}^N$-valued Borel measures. Then $\boldsymbol{\mu} = [\mu_{ij}] \in M(\Omega; \mathbb{S}^N) \Leftrightarrow \mu_{ij} \in C_0'(\Omega)$, $i, j = 1, \dots, N$.

Let $\boldsymbol{\mu}$ and the sequence $\{\boldsymbol{\mu}_k\}_{k\in\mathbb{N}}$ be matrix-valued Radon measures. We say that $\{\boldsymbol{\mu}_k\}_{k\in\mathbb{N}}$ weakly-$*$ converges to $\boldsymbol{\mu}$ in $M(\Omega; \mathbb{S}^N)$ if

$$\lim_{k\to\infty} \int_\Omega \boldsymbol{\varphi} \cdot d\boldsymbol{\mu}_k = \int_\Omega \boldsymbol{\varphi} \cdot d\boldsymbol{\mu} \quad \forall \boldsymbol{\varphi} \in C_0(\Omega; \mathbb{S}^N).$$

In our further considerations we use the following example of such measures

$$d\boldsymbol{\mu}_k = A_k(x)\,dx, \quad d\boldsymbol{\mu} = A(x)\,dx, \tag{19.10}$$

where

$$A_k, A \in \mathfrak{M}_\alpha^\beta(\Omega) \cap L^{p/2}(\Omega; \mathbb{S}^N) \text{ and } A_k \to A \text{ strongly in } L^1(\Omega; \mathbb{S}^N). \tag{19.11}$$

As we will see later (see Lemma 19.1), the sets $\mathfrak{M}_\alpha^\beta(\Omega) \cap L^{p/2}(\Omega; \mathbb{S}^N)$ are sequentially closed with respect to strong convergence in $L^1(\Omega; \mathbb{S}^N)$.

In this section we suppose that the measures $\boldsymbol{\mu}$ and $\{\boldsymbol{\mu}_k\}_{k\in\mathbb{N}}$ are defined by (19.10) and $\boldsymbol{\mu}_k \stackrel{*}{\rightharpoonup} \boldsymbol{\mu}$ in $M(\Omega; \mathbb{S}^N)$. Further, we will use $L^p(\Omega, A\,dx)^N$ to denote the set of measurable vector-valued functions $\mathbf{f} \in \mathbb{R}^N$ on Ω such that

$$\|\mathbf{f}\|_{L^p(\Omega, A\,dx)^N} = \left(\int_\Omega |A^{1/2}(x)\mathbf{f}|^p\,dx\right)^{1/p} = \left(\int_\Omega (\mathbf{f}, A(x)\mathbf{f})_{\mathbb{R}^N}^{p/2}\,dx\right)^{1/p} < +\infty.$$

As follows from estimate (19.8) any vector-valued function of $L^p(\Omega, A\,dx)^N$ is Lebesgue integrable on Ω.

We say that a sequence $\{\mathbf{v}_k \in L^p(\Omega, A_k\,dx)^N\}_{k\in\mathbb{N}}$ is bounded if

$$\limsup_{k\to\infty} \int_\Omega (\mathbf{v}_k, A_k(x)\mathbf{v}_k)_{\mathbb{R}^N}^{p/2}\,dx < +\infty.$$

Definition 19.3 A bounded sequence $\{\mathbf{v}_k \in L^p(\Omega, A_k\,dx)^N\}_{k\in\mathbb{N}}$ is weakly convergent to a function $\mathbf{v} \in L^p(\Omega, A\,dx)^N$ in the variable space $L^p(\Omega, A_k\,dx)^N$ if

$$\lim_{k\to\infty} \int_\Omega (\boldsymbol{\varphi}, A_k(x)\mathbf{v}_k)_{\mathbb{R}^N}\,dx = \int_\Omega (\boldsymbol{\varphi}, A(x)\mathbf{v})_{\mathbb{R}^N}\,dx \quad \forall\,\boldsymbol{\varphi} \in C_0^\infty(\Omega)^N. \tag{19.12}$$

The main property concerning the weak convergence in $L^p(\Omega, A_k\,dx)^N$ can be expressed as follows (see for comparison [14, 26]):

Proposition 19.3 *If a sequence* $\{\mathbf{v}_k \in L^p(\Omega, A_k\,dx)^N\}_{k\in\mathbb{N}}$ *is bounded, then it is compact in the sense of weak convergence in* $L^p(\Omega, A_k\,dx)^N$.

Proof Having set $L_k(\boldsymbol{\varphi}) = \int_\Omega (\boldsymbol{\varphi}, A_k(x)\mathbf{v}_k)_{\mathbb{R}^N}\,dx$ $\forall\,\boldsymbol{\varphi} \in C_0^\infty(\Omega)^N$ and making use the Hölder inequality, we get

$$\begin{aligned} |L_k(\boldsymbol{\varphi})| &\le \left(\int_\Omega |A_k^{1/2}\mathbf{v}_k|_{\mathbb{R}^N}^p\,dx\right)^{1/p} \left(\int_\Omega |A_k^{1/2}\boldsymbol{\varphi}|_{\mathbb{R}^N}^q\,dx\right)^{1/q} \\ &= \left(\int_\Omega (\mathbf{v}_k, A_k\mathbf{v}_k)_{\mathbb{R}^N}^{p/2}\,dx\right)^{1/p} \left(\int_\Omega (\boldsymbol{\varphi}, A_k\boldsymbol{\varphi})_{\mathbb{R}^N}^{q/2}\,dx\right)^{1/q} \end{aligned}$$

$$\leq C\left(\int_\Omega (\boldsymbol{\varphi}, A_k\boldsymbol{\varphi})^{q/2}_{\mathbb{R}^N}\,dx\right)^{1/q} \leq C\left(\int_\Omega (\beta(x))^{q/2}\|\boldsymbol{\varphi}\|^q_{\mathbb{R}^N}\,dx\right)^{1/q}$$
$$\leq C\|\boldsymbol{\varphi}\|_{C(\Omega)^N}\left(\int_\Omega (\beta(x))^{q/2}dx\right)^{1/q} \leq C\|\boldsymbol{\varphi}\|_{C(\Omega)^N}\left(\left(\int_\Omega (\beta(x))^{p/2}\,dx\right)^{q/p}|\Omega|^{(p-q)/p}\right)^{1/q}$$
$$\leq C\|\boldsymbol{\varphi}\|_{C(\Omega)^N}\|\beta\|^{1/2}_{L^{p/2}(\Omega)}|\Omega|^{(p-2)/p} \quad \forall k\in\mathbb{N}. \tag{19.13}$$

Since the set $C_0^\infty(\Omega)^N$ is separable with respect to the norm $\|\cdot\|_{C(\Omega)^N}$ and $\{L_k(\boldsymbol{\varphi})\}_{k\in\mathbb{N}}$ is a uniformly bounded sequence of linear functionals, it follows that there exists a subsequence of positive numbers $\{k_j\}_{j=1}^\infty$ for which the limit (in the sense of pointwise convergence)

$$\lim_{j\to\infty} L_{k_j}(\boldsymbol{\varphi}) = L(\boldsymbol{\varphi}) \tag{19.14}$$

is well defined for every $\boldsymbol{\varphi}\in C_0^\infty(\Omega)^N$. As a result, using (19.11), we have

$$|L(\boldsymbol{\varphi})| \leq C_1\lim_{k\to\infty}\left(\int_\Omega (\boldsymbol{\varphi}, A_k\boldsymbol{\varphi})^{q/2}_{\mathbb{R}^N}\,dx\right)^{1/q}$$
$$\leq C_1\lim_{k\to\infty}\left(\left(\int_\Omega (\boldsymbol{\varphi}, A_k\boldsymbol{\varphi})_{\mathbb{R}^N}\,dx\right)^{q/2}|\Omega|^{(2-q)/2}\right)^{1/q}$$
$$= C_1\left(\int_\Omega (\boldsymbol{\varphi}, A\boldsymbol{\varphi})_{\mathbb{R}^N}\,dx\right)^{1/2}|\Omega|^{(2-q)/2q}$$
$$\leq C_1\left(\left(\int_\Omega (\boldsymbol{\varphi}, A\boldsymbol{\varphi})^{p/2}_{\mathbb{R}^N}\,dx\right)^{2/p}|\Omega|^{(p-2)/p}\right)^{1/2}|\Omega|^{(2-q)/2q}$$
$$= \widetilde{C}_1\left(\int_\Omega (\boldsymbol{\varphi}, A\boldsymbol{\varphi})^{p/2}_{\mathbb{R}^N}\,dx\right)^{1/p}.$$

Hence, $L(\boldsymbol{\varphi})$ is a continuous functional on $L^p(\Omega, A\,dx)^N$ admitting following representation $L(\boldsymbol{\varphi}) = \int_\Omega (\boldsymbol{\varphi}, A(x)\mathbf{v})_{\mathbb{R}^N}\,dx$, where $\mathbf{v}$ is some element of $L^p(\Omega, A\,dx)^N$. Thus, in view of Definition 19.3, $\mathbf{v}$ can be taken as the weak limit of the sequence

$$\left\{\mathbf{v}_k \in L^p(\Omega, A_k\,dx)^N\right\}_{k\in\mathbb{N}}.$$

Remark 19.2 It is easy to see, that when $p\geq 2$ then $1 < q = \dfrac{p}{p-1} \leq 2 \leq p$ and hence $L^p(\Omega, A\,dx)^N \hookrightarrow L^q(\Omega, A\,dx)^N$ with continuous embedding, namely

$$\|\mathbf{v}\|_{L^q(\Omega,A\,dx)^N} = \left(\int_\Omega (\mathbf{v}, A(x)\mathbf{v})_{\mathbb{R}^N}^{q/2} dx\right)^{1/q}$$
$$\leq C\left(\int_\Omega (\mathbf{v}, A(x)\mathbf{v})_{\mathbb{R}^N}^{p/2} dx\right)^{1/p} = C\|\mathbf{v}\|_{L^p(\Omega,A\,dx)^N}.$$

The next property of weak convergence in $L^p(\Omega, A_k\,dx)^N$ shows that the variable L^p-norm is lower semi-continuous with respect to weak convergence.

Proposition 19.4 *If the sequence* $\left\{\mathbf{v}_k \in L^p(\Omega, A_k\,dx)^N\right\}_{k\in\mathbb{N}}$ *converges weakly to* $\mathbf{v} \in L^p(\Omega, A\,dx)^N$*, then*

$$\liminf_{k\to\infty} \int_\Omega (\mathbf{v}_k, A_k(x)\mathbf{v}_k)_{\mathbb{R}^N}^{p/2}\,dx \geq \int_\Omega (\mathbf{v}, A(x)\mathbf{v})_{\mathbb{R}^N}^{p/2}\,dx. \tag{19.15}$$

Proof For the proof we use the well-known Young inequality $ab \leq \dfrac{|a|^p}{p} + \dfrac{|b|^q}{q}$ for $a,\, b \geq 0$ and $p,\, q \geq 1$. Indeed, we have

$$\frac{1}{p}\int_\Omega (\mathbf{v}_k, A_k\mathbf{v}_k)_{\mathbb{R}^N}^{p/2}\,dx = \frac{1}{p}\int_\Omega |A_k^{1/2}\mathbf{v}_k|_{\mathbb{R}^N}^p\,dx$$
$$\geq \int_\Omega \left(A_k^{1/2}\boldsymbol{\varphi}, A_k^{1/2}\mathbf{v}_k\right)_{\mathbb{R}^N}\,dx - \frac{1}{q}\int_\Omega |A_k^{1/2}\boldsymbol{\varphi}|_{\mathbb{R}^N}^q\,dx$$
$$= \int_\Omega (\boldsymbol{\varphi}, A_k\mathbf{v}_k)_{\mathbb{R}^N}\,dx - \frac{1}{q}\int_\Omega (\boldsymbol{\varphi}, A_k\boldsymbol{\varphi})_{\mathbb{R}^N}^{q/2}\,dx \quad \forall \boldsymbol{\varphi} \in C_0^\infty(\Omega)^N,$$
$$\frac{1}{p}\liminf_{k\to\infty}\int_\Omega (\mathbf{v}_k, A_k\mathbf{v}_k)_{\mathbb{R}^N}^{p/2}\,dx \geq \int_\Omega (\boldsymbol{\varphi}, A\mathbf{v})_{\mathbb{R}^N}\,dx - \frac{1}{q}\int_\Omega (\boldsymbol{\varphi}, A\boldsymbol{\varphi})_{\mathbb{R}^N}^{q/2}\,dx. \tag{19.16}$$

It is worth to mention that the limit passage

$$\lim_{k\to\infty}\int_\Omega (\boldsymbol{\varphi}, A_k\boldsymbol{\varphi})_{\mathbb{R}^N}^{q/2}\,dx = \int_\Omega (\boldsymbol{\varphi}, A\boldsymbol{\varphi})_{\mathbb{R}^N}^{q/2}\,dx$$

is guaranteed by the strong L^1-convergence of the sequence $\{A_k\}_{k\in\mathbb{N}} \subset L^{p/2}(\Omega; \mathbb{S}^N)$ to the element $A \in L^{p/2}(\Omega; \mathbb{S}^N)$. Since the inequality (19.16) is valid for all $\boldsymbol{\varphi} \in C_0^\infty(\Omega)^N$ and $C_0^\infty(\Omega)^N$ is a dense subset of $L^q(\Omega, A\,dx)^N$, it holds also true for $\boldsymbol{\varphi} \in L^q(\Omega, A\,dx)^N$. So, taking $\boldsymbol{\varphi} = |A^{1/2}\mathbf{v}|_{\mathbb{R}^N}^{p-2}\mathbf{v}$, we arrive at the following chain of relations

$$\frac{1}{p}\liminf_{k\to\infty}\int_\Omega (\mathbf{v}_k, A_k(x)\mathbf{v}_k)_{\mathbb{R}^N}^{p/2}\,dx$$
$$\geq \int_\Omega \left(|A^{1/2}\mathbf{v}|_{\mathbb{R}^N}^{p-2}\mathbf{v}, A\mathbf{v}\right)_{\mathbb{R}^N} dx - \frac{p-1}{p}\int_\Omega \left(|A^{1/2}\mathbf{v}|_{\mathbb{R}^N}^{p-2}\mathbf{v}, A|A^{1/2}\mathbf{v}|_{\mathbb{R}^N}^{p-2}\mathbf{v}\right)_{\mathbb{R}^N}^{q/2} dx$$
$$= \int_\Omega (\mathbf{v}, A\mathbf{v})_{\mathbb{R}^N}^{p/2}\,dx - \frac{p-1}{p}\int_\Omega \left(|A^{1/2}\mathbf{v}|_{\mathbb{R}^N}^{2(p-1)}\right)^{p/(2(p-1))} dx = \frac{1}{p}\int_\Omega (\mathbf{v}, A(x)\mathbf{v})_{\mathbb{R}^N}^{p/2}\,dx,$$

which completes the proof.

Definition 19.4 A sequence $\left\{\mathbf{v}_k \in L^p(\Omega, A_k\,dx)^N\right\}_{k\in\mathbb{N}}$ is said to be strongly convergent to a function $\mathbf{v} \in L^p(\Omega, A\,dx)^N$ if

$$\lim_{k\to\infty}\int_\Omega (\mathbf{b}_k, A_k(x)\mathbf{v}_k)_{\mathbb{R}^N}\,dx = \int_\Omega (\mathbf{b}, A(x)\mathbf{v})_{\mathbb{R}^N}\,dx \tag{19.17}$$

whenever $\mathbf{b}_k \rightharpoonup \mathbf{b}$ in $L^q(\Omega, A_k\,dx)^N$ as $k \to \infty$.

We have the following property of strong convergence in the variable space $L^p(\Omega, A_k\,dx)^N$.

Proposition 19.5 *Weak convergence of a sequence* $\left\{\mathbf{v}_k \in L^p(\Omega, A_k\,dx)^N\right\}_{k\in\mathbb{N}}$ *to* $\mathbf{v} \in L^p(\Omega, A\,dx)^N$ *and*

$$\lim_{k\to\infty}\int_\Omega (\mathbf{v}_k, A_k(x)\mathbf{v}_k)_{\mathbb{R}^N}^{p/2}\,dx = \int_\Omega (\mathbf{v}, A(x)\mathbf{v})_{\mathbb{R}^N}^{p/2}\,dx \tag{19.18}$$

are equivalent to strong convergence of $\{\mathbf{v}_k\}_{k\in\mathbb{N}}$ *in* $L^p(\Omega, A_k\,dx)^N$ *to the element* $\mathbf{v} \in L^p(\Omega, A\,dx)^N$.

Proof We proceed with the proof following an idea due to V.V. Zhikov (see [27]).

Step 1. First, let us show that strong convergence $\mathbf{v}_k \to \mathbf{v}$ in $L^p(\Omega, A_k\,dx)^N$ implies weak convergence and (19.18). Indeed, if we use $\{\mathbf{b}_k = \boldsymbol{\varphi} \in C_0^\infty(\Omega)^N\}_{k\in\mathbb{N}}$ in (19.17) we immediately get weak convergence of $\{\mathbf{v}_k\}_{k\in\mathbb{N}}$ to $\mathbf{v}$ in $L^p(\Omega, A_k\,dx)^N$.

Now let us set $\mathbf{z}_k = |A_k^{1/2}\mathbf{v}_k|_{\mathbb{R}^N}^{p-2}\mathbf{v}_k\ \forall\, k \in \mathbb{N}$. It is easy to see that the sequence $\{\mathbf{z}_k\}_{k\in\mathbb{N}}$ is bounded in $L^q(\Omega, A_k\,dx)^N$. Indeed, we have

$$\|\,|A_k^{1/2}\mathbf{v}_k|_{\mathbb{R}^N}^{p-2}\,\mathbf{v}_k\|_{L^q(\Omega,A_k\,dx)}^q = \int_\Omega (A_k\mathbf{v}_k, \mathbf{v}_k)_{\mathbb{R}^N}^{p/2}\,dx = \|A_k^{1/2}\mathbf{v}_k\|_{L^p(\Omega,A_k\,dx)}^p \leq C.$$

Then we may assume that there exists an element $\mathbf{z} \in L^q(\Omega, A\,dx)^N$ such that, within a subsequence, $\mathbf{z}_k \to \mathbf{z}$ weakly in $L^q(\Omega, A_k\,dx)^N$. We have

$$\lim_{k\to\infty}\int_\Omega |A_k^{1/2}\mathbf{v}_k|_{\mathbb{R}^N}^p\,dx = \lim_{k\to\infty}\int_\Omega (\mathbf{v}_k, A_k\mathbf{z}_k)_{\mathbb{R}^N}\,dx = \int_\Omega (\mathbf{v}, A(x)\mathbf{z})_{\mathbb{R}^N}. \tag{19.19}$$

It is enough to show that $\mathbf{z} = |A^{1/2}\mathbf{v}|_{\mathbb{R}^N}^{p-2}\mathbf{v}$. To do this, we consider vector-valued functions $g_k(\boldsymbol{t}) = |A_k^{1/2}\boldsymbol{t}\,|_{\mathbb{R}^N}^{p-2}\boldsymbol{t}$ and $g_0(\boldsymbol{t}) = |A^{1/2}\boldsymbol{t}\,|_{\mathbb{R}^N}^{p-2}\boldsymbol{t}$ which obviously satisfy

$$\int_\Omega (g_k(\boldsymbol{t}_1) - g_k(\boldsymbol{t}_2), A_k(\boldsymbol{t}_1 - \boldsymbol{t}_2))_{\mathbb{R}^N}\, dx \geq 2^{p-2} \int_\Omega |A_k^{1/2}(\boldsymbol{t}_1 - \boldsymbol{t}_2)|^p_{\mathbb{R}^N}\, dx \geq 0,\ k = 1, 2, \ldots,$$
$$\int_\Omega (g_0(\boldsymbol{t}_1) - g_0(\boldsymbol{t}_2), A(\boldsymbol{t}_1 - \boldsymbol{t}_2))_{\mathbb{R}^N}\, dx \geq 2^{p-2} \int_\Omega |A^{1/2}(\boldsymbol{t}_1 - \boldsymbol{t}_2)|^p_{\mathbb{R}^N}\, dx \geq 0.$$

Hence, $\int_\Omega ((g_k(\boldsymbol{\varphi}) - g_k(\mathbf{v}_k)), A_k(\boldsymbol{\varphi} - \mathbf{v}_k))_{\mathbb{R}^N}\, dx \geq 0$ for all $\boldsymbol{\varphi} \in C_0^\infty(\Omega)^N$. Passing to the limit in the last relation and taking into account (19.19) we get

$$\int_\Omega ((g_0(\boldsymbol{\varphi}) - \mathbf{z}), A(\boldsymbol{\varphi} - \mathbf{v}))_{\mathbb{R}^N}\, dx \geq 0\ \forall\, \boldsymbol{\varphi} \in C_0^\infty(\Omega)^N.$$

By continuity, this relation takes place for all $\boldsymbol{\varphi} \in L^p(\Omega, A\, dx)^N$ and hence, taking $\boldsymbol{\varphi} = \mathbf{v} + t\mathbf{w}$, where $\mathbf{w} \in L^p(\Omega, A\, dx)^N$ we obtain after passing to the limit as $t \to 0$

$$\int_\Omega ((g_0(\mathbf{v}) - \mathbf{z}), A\mathbf{w})_{\mathbb{R}^N}\, dx = 0\ \forall\, \mathbf{w} \in L^p(\Omega, A\, dx)^N.$$

which implies $\mathbf{z} = g_0(\mathbf{v}) = |A^{1/2}\mathbf{v}|^{p-2}_{\mathbb{R}^N}\mathbf{v}$.

Step 2. For the reader's convenience we prove the following inequality which plays the key role subsequently.

$$0 \leq |\boldsymbol{a} + t\boldsymbol{b}|^p_{\mathbb{R}^N} - |\boldsymbol{a}|^p_{\mathbb{R}^N} - tp\, |\boldsymbol{a}|^{p-2}_{\mathbb{R}^N}(\boldsymbol{a}, \boldsymbol{b})_{\mathbb{R}^N} \leq Ct^2, \tag{19.20}$$

$p > 1$ and the constant C is strictly positive, for each $\boldsymbol{a}, \boldsymbol{b}$. To do this let us consider the function $\Theta(t) = |\boldsymbol{a} + t\boldsymbol{b}|^p_{\mathbb{R}^N}$ and its Taylor formula with a residual term

$$\Theta(t) = \Theta(0) + \Theta'(0)t + \frac{1}{2}\Theta''(\tau)t^2,\ |\tau| \leq |t| \leq 1.$$

Here $\Theta'(t) = p|\boldsymbol{a} + t\boldsymbol{b}|^{p-2}_{\mathbb{R}^N}\left(t|\boldsymbol{b}|^2_{\mathbb{R}^N} + (\boldsymbol{a}, \boldsymbol{b})_{\mathbb{R}^N}\right)$, and

$$\begin{aligned}\Theta''(t) &= p\,(p-2)|\boldsymbol{a} + t\boldsymbol{b}|^{p-4}_{\mathbb{R}^N}\left(t|\boldsymbol{b}|^2_{\mathbb{R}^N} + (\boldsymbol{a}, \boldsymbol{b})_{\mathbb{R}^N}\right)^2 + p\,|\boldsymbol{a} + t\boldsymbol{b}|^{p-2}_{\mathbb{R}^N}|\boldsymbol{b}|^2_{\mathbb{R}^N}\\ &= p\,|\boldsymbol{a} + t\boldsymbol{b}|^{p-2}_{\mathbb{R}^N}|\boldsymbol{b}|^2_{\mathbb{R}^N}\left(1 + (p-2)\frac{\left(t|\boldsymbol{b}|^2_{\mathbb{R}^N} + (\boldsymbol{a}, \boldsymbol{b})_{\mathbb{R}^N}\right)^2}{|\boldsymbol{b}|^2_{\mathbb{R}^N}|\boldsymbol{a} + t\boldsymbol{b}|^2_{\mathbb{R}^N}}\right).\end{aligned}$$

It is easy to see that setting $\dfrac{\boldsymbol{a}}{|\boldsymbol{b}|_{\mathbb{R}^N}} = \boldsymbol{\xi}$, $\dfrac{\boldsymbol{b}}{|\boldsymbol{b}|_{\mathbb{R}^N}} = \boldsymbol{\eta}$, we have

$$0 \leq \frac{\left(t|\boldsymbol{b}|^2_{\mathbb{R}^N} + (\boldsymbol{a}, \boldsymbol{b})_{\mathbb{R}^N}\right)^2}{|\boldsymbol{b}|^2_{\mathbb{R}^N}|\boldsymbol{a} + t\boldsymbol{b}|^2_{\mathbb{R}^N}} = \frac{|\boldsymbol{b}|^4_{\mathbb{R}^N}\left(t^2 + 2t(\boldsymbol{\xi}, \boldsymbol{\eta})_{\mathbb{R}^N} + (\boldsymbol{\xi}, \boldsymbol{\eta})^2_{\mathbb{R}^N}\right)}{|\boldsymbol{b}|^4_{\mathbb{R}^N}(t^2 + 2t(\boldsymbol{\xi}, \boldsymbol{\eta})_{\mathbb{R}^N} + |\boldsymbol{\xi}|^2_{\mathbb{R}^N})} \leq 1,$$

since $(\boldsymbol{\xi}, \boldsymbol{\eta})^2_{\mathbb{R}^N} \le |\boldsymbol{\xi}|^2_{\mathbb{R}^N}$. Hence, for $p \ge 2$ we have

$$0 \le \Theta''(\tau) \le p\,(p-1)|\boldsymbol{a} + \tau \boldsymbol{b}|^{p-2}_{\mathbb{R}^N} |\boldsymbol{b}|^2_{\mathbb{R}^N} = C_1,\ C_1 > 0\,\forall\, \boldsymbol{a}, \boldsymbol{b} \neq \mathbf{0}.$$

and for $1 < p < 2$ we have

$$0 \le p\,(p-1)|\boldsymbol{a}+\tau \boldsymbol{b}|^{p-2}_{\mathbb{R}^N} |\boldsymbol{b}|^2_{\mathbb{R}^N} \le \Theta''(\tau) \le p\,|\boldsymbol{a}+\tau \boldsymbol{b}|^{p-2}_{\mathbb{R}^N} |\boldsymbol{b}|^2_{\mathbb{R}^N} = \frac{1}{p-1} C_1 = C_2.$$

Hence, (19.20) follows from $0 \le \Theta(t) - \Theta(0) - t\Theta'(0) = \frac{1}{2}\Theta''(\tau)t^2 \le \frac{1}{2} C_k t^2$, $k = 1, 2$.

Step 3. Now we prove the following statement: if $\mathbf{v_k} \to \mathbf{v}$ weakly in the space $L^p(\Omega, A_k\,dx)^N$ and (19.18) holds, then $\mathbf{z_k} = |A_k^{1/2}\mathbf{v}_k|^{p-2}_{\mathbb{R}^N}\mathbf{v}_k \to |A^{1/2}\mathbf{v}|^{p-2}_{\mathbb{R}^N}\mathbf{v}$ weakly in $L^q(\Omega, A_k\,dx)^N$. For $\boldsymbol{\varphi} \in C_0^\infty(\Omega)^N$ and $|t| \le 1$ with respect to (19.20) we have

$$\int_\Omega |A_k^{1/2}(\mathbf{v}_k + t\,\boldsymbol{\varphi})|^p_{\mathbb{R}^N}\,dx \le \int_\Omega |A_k^{1/2}\mathbf{v}_k|^p_{\mathbb{R}^N}\,dx + p\,t \int_\Omega (\mathbf{z}_k, A_k\boldsymbol{\varphi})_{\mathbb{R}^N}\,dx + \widetilde{C}t^2.$$

Here we pass to the limit taking into account (19.18) in the right-hand side and the lower semi-continuity property (19.15) in the left-hand side

$$\int_\Omega |A^{1/2}(\mathbf{v} + t\,\boldsymbol{\varphi})|^p_{\mathbb{R}^N}\,dx \le \int_\Omega |A^{1/2}\mathbf{v}|^p_{\mathbb{R}^N}\,dx + p\,t \int_\Omega (\mathbf{z}, A\boldsymbol{\varphi})_{\mathbb{R}^N}\,dx + \widetilde{C}t^2, \tag{19.21}$$

where $\mathbf{z}$ is a weak limit of $\{\mathbf{z}_k\}_{k\in\mathbb{N}}$. On the other hand, inequality (19.20) also implies

$$\int_\Omega |A^{1/2}(\mathbf{v} + t\,\boldsymbol{\varphi})|^p_{\mathbb{R}^N}\,dx \ge \int_\Omega |A^{1/2}\mathbf{v}|^p_{\mathbb{R}^N}\,dx + p\,t \int_\Omega (|A^{1/2}\mathbf{v}|^{p-2}_{\mathbb{R}^N}\mathbf{v}, A\boldsymbol{\varphi}))_{\mathbb{R}^N}\,dx. \tag{19.22}$$

Hence, combining (19.21) and (19.22) we obtain

$$p\,t \int_\Omega (|A^{1/2}\mathbf{v}|^{p-2}_{\mathbb{R}^N}\mathbf{v}, A\boldsymbol{\varphi}))_{\mathbb{R}^N} \le p\,t \int_\Omega (\mathbf{z}, A\boldsymbol{\varphi})_{\mathbb{R}^N}\,dx + \widetilde{C}_1 t^2,$$

which gives $\mathbf{z} = |A^{1/2}\mathbf{v}|^{p-2}_{\mathbb{R}^N}\mathbf{v}$.

Step 4. Finally, let us establish the desired strong convergence $\mathbf{v_k} \to \mathbf{v}$ in $L^p(\Omega, A_k\,dx)^N$ on the basis of weak convergence $\mathbf{v_k} \to \mathbf{v}$ in $L^p(\Omega, A_k\,dx)^N$ and (19.18). Assume that $\mathbf{b}_k \to \mathbf{b}$ strongly in $L^q(\Omega, A_k\,dx)^N$ and within a subsequence

$$\lim_{k\to\infty} \int_\Omega (\mathbf{v}_k, A_k(x)\mathbf{b}_k)_{\mathbb{R}^N}\,dx = \gamma.$$

We must show that $\gamma = \int_\Omega (\mathbf{v}, A(x)\mathbf{b})_{\mathbb{R}^N}\,dx$. To do it we use inequality (19.20) once more for the dual exponent $q = p/(p-1)$. If $\mathbf{z}_k = |A_k^{1/2}\mathbf{v}_k|_{\mathbb{R}^N}^{p-2}\mathbf{v}_k$, then

$$\begin{aligned}\int_\Omega |A_k^{1/2}(\mathbf{z}_k + t\mathbf{b}_k)|_{\mathbb{R}^N}^q\,dx \\ \le \int_\Omega |A_k^{1/2}\mathbf{z}_k|_{\mathbb{R}^N}^q\,dx + q\,t\int_\Omega (|A_k^{1/2}\mathbf{z}_k|_{\mathbb{R}^N}^{q-2}\mathbf{z}_k, A_k\mathbf{b}_k)_{\mathbb{R}^N}\,dx + \widetilde{C}_2 t^2 \\ = \int_\Omega |A_k^{1/2}\mathbf{v}_k|_{\mathbb{R}^N}^p\,dx + q\,t\int_\Omega (\mathbf{v}_k, A_k\mathbf{b}_k)_{\mathbb{R}^N}\,dx + \widetilde{C}_2 t^2,\end{aligned}$$

since $|A_k^{1/2}\mathbf{z}_k|_{\mathbb{R}^N}^q = |A_k^{1/2}\mathbf{v}_k|_{\mathbb{R}^N}^p$ and $|A_k^{1/2}\mathbf{z}_k|_{\mathbb{R}^N}^{q-2}\mathbf{z}_k = \mathbf{v}_k$. Passing to the limit in the previous relation gives

$$\begin{aligned}\int_\Omega |A^{1/2}(\mathbf{z} + t\mathbf{b})|_{\mathbb{R}^N}^q\,dx \le \int_\Omega |A^{1/2}\mathbf{v}|_{\mathbb{R}^N}^p\,dx + q\,t\gamma + \widetilde{C}_2 t^2 \\ = \int_\Omega |A^{1/2}\mathbf{z}|_{\mathbb{R}^N}^q\,dx + q\,t\gamma + \widetilde{C}_2 t^2,\end{aligned} \tag{19.23}$$

since $\mathbf{z} = |A^{1/2}\mathbf{v}|_{\mathbb{R}^N}^{p-2}\mathbf{v}$, as we showed in Step 3, and hence $|A^{1/2}\mathbf{z}|_{\mathbb{R}^N}^q = |A^{1/2}\mathbf{v}|_{\mathbb{R}^N}^p$. On the other hand, we have

$$\begin{aligned}\int_\Omega |A^{1/2}(\mathbf{z} + t\mathbf{b})|_{\mathbb{R}^N}^q\,dx \ge \int_\Omega |A^{1/2}\mathbf{z}|_{\mathbb{R}^N}^q\,dx + q\,t\int_\Omega (|A^{1/2}\mathbf{z}|_{\mathbb{R}^N}^{q-2}\mathbf{z}, A\mathbf{b})_{\mathbb{R}^N}\,dx \\ = \int_\Omega |A^{1/2}\mathbf{z}|_{\mathbb{R}^N}^q\,dx + q\,t\int_\Omega (\mathbf{v}, A\mathbf{b})_{\mathbb{R}^N}\,dx.\end{aligned} \tag{19.24}$$

As a result of combining (19.23) and (19.24) we obtain

$$q\,t\int_\Omega (\mathbf{v}, A(x)\mathbf{b})_{\mathbb{R}^N}\,dx \le q\,t\gamma + \widetilde{C}_2 t^2,$$

which immediately implies the equity $\gamma = \int_\Omega (\mathbf{v}, A(x)\mathbf{b})_{\mathbb{R}^N}\,dx$ and completes the proof.

19.4 Auxiliary Results

Definition 19.5 We say that a bounded sequence

$$\left\{(A_k, y_k) \in L^{p/2}(\Omega; \mathbb{S}^N) \times W_{A_k}(\Omega)\right\}_{k\in\mathbb{N}} \tag{19.25}$$

w-converges to $(A, y) \in L^{p/2}(\Omega, \mathbb{S}^N) \times W^{1,1}(\Omega)$ as $k \to \infty$ if

$$A_k \to A \quad \text{strongly in } L^1(\Omega; \mathbb{S}^N), \tag{19.26}$$

$$y_k \to y \quad \text{weakly in } L^p(\Omega), \tag{19.27}$$

$$\nabla y_k \to \nabla y \quad \text{weakly in the variable space } L^p(\Omega, A_k\, dx)^N, \tag{19.28}$$

therefore,

$$\lim_{k\to\infty} \int_\Omega A_k \cdot \boldsymbol{\eta}\, dx = \int_\Omega A \cdot \boldsymbol{\eta}\, dx \quad \forall \boldsymbol{\eta} \in L^\infty(\Omega; \mathbb{S}^N), \tag{19.29}$$

$$\lim_{k\to\infty} \int_\Omega y_k \lambda\, dx = \int_\Omega y\lambda\, dx \quad \forall \lambda \in L^q(\Omega), \tag{19.30}$$

$$\lim_{k\to\infty} \int_\Omega (\boldsymbol{\xi}, A_k \nabla y_k)_{\mathbb{R}^N}\, dx = \int_\Omega (\boldsymbol{\xi}, A\nabla y)_{\mathbb{R}^N}\, dx \quad \forall \boldsymbol{\xi} \in C_0^\infty(\Omega)^N. \tag{19.31}$$

In order to motivate this definition, we give the following result.

Lemma 19.1 *Let* $\left\{(A_k, y_k) \in L^{p/2}(\Omega; \mathbb{S}^N) \times W_{A_k}(\Omega)\right\}_{k\in\mathbb{N}}$ *be a sequence such that*

(i) the sequence $\left\{y_k \in W_{A_k}(\Omega)\right\}_{k\in\mathbb{N}}$ *is bounded, i.e.,*

$$\sup_{k\in\mathbb{N}} \int_\Omega \left(|y_k|^p + (\nabla y_k, A_k \nabla y_k)_{\mathbb{R}^N}^{p/2}\right) dx < +\infty; \tag{19.32}$$

(ii) $\{A_k\}_{k\in\mathbb{N}} \subset \mathfrak{M}_\alpha^\beta(\Omega)$ *and there exists a matrix-valued function* $A(x) \in \mathbb{S}^N$ *such that*

$$A_k \to A \ \text{ and } \ A_k^{-1} \to A^{-1} \ \text{ in } L^1(\Omega; \mathbb{S}^N) \ \text{ as } k \to \infty. \tag{19.33}$$

Then, $A \in \mathfrak{M}_\alpha^\beta(\Omega) \cap L^{p/2}(\Omega; \mathbb{S}^N)$ *and the original sequence is relatively compact with respect to w-convergence. Moreover, each w-limit pair* (A, y) *belongs to the space* $L^{p/2}(\Omega; \mathbb{S}^N) \times W_A(\Omega)$.

Proof We note that (19.32), (19.33), (19.7), (19.8), and (19.4) immediately imply the boundedness of the original sequence in $L^{p/2}(\Omega; \mathbb{S}^N) \times W^{1,1}(\Omega)$. Moreover, due to (19.33), we have:

$$d\boldsymbol{\mu}_n := A_n\, dx \stackrel{*}{\rightharpoonup} A\, dx =: d\boldsymbol{\mu} \ \text{ in } M(\Omega; \mathbb{S}^N).$$

Thus, the compactness criterium for weak convergence in variable spaces (see Proposition 19.3) and (19.32) imply the existence of a pair $(y, \mathbf{v}) \in L^p(\Omega) \times L^p(\Omega, A\, dx)^N$ such that, within a subsequence of $\{y_k\}_{k\in\mathbb{N}}$,

$$y_k \to y \text{ weakly in } L^p(\Omega), \tag{19.34}$$

$$\nabla y_k \to \mathbf{v} \text{ weakly in variable space } L^p(\Omega, A_k\,dx)^N. \tag{19.35}$$

Our aim is to show that $A \in \mathfrak{M}_\alpha^\beta(\Omega)$, $\mathbf{v} = \nabla y$, and $y \in W_A(\Omega)$. It is clear that $A(x) \in \mathbb{S}^N$ and this matrix satisfies (19.3). Thus, $A \in \mathfrak{M}_\alpha^\beta(\Omega)$ and the limit matrix $A(x) \in \mathbb{S}^N$ satisfies (19.4) and (19.5).

For our further analysis we fix any test function $\boldsymbol{\varphi} \in C_0^\infty(\Omega)^N$ and observe

$$\lim_{k\to\infty} \int_\Omega \left(A_k^{-1}\boldsymbol{\varphi}, A_k\boldsymbol{\psi}\right)_{\mathbb{R}^N} dx = \int_\Omega (\boldsymbol{\varphi}, \boldsymbol{\psi})_{\mathbb{R}^N}\,dx = \int_\Omega \left(A^{-1}\boldsymbol{\varphi}, A\boldsymbol{\psi}\right)_{\mathbb{R}^N} dx, \tag{19.36}$$

which is obviously true for each $\boldsymbol{\psi} \in C_0^\infty(\Omega)^N$ and for all $k \in \mathbb{N}$. Since

$$\begin{aligned}
\limsup_{k\to\infty} \int_\Omega \left(A_k^{-1}\boldsymbol{\varphi}, A_k A_k^{-1}\boldsymbol{\varphi}\right)_{\mathbb{R}^N}^{q/2} dx &= \limsup_{k\to\infty} \int_\Omega \left(\boldsymbol{\varphi}, A_k^{-1}\boldsymbol{\varphi}\right)_{\mathbb{R}^N}^{q/2} dx \\
&\le \int_\Omega \zeta_{ad}^{-q/2} |\boldsymbol{\varphi}|_{\mathbb{R}^N}^q\,dx \le \|\boldsymbol{\varphi}\|_{C(\Omega)^N}^q \int_\Omega \zeta_{ad}^{-q/2}\,dx \\
\le \|\boldsymbol{\varphi}\|_{C(\Omega)^N}^q \left(\int_\Omega \zeta_{ad}^{-1} dx\right)^{q/2} &|\Omega|^{(2-q)/2} \le C\|\boldsymbol{\varphi}\|_{C(\Omega)^N}^q \|\zeta_{ad}^{-1}\|_{L^1(\Omega)}^{q/2} < +\infty,
\end{aligned}$$

it follows that the sequence $\left\{A_k^{-1}\boldsymbol{\varphi} \in L^q(\Omega, A_k\,dx)^N\right\}_{k\in\mathbb{N}}$ is bounded. Consequently, combining this fact with (19.36), we conclude $A_k^{-1}\boldsymbol{\varphi} \to A^{-1}\boldsymbol{\varphi}$ weakly in the variable space $L^q(\Omega, A_k\,dx)^N$ (see Definition 19.3). At the same time, strong convergence in (19.33) implies the relation

$$\begin{aligned}
\lim_{k\to\infty} \int_\Omega \left(A_k^{-1}\boldsymbol{\varphi}, A_k A_k^{-1}\boldsymbol{\varphi}\right)_{\mathbb{R}^N}^{q/2} dx &= \lim_{k\to\infty} \int_\Omega \left(\boldsymbol{\varphi}, A_k^{-1}\boldsymbol{\varphi}\right)_{\mathbb{R}^N}^{q/2} dx \\
&= \int_\Omega (\boldsymbol{\varphi}, A^{-1}\boldsymbol{\varphi})_{\mathbb{R}^N}^{q/2}\,dx = \int_\Omega (A^{-1}\boldsymbol{\varphi}, AA^{-1}\boldsymbol{\varphi})_{\mathbb{R}^N}^{q/2}\,dx.
\end{aligned}$$

Indeed, strong convergence of the sequence $\{\left(\boldsymbol{\varphi}, A_k^{-1}\boldsymbol{\varphi}\right)_{\mathbb{R}^N}^{q/2}\}_{k\in\mathbb{N}}$ in $L^1(\Omega)$ follows from its equi-integrability and convergence a.e. in Ω, implied by (19.33). Hence (see Proposition 19.5),

$$A_k^{-1}\boldsymbol{\varphi} \to A^{-1}\boldsymbol{\varphi} \text{ strongly in } L^q(\Omega, A_k\,dx)^N \quad \forall\,\boldsymbol{\varphi} \in C_0^\infty(\Omega)^N. \tag{19.37}$$

Further, we note that for every measurable subset $K \subset \Omega$, the estimate

$$\int_K \|\nabla y_k\|_{\mathbb{R}^N}\, dx \le \Big(\int_K \|\nabla y_k\|^p_{\mathbb{R}^N} \zeta_{ad}^{p/2}\, dx\Big)^{1/p}\Big(\int_K \zeta_{ad}^{-q/2}\, dx\Big)^{1/q}$$
$$\le \Big(\int_\Omega (\nabla y_k, A_k(x)\nabla y_k)^{p/2}_{\mathbb{R}^N}\, dx\Big)^{1/p}\Big(\int_K \zeta_{ad}^{-q/2}\, dx\Big)^{1/q} \le C\Big(\int_K \zeta_{ad}^{-1}\, dx\Big)^{1/2}$$

implies equi-integrability of the family $\big\{\|\nabla y_k\|_{\mathbb{R}^N}\big\}_{k\in\mathbb{N}}$. Hence, $\big\{\|\nabla y_k\|_{\mathbb{R}^N}\big\}_{k\in\mathbb{N}}$ is weakly compact in $L^1(\Omega)$, which means the weak compactness of the vector-valued sequence $\{\nabla y_k\}_{k\in\mathbb{N}}$ in $L^1(\Omega;\mathbb{R}^N)$. As a result, by the properties of the strong convergence in variable spaces, we obtain

$$\int_\Omega (\boldsymbol{\xi}, \nabla y_k)_{\mathbb{R}^N}\, dx = \int_\Omega \Big(A_k^{-1}\boldsymbol{\xi}, A_k\nabla y_k\Big)_{\mathbb{R}^N}\, dx$$
$$\overset{\text{by (19.17), (19.35), and (19.37)}}{\longrightarrow} \int_\Omega \Big(A^{-1}\boldsymbol{\xi}, A\mathbf{v}\Big)_{\mathbb{R}^N}\, dx$$
$$= \int_\Omega (\boldsymbol{\xi}, \mathbf{v})_{\mathbb{R}^N}\, dx \quad \forall \boldsymbol{\xi} \in C_0^\infty(\Omega)^N.$$

Thus, in view of the weak compactness property of $\{\nabla y_k\}_{k\in\mathbb{N}}$ in $L^1(\Omega;\mathbb{R}^N)$, we conclude

$$\nabla y_k \to \mathbf{v} \text{ weakly in } L^1(\Omega;\mathbb{R}^N) \quad \text{as } k\to\infty. \tag{19.38}$$

Since $y_k \in W_0^{1,1}(\Omega)$ for all $k\in\mathbb{N}$ and the Sobolev space $W_0^{1,1}(\Omega)$ is complete, (19.34) and (19.38) imply $\nabla y = \mathbf{v}$, and consequently $y \in W_0^{1,1}(\Omega)$. To end the proof, it remains to observe that (19.34) and (19.35) guarantee the finiteness of the norm $\|y\|_A$ (see (19.6)). Hence, $y \in W_A(\Omega)$ and this concludes the proof.

19.5 Setting of the Optimal Control Problem

Let $M \in \mathbb{S}^N$ be a given constant matrix satisfying the condition

$$(M\xi,\xi)_{\mathbb{R}^N} \ge m\|\xi\|^2_{\mathbb{R}^N} \quad \text{for some } m>0.$$

Let $f \in L^q(\Omega)$ be a given function. We consider the following boundary value problem

$$-\operatorname{div}\Big(|(A(x)\nabla y, \nabla y)_{\mathbb{R}^N}|^{(p-2)/2} A(x)\nabla y\Big) = f \quad \text{in } \Omega, \tag{19.39}$$
$$y = 0 \text{ on } \partial\Omega. \tag{19.40}$$

To introduce the class of admissible controls in coefficients, we adopt the following concept:

Definition 19.6 We say that a matrix-valued function $A = A(x) \in \mathbb{S}^N$ is an admissible control for the boundary value problem (19.39)–(19.40) (it is written as $A \in \mathfrak{A}_{ad}$) if

$$A \in BV(\Omega; \mathbb{S}^N), \quad \int_\Omega A(x)\,dx = M, \tag{19.41}$$

$$A \in \mathfrak{M}_\alpha^\beta(\Omega). \tag{19.42}$$

Hereinafter we assume that the set $\mathfrak{A}_{ad}$ is nonempty.

Remark 19.3 In view of (19.42) and $(19.3)_1$ (see also Remark 19.1), we deal with a boundary value problem for the degenerate elliptic equation. It means that for some admissible controls $A \in \mathfrak{A}_{ad}$ the boundary value problem (19.39) and (19.40) can exhibit the Lavrentiev phenomenon and nonuniqueness of the weak solutions.

Definition 19.7 We say that a function $y = y(A, f)$ is a weak solution (in the sense of Minty) to boundary value problem (19.39) and (19.40) for a fixed control $A \in \mathfrak{A}_{ad}$ and given function $f \in L^q(\Omega)$ if $y \in W_A(\Omega)$ and the inequality

$$\int_\Omega |A^{1/2}\nabla\varphi|_{\mathbb{R}^N}^{p-2}(A\nabla\varphi, \nabla\varphi - \nabla y)_{\mathbb{R}^N}\,dx \geq \int_\Omega f(\varphi - y)\,dx \tag{19.43}$$

holds for any $\varphi \in C_0^\infty(\Omega)$.

Remark 19.4 Another definition of the weak solution to the considered boundary value problem appears more natural:

$$y \in W_A(\Omega),$$
$$\int_\Omega |A^{1/2}\nabla y|_{\mathbb{R}^N}^{p-2}(A\nabla y, \nabla\varphi)_{\mathbb{R}^N}\,dx = \int_\Omega f\varphi\,dx \;\forall\,\varphi \in C_0^\infty(\Omega).$$

However, as was shown by Pastukhova [22], both concepts for the weak solutions coincide in the case when the subspace of smooth functions $C_0^\infty(\Omega)$ is dense in the weighted space $W_A(\Omega)$, which is not true for the case of "typical" degenerate matrix-valued weight function A.

Also, it is worth to notice that the original boundary value problem (19.39) and (19.40) is ill-possed, in general. Indeed, this problem may not admit a weak solution $y \in W_A(\Omega)$ in the sense of Definition 19.7 for every admissible initial data $f \in L^q(\Omega)$ and $A \in \mathfrak{A}_{ad}$. So, it is not possible to write in this case $y = y(A, f)$. Moreover, it should be emphasized that, to the best knowledge of authors, the existence of a weak solution to (19.39) and (19.40) for fixed $A \in \mathfrak{A}_{ad}$ and $f \in L^q(\Omega)$ is an open question. On the other hand, even if a weak solution to the above problem exists, the question about its uniqueness leads us again to the problem of density of the subspace of smooth functions $C_0^\infty(\Omega)$ in $W_A(\Omega)$. However, as was indicated in [28], there exists a diagonal matrix-valued function $A(x) = \rho(x)I$ with $\rho \geq \zeta_{ad}$ a.e. in Ω such that the subspace $C_0^\infty(\Omega)$ is not dense in $W_A(\Omega)$, and, hence, there is no uniqueness of weak solutions (for more details and other types of solutions we refer to [2, 25, 28]). Thus, the mapping $A \mapsto y(A, f)$ can be multivalued, in general.

To avoid this situation in our analysis, we introduce the set of admissible solutions to the original optimal control problem as follows:

$$\Xi_w = \big\{(A, y) \mid A \in \mathfrak{A}_{ad},\ y \in W_A(\Omega),\ (A, y) \text{ are related by (19.43)}\big\}. \quad (19.44)$$

The optimal control problem we consider here is to minimize the discrepancy (tracking error) between a given distribution $y_d \in L^p(\Omega)$ and the solution of boundary valued problem (19.39) and (19.40) by choosing an appropriate coefficients matrix $A \in \mathfrak{A}_{ad}$. More precisely, we are concerned with the following optimal control problem

$$\text{Minimize} \left\{ I(A, y) = \int_\Omega |y(x) - y_d(x)|^p\, dx + \int_\Omega (\nabla y(x), A(x)\nabla y(x))^{p/2}_{\mathbb{R}^N}\, dx + \sum_{i,j=1}^N \int_\Omega |D\, a_{ij}(x)| \right\} \quad (19.45)$$

subject to the constraints (19.39)–(19.42).

Remark 19.5 The second term in (19.45) plays a special role in this problem. Its appearance in the cost function (19.45) is motivated by the fact that there are no appropriate a priori estimates in the $W_A(\Omega)$-norm for weak solutions $y = y(A, f)$ of the degenerate boundary value problem (19.39) and (19.40) (in the sense of Definition 19.7). Hence, the term $\int_\Omega (\nabla y(x), A(x)\nabla y(x))^{p/2}_{\mathbb{R}^N}\, dx$ together with the first one in (19.45) ensures the coercivity of the cost function on the space of weak solutions $W_A(\Omega)$.

Remark 19.6 Note that due to (19.7) and (19.8), we have the following obvious inclusion for the set of admissible solutions

$$\Xi_w \subset L^{p/2}(\Omega; \mathbb{S}^N) \times W^{1,1}_0(\Omega).$$

However, the characteristic feature of this set is the fact that for different admissible controls $A \in \mathfrak{A}_{ad}$ the corresponding admissible solutions y of optimal control problem (19.39)–(19.42) and (19.45) belong to different weighted spaces. It is a nontypical situation from the point of view of classical optimal control theory.

It is worth noticing that for any admissible initial data $f \in L^q(\Omega)$, verification of $\Xi_w \neq \emptyset$ is a nontrivial matter, in general. In the particular case, when the set of admissible controls $\mathfrak{A}_{ad}$ possesses the property:

$$A \in L^\infty(\Omega; \mathbb{S}^N),\ A(x) \geq \nu I \text{ a.e. in } \Omega \quad \forall A \in \mathfrak{A}_{ad},$$

for some $\nu > 0$, it is obvious that $\Xi_w \neq \emptyset$ since the corresponding boundary value problem (19.39) and (19.40) has a unique weak solution $y = y(A)$. Therefore, we adopt the following hypothesis, which is mainly motivated by Remark 19.4.

Hypothesis A. *The set of admissible solutions Ξ_w is nonempty, that is, the minimization problem $\inf_{(A,y)\in\Xi_w} I(A, y)$ is regular.*

Definition 19.8 We say that a pair $(A^0, y^0) \in L^{p/2}(\Omega; \mathbb{S}^N) \times W_A(\Omega)$ is weakly optimal for problem (19.39)–(19.42) and (19.45) if

$$(A^0, y^0) \in \Xi_w \text{ and } I(A^0, y^0) = \inf_{(A,y)\in\Xi_w} I(A, y). \tag{19.46}$$

19.6 Existence of Weak Optimal Solutions

Since our prime interest is the solvability of optimal control problem (19.39)–(19.42) and (19.45), we begin with the study of the topological properties of the set of admissible solutions Ξ_w. To do so, we give some auxiliary results.

Definition 19.9 We say that a sequence $\{(A_k, y_k) \in \Xi_w\}_{k\in\mathbb{N}}$ is bounded if

$$\sup_{k\in\mathbb{N}} \left[\|A_k\|_{BV(\Omega;\mathbb{S}^N)} + \|y_k\|_{A_k} \right] < +\infty.$$

Lemma 19.2 *Let $\{(A_k, y_k) \in \Xi_w\}_{k\in\mathbb{N}}$ be a bounded sequence in the sense of Definition 19.9. Then there exists a pair $(A, y) \in L^{p/2}(\Omega; \mathbb{S}^N) \times W_0^{1,1}(\Omega)$ such that, up to a subsequence,*

$$(A_k, y_k) \xrightarrow{w} (A, y), \ A \in \mathfrak{A}_{ad}, \text{ and } y \in W_A(\Omega). \tag{19.47}$$

Proof By the compactness BV-functions (see Proposition 19.2), there exists a subsequence of $\{A_k\}_{k\in\mathbb{N}}$, still denoted by the same indices, and a matrix $A \in BV(\Omega; \mathbb{S}^N) \cap L^{p/2}(\Omega; \mathbb{S}^N)$ such that $A_k \to A$ weakly in $L^{p/2}(\Omega; \mathbb{S}^N)$ and $A_k \to A$ strongly in $L^1(\Omega; \mathbb{S}^N)$. Thus,

$$A \in BV(\Omega; \mathbb{S}^N), \tag{19.48}$$

$$\int_\Omega A(x)\,dx = \lim_{k\to\infty} \int_\Omega A_k(x)\,dx = M, \tag{19.49}$$

and the condition (19.26) of Definition 19.5 holds true. In order to check the remaining conditions (19.27) and (19.28) of this definition and to show that $A \in \mathfrak{A}_{ad}$, we make the following observation.

We have $(A_k, y_k) \in \Xi_w$ for all $k \in \mathbb{N}$. Hence, in view of strong L^1-convergence $A_k \to A$, we may assume that $A_k^{-1} \to A^{-1}$ almost everywhere in Ω. Since $A_k(x) \geq \zeta_{ad} I$ a. e. in Ω, it follows that

$$\int_K \left(\boldsymbol{\xi}, A_k^{-1}\boldsymbol{\xi}\right)_{\mathbb{R}^N} dx \leq \int_K \zeta_{ad}^{-1}\, dx \|\boldsymbol{\xi}\|^2_{\mathbb{R}^N} \quad \forall k \in \mathbb{N}$$

for any subset $K \subset \Omega$. Hence, the sequence $\{A_k^{-1}\}_{k\in\mathbb{N}}$ is equi-integrable. Then, by Lebesgue's Theorem (see Theorem 19.1) we obtain $A_k^{-1} \to A^{-1}$ strongly in $L^1(\Omega; \mathbb{S}^N)$ as $k \to \infty$. As a result, $A \in \mathfrak{M}_\alpha^\beta(\Omega) \cap L^{p/2}(\Omega; \mathbb{S}^N)$ by Lemma 19.1. Combining this fact with properties (19.48) and (19.49), we conclude $A \in \mathfrak{A}_{ad}$.

To complete the proof, it remains to observe that the remaining conditions (19.27) and (19.28) of Definition 19.5 and $y \in W_A(\Omega)$ for the w-limiting component (A, y) of the sequence $\{(A_k, y_k)\}_{k\in\mathbb{N}}$, are ensured by Lemma 19.1. This concludes the proof.

Our next step deals with the study of topological properties of the set of admissible solutions Ξ_w to the problem (19.39)–(19.42) and (19.45). The following theorem is crucial for our analysis.

Theorem 19.2 *Assume that the Hypothesis A is valid. Then for any admissible initial data $f \in L^q(\Omega)$ the set of admissible solutions Ξ_w is sequentially closed with respect to w-convergence.*

Proof Let $\{(A_k, y_k) \in \Xi_w\}_{k\in\mathbb{N}}$ be a bounded w-convergent sequence of admissible solutions to the optimal control problem (19.39)–(19.42) and (19.45). Let $(\widehat{A}, \widehat{y})$ be its w-limit. Our aim is to prove that $(\widehat{A}, \widehat{y}) \in \Xi_w$. By Lemma 19.2, we have: $\widehat{A} \in \mathfrak{A}_{ad}$ and $\widehat{y} \in W_{\widehat{A}}(\Omega)$. More precisely, the following convergence takes place

(i) $A_k \to \widehat{A}$ strongly in $L^1(\Omega; \mathbb{S}^N)$;
(ii) $y_k \to \widehat{y}$ weakly in $L^p(\Omega)$;
(iii) $\nabla y_k \to \nabla\widehat{y}$ weakly in $L^p(\Omega, A_k\, dx)$, $\nabla\widehat{y} \in L^p(\Omega, \widehat{A}\, dx)$.

It remains to show that the pair $(\widehat{A}, \widehat{y})$ is related by (19.43) for all $\varphi \in C_0^\infty(\Omega)$. To do this we must pass to the limit as $k \to \infty$ in the following relation

$$\int_\Omega |A_k^{1/2}\nabla\varphi|^{p-2}_{\mathbb{R}^N} (A_k\nabla\varphi, \nabla\varphi - \nabla y_k)_{\mathbb{R}^N}\, dx \geq \int_\Omega f(\varphi - y_k)\, dx \;\forall \varphi \in C_0^\infty(\Omega)$$

or we may also rewrite it in the form

$$\int_\Omega |A_k^{1/2}\nabla\varphi|^p_{\mathbb{R}^N}\, dx - \int_\Omega (|A_k^{1/2}\nabla\varphi|^{p-2}_{\mathbb{R}^N}\nabla\varphi, A_k\nabla y_k)_{\mathbb{R}^N}\, dx \geq \int_\Omega f(\varphi - y_k)\, dx. \tag{19.50}$$

As for the right-hand side of the inequality (19.50), from (ii) we obviously have

$$\lim_{k\to\infty} \int_\Omega f(\varphi - y_k)\, dx = \int_\Omega f(\varphi - \widehat{y})\, dx.$$

Let us show now, that the sequence $\{\mathbf{z}_k = |A_k^{1/2}\nabla\varphi|^{p-2}_{\mathbb{R}^N}\nabla\varphi\}_{k\in\mathbb{N}}$ converges strongly in $L^q(\Omega, A_k\, dx)$ to the element $\mathbf{z} = |\widehat{A}^{1/2}\nabla\varphi|^{p-2}_{\mathbb{R}^N}\nabla\varphi$. To do that, according to the proof of Proposition 19.5 (see Step 3) it is enough to make sure that the sequence $\{\mathbf{v}_k \equiv \nabla\varphi\}_{k\in\mathbb{N}}$ converges weakly to $\mathbf{v} = \nabla\varphi$ in $L^p(\Omega, A_k\,dx)$ and that

$$\lim_{k\to\infty}\int_\Omega |A_k^{1/2}\mathbf{v}_k|^p_{\mathbb{R}^N}\,dx = \lim_{k\to\infty}\int_\Omega (A_k\nabla\varphi, \nabla\varphi)^{p/2}_{\mathbb{R}^N}\,dx$$
$$= \int_\Omega (\widehat{A}\nabla\varphi, \nabla\varphi)^{p/2}_{\mathbb{R}^N}\,dx = \int_\Omega |\widehat{A}^{1/2}\mathbf{v}|^p_{\mathbb{R}^N}\,dx \tag{19.51}$$

It is easy to see, that strong convergence $A_k \to \widehat{A}$ in $L^1(\Omega; \mathbb{S}^N)$ immediately implies weak convergence of $\{\mathbf{v}_k\}_{k\in\mathbb{N}}$ to $\nabla\varphi$ in $L^p(\Omega, A_k\,dx)$ as well as convergence almost everywhere in Ω for the sequence $\{(A_k\nabla\varphi, \nabla\varphi)^{p/2}_{\mathbb{R}^N}\}_{k\in\mathbb{N}}$ to the element $(\widehat{A}\nabla\varphi, \nabla\varphi)^{p/2}_{\mathbb{R}^N}$. For the latter we also have an equi-integrability property

$$\int_K (A_k\nabla\varphi, \nabla\varphi)^{p/2}_{\mathbb{R}^N}\,dx \le \int_K \left(\beta(x)|\nabla\varphi|^2_{\mathbb{R}^N}\right)^{p/2}\,dx \le \|\nabla\varphi\|^p_{C(\Omega;\mathbb{R}^N)}\int_K \beta^{p/2}(x)\,dx$$

satisfied. Therefore, by Theorem 19.1, the sequence $\{(A_k\nabla\varphi, \nabla\varphi)^{p/2}_{\mathbb{R}^N}\}_{k\in\mathbb{N}}$ converges to $(\widehat{A}\nabla\varphi, \nabla\varphi)^{p/2}_{\mathbb{R}^N}$ strongly in $L^1(\Omega, \mathbb{S}^N)$ and this fact justifies the limit passage in the first term of relation (19.50). Weak convergence of $\{\mathbf{v}_k\}_{k\in\mathbb{R}^N}$ in variable space $L^p(\Omega, A_k\,dx)$ together with convergence of norms (19.51) imply weak convergence $\mathbf{z_k} \to \mathbf{z}$ in $L^q(\Omega, A_k\,dx)$. It is left to notice here, that

$$\lim_{k\to\infty}\int_\Omega |A_k^{1/2}\mathbf{z}_k|^q_{\mathbb{R}^N}\,dx = \lim_{k\to\infty}\int_\Omega |A_k^{1/2}\nabla\varphi|^p_{\mathbb{R}^N}\,dx$$
$$= \int_\Omega |\widehat{A}^{1/2}\nabla\varphi|^p_{\mathbb{R}^N}\,dx = \int_\Omega |\widehat{A}^{1/2}\mathbf{z}|^q_{\mathbb{R}^N}\,dx.$$

Hence, $|A_k^{1/2}\nabla\varphi|^{p-2}_{\mathbb{R}^N}\nabla\varphi \to |\widehat{A}^{1/2}\nabla\varphi|^{p-2}_{\mathbb{R}^N}\nabla\varphi$ strongly in $L^q(\Omega, A_k\,dx)$ and we can pass to the limit in the second term of relation (19.50), since a product appears of strongly and weakly convergent sequences in dual variable spaces. Therefore, the limit relation takes the desired form

$$\int_\Omega |\widehat{A}^{1/2}\nabla\varphi|^p_{\mathbb{R}^N}\,dx - \int_\Omega (|\widehat{A}^{1/2}\nabla\varphi|^{p-2}_{\mathbb{R}^N}\nabla\varphi, \widehat{A}\nabla\widehat{y})_{\mathbb{R}^N}\,dx \ge \int_\Omega f(\varphi - \widehat{y})\,dx,$$

or

$$\int_\Omega (|\widehat{A}^{1/2}\nabla\varphi|^{p-2}_{\mathbb{R}^N}\widehat{A}\,\nabla\varphi, \nabla\varphi - \nabla\widehat{y})_{\mathbb{R}^N}\,dx \ge \int_\Omega f(\varphi - \widehat{y})\,dx,$$

Hence $\widehat{y} \in W_{\widehat{A}}(\Omega)$ is a weak solution to (19.39) and (19.40) under $A = \widehat{A}$ in the sense of Definition 19.7. Thus, the w-limit pair $(\widehat{A}, \widehat{y})$ belongs to set Ξ_w, and this concludes the proof.

We are now in a position to state the existence of weak optimal solution to the problem (19.39)–(19.42) and (19.45).

Theorem 19.3 *Let $f \in L^q(\Omega)$ and $y_d \in L^p(\Omega)$ be given functions. Assume that the Hypothesis A is valid. Then the optimal control problem* (19.39)–(19.42) *and* (19.45) *admits at least one solution* $(A^0, y^0) \in L^{p/2}(\Omega; \mathbb{S}^N) \times W_{A^0}(\Omega)$.

Proof Since the cost functional $I = I(A, y)$ is bounded below and $\Xi_w \neq \emptyset$, it provides the existence of a minimizing sequence $\{(A_k, y_k) \in \Xi_w\}_{k\in\mathbb{N}}$ to the problem (19.46). Then,

$$\inf_{(A,y)\in\Xi_w} I(A, y) = \lim_{k\to\infty} I(A_k, y_k) = \lim_{k\to\infty} \Big[\int_\Omega |y_k(x) - y_d(x)|^p \, dx + \int_\Omega (\nabla y_k(x), A_k(x)\nabla y_k(x))_{\mathbb{R}^N}^{p/2} \, dx + \sum_{i,j=1}^{N} \int_\Omega |D a_{ij}^k(x)| \Big] < +\infty \tag{19.52}$$

implies the existence of a constant $C > 0$ such that

$$\sup_{k\in\mathbb{N}} \|\nabla y_k\|_{L^p(\Omega, A_k\,dx)^N} \le C, \quad \sup_{k\in\mathbb{N}} \|y_k\|_{L^p(\Omega)} \le C, \tag{19.53}$$

$$\sup_{k\in\mathbb{N}} \|A_k\|_{BV(\Omega;\, \mathbb{S}^N)} \le C. \tag{19.54}$$

Hence, the minimizing sequence $\{(A_k, y_k) \in \Xi_w\}_{k\in\mathbb{N}}$ is bounded in the sense of Definition 19.9 by Lemma 19.2 there exist functions $A^0 \in L^{p/2}(\Omega; \mathbb{S}^N)$ and $y^0 \in W_{A^0}(\Omega)$ such that, up to a subsequence, $(A_k, y_k) \xrightarrow{w} (A^0, y^0)$. Since the set Ξ_w is sequentially closed with respect to the w-convergence (see Theorem 19.2), it follows that the w-limit pair (A^0, y^0) is an admissible solution to (19.39)–(19.42) and (19.45) (i.e., $(A^0, y^0) \in \Xi_w$). To conclude the proof it is enough to observe that the cost functional I is sequentially lower w-semi-continuous. Hence,

$$I(A^0, y^0) \le \liminf_{k\to\infty} I(A_k, y_k) = \inf_{(A,y)\in\Xi_w} I(A, y),$$

i.e., (A^0, y^0) is an optimal solution. The proof is complete.

Acknowledgments Research funded by the DFG-cluster CE315: Engineering of Advanced Materials

References

1. Alessandrini, G., Sigalotti, M.: Geometric properties of solutions to the anistropic p-Laplace equation in dimension two. Annal. Acad. Scient. Fen. Mat. **21**, 249–266 (2001)

2. Boccardo, L., Gallouët, T., Marcellini, P.: Anisotropic equations in L^1. Differ. Integral Equ. **9**, 209–212 (1996)
3. Bouchitte, G., Buttazzo, G.: Characterization of optimal shapes and masses through Monge-Kantorovich equation. J. Eur. Math. Soc. **3**, 139–168 (2001)
4. Buttazzo, G., Kogut, P.I.: Weak optimal controls in coefficients for linear elliptic problems. Revista Matematica Complutense (2010). doi:10.1007/s13163-010-0030-y
5. Buttazzo, G., Varchon, N.: On the optimal reinforcement of an elastic membrane. Riv. Mat. Univ. Parma. **7**(4), 115–125 (2005)
6. Caldiroli, P., Musina, R.: On a variational degenerate elliptic problem. Nonlinear Differ. Equ. Appl. **7**, 187–199 (2000)
7. Calvo-Jurado, C., Casado-Díaz, J.: Optimization by the homogenization method for nonlinear elliptic Dirichlet problems. Mediterr. J. Math. **4**, 53–63 (2007)
8. Chabrowski, J.: Degenerate elliptic equation involving a subcritical Sobolev exponent. Portugal Math. **53**, 167–177 (1996)
9. Chiadò Piat, V., Serra Cassano, F.: Some remarks about the density of smooth functions in weighted Sobolev spaces. J. Convex Anal. **2**(1), 135–142 (1994)
10. Dautray, R., Lions, J.-L.: Mathematical Analysis and Numerical Methods for Science and Technology. Physical Origins and Classical Methods, vol. 1. Springer, Berlin (1985)
11. Giusti, E.: Minimal Surfaces and Functions of Bounded Variation. Birkhäuser, Boston (1984)
12. Greenleaf, A., Kurylev, Y., Lassas, M., Uhlmann, G.: Cloaking devices, electromagnetic wormholes, and transformation optics. SIAM Rev. **1**(51), 3–33 (2009)
13. Kenig, C.E., Salo, M., Uhlmann, G.: Inverse problems for the anisotropic Maxwell equations. Duke Math. J. **2**(157), 369–419 (2011)
14. Kogut, P.I., Leugering, G.: Matrix-valued L1-optimal control in the coefficients of linear elliptic problems. ZAA (Zeitschrift für Analysis und ihre Anwendungen) **4**(32), 433–456 (2013)
15. Kogut, P.I., Leugering, G.: Optimal L^1-Control in Coefficients for Dirichlet Elliptic Problems: H-Optimal Solutions, Zeitschrift für Analysis und ihre Anwendungen (2011)
16. Kogut, P.I., Leugering, G.: Optimal L^1-Control in Coefficients for Dirichlet Elliptic Problems: W-Optimal Solutions. J. Optim. Theory Appl. **2**(150), 205–232 (2011)
17. Kupenko, O.P., Manzo, R.: On an optimal L^1-control problem in coefficients for linear elliptic variational inequality. Abstr. Appl. Anal. 1–13 (2013), Article ID 821964, doi:10.1155/2013/821964
18. Leonhardt, U.: Optical Conformal Mapping, Science 312 1777–1780 (2006) (23 June, 2006)
19. Lions, J.-L., Magenes, E.: Non-Homogeneous Boundary Value Problems and Applications. Springer, Berlin (1972)
20. Murat, F.: Un contre-exemple pour le prolème de contrôle dans les coefficients, C.R.A.S. Paris, Sér. A, 273, 708–711 (1971)
21. Murthy, M.K.V., Stampacchia, V.: Boundary problems for some degenerate elliptic operators. Ann. Mat. Pura e Appl. **4**(5), 1–122 (1968)
22. Pastukhova, S.E.: Degenerate equations of monotone type: Lavrentév phenomenon and attainability problems. Sb.: Math. **10**(198), 1465–1494 (2007)
23. Pendry, J.: Negative refraction makes a perfect lens. Phys. Rev. Lett. **8**, 3966–3969 (2000)
24. Uhlmann G. (ed.): Inside out: Inverse problems and applications, Reprint of the 2003 hardback ed. (English) Mathematical Sciences Research Institute Publications 47. Cambridge: Cambridge University Press (ISBN 978-0-521-16874-8/pbk), 400 p. (2011)
25. Zhikov, V.V.: Weighted Sobolev spaces. Sb.: Math. **8**(189), 27–58 (1998)
26. Zhikov, V.V.: On an extension of the method of two-scale convergence and its applications. Sb.: Math. **7**(191), 973–1014 (2000)
27. Zhikov, V.V.: On two-scale convergence. J. Math. Sci. **3**(120), 1328–1352 (2004)
28. Zhikov, V.V., Pastukhova, S.E.: Homogenization of degenerate elliptic equations. Sib. Math. J. **1**(49), 80–101 (2006)

Part IV
Fundamental and Computational Mechanics

Chapter 20
Uniform Approach to Construction of Nonisothermal Models in the Theory of Constitutive Relations

B.E. Pobedria and D.V. Georgievskii

Abstract A uniform methodical approach for the construction of closed-coupled systems simulating nonisothermal processes in various continual media is described. Differential consequences of five general postulates of continuum mechanics are completed by constitutive relations taking into account both mechanical and thermodynamic matter. Classic models such as ideal liquid (gas), Newtonian viscous fluid, and linear anisotropic elastic solid are considered in this paper.

20.1 Postulates of Continuum Mechanics

Classic axiomatics of continuum mechanics is based on five principal postulates which actually represent the laws in mechanics. Each of the five postulates have both the integral statement (for any individual moving material volume) and the differential consequence (at any point of material continuum). Write out below these five well-known consequences [1–5].

The equation of continuity

$$\frac{d\rho}{dt} + \rho \mathrm{div}\, \mathbf{v} = 0 \tag{20.1}$$

follows from the Postulate (I) of conservation of mass.

The equations of motion for continuum

$$\rho \frac{d\mathbf{v}}{dt} = \nabla_i \mathbf{P}^i + \rho \mathbf{F} \tag{20.2}$$

are the consequence of the Postulate (II) of change of motion amount (or impulse).

B.E. Pobedria (✉) · D.V. Georgievskii
Moscow State University, Moscow 119991, Russia
e-mail: pobedria@mech.math.msu.su

D.V. Georgievskii
e-mail: georgiev@mech.math.msu.su

© Springer International Publishing Switzerland 2015

V.A. Sadovnichiy and M.Z. Zgurovsky (eds.), *Continuous and Distributed Systems II*, Studies in Systems, Decision and Control 30,
DOI 10.1007/978-3-319-19075-4_20

It follows from the Postulate (III) of change of impulse moment (or moment of momentum) that Cauchy stress tensor $\mathbf{P} = \mathbf{E}_i \otimes \mathbf{P}^i = P^{ij}\,\mathbf{E}_i \otimes \mathbf{E}_j$ is symmetric:

$$P^{ij} = P^{ji} \tag{20.3}$$

The local equation of energy

$$\rho \frac{de}{dt} = \rho q - \operatorname{div} \mathbf{q} + \mathrm{P}^{ij}\mathrm{D}_{ij} \tag{20.4}$$

is the differential consequence of the Postulate (IV) of change of internal energy which is also known as the first law of thermodynamics.

At last the local equation of entropy

$$\rho T \frac{ds}{dt} = \rho q - \operatorname{div} \mathbf{q} + \mathrm{w}^* \tag{20.5}$$

represents the differential consequence of the Postulate (V) of change of entropy (the second law of thermodynamics).

As is obvious from what has been said, the Postulates (I), (IV), and (V) have scalar forms while the Postulates (II) and (III) have vector statements.

The following notation are used in (1)–(5): ρ is mass density; $\mathbf{v}$ is velocity vector; $\mathbf{F}$ is mass force; $\mathbf{P}^i$ are stress vectors; P^{ij} and D_{ij} are components of Cauchy stress tensor and strain rate tensor; e is mass density of internal energy; $\mathbf{q}$ is heat flux vector; q is power of heat sources; T is temperature; s is mass density of entropy; and w^* is density of scattering.

20.2 Ideal Liquid and Gas

At first let us consider the model of ideal liquid that is convertible medium ($w^* = 0$) possessing the spherical stress tensor [2]:

$$P^{ij} = -pG^{ij} \tag{20.6}$$

where p is pressure, G^{ij} are components of metric tensor.

Then the change of work $\delta \mathsf{A}^{(\mathrm{int})}$ of internal forces inside some individual volume V is equal to

$$\delta \mathsf{A}^{(\mathrm{int})} = -dt \int\limits_V P^{ij} D_{ij}\, dV \equiv \int\limits_V \delta \mathsf{a}^{(\mathrm{int})}\, dV \tag{20.7}$$

where

$$\delta \mathsf{a}^{(\mathrm{int})} = pG^{ij} D_{ij}\, dt = p \operatorname{div} \mathbf{v}\, dt = -\frac{\mathrm{p}}{\rho}\, \mathrm{d}\rho = \mathrm{p}\rho\, \mathrm{d}\frac{1}{\rho} \tag{20.8}$$

The equations of motion (20.2) for ideal liquid are said to be the Euler equations

$$\frac{d\mathbf{v}}{dt} = -\frac{1}{\rho}\,\mathrm{grad}\, p + \mathbf{F} \tag{20.9}$$

The local equations of energy (20.4) and entropy (20.5) according to (20.7) and (20.8) may be written in the following form

$$\rho\frac{de}{dt} = \rho q - \mathrm{div}\,\mathbf{q} + \frac{p}{\rho}\frac{\mathrm{d}\rho}{dt} \tag{20.10}$$

$$\rho T\frac{ds}{dt} = \rho q - \mathrm{div}\,\mathbf{q} \tag{20.11}$$

For a perfect gas, i.e., for an ideal medium with the Clapeyron constitutive relation $p = \rho RT$ (R is a gas constant), the density of internal energy e is following

$$\rho e = \rho c_v T + \mathrm{const} \tag{20.12}$$

where c_v is mass density of heat capacity by constant volume. Substituting (20.12) to the Eq. (20.10) we derive

$$\rho c_v\frac{dT}{dt} = \rho q - \mathrm{div}\,\mathbf{q} + \frac{p}{\rho}\frac{\mathrm{d}\rho}{dt} \tag{20.13}$$

In case of incompressibility ($d\rho/dt = 0$) taking into account the Fourier law for isotropic medium

$$\mathbf{q} = -\Lambda \mathrm{grad}\, T \tag{20.14}$$

we obtain from (20.13) the equation of heat conduction

$$\rho c_v\frac{dT}{dt} = \rho q + \Lambda\Delta T \tag{20.15}$$

On the other hand the Eq. (20.11) may be written on the basis of (20.14) as

$$\rho T\frac{ds}{dt} = \rho q + \Lambda\Delta T \tag{20.16}$$

Comparison of (20.15) and (20.16) for incompressible medium results in

$$\rho T\frac{ds}{dt} = \rho c_v\frac{dT}{dt} \tag{20.17}$$

so the density of entropy has logarithmic form

$$s = c_v \ln T + \text{const} \tag{20.18}$$

If an ideal liquid (gas) is compressible then its density of internal energy depends on two state parameters:

$$e = e(s, \rho), \quad \rho \frac{de}{dt} = \rho \frac{\partial e}{\partial s} \frac{ds}{dt} + \rho \frac{\partial e}{\partial \rho} \frac{d\rho}{dt} \tag{20.19}$$

where the function $e(s, \rho)$ must be given for every concrete liquid (gas). Comparing the equalities (20.10), (20.11), and (20.19), we obtain

$$de = T\, ds + \frac{p}{\rho^2}\, d\rho \tag{20.20}$$

so

$$\left(\frac{\partial e}{\partial s}\right)_{\rho=\text{const}} = T, \quad \left(\frac{\partial e}{\partial \rho}\right)_{s=\text{const}} = \frac{p}{\rho^2} \tag{20.21}$$

Thus the closed system of equations for perfect gas by nonisothermal processes consists of three equations of motion (20.9), two state equations (20.21), the equation of continuity (20.1), and the equation of heat influx (20.16). Seven mentioned equations include seven unknown variables: $\mathbf{v}$, p, ρ, s, T.

It is known that total entropy S in some individual volume V of perfect gas is equal to

$$S = c_V \ln\left(T V^{\gamma-1}\right) + \text{const} \tag{20.22}$$

Assuming that thermodynamic parameters for the certain state are fixed: S_0, T_0, V_0, we write (20.22) in the following way

$$\frac{S - S_0}{c_V} = \ln\left[\frac{T}{T_0}\left(\frac{V}{v_0}\right)^{\gamma-1}\right], \quad c_V = \int\limits_V \rho c_v\, dV \tag{20.23}$$

or for the density of entropy

$$\frac{s - s_0}{c_v} = \ln\left[\frac{T}{T_0}\left(\frac{\rho_0}{\rho}\right)^{\gamma-1}\right] \tag{20.24}$$

Then temperature T may be expressed from (20.25):

$$T = T_0 \left(\frac{\rho}{\rho_0}\right)^{\gamma-1} \exp\left(\frac{s - s_0}{c_v}\right) \tag{20.25}$$

The density of internal energy (20.12) with regard for (20.25) is transformed into the following form

$$e = c_v T_0 \left(\frac{\rho}{\rho_0}\right)^{\gamma-1} \exp\left(\frac{s - s_0}{c_v}\right) + \text{const} \tag{20.26}$$

20.3 Newtonian Viscous Fluid

Now we define the model on Newtonian viscous fluid [2] as a irreversible medium for which: (a) the density f of Helmholtz free energy depends on two state parameters:

$$f = f(T, \rho) \tag{20.27}$$

and (b) Cauchy stress tensor is a sum of $-pG^{ij}$ as in (20.6) and tensor of viscous stresses:

$$P^{ij} = -pG^{ij} + \tau^{ij} \tag{20.28}$$

$$\tau^{ij} = \lambda_1 \text{div}\,\mathbf{v}\,G^{ij} + 2\mu_1 G^{ik} G^{jl} D_{kl} \tag{20.29}$$

The term "Newtonian" means that the tensor function (20.29) is linear function of its argument $\mathbf{D}$ with two material constants λ_1 (volume viscosity) and μ_1 (shear viscosity).

It follows from (20.8), (20.27)–(20.29) that the change of density of work $\delta \mathsf{a}^{(\text{int})}$ of internal forces for Newtonian viscous fluids has the form

$$\delta \mathsf{a}^{(\text{int})} = -\frac{p}{\rho} d\rho - \tau^{ij} D_{ij}\, dt = p\rho\, d\frac{1}{\rho} + \lambda_1 (\text{div}\,\mathbf{v})^2 + 2\mu_1 \text{tr}\,\mathbf{D}^2 \tag{20.30}$$

where

$$\text{tr}\,\mathbf{D}^2 = G^{ik} G^{jl} D_{ij} D_{kl} \tag{20.31}$$

Representing the strain rate tensor $\mathbf{D}$ as a sum of spherical part and deviator $\bar{\mathbf{D}}$:

$$D_{ij} = \bar{D}_{ij} + \frac{1}{3} \text{div}\,\mathbf{v}\,G_{ij} \tag{20.32}$$

one can obtain

$$\text{tr}\,\mathbf{D}^2 = D^2_{\text{inten}} + \frac{1}{3} (\text{div}\,\mathbf{v})^2 \tag{20.33}$$

where D_{inten} is an intensity of the strain rate tensor

$$D_{\text{inten}} = \sqrt{\text{tr}\,\bar{\mathbf{D}}^2} \tag{20.34}$$

The known connections of various thermodynamic potentials result in the following relation

$$dF + S\,dT = -\delta \mathsf{A}^{(\mathrm{int})} - W^*\,dt \tag{20.35}$$

which may be written in terms of densities

$$\rho\,df + \rho s\,dT = -\delta \mathsf{a}^{(\mathrm{int})} - w^*\,dt \tag{20.36}$$

The Eqs. (20.30) and (20.36) lead to the following

$$\rho\,df + \rho s\,dT = \frac{p}{\rho}\,d\rho + \tau^{ij} D_{ij}\,dt - w^*\,dt \tag{20.37}$$

Taking into account the model of viscous fluid (20.29) we conclude from (20.37):

$$\left(\frac{\partial f}{\partial T}\right)_{\rho=\mathrm{const}} = -s, \quad \left(\frac{\partial f}{\partial \rho}\right)_{T=\mathrm{const}} = \frac{p}{\rho^2} \tag{20.38}$$

$$w^* = \tau^{ij} D_{ij} = \left(\lambda_1 + \frac{2\mu_1}{3}\right)(\operatorname{div}\mathbf{v})^2 + 2\mu_1 D^2_{\mathrm{inten}} \tag{20.39}$$

The closed system of equations for Newtonian viscous fluid consists of three equations of motion (consequences of (20.2), (20.28), and (20.29))

$$\rho \frac{d\mathbf{v}}{dt} = -\operatorname{grad} p + (\lambda_1 + \mu_1)\operatorname{grad}\operatorname{div}\mathbf{v} + \mu_1 \Delta \mathbf{v} + \rho \mathbf{F} \tag{20.40}$$

as well as two state equations (20.38), the equation of continuity (20.1) and the equation of heat influx (20.15) which may be written as

$$\rho T \frac{ds}{dt} = \rho q + \Lambda \Delta T + w^* \tag{20.41}$$

with regard for the Fourier law for isotropic medium. In addition, the density of scattering w^* is expressed by the formula (20.39) and the strain rate tensor components are expressed in terms of the velocity components:

$$D_{ij} = \frac{1}{2}(\nabla_i v_j + \nabla_j v_i) \tag{20.42}$$

So the density of scattering w^* in (20.41) may be expressed in terms of the velocity components and we have (just as for ideal liquid) seven equations with respect to the same seven unknown variables: $\mathbf{v}$, p, ρ, s, T.

As is obvious from (20.39) that w^* is a positive definite scalar value if

$$\lambda_1 + \frac{2\mu_1}{3} > 0, \quad \mu_1 > 0 \tag{20.43}$$

If a fluid is incompressible, then three equations of motion (20.40) are transformed to Navier–Stokes form

$$\frac{d\mathbf{v}}{dt} = -\frac{1}{\rho}\,\mathrm{grad}\, p + \eta \Delta \mathbf{v} + \mathbf{F}, \quad \eta = \frac{\mu_1}{\rho} \tag{20.44}$$

where η is kinematic viscosity. The function w^* in (20.39)

$$w^* = 2\mu_1 D^2_{\mathrm{inten}} \tag{20.45}$$

is positive definite if $\mu_1 > 0$ (or $\eta > 0$).

20.4 Linear Anisotropic Elastic Solid

Let us define a model of linear elastic solid by nonisothermal processes [3–7] as a reversible medium ($w^* = 0$) where the density f of Helmholtz free energy depends on temperature T and strain tensor ε:

$$f = f(\varepsilon, T) \tag{20.46}$$

We use the Duhamel-Neumann hypothesis which presupposes that the following combination of mechanical deformation and temperature may be argued in (20.46):

$$\varepsilon^T_{ij} = \varepsilon_{ij} - \alpha_{ij}\vartheta \tag{20.47}$$

where α_{ij} are the components of symmetric thermal extension tensor, ϑ is temperature overfall, i.e., difference of the current temperature T and some constant T_0:

$$\vartheta = T - T_0 \tag{20.48}$$

This constant T_0 should be introduced because of inaccessibility of absolute zero $T = 0$.

We represent the function f in the form

$$f = f_0(T) + \tilde{f}(\varepsilon^T) \tag{20.49}$$

with additive term $f_0(T)$ depending only on temperature. Below (with a view to brevity), we should omit upper tilde in the second term in (20.49). In order that constitutive relations of elastic medium would be linear it is natural to choose the density of free energy as a quadratic function of "temperature strain":

$$\rho f = \rho f_0(T) + \frac{1}{2} C^{ijkl} \varepsilon_{ij}^T \varepsilon_{kl}^T \qquad (20.50)$$

where C^{ijkl} are the components of fourth rank tensor possessing the following symmetries:

$$C^{ijkl} = C^{jikl} = C^{ijlk} = C^{klij} \qquad (20.51)$$

The formula (20.7) may be written as

$$\delta \mathsf{A}^{(\mathrm{int})} = -\int\limits_V P^{ij}\, d\varepsilon_{ij}\, dV \equiv \int\limits_V \delta \mathsf{a}^{(\mathrm{int})}\, dV \qquad (20.52)$$

With a view to simplicity, we take the Cartesian coordinate system. Then $\mathbf{P} = \sigma$ and the relations (20.8) and (20.52) result in

$$\delta \mathsf{a}^{(\mathrm{int})} = -dt\, \sigma_{ij} \dot{\varepsilon}_{ij} = -\sigma_{ij}\, d\varepsilon_{ij} \qquad (20.53)$$

The Eq. (20.36) for reversible elastic medium is the following

$$\rho\, df + \rho s\, dT = \sigma_{ij}\, d\varepsilon_{ij} \qquad (20.54)$$

Substituting the expression (20.49)–(20.54) we obtain

$$\rho \frac{\partial f_0}{\partial T} + \rho \frac{\partial f}{\partial \varepsilon_{ij}^T} (d\varepsilon_{ij} - \alpha_{ij}\, dT) + \rho s\, dT = \sigma_{ij}\, d\varepsilon_{ij} \qquad (20.55)$$

Equalization of the coefficients in independent differentials dT and $d\varepsilon_{ij}$ in (20.55) carries into

$$\rho \frac{\partial f_0}{\partial T} - \rho \alpha_{ij} \frac{\partial f}{\partial \varepsilon_{ij}^T} = -\rho s \qquad (20.56)$$

$$\rho \frac{\partial f}{\partial \varepsilon_{ij}^T} = \sigma_{ij} \qquad (20.57)$$

It is evident that relations (20.56) and (20.57) are correct for the density f of Helmholtz free energy arbitrary depending on tensor ε^T. Using (20.50) we can obtain from (20.57)

$$\sigma_{ij} = C_{ijkl}(\varepsilon_{kl} - \alpha_{kl} \vartheta) \qquad (20.58)$$

as well as from (20.56) to (20.57)

$$\rho s = -\rho \frac{\partial f_0}{\partial T} + \alpha_{ij}\sigma_{ij} \tag{20.59}$$

i.e., constitutive relations for linear thermoelastic solid.

Now we denote (by analogy with f_0) the additive terms of the densities of both entropy and free energy depending only on temperature by s_0 and e_0 correspondingly. The following limit by constant strain

$$c_V = \lim_{\Delta T \to 0} \left(\frac{\Delta Q}{\Delta T} \right)_{\varepsilon = \text{const}} \equiv \left(\frac{\partial Q}{\partial T} \right)_{\varepsilon = \text{const}} \tag{20.60}$$

is said to be a heat capacity by constant strain.

Taking account of the first law of thermodynamics

$$dE = \delta Q - \delta \mathsf{A}^{(\text{int})} \tag{20.61}$$

the definition (20.61) results in

$$c_V = \left(\frac{\partial E}{\partial T} \right)_{\varepsilon = \text{const}} \tag{20.62}$$

We also introduce a heat capacity by constant stress

$$c_P = \frac{\partial E_0}{\partial T} \tag{20.63}$$

With regard for the connection

$$f = e - Ts \tag{20.64}$$

as well as relations

$$s = -\frac{\partial f}{\partial T}, \quad s_0 = -\frac{\partial f_0}{\partial T} \tag{20.65}$$

we can write

$$c_v = \frac{\partial e}{\partial T} = T \frac{\partial s}{\partial T} = -T \frac{\partial^2 f}{\partial T^2} \tag{20.66}$$

$$c_p = \frac{\partial e_0}{\partial T} = T \frac{\partial s_0}{\partial T} = -T \frac{\partial^2 f_0}{\partial T^2} \tag{20.67}$$

Then the Eq. (20.59) results in

$$\rho c_v = \rho c_p - T\alpha_{ij} C_{ijkl}\alpha_{kl} = \rho c_p - T\alpha_{ij}\beta_{ij} \tag{20.68}$$

where

$$\beta_{ij} = C_{ijkl}\alpha_{kl} \tag{20.69}$$

In this way using (20.67) we obtain

$$-\frac{\partial f_0}{\partial T} = \int\limits_{T_0}^{T} \frac{c_p}{T}\, dT = c_p \ln \frac{T}{T_0} = c_p \ln \left(1 + \frac{\vartheta}{T_0}\right) \tag{20.70}$$

The Dulong-Petit law has been taken into account here: a heat capacity of solids is assumed to be constant by temperatures exceeding so called Debye temperature (not more than 100–200° K for most crystals).

Therefore, mass density of entropy for elastic solids has the form

$$\rho s = \rho c_p \ln \frac{T}{T_0} + \alpha_{ij}\sigma_{ij} = \rho c_p \ln \frac{T}{T_0} + \beta_{ij}(\varepsilon_{ij} - \alpha_{ij}\vartheta) \tag{20.71}$$

If temperature overfall is not large ($\vartheta \ll T$) then both the equations (20.70) and (20.71) result in

$$\rho s = \rho c_p \frac{\vartheta}{T_0} + \alpha_{ij}\, C_{ijkl}\varepsilon_{kl} - C_{ijkl}\alpha_{ij}\alpha_{kl}\vartheta = \rho c_v \frac{\vartheta}{T_0} + \beta_{ij}\varepsilon_{ij} \tag{20.72}$$

Since the equation of heat influx (20.41) for anisotropic solid may be represented as

$$\rho T \frac{ds}{dt} = \rho q + \Lambda_{ij} T_{,ij} + w^* \tag{20.73}$$

as well as an elastic medium is reversible then bearing in mind the relation

$$\rho \frac{ds}{dt} = \frac{\rho c_p}{T} \frac{dT}{dt} + (\alpha_{ij}\sigma_{ij})^{\cdot} \tag{20.74}$$

which may be obtained by derivation (20.71) with respect to time, we write

$$\rho c_p \frac{\partial T}{\partial t} = \rho q + \Lambda_{ij} T_{,ij} - T(\alpha_{ij}\sigma_{ij})^{\cdot} \tag{20.75}$$

or

$$\rho c_p \frac{\partial T}{\partial t} = \rho q + \Lambda_{ij} T_{,ij} - T\alpha_{ij}\, C_{ijkl}(\dot{\varepsilon}_{kl} - \alpha_{kl}\dot{T}) \tag{20.76}$$

$$\rho c_v \frac{\partial T}{\partial t} = \rho q + \Lambda_{ij} T_{,ij} - T\beta_{ij}\dot{\varepsilon}_{ij} \tag{20.77}$$

The equations of motion for elastic medium

$$\rho \frac{\partial^2 u_i}{\partial t^2} = \sigma_{ij,j} + \rho F_i \tag{20.78}$$

have the following form with regard for (20.58)

$$\rho \frac{\partial^2 u_i}{\partial t^2} = C_{ijkl} u_{kl,j} - \beta_{ij} T_{,j} + \rho F_i \tag{20.79}$$

So the closed system of equations for anisotropic thermoelastic medium consists of three equations of motion (20.79) and the equation of heat influx (20.77) (the last term in (20.77) should be written as $-T\beta_{ij}\dot{u}_{i,j}$) with respect to four unknown variables: u_i and T. In most cases, one can disregard the last term in (20.77) because of smallness of dimensionless values $T\alpha_{ij}$. Thus the equation of heat influx, for example, in conformity to isotropic medium represents a linear not uniform equation by parabolic type:

$$\rho c_v \frac{\partial T}{\partial t} = \rho q + \Lambda \Delta T \tag{20.80}$$

The "thermal" equation (20.80) may be solved separately from others (using the corresponding initial and boundary conditions). After that, the Eq. (20.79) become the closed system of not uniform (with the known temperature T) equations with respect to three components u_i. In this case the problem of thermoelasticity is said to be not coupled.

For isotropic medium, the equations of motion (20.79) are following

$$\rho \frac{\partial^2 u_i}{\partial t^2} = \lambda u_{k,ik} + \mu \Delta u_i - 3\alpha K T_{,i} + \rho F_i \tag{20.81}$$

because of

$$\beta_{kl} = \alpha_{ij} C_{ijkl} = \alpha \delta_{ij} [\lambda \delta_{ij} \delta_{kl} + \mu (\delta_{ik} \delta_{jl} + \delta_{jk} \delta_{il})] = \alpha (3\lambda + 2\mu) \delta_{kl} = 3\alpha K \delta_{kl} \tag{20.82}$$

for isotropic case.

It should be noted that in problems of thermoelasticity the adiabatic material constants are inconsistent with the isothermal ones. In (20.58) namely isothermal values C_{ijkl} are present as they are determined in experiment by constant temperature. In this case the relations (20.50) may be rewritten as

$$\rho f = \rho f_0(T) + W \tag{20.83}$$

with elastic potential W:

$$W = \rho \tilde{f} = \frac{1}{2} C_{ijkl} \varepsilon_{ij}^T \varepsilon_{kl}^T \tag{20.84}$$

In order to obtain the adiabatic elastic constants C^{ad}_{ijkl} it is necessary to pass from the pair of thermodynamic state parameters T, ε to the pair s, ε. To this end, we introduce according to (20.64) the density of internal energy:

$$\rho e = \frac{1}{2} C_{ijkl}(\varepsilon_{ij} - \alpha_{ij}\vartheta)(\varepsilon_{kl} - \alpha_{kl}\vartheta) - \rho c_p T \ln \frac{T}{T_0} - \beta_{ij} T(\varepsilon_{ij} - \alpha_{ij}\vartheta) \quad (20.85)$$

then express temperature overfall from (20.72):

$$\vartheta = \frac{T_0}{\rho c_v}(\rho s - \beta_{ij}\varepsilon_{ij}) \quad (20.86)$$

and substitute to the generalized Hooke' law (20.58):

$$\sigma_{ij} = C_{ijkl}\varepsilon_{kl} - \beta_{ij}\frac{T_0}{\rho c_v}(\rho s - \beta_{kl}\varepsilon_{kl}) \quad (20.87)$$

The values of adiabatic constants are determined from comparison of (20.58) and (20.87):

$$C^{\mathrm{ad}}_{ijkl} = C_{ijkl} + \frac{T_0}{\rho c_v}\beta_{ij}\beta_{kl} \quad (20.88)$$

References

1. Germain, P.: Cours de Mécanique des Milieux Continus. T. 1. Théorie Générale. Masson Éditeurs, Paris (1973)
2. Sedov, L.I.: Mechanics of Continuous Media, vol. I, II. World Scientific Publishing, Singapore (1997)
3. Ilyushin, A.A.: Mechanics of Continuous Media. Moscow State University Publ, Moscow (1990) (in Russian)
4. Pobedria, B.E.: Numerical Methods in Elasticity and Plasticity. Moscow State University Publ, Moscow (1995) (in Russian)
5. Pobedria, B.E., Georgievskii, D.V.: Foundations of Mechanics of Continuous Media. Fizmatlit, Moscow (2006) (in Russian)
6. Karnaukhov, V.G.: Coupled Problems of Thermoviscoelasticity. Naukova Dumka, Kiev (1982) (in Russian)
7. Dimitrienko, Yu.D: Nonlinear Mechanics of Continuous Media. Fizmatlit, Moscow (2009) (in Russian)

Chapter 21
Active Near-Wall Flow Control via a Cross Groove with Suction

I.M. Gorban and O.V. Khomenko

Abstract Theoretical model of the nonlinear active near-wall flow control that uses a vortex trapped in the cross groove and suction of fluid is developed. The system parameters are evaluated from the equation of vortex equilibrium and the Kutta condition in the groove edges. Dynamical system analysis is used to explore the performance of the control strategy. The suction is shown to change the flow topology as compared with uncontrolled case. The equilibrium vortex satisfies now either stable or unstable focus that depends on suction parameters. The optimal characteristics of the control system, when a stable vortex is supported with minimal energy costs, are obtained in several groove configurations.

Keywords Cross groove · Flow control · Vortex · Stability · Fluid ejection

21.1 Introduction

Transformation of a near-wall turbulent flow to the regular vortical pattern is one of the control strategies that has been successfully used, for example, for improving hydrodynamic characteristics of bluff bodies [1–3]. The local vortical zones in near-wall region may be generated by artificial change of the body configuration, with help of bulges, grooves or ribs, as well as by body vibrations or surface deformation. Because of sensitivity of a circulation flow to external perturbations [4], the practical realization of such type control algorithms in a wide range of the Reynolds number requires the development of active (with supplying external energy) schemes for stabilization of the vortices.

I.M. Gorban
Institute of Hydromechanics, National Academy of Sciences of Ukraine,
Zheliabova St. 8/4, Kyiv 03680, Ukraine
e-mail: ivgorban@gmail.com

O.V. Khomenko (✉)
Institute for Applied System Analysis, National Technical University of Ukraine
"Kyiv Polytechnic Institute", Peremogy Ave. 37, Build 35, Kyiv 03056, Ukraine
e-mail: olghomenko@mail.ru

© Springer International Publishing Switzerland 2015

V.A. Sadovnichiy and M.Z. Zgurovsky (eds.), *Continuous and Distributed Systems II*, Studies in Systems, Decision and Control 30,
DOI 10.1007/978-3-319-19075-4_21

The problem of active control for unsteady fluid flows is nonlinear one. As far as it has not have the general solution, the model of fluid that is used for constructing a control algorithm is very important. The controllers based on the model of viscous fluid are very complex because they use the full Navier–Stokes equations and has to take into account information about a large number of physical parameters. On the other hand, the simple linear models might generate a controller unable to achieve the desired control objective. A reasonable compromise is a reduced model based on topological information obtained from the dynamical analysis of the flow field [5, 6]. Modern approaches in this area are connected with generating the necessary flow topology at which the desired result is achieved with minimal energy losses.

Investigations of the dynamic properties of vortical flows are mainly based on the nonviscous model of point vortices, in which the vorticity field is represented by a discrete set of isolated singular elements whose axes are perpendicular to the flow plane [7]. Since, a point vortex is the "weak" solution of the Euler equations, the flow field in this case is reduced to the finite system of vortices that move along the trajectories of fluid particles. The vortex model has ensured many important results about the behavior of coherent structures in turbulent flows [8] and in the field of flow control [6]. A detailed overview of the researches devoted to chaotization of vortex systems and its relation with two-dimensional turbulence is presented in paper [9]. The examples of flows for which the computational results obtained by the idealized model of point vortices are very close to the experimental data are shown in paper [10].

It has been mentioned above, one of the effective ways to generate large-scale vortices near the body surface is installation of cross grooves there. Those are applied to concentrate the vorticity from the boundary layer in the isolated recirculation zones. For the first time, this technology was proposed by Ringleb to reduce the hydraulic losses in diffusers [11]. He developed the model of trapped vortex in the special groove having the form of natural snow cornice. It is based on analysis of the flow critical points and allows to derive the groove optimal geometrical parameters that ensure the trapped vortex without expenditure of external energy. The Kasper wing is an example of the successful realization of the trapped vortex conception in aerodynamics [3]. The theoretical model developed in [12] expands the class of the surfaces near which the large-scale stable vortices are possible.

Analysis of the dynamic behavior of trapped vortices in cylindrical grooves shows that the vortices are located on the axis of the groove and have neutral stability [5, 13]. As a result, they exhibit selective sensitivity to external perturbation, reacting especially to the perturbations with a periodic component. In the periodically perturbed flow the trapped vortex deviates from its equilibrium point and demonstrates the resonant behavior when the frequency perturbation is close to the vortex eigenfrequency. In real conditions, vortex zone pulsations lead to leakage vorticity from the groove that reduces essentially the effectiveness of control. In order to stabilize the circulating flow in the groove, active control that uses external energy—fluid suction, rotors for additional impulse—are applied [14, 15]. Its efficiency and feasibility of application depends on the required energy. Therefore, the development of active control scheme has to include solving optimization and stabilization problems with regard to controller, which is proposed [6, 16].

In this paper, the complex near-wall flow control scheme that uses standing (trapped) vortex in the cross groove and its stabilization by fluid ejection (or injection) is developed. The scheme justification is based on the simplified dynamic model with one degree of freedom in which the recirculation zone in the groove is replaced by a point vortex that locates in the vorticity center. The purpose of the control is to eliminate the generation of vorticity in the groove corners. So, the controller is built in such a way as to satisfy the Kutta condition in the sharp groove edges. In addition, we found the optimal parameters of control device, at which the stable flow configuration is supported with minimal energy losses.

21.2 Mathematical Formulation of the Problem

The two-dimensional flow of ideal incompressible fluid bounded by the wall with a cylindrical groove is considered (Fig. 21.1). The groove linear parameters are assumed to be much higher than the thickness of the boundary layer developed on the wall. The axis Ox of the coordinate system is directed along the wall, the vertical axis Oy passes through the groove center. The geometry of the groove is described by its semichord a and angle β between the axis Ox and the tangent line to the groove surface (Fig. 21.1). Depending on the groove depth, its center (x_c, y_c) may be located either above or below the wall. The lateral flow velocity in general case consists of an uniform flow velocity U_∞ and nonstationary component $u(t)$: $U(t) = U_\infty + u(t)$. The circulating zone, generated in the groove owing to boundary layer separation, is replaced by a point vortex with circulation Γ_v and coordinates (x_v, y_v). The fluid ejection, which is applied to stabilize the vortex, is modeled by a sink with power Q and coordinates x_q, y_q, that locates on the groove wall. Its position is defined identically by an angular coordinate α (Fig. 21.1): $x_q = r \sin\alpha$, $y_q = -r\cos\alpha + y_c$.

The practical goal of control is to create and maintain such circulating flow in the groove which will prevent the generation of vorticity in sharp corners. Therefore, theoretical modeling of the process consists in determining parameters Q, α of the ejection device providing existence of the stable standing vortex in the groove under condition that Kutta theorem on finiteness of the flow velocity holds in the sharp edges.

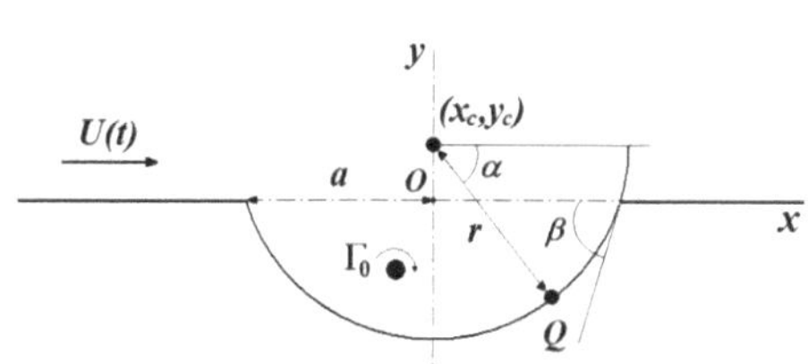

Fig. 21.1 The geometry of the flow region

The dynamical system under consideration has one degree of freedom. Therefore, its evolution is described by the following nonlinear differential equation in R^2:

$$\frac{d\overrightarrow{X}(t)}{dt} = f(\overrightarrow{X}), \tag{21.1}$$

where $\overrightarrow{X} \in R^2$ is the vector of vortex coordinates, the vector function $f : R^2 \to R^2$ sets the vortex velocity generated by the lateral flow and fluid ejection. If the flow velocity does not change in time, Eq. (21.1) is autonomous one. Its right part is expressed by the stream function, which is the Hamiltonian of the system [9]. As the vortex moves along the streamlines, the phase space of the dynamical system coincides with the flow region, in the sense that the Cartesian coordinates of the vortex (x_v, y_v) represent the conjugate variables.

Positions of the vortices in equilibrium coincide with flow critical points. Those may be derived from the equation:

$$f(\overrightarrow{X}) = 0, \tag{21.2}$$

It means, the modes of behavior of the standing vortex are specified by the critical point type that is determined from the eigenvalues analysis of the Jacobian of linearized system:

$$\frac{d\overrightarrow{X}'}{dt} = A\overrightarrow{X}', \tag{21.3}$$

where $\overrightarrow{X}'(t)$ is the vector of perturbations of the equilibrium solution $\overrightarrow{X}_0$, $A = \nabla f(\overrightarrow{X}_0)$.

It follows from the Kutta theorem, the groove corners will be flowed smoothly if the following equations are satisfied:

$$\overrightarrow{V}(z_1^*) = C_1, \quad \overrightarrow{V}(z_2^*) = C_2 \tag{21.4}$$

Here $\overrightarrow{V}$ is the flow velocity, z_1^*, z_2^* are the complex coordinates of sharp edges, C_1, C_2 are the arbitrary constants. The set of Eqs. (21.2)–(21.4) describes the control problem in the sense that it allows to determine uniquely the characteristics of the circulation flow and parameters of fluid ejection, which provide unseparated flow in the groove corners.

For an ideal incompressible fluid the hydrodynamic problem under consideration is governed by the Euler's equation with the boundary condition of nonleaking on the wall.

$$\frac{\partial \overrightarrow{V}}{\partial t} + (\overrightarrow{V} \cdot \nabla)\overrightarrow{V} = 0 \quad in \quad S \times (0, T), \tag{21.5}$$

$$\overrightarrow{V} \cdot \overrightarrow{n}|_{\partial S} = 0, \tag{21.6}$$

$$\overrightarrow{V}|_{t=0} = U_\infty, \tag{21.7}$$

where S, ∂S are the flow region and its boundary, respectively, T is the time period under consideration.

One of the advantages of the vortex models is potentiality of the flow in the whole region with the exception of the points where vortices are located. This allows to apply the theory of complex variable functions to solve the problem (21.5)–(21.7). In this work, boundary condition (21.6) is satisfied by conformal mapping of the flow field in the physical plane $z(x, y)$ into an upper half-plane of the auxiliary plane $\zeta(\xi, \eta)$, where the Green's function for vortex builds by its mirror image with respect to the wall. The function which implements aforesaid transformation for the half-plane with the cut cylindrical groove has the following form [17]:

$$f(z) = a\gamma \frac{1 + \left(\frac{z-a}{z+a}\right)^{\gamma}}{1 - \left(\frac{z-a}{z+a}\right)^{\gamma}}, \quad \gamma = \frac{\beta}{\pi - \beta} \tag{21.8}$$

The flow configurations in physical and canonical planes are shown in Fig. 21.2.

The complex flow potential in the canonical plane is constructed by superposition of particular flows:

$$W(\zeta) = U_{\infty\zeta}\zeta + \frac{\Gamma_v}{2\pi i}\left[\ln(\zeta - \zeta_v) - \ln(\zeta - \overline{\zeta_v})\right] + Q(t)\ln(\zeta - \zeta_q), \tag{21.9}$$

where $U_{\infty\zeta}$, $\zeta_v(\xi_v, \eta_v)$, $\zeta_q(\xi_q, 0)$ is the velocity of lateral flow and complex coordinates of vortex and sink in the plane ζ, respectively.

It follows from the invariance of the complex potential under conformal mapping and properties of function (21.8):

$$U_\infty = \lim_{z\to\infty}\frac{dW}{dz} = \lim_{\zeta\to\infty}\frac{dW}{d\zeta}\frac{d\zeta}{dz} = U_\infty\zeta. \tag{21.10}$$

Then the complex conjugate velocity in the physical region is given by the following expression:

$$\overline{V}(x, y) = \frac{dW}{d\zeta}\frac{df}{dz} = \left(U_\infty + \frac{\Gamma_v}{2\pi i}\left[\frac{1}{\zeta - \zeta_v} - \frac{1}{\zeta - \overline{\zeta}_v}\right] + \frac{Q(t)}{\zeta - \xi_q}\right)\frac{df}{dz}. \tag{21.11}$$

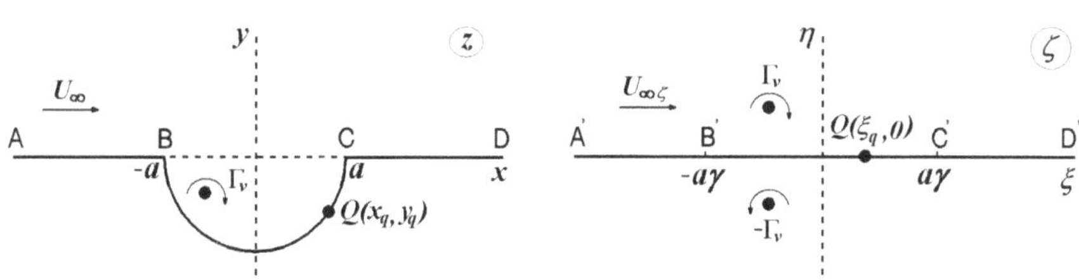

Fig. 21.2 Physical and canonical planes for a cross groove

Taking into account the Routh rule [18] from (21.11) we have:

$$\overline{V}(x_v, y_v) = \left(U_\infty + \frac{\Gamma_v}{4\pi\eta_v} + \frac{Q(t)}{\zeta - \xi_q}\right)\frac{df}{dz}\bigg|_{\zeta=\zeta_v} + \frac{\Gamma_v}{4\pi i}\left(\frac{d^2 f}{dz^2}\bigg/\frac{df}{dz}\right)\bigg|_{\zeta=\zeta_v}. \tag{21.12}$$

To construct the equations for determining the parameters of the standing vortex and control device, we segregate the real and imaginary parts of (21.12). Then we put that those are equal to zero and take into account conditions (21.4) in sharp edges $z^*_{1,2}$. Since the velocity has a singularity at the points $z^*_{1,2}$, from the Kutta theorem we obtain: $\left.\frac{dW}{d\zeta}\right|_{\zeta=\zeta^*_{1,2}} = 0$ or

$$U(t) + \frac{1}{\pi}\frac{\Gamma_v \eta_v}{(\xi^*_{1,2} - \xi_v)^2 + \eta_v^2} + \frac{Q(t)}{\xi^*_{1,2} - \xi_q} = 0 \tag{21.13}$$

where $\zeta^*_{1,2}(\xi^*_{1,2}, 0)$ are the coordinates of the groove edges in the plane ζ.

So, we have four transcendental equation for calculating standing vortex coordinates (x_v, y_v), its circulation Γ_v, the intensity Q and angular coordinate α of the sink. In order to this set of equations would be closed, one of these parameters need to be fixed. If we set the coordinate x_v, we will obtain the curve, on which standing vortices lie. Because of the complexity of the obtained equations they are solved numerically. Note the problem is considered in dimensionless form, where the groove semichord a and the unperturbed flow velocity U_∞ are the characteristic parameters, so $\overline{x} = \frac{x}{a}$, $\overline{y} = \frac{y}{a}$, $\overline{t} = tU_\infty$ (further the dashes denoting dimensionless quantities will be omitted).

21.3 Standing Vortex Within the Groove in the Stationary Flow

The stationary lateral near-wall flow of velocity U_∞ is considered. If ejection of fluid is absent, the coordinates of the critical point will be calculated from Eq. (21.2) and the standing vortex circulation will be uniquely determined from the Kutta condition in one of the sharp edges of the groove [13]. The results of calculations discovered that the flow critical point lies on the groove axis and is characterized by a pair of conjugated imaginary eigenvalues. It means the standing vortex corresponding to this point is of neutral stability (21.3). Such a vortex rotates in the small neighborhood of the critical point and the frequency of its rotation is equal to eigenvalue, so it may be considered as the vortex frequency ω_0.

The dependencies of the vertical coordinate, circulation, and eigenfrequency of the standing vortex against the groove depth have been represented in [13]. Note if the groove is shallow ($\beta \leq 80°$) the standing vortex will lie above the wall. This fact allows to apply such grooves for creating so-called "vortex lubrication" of the wall when the turbulent boundary layer is replaced by a system of regular vortices. The standing vortex circulation is approximately constant in shallow and medium grooves and grows rapidly in deep grooves. The highest frequency of vortex is observed in the medium grooves (with $\beta = 90°$), and it decreases in both shallow and deep grooves [13].

The motion of fluid particles in the potential flow is governed by the Hamiltonian dynamical system:

$$\frac{dx}{dt} = \frac{\partial \psi}{\partial y}, \quad \frac{dy}{dt} = -\frac{\partial \psi}{\partial x}, \tag{21.14}$$

where the stream function ψ represents the Hamilton function and the coordinates x, y are canonical variables. While the vortex is in equilibrium, system (21.14) is autonomous and integrated. Its right part is derived from the complex conjugate velocity (21.11) with $Q = 0$. The pattern of streamlines points out a regular motion of fluid particles in this case (Fig. 21.3a). On the other hand, there is a line on this picture which separates trajectories of different types. The presence of such heteroclinic orbit is a precondition of occurrence of chaotic mixing of fluid particles in the perturbed system.

If the standing vortex is deflected from the equilibrium, its trajectory in the phase space (x_v, y_v) will be derived by integrating Eq. (21.1) where the right part is calculated from expression (21.12). The obtained portrait of standing vortex trajectories in the groove (Fig. 21.3b) demonstrates lines of different types. They are connected with the stable critical point that lies on the axis and two hyperbolic "saddle" points located near the groove corners. Figure 21.3b shows if initial deviation of the standing vortex from equilibrium point is small, one will move around its stationary position. But when the initial disturbance is large enough, the vortex may pass the separatrix between the different trajectories and "wash out" from the groove in the near-wall flow. Ejection of fluid from the groove, which will enter to the control scheme fur-

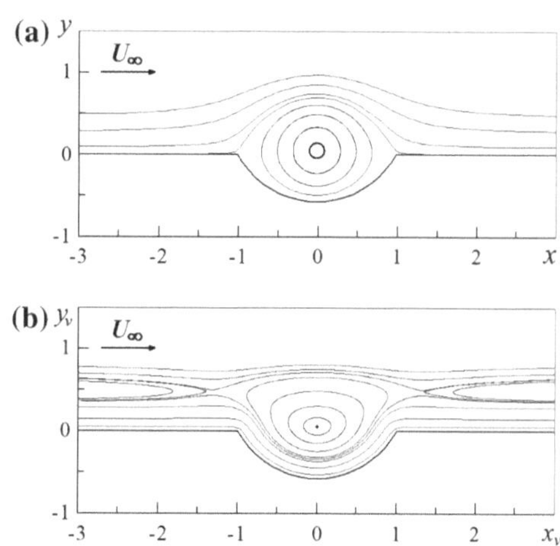

Fig. 21.3 **a** Picture of streamlines around standing vortex; **b** phase portrait of standing vortex trajectories above the groove, $\beta = 60°$

ther should make the standing vortex more stable and return it to equilibrium at large deviations.

Consider as the flow topology will change when fluid suction (ejection) is realized on the groove wall. If one gives the horizontal coordinate x_v of the standing vortex, its another coordinate y_v and circulation Γ_v as well as the power Q and angular coordinate α of the ejection device are calculated uniquely from Eqs. (21.12), (21.13). The stationary curves calculated for different groove shapes are shown in Fig. 21.4. Each point of the curve corresponds to the vortex, which is stationary one and satisfies the Kutta condition on the groove edges. Dependencies of the standing vortex circulation Γ_v and suction power Q from the horizontal coordinate are shown in Fig. 21.5a, b, respectively.

Eigenvalue analysis of the Jacobian of linearized system (21.3) reveals the following topological modes of the standing vortex in the groove with fluid ejection:

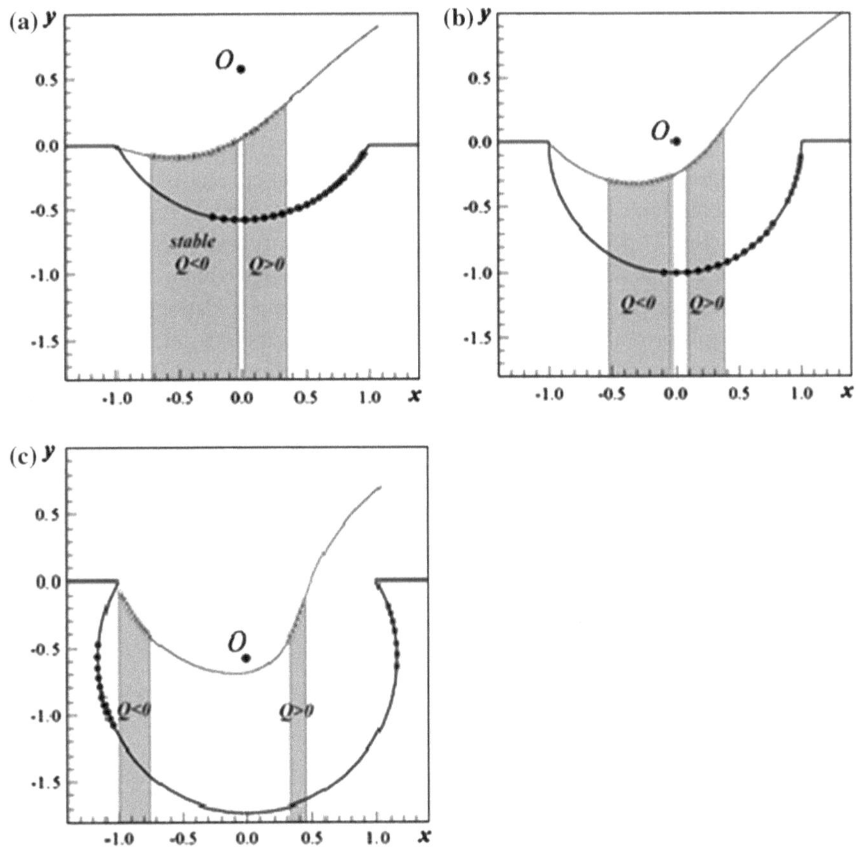

Fig. 21.4 *Curves* of standing vortices in the grooves of different depths: **a** $\beta = 60°$, **b** $\beta = 90°$, **c** $\beta = 120°$

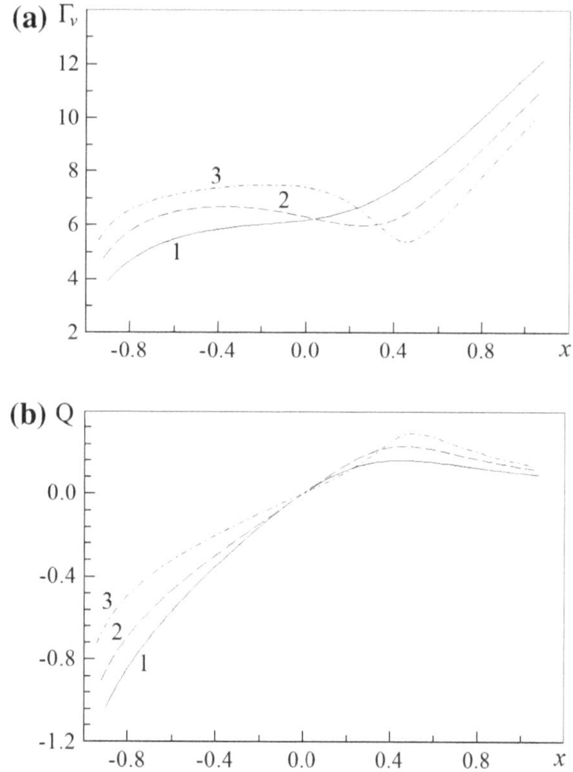

Fig. 21.5 **a** Standing vortex circulation $\Gamma(x)$, **b** strength of fluid suction $Q(x)$ in the grooves of different depths: (*1*) $\beta = 60°$, (*2*) $\beta = 90°$, (*3*) $\beta = 120°$

(1) stable focus, which corresponds to a conjugate pair of eigenvalues with negative real part;
(2) unstable focus, when the real parts of both eigenvalues are positive;
(3) saddle, when the eigenvalues are real and of opposite sign.

Zones of vortex stability in stationary curves are shaded (Fig. 21.4). It is seen that in shallow grooves those are wider and localize in the groove central part (Fig. 21.4a, b). In the deep grooves the zones of stationary vortices are narrow and move to the edges (Fig. 21.4c). It follows from these results the shallow grooves are more perspective ones for creating stable circulation zones in near-wall flow. Each vortex of the stationary curve realizes at a specific position and power of fluid suction. It is seen in Fig. 21.5b that stable circulation zone may be achieved not only by suction, but blowing (injection) fluid in the groove. In Fig. 21.4, the calculated locations of ejection/injection device corresponding to stable standing vortices are shown by markers on the groove borders (•—$Q < 0$, ⋆—$Q > 0$). Pictures of streamlines with a standing vortex and ejection/injection of fluid are shown in Fig. 21.6a, b, respectively.

Figure 21.6a indicates the presence of sufficiently broad layer of fluid that is sucked by the control device. This layer separates the circulating zove and external flow and supports stable vortex configuration in the region. The extension of

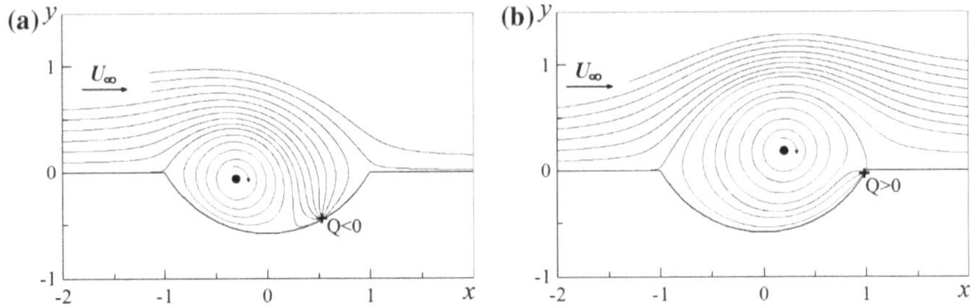

Fig. 21.6 Streamlines pictures in the shallow cross groove ($\beta = 60°$) **a** with fluid suction, **b** with fluid injection

the recirculation zone is significantly reduced compared to the uncontrolled flow (Fig. 21.3a). Figure 21.6b shows that the fluid that is blown into the groove, twists around the vortex. This leads to gain of the vortex circulation and increasing the size of the circulation zone compared to the uncontrolled case.

The intensity of the attraction or repulsion of the vortex with respect to the flow critical point is determined by the real part of the eigenvalues. The calculated values λ_r for the stable focus are of order 10^{-2} when suction and of 10^{-4} when fluid injection. This fact, as well as analysis of Fig. 21.6 points out that the suction provides more stable vortex configuration in the groove in comparison with injection. Therefore, fluid suction is more convenient for the control.

Figure 21.7 demonstrates the trajectories on which the vortex will move in the control scheme with fluid suction in case of its deviation from equilibrium position. The results confirm that suction prevents washing of the vortex from the groove downstream. But Fig. 21.7 demonstrates not only closed rotational vortex trajectories around the stable critical point but and parabolic curves caused by the presence of unstable singularity over front angle groove. Finding such a trajectory the vortex moves upstream from the groove. To stabilize the vortex in the neighborhood of stable critical point and not allow him pass through a separatrix that divides the trajectory of various types, active feedback control has to be applied. Then the parameters of control device will be chosen depending on the changes in the flow.

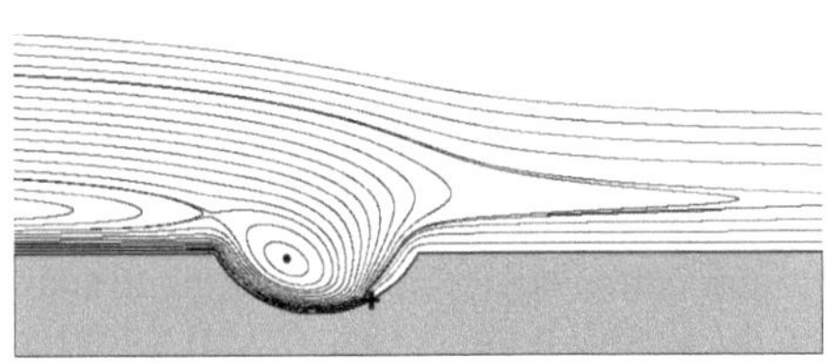

Fig. 21.7 The phase portrait of standing vortex trajectories in the groove with fluid suction ($\beta = 60°$): •—stable equilibrium point, +—position of fluid suction

21.4 Standing Vortex in the Groove in Periodically Perturbed Flow

In practice, near-wall flows are heterogeneous. There are many factors that cause nonstationarity of an external flow, for example, body vibrations, migration of turbulent spots and passing external large-scale vortices. Often those have a periodic component that leads to resonant hydrodynamic loads on the system under consideration. Reaction of a standing vortex to periodic flow perturbations in the groove with fluid ejection/injection is defined by the imaginary part of the eigenvalues. Depending on the vortex location on the stationary curve they correspond to stable or unstable focus. This parameter characterizes the standing vortex rotation in a small neighborhood of the critical point. As in [13], we call it the eigenfrequency of the vortex ω_0. The results of calculations (Fig. 21.8) show that fluid suction causes significant rise of ω_0. With fluid injection ω_0 decreases.

The role of standing vortex eigenfrequency becomes evident in periodically perturbed flow. Let the velocity of external flow has a small periodic component:

$$U(t) = U_\infty(1 + \varepsilon \sin \Omega t), \quad \varepsilon \ll 1, \tag{21.15}$$

where ε, Ω are the amplitude and frequency of perturbations, respectively.

Velocity fluctuations cause deviations of standing vortex from the equilibrium position. If the amplitude of perturbations is small, these deviations will be insignificant, and as shown in Fig. 21.7, the vortex remains in the neighborhood of the critical point. Motion of standing vortex in the oscillating flow is characterized by the function:

$$R(t) = \sqrt{(x_\nu(t) - x_0)^2 + (y_\nu(t) - y_0)^2},$$

where x_0, y_0 are the coordinates of the flow critical point. Then the maximum deviation $R_{max} = max\{R(t), t = (0, T)\}$, where $T \to \infty$, determines the amplitude of vortex motion in the perturbed flow.

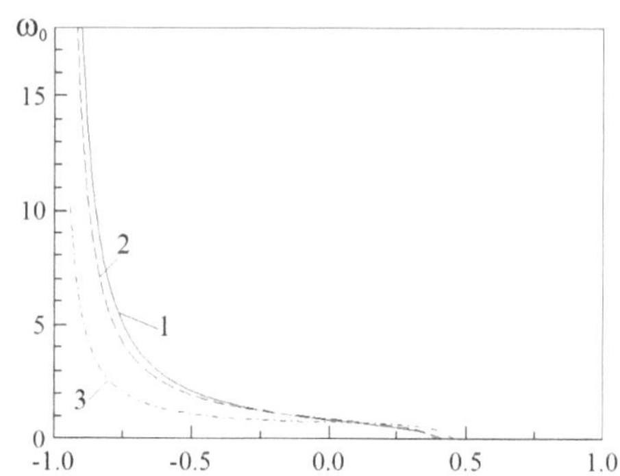

Fig. 21.8 The frequency of standing vortex $\omega_0(x)$ against the vortex position in grooves of different depth at fluid suction (injection): (*1*) $\beta = 60°$, (*2*) $\beta = 90°$, (*3*) $\beta = 120°$

The calculations show the vortex motion in the perturbed flow depends significantly on the ratio of the frequency of external perturbations Ω to the eigenfrequency of vortex ω_0. If Ω differs significantly from ω_0, the vortex will move in a small neighborhood of the critical point along periodical trajectories. The trajectories are changed with frequencies that are proportional to the excitation frequency Ω, in other words those are subharmonics of Ω.

In general, this process is periodic one, it means that there is a long period after which the process repeats. Since the vortex weakly deviates from the critical point in this case, such perturbations are not dangerous for system stability.

If $\Omega \to \omega_0$, the vortex precession amplitude will increase sharply and vortex trajectories will become more complicated, so this case requires detailed study. Figure 21.9 presents trajectories of two different vortices from the stationary curve in the shallow groove ($\beta = 60°$) calculated at $\frac{\Omega}{\omega_0} = 1$. One can see that the vortices move along the closed periodic trajectories with the start and the end at the critical point. The maximum deviation R_{max} of vortex from the critical point is inversely proportional to the power of fluid suction that corresponds to the given initial vortex coordinates (Fig. 21.5b). Small trajectory in Fig. 21.9 corresponds to the power $Q = -0,55$, large trajectory is obtained for $Q = -0,25$. Frequency characteristics of vortex motion in the resonance perturbed flow may be obtained from the analysis of function $R(t)$. As shown in Fig. 21.10, high-frequency pulsations whose frequency is 2Ω impose on the main trajectory here.

The low frequency which corresponds to the vortex oscillation with a large amplitude associates with the frequency of external action by ratio $\frac{\Omega}{50}$. The obtained estimates show the frequency characteristics of the perturbed flow are the same for all vortices from the stationary curve which corresponds to the given groove depth. Thus, the small resonant perturbations of external flow generate low-frequency vortex motion with large amplitude and high-frequency vibrations. Both the first and the second are related by some ratio with the frequency of external action, those are its subharmonics. It is obvious that generation of subharmonics in the oscillatory

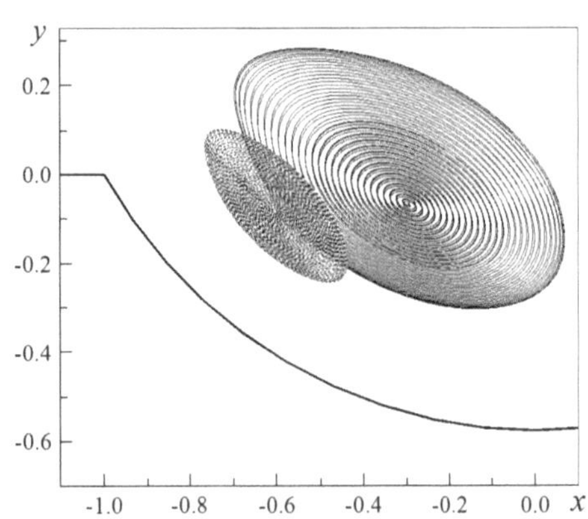

Fig. 21.9 Trajectories of standing vortex in perturbed flow for shallow groove ($\beta = 60°$): $\varepsilon = 0,01$, $\frac{\Omega}{\omega_0} = 1$, $x_v = -0,3$, $x_v = -0,6$

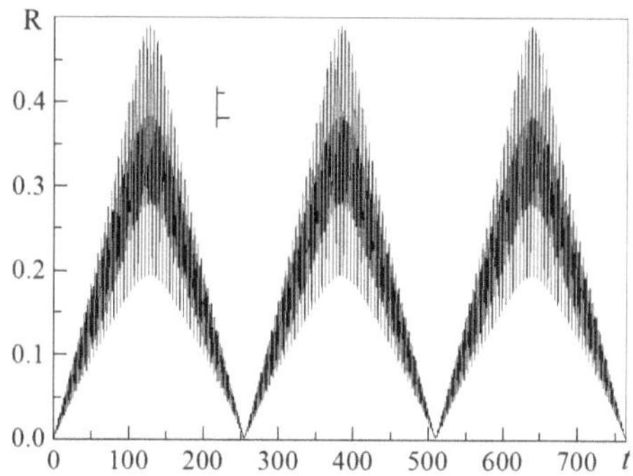

Fig. 21.10 Deviations of standing vortex from the equilibrium position in the perturbed flow: $\frac{\Omega}{\omega_0} = 1$, $x_v = -0,3$, $\varepsilon = 0,01$

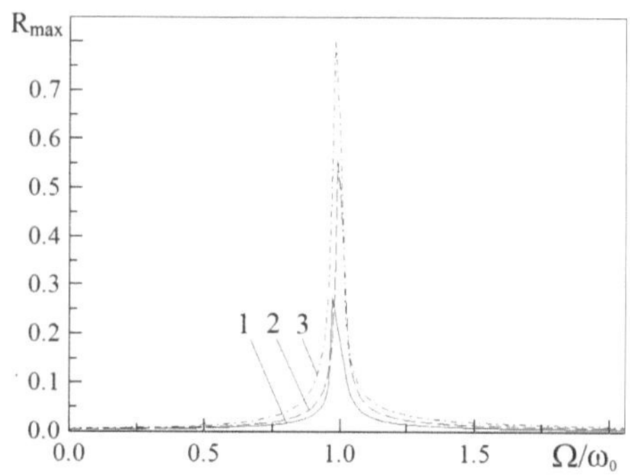

Fig. 21.11 Dependence of the vortex motion amplitude R_{max} from the frequency of external perturbation in various power of fluid suction: *(1)* $x_v = -0,6$, $Q = -0,55$; *(2)* $x_v = -0,3$, $Q = -0,25$; *(3)* $x_v = 0$, $Q = 0$

system is caused by its internal mechanisms, which follow from nonlinear nature of the equations of vortex dynamics in near-wall regions.

Figure 21.11 presents the dependence of maximum vortex deviation R_{max} in perturbed flow from the ratio of forced frequency Ω to the natural frequency of the system ω_0 at different configurations of the flow field in the groove (without fluid suction and with suction of different intensity).

One can see that all the three curves have the sharp resonance character. But if the amplitude of vortex oscillations R_{max} is commensurate with the size of the groove in the uncontrolled flow (curve 3), then fluid suction leads to its significant reduction (curves 1, 2).

To define the character of motion of standing vortex in the perturbed flow, the corresponding Puincare sections are computed when positions of vortex are calculated at the discrete time points: $t_n = \frac{2\pi n}{\Omega}$, $n = 1, 2, \ldots$ Figure 21.12 shows the Poincare sections of two standing vortices with $x_v = -0,3$ and $x_v = -0,6$ calculated at $\Omega = \omega_0$.

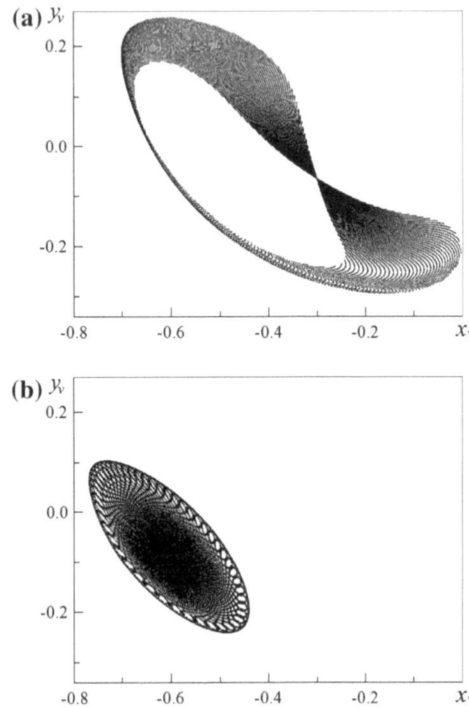

Fig. 21.12 Poincare sections of standing vortex trajectories in shallow groove in perturbed flow: $\varepsilon = 0,01$, $\frac{\Omega}{\omega_0} = 1$, **a** $x_v = -0,3$, **b** $x_v = -0,6$

21.5 Summary

The theoretical model of the nonlinear active near-wall flow control that uses the vortex trapped in a cross groove and suction of fluid is developed. Its analysis is based on the simplified model in which circulating flow is replaced by a point vortex, and the fluid suction is modeled by the hydrodynamic fluid sink. The nonlinear near-wall flow controller constructed contains the equation of vortex equilibrium and the Kutta condition in the groove sharp edges. It is applied for calculating the parameters of the control system which provide the existence of the stable standing vortex and unseparated flow in the groove corners. Dynamic analysis revealed the critical points in the groove with fluid suction are either stable or unstable foci or saddles. The region of stability of vortices in shallow grooves is wider than in deep grooves, so they are more perspective for near-wall flow control.

The reaction of the system to small external periodic perturbations is defined by the ratio of the frequency of these perturbations to the eigenfrequency of standing vortex, which characterizes the rotation of the vortex in the neighborhood of the critical point. Resonant perturbations generate the low-frequency vortex motion with large amplitude and high-frequency vibrations that lead to violation of the demand of unseparated flow in groove corners. Fluid suction reduces significantly the resonant loads in the system and increases the stability of the flow configuration.

References

1. Gad-el-Hak, M., Bushnell, D.M.: Separation control: review. J. Fluids Eng. **113**(1), 5–30 (1991)
2. Protas, B., Wesfreid, J.E.: Drag force in the open-loop control of the cylinder wake in the laminar regime. Phys. Fluids **14**(2), 810–826 (2002)
3. Wu, J.Z., Vakili, A.D., Wu, J.M.: Review of the physics of enhancing vortex lift by unsteady excitation. Prog. Aerosp. Sci. **28**(2), 73–131 (1991)
4. Roos, F.W., Kegelman, J.T.: Control of coherent structures in reattaching laminar and turbulent shear layers. AIAA J. **24**(12), 1956–1963 (1986)
5. Gorban, V., Gorban, I.: Dynamics of vortices in near-wall flows: eigenfrequencies, resonant properties, algorithms of control. AGARD Rep. **827**, 15-1–15-11 (1998)
6. Protas, B.: Vortex dynamics models in flow control problems. Nonlinearity **21**, 1–54 (2008)
7. Batchelor, J.: Introduction to Fluid Dynamics [in Russian]. Mir, Moscow (1973)
8. Chorin, A.J.: Vorticity and Turbulence. Springer, New York (1994)
9. Aref, H., Kadtke, J.B., Zawadski, I.: Point vortex dynamics: recent results and open problems. J. Fluid Dyn. Res. **3**, 63–74 (1988)
10. Meleshko, V.V., van Heijst, G.J.F.: Interacting two-dimensional vortex structures: point vortices, contour kinematics and stirring properties. Chaos Solut. Fractals **4**, 977–1010 (1994)
11. Ringleb, F.O.: Two-dimensional flow with standing vortex in ducts and diffusers. J. Fluids Eng. **82**(4), 921–927 (2011)
12. Bunyakin, A.V., Chernyshenko, S.I., Stepanov, G.Y.: High-Reynolds-number Bftchelor-model asymptotics of a flow past an aerofoil with a vortex trapped in a cavity. J. Fluid Mech. **358**, 283–297 (1998)
13. Gorban, I.M., Homenko, O.V.: In: Zgurovsky, M.Z., Sadovnichiy, V.A. (eds.) Dynamics of vortices in near-wall flows with irregular boundaries. Continuous and Distributed Systems: Theory and Applications. Solid Mechanics and Its Applications, pp. 115–128 (2014)
14. Cortelezzi, L., Leonard, A., Doyle, J.: An example of active circulation control of the unsteady separated flow past a semi-infinite plate. J. Fluid Mech. **260**, 127–154 (1994)
15. Chernyshenko, S.I.: Stabilization of trapped vortices by alternating blowing suction. Phys. Fluids **7**(4), 802–807 (1995)
16. Iollo, A., Zanetti, L.: Trapped vortex optimal control by suction and blowing at the wall. Eur. J. Mech. B-Fluids **20**(1), 7–24 (2001)
17. Filchakov, P.F.: Approximate methods of conformal mappings [in Russian]. K. Naukova Dumka, Kiev (1964)
18. Clements, R.R.: An inviscid model of two-dimensional vortex shedding. J. Fluid Mech. **57**, 321–336 (1973)
19. Cortelezzi, L.: Nonlinear feedback control of the wake past a plate with a suction point on the downstream wall. J. Fluid Mech. **327**, 303–324 (1996)

Chapter 22
A Numerical Study of Solitary Wave Interactions with a Bottom Step

I.M. Gorban

Abstract The numerical scheme for simulation of viscous nonlinear interactions between a solitary wave and a nonregular bottom is developed. It combines the boundary integral method for description of free-surface deformations, conformal mapping used to satisfy the nonleaking boundary condition on the bottom and the vortex method for integrating the fluid dynamic equations. A series of simulations were performed to study free-surface transformations and vortical flow patterns when propagating the solitary wave over a submerged step. Types of both the reflected and transmitted waves are shown to depend on the ratio of the incident wave amplitude to the water depth over the top step wall. The obtained critical value of this coefficient, at which the transmitted wave will be always breaking, is about 0.8 that is in congruence with the experimental data. The detailed investigation of the vortical patterns generated by a solitary wave near the step edge detected two large opposite vortices shedding in both the upstream and downstream directions. Interaction of those specifies the fluid dynamics and turbulent processes in the region.

22.1 Introduction

The study of interactions between surface waves and sea bottom irregularities, such as a step, is important for understanding factors that may lead to dangerous processes in coastal zones. Frequently, the bottom step is considered as the model that describes transfer of the sea bottom to continental shelf [1, 2]. On the other hand, solitary waves present a limiting condition for the run-up of long waves [3]. So evolution of a soliton over a submerged step simulates interaction of long waves with shelf.

Because of great importance of the problem, there are a lot of papers devoted to study of a solitary wave when it propagates from deep water into shallower water. Earlier researchers were based on the nonlinear shallow-water wave theories such

I.M. Gorban (✉)
Institute of Hydromechanics, National Academy of Sciences of Ukraine,
Zheliabova St. 8/4, Kyiv 03680, Ukraine
e-mail: ivgorban@gmail.com

© Springer International Publishing Switzerland 2015

V.A. Sadovnichiy and M.Z. Zgurovsky (eds.), *Continuous and Distributed Systems II*, Studies in Systems, Decision and Control 30,
DOI 10.1007/978-3-319-19075-4_22

as Boussinesq and the KdV equations [1, 2]. Within its applicability, the models predict accurately the reflected and transmitted waves as well as generation of high harmonics on the free surface when the solitary wave passes the step. Since these approximations are derived in assumptions of weak nonlinearity and weak dispersion, they calculate higher wave amplitudes as compared with experimental data for strongly nonlinear problems [2, 4]. That is due to nonlinear and dispersive effects associated, in particular, with the intensive vertical water flow generated by the wave that is ignored by the shallow-water theory.

In order to overcome this defect, both the improvement of the Boussinesq equations taking into account dispersive processes [5] and the boundary element method applied to solve the Laplace equation at the nonlinear free-surface boundary conditions were developed [6]. Using those, decomposition phenomenon of waves passing over a step without breaking was studied. The obtained wave amplitudes were in good agreement with experimental data. In paper [7], both nonbreaking and breaking solitary waves over a step were numerically investigated by the COBRAS model based on the Reynolds-averaged Navier-Stokes (RANS) equations with a turbulence model. A complex study of soliton transformations at the bottom step including the theoretical analysis based on the KdV and Boussinesq-like systems and computational RANS modeling was carried out in [8]. Comparison of the theoretical predictions with numerical results revealed good agreement for the soliton amplitudes but the difference in the travel time, that characterizes the soliton speed, was large enough. The latest fact follows from disability of asymptotic and potential theories to predict correctly the velocity field in the vicinity of bottom irregularities due to the vortical and turbulent nature of the flow there.

The wave motion causes a fluid flow under the free surface. Its separation in bottom sharp edges induces large-scale vortices which strongly affect the wave energy and stress distribution on the bottom. At periodic wave processes, the last causes the bottom erosion which may seriously affect the safety and the stability of underwater constructions. So the study of separation processes near bottom irregularities is important for both advanced estimations of wave evolution and prediction of possible dangers to the sea bottom and underwater technical systems.

For this reason, numerical models for investigation of the viscous wave-bottom interactions have been developed. Most of them solve the Navier-Stokes equations for laminar flows or the Reynolds-averaged Navier-Stokes equations with turbulence closure for turbulent flows using grid-based schemes [9–12]. The main patterns of the vortical fields generated around submerged structures, such as rectangular and semi-circular dikes, were derived using those. Because of accurate modeling of the vortex generation mechanism requires an extremely fine resolution; the grid-based schemes are costly with a view to computer power. Besides, the great disadvantages of these schemes are the difficulty in tracking the fluid interface and implementation of free-surface boundary conditions.

The alternate algorithm is a Lagrangian-type numerical scheme which combines the boundary integral method used for modeling free surface transformations and the vortex method for calculation of the vorticity field. It has been confirmed by simulating a solitary wave traveling over a submerged rectangular obstacle [13].

Complexity and importance of the problem under consideration make for necessity of development of new numerical schemes which are capable to simulate accurately wave-structure interactions. The numerical model proposed in this study is based on decomposition of the flow field into two components: the first is related with a free surface and the second defines the vorticity field generated near a submerged structure. In this statement, the effects of viscosity and vorticity generation at the free surface are ignored; at the same time, the flow field far from the structure is supposed to be irrotational. These assumptions permit to apply the potential-flow analysis for calculating the free-surface evolution. The flow field under the free surface is described by the total system of fluid dynamics equations. It should be noted that the two components are not solved independently because they affect each other.

So the numerical scheme developed combines the boundary integral method used to derive free-surface deformations with a vortex algorithm for integrating the Navier-Stokes equations [14, 15]. To satisfy the nonleaking condition at the bottom, the conformal mapping of the flow field into the upper half-plane of canonical domain is applied. The approach is employed to investigate the interaction of a solitary wave and a submerged step under nonbreaking wave conditions. The effect of wave height on both the water surface deformations and the vortical flow near the step is studied. The results are systematized with respect to the value of interaction coefficient, which is given by the ratio of the incident wave amplitude to the water depth over the top step wall.

22.2 Problem Statement

A solitary wave passing over a bottom step in viscous incompressible fluid is considered. A Cartesian coordinate system is fixed such that its origin is connected with the step, the x-axis lies in the bottom and the y-axis points vertically upward (Fig. 22.1). The still water depth is h and the amplitude of the incident wave is A_i. The height of the step is depicted by d; so the shallow water depth above the step is $h_1 = h - d$.

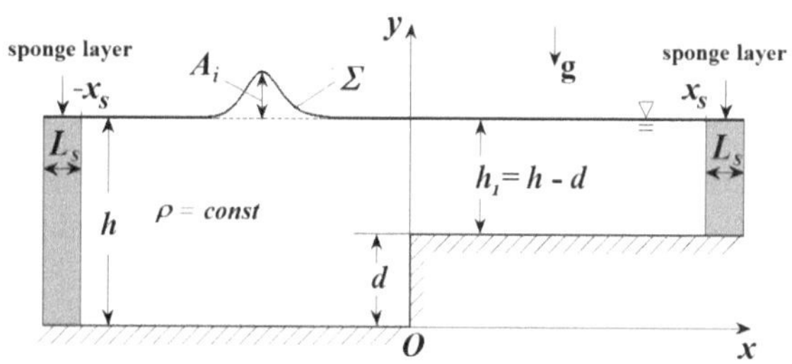

Fig. 22.1 Schematic diagram of a solitary wave passing over a submerged step

The fluid is assumed to be incompressible with constant density and viscosity. Its motion is governed by the continuity equation and the Navier-Stokes equations:

$$\nabla \cdot \mathbf{V} = 0, \tag{22.1}$$

$$\frac{\partial \mathbf{V}}{\partial t} + (\mathbf{V} \cdot \nabla)\mathbf{V} = -\nabla P + \frac{1}{Re}\nabla^2 \mathbf{V}, \tag{22.2}$$

where the hydrodynamic pressure $P = p + y/Fr^2$ is the total pressure p minus the hydrostatic pressure, $\mathbf{V} = (v_x, v_y)$ is the velocity and t is the time. The variables in Eqs. (22.1) and (22.2) are scaled by the undisturbed water depth h and the phase speed of linear long-wave $\sqrt{gh}$, where g is the acceleration due to gravity. So, the Reynolds number is defined as $Re = h\sqrt{gh}/\nu$, where ν is the kinematic viscosity of water. The Froude number specifies the ratio of the phase speed of the solitary wave $\sqrt{g(A_i + h)}$ to $\sqrt{gh} \cdot (Fr = \sqrt{g(A_i + h)}/\sqrt{gh})$, while $t = \sqrt{gh/h}$ is chosen to non-dimensionalize the time.

On the fluid boundary, the dynamic and kinematic boundary conditions must be required. The kinematic free-surface boundary condition states that fluid particles at a free surface remain on the surface; it can be expressed as

$$\frac{\partial \eta}{\partial t} + v_x \frac{\partial \eta}{\partial x} = v_y \quad at \quad y = h + \eta(x, t), \tag{22.3}$$

where $\eta = \eta(x, t)$ is the free surface equation.

For dynamic free-surface boundary condition, continuity of stress components must be satisfied. The surface tension on the fluid boundary is neglected in the present study and then the condition for the most general form can be expressed as

$$(-pI + \sigma) \cdot \mathbf{n}\big|_{\Sigma} = 0, \tag{22.4}$$

where $\mathbf{n}$ is the outward unit vector normal to Σ, I is the identity tensor, $\sigma = \mu[\nabla\mathbf{V} + (\nabla\mathbf{V})^T]$ is the deviatoric stress, μ is the molecular viscosity. Since for most problems in water-wave mechanics, the free-surface viscosity is negligible, we ignore the effect of viscosity at the free surface. Then condition (22.4) takes on form:

$$p_{-} = 0, \tag{22.5}$$

where "_" denotes the limit of the total pressure under the fluid boundary.

On the solid bottom, the nonleaking and no-slip boundary conditions must be required

$$\mathbf{V}(\mathbf{r_b}, t) \cdot \mathbf{n} = 0, \tag{22.6}$$

$$\mathbf{V}(\mathbf{r_b}, t) \cdot \boldsymbol{\tau} = 0, \tag{22.7}$$

where $\mathbf{r}_b$ denotes the radius-vector of bottom points, $\mathbf{n}, \boldsymbol{\tau}$ are the normal and tangential unit vectors on the bottom, respectively.

To avoid wave reflection on the lateral boundaries of the computational domain, wave damping is introduced following the method proposed in [16]. For this reason, numerical sponge layers are placed at both lateral boundaries to absorb the outward-traveling waves.

22.3 Numerical Method

22.3.1 General Principles

Following the Helmholtz decomposition principle [17], the flow under consideration is decomposed into the irrotational part in the thin layer near a free surface and the rotational part everywhere except the fluid boundary. Then the velocity field may be represented as

$$\mathbf{V}(\mathbf{r}, t) = \nabla\varphi + \nabla \times (\Phi\mathbf{k}), \tag{22.8}$$

where $\mathbf{r}$ denotes the radius-vector of a point, φ and $\Phi\mathbf{k}$ are scalar and vector potentials, respectively, $\mathbf{k}$ is the unit vector out of the page.

The rotation field in the computational domain S is described by the function of vorticity $\omega = \mathbf{k} \cdot \nabla \times \mathbf{V}$ and the free-surface Σ is modeled by a vortex sheet whose strength γ is induced by the jump in tangential velocity across the sheet. Then, expression (22.8) takes the form

$$\mathbf{V}(\mathbf{r}, t) = \int_\Sigma \gamma(\mathbf{r}', t)\mathbf{k} \times \nabla G(\mathbf{r}, \mathbf{r}')dl(\mathbf{r}') + \int_S \omega(\mathbf{r}', t)\mathbf{k} \times \nabla G(\mathbf{r}, \mathbf{r}')ds(\mathbf{r}') \tag{22.9}$$

where $G(\mathbf{r}, \mathbf{r}')$ is the vortex function in the region under consideration.

To find the function G, which will satisfy nonleaking of solid bottom, the conformal mapping of the flow field in the physical plane $z(x, y)$ into an upper half-plane of the auxiliary plane $\zeta(\xi, \lambda)$ is performed. At this transformation, the bottom step maps into the flat wall that coincides with the axis ξ and the free-surface $\eta(x, t)$ transfers to the conditional boundary $\eta^*(\xi, t)$. The function that realizes the present mapping is built with applying the Schwarz-Christoffel transformation [18]:

$$z = \frac{d}{\pi}\left[\sqrt{\zeta^2 - 1} - \ln(\zeta + \sqrt{\zeta^2 - 1})\right].$$

The inverse transformation $\zeta = \zeta(z)$ is determined numerically. The compliance of the points in physical and canonical domains is depicted in Fig. 22.2.

For convenience, the complex notations $z = x + iy$ and $\zeta = \xi + i\lambda$ are used to describe the complex field points in physical and auxiliary planes, respectively. The fundamental solution of Laplace's equation for a vortex in the half-plane is known

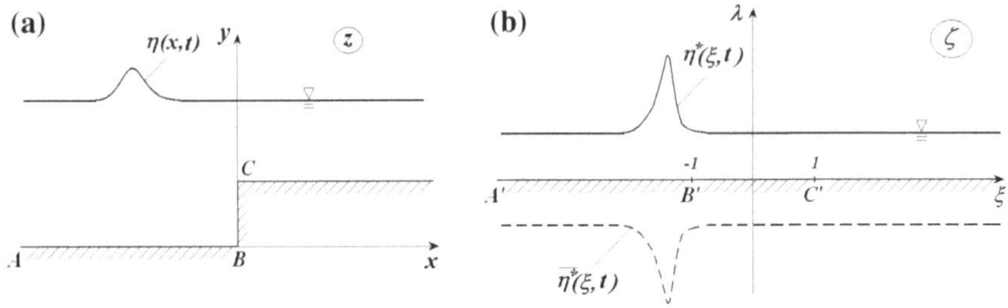

Fig. 22.2 Coordinate systems in the physical plane z (**a**) and in the auxiliary plane ζ (**b**)

to be built by mirror image of the vortex relative to the wall. So function G takes the form

$$G(\zeta,\zeta_v)=\frac{1}{2\pi i}\Big[\ \ln(\zeta-\zeta_v)-\ln(\zeta-\overline{\zeta_v})\Big], \tag{22.10}$$

where ζ_v is the vortex coordinate in the auxiliary plane and the bar designates the complex conjugation.

The free-surface Σ and the vortex distribution γ on it are parametrized in terms of the time t and a space parameter e as $z_\Sigma(e,t)$ and $\gamma(e,t)$. The vorticity field ω is approximated by a set of vorticity carrying particles as proposed in [19]:

$$\omega(z,t)\approx\sum_{j=1}^{N}\Gamma_j(t)f_\delta(z-z_j), \tag{22.11}$$

where Γ_j and z_j are circulation and coordinate of the jth vortex, respectively, f_δ is the distribution function of vortex [19]. Taking into account that vortex circulation is conserved when conformal mapping, we obtain the following expression for the velocity field

$$V(z,t)=\Bigg(\frac{1}{2\pi i}\int_\Sigma\gamma(e',t)\Bigg[\frac{1}{\zeta(z)-\zeta_{\Sigma^*}(z(e'))}-\frac{1}{\zeta(z)-\overline{\zeta}_{\Sigma^*}(z(e'))}\Bigg]de'+ \\ \frac{1}{2\pi i}\sum_{j=1}^{N}\Gamma_j\Bigg[\frac{1}{\zeta(z)-\zeta(z_j)}-\frac{1}{\zeta(z)-\overline{\zeta}(z_j)}\Bigg]\Bigg)\frac{d\zeta}{dz}, \tag{22.12}$$

where Σ^* is the imagery of the free surface in ζ -plane.

22.3.2 Free-Surface Modeling

The free-surface points $z_\Sigma(e,t)=x_\Sigma(e,t)+iy_\Sigma(e,t)$ and the vortex sheet strength $\gamma(e,t)$ are determined by imposing the free-surface boundary conditions. Kinematic boundary condition (22.3) is satisfied automatically if one assumes that free surface follows trajectories of fluid particles lying on the free surface. The latest are determined from the evolutionary equation:

$$\frac{dz_{\Sigma}(e,t)}{dt} = \mathbf{V}_{\Sigma}(e,t), \tag{22.13}$$

where $\mathbf{V}_{\Sigma}$ is the velocity of the free-surface Lagrangian points. It is determined by formula (22.12). To remove the singularity in the first term of (22.12), the principal value of the integral is considered.

To calculate the strength of the vortex sheet distributed along the free surface, dynamics condition (22.4) is applied. Following the algorithm proposed in [20], we evaluate the Bernoulli's equation on the lower side of the interface. Then boundary condition (22.4) takes the form

$$\frac{\partial \varphi_{-}(e,t)}{\partial t} + \frac{\mathbf{V}_{-}^{2}(e,t)}{2} + \frac{y_{\Sigma}(e,t)}{Fr^{2}} = 0, \tag{22.14}$$

where $\mathbf{V}_{-}$, φ_{-} are the limit values of the velocity and of its potential under the free surface, respectively.

Taking into account that the substantial derivative of potential $d\varphi/dt$ in the point moving with velocity $\overrightarrow{V}_{\Sigma}$ is [17]

$$\frac{d\varphi}{dt} = \frac{\partial \varphi}{\partial t} + \mathbf{V}_{\Sigma} \cdot \nabla \varphi,.$$

one may rewrite Eq. (22.14) as

$$\frac{d\varphi_{-}}{dt} + \frac{\mathbf{V}_{-}^{2}}{2} - \mathbf{V}_{\Sigma} \cdot \mathbf{V}_{-} + \frac{y_{\Sigma}}{Fr^{2}} = 0, \tag{22.15}$$

The jump of velocity across the vortex sheet is known to be $\gamma/2$ [17]. So, the velocity of free-surface points $\mathbf{V}_{\Sigma}$ and velocity under the free-surface $\mathbf{V}_{-}$ are connected with the following relation

$$\mathbf{V}_{-} = \mathbf{V}_{\Sigma} + \frac{1}{2}\gamma \cdot \boldsymbol{\tau}, \tag{22.16}$$

where $\boldsymbol{\tau}$ is the tangential unit vector to the free surface. Substituting (22.16) into equation (22.15) we obtain the evolution equation for the velocity potential:

$$\frac{d\varphi_{-}}{dt} = \frac{\mathbf{V}_{\Sigma}^{2}}{2} - \frac{1}{8}\gamma^{2} - \frac{y_{\Sigma}}{Fr^{2}}. \tag{22.17}$$

Equation (22.17) is the detailed dynamical condition on the interface. Further, it will be used for calculating the strength of the vortex layer simulating the free surface.

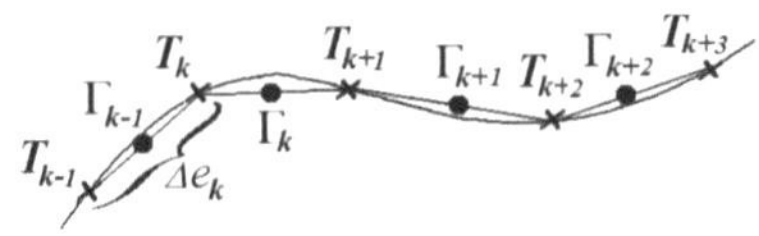

Fig. 22.3 Sketch of the computational domain: • — discrete vortices, **x** — control points

In the numerical implementation of the method, the free surface is represented by a set of straight segments (panels) with a piecewise distribution of the vortex intensity. The end coordinates and the linear strength of the k-th segment are estimated as $z_{\Sigma_k} = z_\Sigma(e_k)$ and $\gamma_k = \gamma(e_k)$, where e_k is the discrete parametrical coordinate. Further, each segment is replaced by a discrete vortex placed in the middle of the segment. So, the discrete vortex circulation is $\Gamma_k^\Sigma = \gamma_k \Delta e_k$, where Δe_k is the k-th panel length ($k = 1, 2, ..., N_\Sigma$). The collocation points (control points), in which the dynamics boundary condition is satisfied, are located in the ends of the segments (Fig. 22.3).

Such discretization means that the integral along the free surface Σ in (22.12) is replaced by a sum whose components are based on trapezoidal quadrature. So, expression (22.12) takes the form

$$V(z,t) = \left(\frac{1}{2\pi i} \sum_{k=1}^{N_\Sigma} \Gamma_k^\Sigma \left[\frac{1}{\zeta(z) - \zeta(z_{\Sigma k})} - \frac{1}{\zeta(z) - \overline{\zeta}(z_{\Sigma k})} \right] + \frac{1}{2\pi i} \sum_{j=1}^{N} \Gamma_j \left[\frac{1}{\zeta(z) - \zeta(z_j)} - \frac{1}{\zeta(z) - \overline{\zeta}(z_j)} \right] \right) \frac{d\zeta}{dz} \quad (22.18)$$

If one considers a positive vortex rotating counterclockwise, the limit value of the tangential velocity $V_\tau = \partial\varphi/\partial\tau$ under the vortex layer may be calculated from the following relation

$$\frac{\partial \varphi_-}{\partial \tau} = V_{\Sigma\tau} + \frac{\gamma}{2}. \quad (22.19)$$

Then for two adjacent control points T_k and T_{k+1} on the free surface, we have

$$\frac{1}{2}\left[(V_{\Sigma\tau})_k + (V_{\Sigma\tau})_{k+1} \right] \Delta e_k + \frac{\Gamma_k^\Sigma}{2} = \varphi_{k+1} - \varphi_k, \quad k = 1, 2, ..., N_\Sigma, \quad (22.20)$$

where $(V_{\Sigma\tau})_k$ is determined from (22.18) noted at the control points. If to retain in the left part of (22.20) the terms containing the circulation of free-surface vortices only, the system of linear algebraic equations relative to the discrete vortex circulations Γ_k^Σ will be derived. In the right part of the system, the velocity potentials and the velocities induced by the viscous vortices at the points under consideration are settled.

Following [13, 21], we use a fourth-order Adams-Bashforth-Moulton (ABM) predictor-corrector scheme for time integrating equations (22.13), (22.17), which describe evolution of the free-surface and the velocity potential, respectively. If one considers the ordinarily differential equation $dz/dt = f(t)$, the ABM scheme for its decision is expressed as

$$\begin{aligned} z_1^{pr} &= z_0 + \frac{\Delta t}{24}(55 f_0 - 59 f_{-1} + 37 f_{-2} - 9 f_{-3}) \\ z_1^{corr} &= z_0 + \frac{\Delta t}{24}(9 f_1^{pr} + 19 f_0 - 5 f_{-1} + f_{-2}) \end{aligned} \tag{22.21}$$

where Δt is the time step.

Note Eq. (22.13) is considered relative to the discrete vortices modeling the free surface. Velocities of those are determined from (22.18) with allowing for the Routh correction occurring due to conformal mapping [22]. To remove self-intersection of the free surface, redistribution of the vortices is executed at the end of each time step. The procedure includes recalculating both positions and circulation of the vortices along the surface in order to achieve the equal spacing between those. Regularization of the vortex sheet improves considerably stability of the computations. To avoid significant wave reflections at the lateral boundaries of the computational domain, the sponge layer technique is applied. For the first time, the technique was proposed in [23] and since then it has been successfully used when integrating the problems with a free-surface [13, 21, 24]. The sponge layers of length L_s are located at the left and right boundaries (Fig. 22.1) and damped members are introduced into equations (22.13), (22.17), such as

$$\begin{aligned} \frac{dz_\Sigma(e,t)}{dt} &= V_\Sigma(e,t) - iD_\Sigma\left|\left(|x_\Sigma(e)| - x_s\right)\right| y_\Sigma(e), \\ \frac{d\varphi_-}{dt} &= R_\varphi - D_\Sigma\left|\left(|x_\Sigma(e)| - x_s\right)\right| y_\Sigma(e), \end{aligned} \tag{22.22}$$

where R_φ is the right-hand side of (22.17), D_Σ is the damping coefficient: $D_\Sigma = 0$ for $|x| < x_s$ and $D_\Sigma = S/L_s$ for $|x| > x_s$. To determine the optimal values of L_s and S, a series of numerical tests was performed. It was given in the present calculations: $L_s - 5$, $S - 2$. The presented realization of the Lagrangian method for free surface modeling differs from that developed in [16, 23, 24], where the strength of the vortex sheet distributed along a fluid boundary is govern by a Fredholm integral equation of the second kind. The present algorithm, which is based on results of paper [20], comes to the system of linear algebraic equations relative to the discrete analog of function γ. In spite of this difference, the results of test calculations point out the total identity of both approaches. The problem of interaction of a pair of counter-rotating point vortices with a free surface in ideal fluid was considered as of the test case. The flow parameters were specified as such in paper [24]: $1/Fr^2 = 0.02$, the circulations and initial positions of the vortices were $\Gamma_{1,2} = \pm 1$, $x_{1,2} = \pm 0.5$, $y_{1,2} = -5$. The results of calculations at $t = 32$ are depicted in Fig. 22.4. Both the vortex trajectories and the free surface configuration obtained coincide absolutely with those presented in [24].

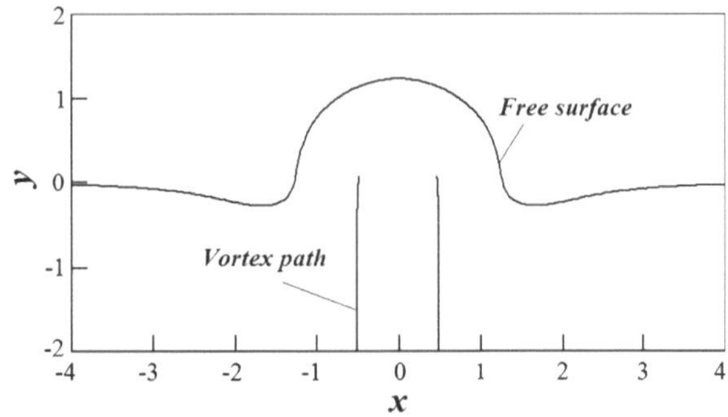

Fig. 22.4 Vortex paths and the free surface for $1/Fr^2 = 0.02$ at $t = 32$

22.3.3 *The Vortex Method for 2-D Flows*

The viscous flow field under the free surface is calculated via the generalized vortex method. It has been described in details in papers [14, 15], therefore, generalities of the method are only considered here. The motion of fluid is governed by the vorticity transport equation, which is derived from Eqs. (22.1) and (22.2) via the procedure "curl" applied to each term of the equations

$$\frac{\partial \omega}{\partial t} + \left(\mathbf{V} \cdot \nabla\right)\omega = \frac{1}{Re}\Delta\omega. \tag{22.23}$$

So the flow field is here described in terms of "vorticity" and "velocity." This approach is preferential due to absence of the pressure, automatic implementation of the continuity equation, and adaptability because of the regions of concentrated vorticity are only considered.

Equation (22.23) is solved using a fractional step procedure [19] when that is split into convective and diffusive parts, which are solved separately. The spatial derivative in the diffusion equation is approximated by the finite-difference scheme on the orthogonal grid put on the calculation domain. The convective transfer of vorticity is simulated by the method of finite volumes when the vorticity flows across boundaries of given elementary volumes are controlled. The volumes are connected with node points of the orthogonal grid. The vorticity is assumed to distribute evenly inside each volume element. To integrate the process in time, the explicit scheme of the second order with correction of the variables after each performed operator is applied.

The velocity field in the region is calculated by formula (22.18), where the strengths of the vortices modeling the free surface Γ_k^{Σ} are known from decision of the potential problem and circulations Γ_j of the viscous vortices are determined as

$$\Gamma_j(t) = \omega(z_j, t)\Delta s_j, \tag{22.24}$$

where Δs_j is the area of the considered elementary volume. Relation (22.24) means the vorticity distributed in the volume element is converted into the point vortex locating in the middle of the element.

Note that the grid vertical size is not fixed because we deal with permanently changing fluid boundary. At each time step, the top boundary of the calculation domain is approximated by the step function.

To determine the vortex flux on the solid boundary (bottom), the Lighthill vorticity creation mechanism when the spurious slip velocity is turned into the vorticity flux is applied [19]. The no-penetration condition is automatically fulfilled owing to the conformal map and method of images. As results of numerous test calculations have shown, it is preferable use the Kutta condition to compute the vorticity generation in sharp boundary edges.

Validation of this algorithm was examined for a problem of viscous flow past a square prism [14, 15]. The comparison of calculated prism characteristics (force coefficients and Strouhal number) with the known experimental data and numerical results has affirmed the efficiency of the algorithm with acceptable parameters of discrete scheme.

22.4 Results and Disscussion

This section presents the computational results concerning both free-surface transformations and vortical flow evolution generated by a solitary wave propagating above a submerged step. The initial wave data are obtained using the MatLab implementation of the algorithm proposed in [25]. Since the scheme deals with the Euler equations, the solution describes as horizontal as vertical fluid motions generated by the wave.

To estimate calculation ability of the developed numerical algorithm with respect to a solitary wave, its propagation in the channel of constant water depth is first considered. The wave is supposed to have the normalized initial amplitude $A_i = 0.15$ and move from left to right in the channel of length 70 h. The problem is simulated with the following scheme parameters: the time step is $\Delta t = 0.01$ and the initial length of the vortex segments along the free surface is calculated under condition that $\Delta x = 0.05$. The calculated free-surface profiles during wave propagation are depicted in Fig. 22.5. The wave is seen to preserve its permanent shape during a long time that points out the stability of the computational scheme at given discretization parameters.

Evolution of the solitary wave propagating above a submerged step depends on both the amplitude of the incident wave A_i and the step height d. It has been experimentally shown in paper [26] that the process may be controlled only by the quantitative parameter $K_{int} = A_i/(h - d)$. It was called the coefficient of interaction. Depending on its value, the incident solitary wave will either remain nonbreaking throughout the entire evolution process or break in front of the step or break above the step. In this study, nonbreaking solitary waves are only computed. The values

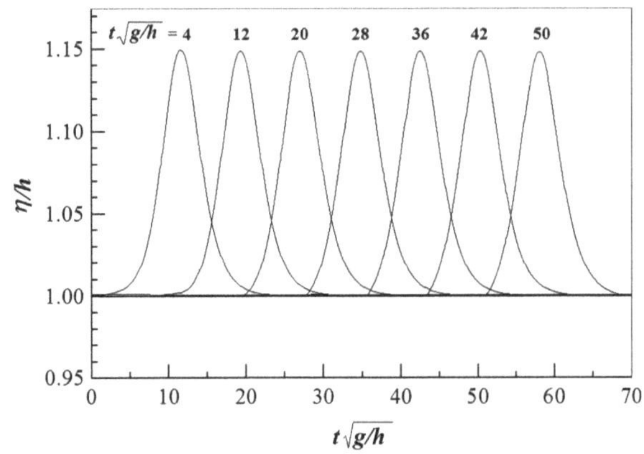

Fig. 22.5 Calculated evolution of the solitary wave propagating on constant water depth

of parameter K_{int} corresponding to different modes of interaction between the wave and the step are evaluated.

In all computations, the channel with domain $-50 \leq x \leq 50$ is used and the step is located at $x = 0$. The incident wave amplitude and the step height are changed in ranges $A_i = 0.1 \div 0.4$, $d = 0.2 \div 0.8$, respectively. The free surface is initially discretized with the step $\Delta x = 0.05$, so, 2000 panels on the free surface are given. The time step is $\Delta t = 0.01$. The grid parameters of the viscous problem are $\Delta x = \Delta y = 0.02$. The Reynolds number is $Re = 3 \cdot 10^6$. Note this evaluation is true when the depth of channel is about 1 m.

Transformations of the solitary wave of normalized amplitude $A_i = 0.2$ passing over submerged steps at different values of interaction coefficient K_{int} are shown in Fig. 22.6.

Here $t = 0$ represents the moment when the wave crest arrives at the edge of the obstacle. When the wave front is seen to approach the obstacle, part of the wave energy is reflected back, forming the wave reflection, and the rest passes the obstacle and continues to propagate forward. Transmitted wave transformations over the step are determined by the water depth as well as by the vortical and turbulent processes in the region. Losada et al. [4] classified the solitary wave evolution at an abrupt junction into four modes, namely: (1) propagation with weak distortion, (2) fission of the wave in solitons, (3) fission in solitons and peaking of the first soliton, and (4) plunging of the wave and evolution of the subsequent bore. The transmitted wave in Fig. 22.6a, when $K_{int} = 0.3$, seems to be the first mode that describes weak interaction of a solitary wave with a step. The transmitted wave in Fig. 22.6b ($K_{int} = 0.4$) belongs to mode 2 because it splits into subharmonics and the height of the first soliton grows considerably. The first soliton in Fig. 22.6c ($K_{int} = 0.8$) stretches as much as possible. Besides, the intense secondary solitons and dispersive wave chain are observed. This case is an example of the strong interaction between the wave and the step and belongs to mode 3. Further augmentation of the step height leads to break of the wave yet before the obstacle. This mode cannot be discussed here because simulation of breaking waves is beyond the scope of this study. It

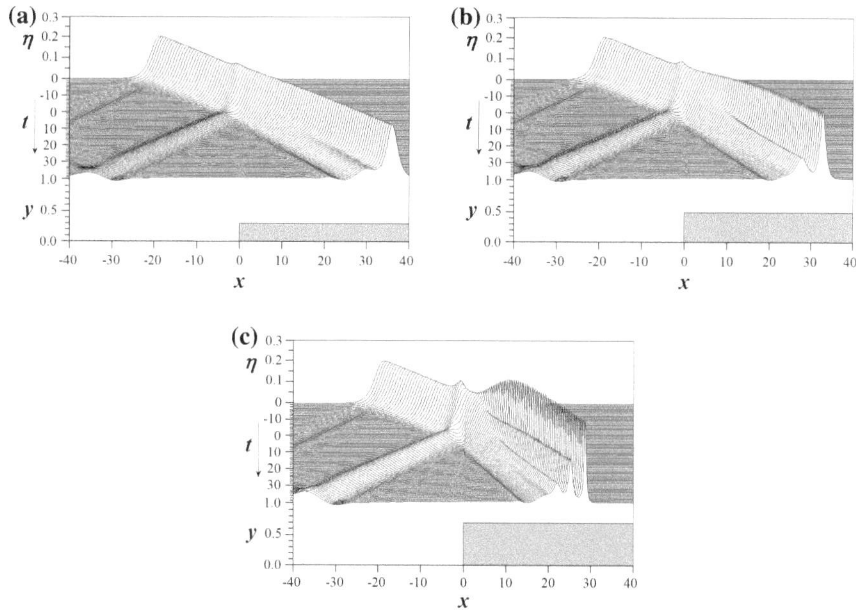

Fig. 22.6 Evolution of free-surface profiles above a step: **a** $-K_{int} = 0.3$, **b** $-K_{int} = 0.4$, **c** $-K_{int} = 0.8$, $A_i = 0.2$

follows from the calculation results that the interaction coefficient value $K_{int} = 0.8$ is maximal at which the solution does not yet break. The conclusion is reasonable for solitary waves of different amplitudes from the declared range. So $K_{int} = 0.8$ may be considered as the critical value after which a solitary wave is always breaking above a submerged step. Note this value coincides with that obtained in the experimental researches [26], where the analogous wave patterns including the dispersion chain have been observed.

It is seen in Fig. 22.6 that the transmitted wave always disintegrates at an abrupt channel junction; so at least one secondary soliton generates next to the leading wave. The amplitude of the first transmitted soliton is always higher than that of the incident wave. Following paper [2], we drew the relative amplitudes both of the transmitted waves A_t/A_i and of the reflected wave A_r/A_i against the normalized amplitude of incident wave A_i/h at $d = 0.5\,\mathrm{h}$ (Fig. 22.7).

The amplitudes obtained are compared with the experimental data [2] and with the results of asymptotic theory [1]. The present results are seen to be closer to the experimental data than to theoretical ones. The dashed vertical line in Fig. 22.7a indicates the upper limit for A_i/h after which the first transmitted wave breaks in the analytic theory. The critical coefficient of interaction K_{int}^{cr} evaluated with this value A_i/h is about 0.4 while $K_{int}^{cr} = 0.8$ as in our calculations as in experimental researches [26]. This difference may be conditioned by action both of nonlinear and of dissipate effects, which the asymptotic theory does not take into account.

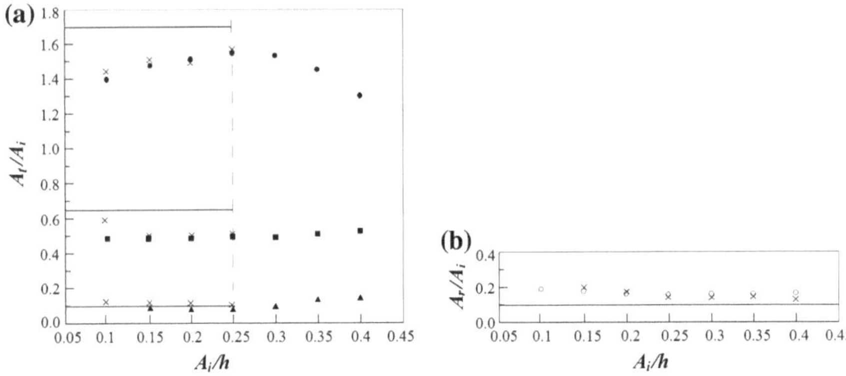

Fig. 22.7 Relative amplitudes of transmitted (**a**) and reflected (**b**) waves against the normalized amplitude of incident wave at $d = 0.5$: *black circle*—first, *black square*—second and *triangle*—third transmitted waves, clear circle—reflected wave,— -results of the long-wave theory [1], $\times\times$-experimental data [2]

The comparative analysis of the results presented in Fig. 22.7 shows that the numerical model applied can produce accurate simulation of fission process when a solitary wave passes an underwater step. Physical interpretation of the results is important to predict the dangers that may be caused by long waves propagating above the continental shelf. The amplitude of the leading soliton is seen to grow significantly as compared with that of the incident wave.

Figure 22.8 systematizes the obtained amplitude data with respect to the interaction coefficient K_{int}. The relative amplitudes of both the leading transmitted soliton A_{t_1}/A_i and the reflected wave A_r/A_i against the parameter K_{int} are presented for two values of incident wave amplitude: $A_i = 0.15$ and $A_i = 0.3$. The solution is seen to grow much more in the first case. In both cases, the parameter A_{t_1}/A_i reaches the highest value at $K_{int} \approx 0.5$ and A_r/A_i grows as the linear function.

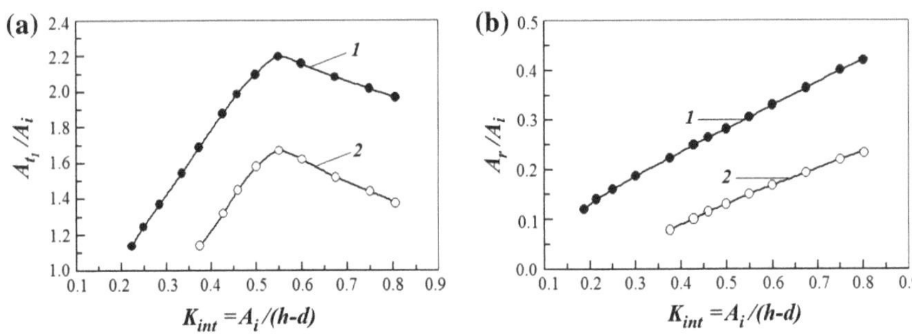

Fig. 22.8 Relative amplitudes of the leading transmitted soliton (**a**) and the reflected wave (**b**) against the interaction coefficient K_{int}: $1 - A_i = 0.15$, $2 - A_i = 0.3$

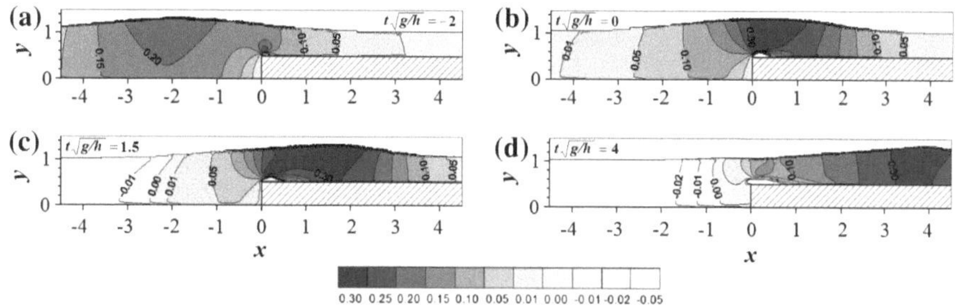

Fig. 22.9 Contours of the horizontal velocity generated by the solitary wave of $A_i = 0.3$ near the step of $d = 0.5$

Traveling the solitary wave causes the intense enough fluid flow under the wave. When the wave front approaches to the step edge, the flow will transform due to channel constriction. To describe the flow evolution, the strong nonlinear wave of $A_i = 0.3$ and the step of $d = 0.5$ are considered. Figure 22.9 demonstrates changes of the horizontal velocity during passing the wave above the step. One can see the backflow and local regions of flow acceleration. Those are obvious to be specified by the flow separation in the step sharp edge. Besides, the flow generated by the wave in the shallow water has much stronger velocity than before the step.

To understand how a vortex flow develops, velocity fields near the step edge at six different time instances are shown in Fig. 22.10a–f. The shear flow above the step is obtained to be generated when a wave crest locates at $x = -10$. As the wave front approaches the step, the shear flow develops into a clockwise vortex (Fig. 22.10a). The vortex grows gradually (Fig. 22.10b) and reaches the maximal strength when a wave crest passes the step edge (Fig. 22.10c). Further, the vortex generated in Figs. 22.10a–c sheds out and continues to grow in size. After the vortex has grown almost to the water depth (Fig. 22.10d), it is confined by the bottom and free surface. To this instant, the wave has passed the step; as a result, the translatory motion of fluid particles almost completely ceases. It forces the water to move upstream in the thin layer adjacent to the step wall. This creates a counterclockwise vortex in the step edge (Fig. 22.10e). The vortex moves upward and upstream and continues to grow. Subsequently, the main clockwise and counterclockwise vortices reach the free surface and cause a small bulge on the free surface (Fig. 22.10f).

The vortex processes induced by a solitary wave near a submerged step were also studied in terms of circulation. According the vorticity definition, the clockwise vortex is characterized by a negative circulation and the counterclockwise vortex is of a positive circulation. The clockwise and counterclockwise circulations are then determined by integrating the negative and positive point vortices over the regions of the vortex motion, which are chosen a posteriori from the velocity field figures. The temporal variations of dimensionless circulations generated by the waves of amplitudes $A_i = 0.15$ and $A_i = 0.3$ near a submerged step of height $d = 0.5$ are presented in Fig. 22.11. The solid lines in Fig. 22.11 correspond to main (clockwise) vortices and the dashed curves describe secondary (counterclockwise) vortices. The

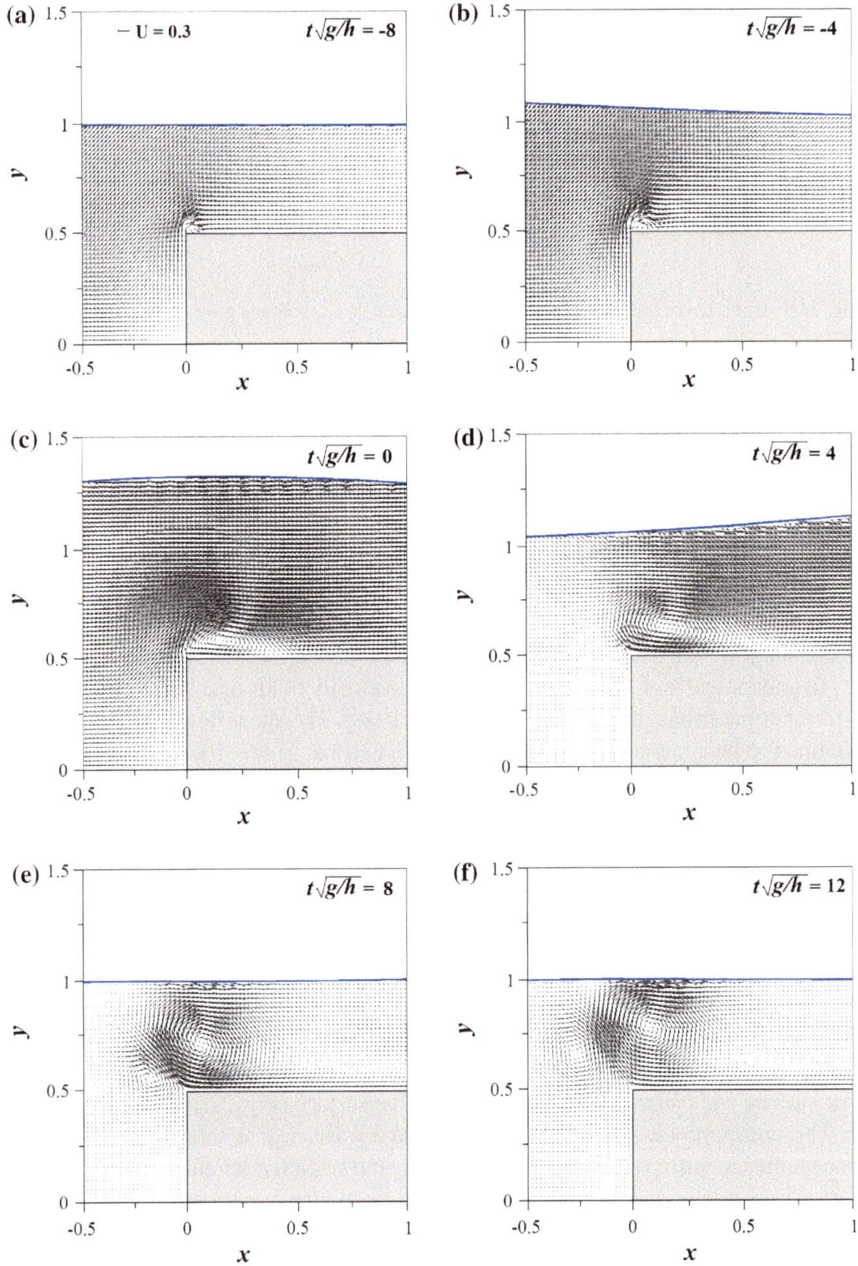

Fig. 22.10 Velocity fields at different times as the solitary wave of $A_i = 0.3$ passes over the step of $d = 0.5$

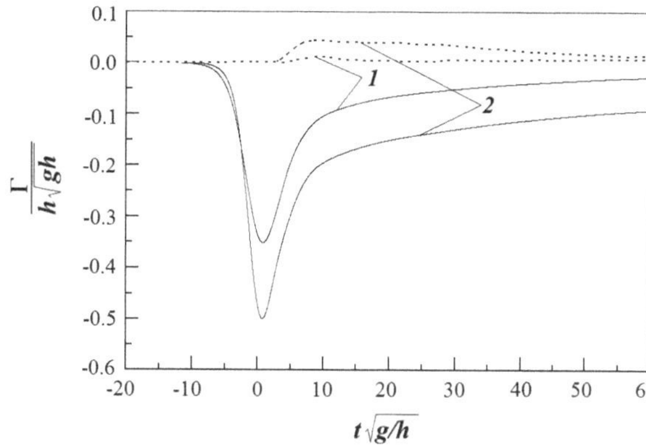

Fig. 22.11 Vortex paths and the free surface for $1/Fr^2 = 0.02$ at $t = 32$

primary circulation is seen to reach its maximum value at $t\sqrt{g/h} \approx 0$ and it decays fast enough due to the strong velocity gradients within the vortex, which can be seen from Fig. 22.9. The secondary circulation is small compared to the main vortex.

22.5 Conclusions

In coastal engineering, a solitary wave is used to represent certain characteristics of tsunamis, storm surges, and other long free-surface waves. On the other hand, the bottom step is often considered as the model that describes transfer of the sea bottom to the continental shelf. So study of interactions of a solitary wave with a submerged step is important for interpretation of catastrophic processes in coastal zones.

This paper provides a systematic study of the viscous interaction between a non-breaking solitary wave and a submerged step. The numerical scheme developed for simulation of viscous flows in the free-surface channels with a nonregular bottom is applied. The scheme combines the boundary integral method for description free-surface deformations, conformal mapping using to satisfy the nonleaking boundary condition on the bottom, and the vortex method for integrating the fluid dynamic equations. Both free surface transformations and vortical flow characteristics are obtained in a wide range of incident wave amplitudes and step heights.

The free-surface results are systematized by the interaction coefficient K_{int}, which is the ratio of the incident wave amplitude to the water depth over the top wall of step. The calculated evolutionary modes for the transmitted wave coincide with those observed in the experimental researches. Either weak distortion or fission in solitons and peaking of the first soliton are realized depending on the parameter K_{int}. The interaction coefficient, at which the solitary wave over an underwater step will be always breaking, is considered as the critical parameter K_{int}^{cr} of the process. Its value is found to be 0.8 that is in congruence with the experimental data.

The detailed investigation of the vortical patterns induced by a solitary wave near the bottom detected two large-scale opposite vortices shedding from the step edge in both upstream and downstream directions. Interaction of those specifies the fluid dynamics and turbulent processes in the region.

References

1. Mei, C.C.: Scattering of solitary wave at abrupt junction. J. Waterw. Port Coast. Ocean Eng. **111**(2), 319–328 (1985)
2. Seabra-Santos, F.J., Renouard, D.P., Temperville, A.M.: Numerical and experimental study of the transformation of a solitary wave over a shelf or isolated obstacle. J. Fluid Mech. **176**, 117–134 (1987)
3. Synolakis, C.E.: The run-up of solitary waves. J. Fluid Mech. **185**, 523–545 (1987)
4. Losada, M.A., Vidal, C., Medina, R.: Experimental study of the evolution of a solitary wave at an abrupt junction. J. Geophys. Res. **94**, 557–566 (1989)
5. Beji, S., Battjes, J.A.: Numerical simulation of nonlinear wave propagation over a bar. J. Coast. Eng. **23**, 1–16 (1994)
6. Ohyama, T., Nadaoka, K.: Transformation of non-linear wave train passing over a submerged shelf without breaking. J. Coast. Eng. **24**, 1–22 (1994)
7. Liu, P.L.-F., Cheng, Y.: A numerical study evolution of a solitary wave over a shelf. J. Phys. Fluids **13**(6), 1660–1667 (2001)
8. Pelinovsky, E., Choi, B.H., Talipova, T., Woo, S.B., Kim, D.C.: Solitary wave transformation on the underwater step: asymptotic theory and numerical experiments. J. Appl. Math. Comput. **217**, 1704–1718 (2010)
9. Tang, C.J., Chang, J.H.: Flow separation during solitary wave passing over submerged obstacle. J. Hydraul. Eng. **124**, 742–749 (1998)
10. Huang, C.-J., Dong, C.-M.: On the interaction of a solitary wave and a submerged dike. J. Coast. Eng. **43**, 265–286 (2001)
11. Hsu, T.W., Hsien, C.M., Hwang, R.R.: Using RANS to simulate generation and dissipation around impermeable submerged double breakwaters. J. Coast. Eng. **51**, 557–579 (2004)
12. Klettner, C.A., Eames, I.: Momentum and energy of a solitary wave interacting with a submerged semi-circular cylinder. J. Fluid Mech. **708**, 576–595 (2012)
13. Lin, M.Y., Huang, L.H.: Vortex shedding from a submerged rectangular obstacle attacked by a solitary wave. J. Fluid Mech. **651**, 503–518 (2010)
14. Gorban, V., Gorban, I.: Vortical flow structure near a square prism: numerical model and algorithms of control [in Ukrainian]. J. Appl. Hydromech. **7**, 8–26 (2005)
15. Zgurovsky, M., Mel'nik, V., Kasyanov, P.: Evolution Inclusions and Variation Inequalities for Earth Data Processing. Advances in Mechanical and Mathematics. Springer, Berlin (2011)
16. Baker, G.R., Meiron, D.J., Orszag, S.A.: Generalized vortex methods for free-surface flow problems. J. Fluid Mech. **123**, 477–501 (1982)
17. Lamb, H.: Hydrodynamics. Cambridge University Press, Cambridge (1932)
18. Lavric, V.I., Savenkov, V.N.: Handbook of conformal transformations [in Russian]. naukova dumka, Kyiv (1970)
19. Cottet, G.-H., Koumoutsakos, P.: Vortex Methods: Theory and Practice. Cambridge University Press, London (2000)
20. Molyakov, N.M.: Unsteady flow around a profile under a free surface separating fluids of different densities [in Russian]. Trans. Zhoukovsky Air Force Acad. **1313**, 336–347 (1985)
21. Lin, M.Y., Huang, L.H.: Study of water waves with submerged obstacles using a vortex method with Helmholtz decomposition. J. Numer. Methods Fluids **60**, 119–148 (2009)
22. Clements, R.R.: An inviscid model of two-dimensional vortex shedding. J. Fluid Mech. **57**, 321–336 (1973)

23. Israeli, M., Orszag, S.A.: Approximation of radiation boundary conditions. J. Comput. Phys. **41**, 115–135 (1981)
24. Telste, J.G.: Potential flow about two counter-rotating vortices approaching a free surface. J. Fluid Mech. **201**, 259–278 (1989)
25. Clamond, D., Dutykh D.: Fast accurate computation of the fully nonlinear solitary surface gravity waves. J. Comput. Fluids. **3** (2013)
26. Kotelnikova, A.S., Nikishov, V.I., Srebnyuk, S.M.: Interaction of surface solitary waves with submerged obstacles. Dopovidi NAC Ukr. **7**, 54–59 (2013)

MIX
Papier aus verantwortungsvollen Quellen
Paper from responsible sources
FSC® C105338

If you have any concerns about our products,
you can contact us on
ProductSafety@springernature.com

In case Publisher is established outside the EU,
the EU authorized representative is:
Springer Nature Customer Service Center GmbH
Europaplatz 3, 69115 Heidelberg, Germany

Printed by Libri Plureos GmbH
in Hamburg, Germany